THE SCIENCE OF ELECTRONICS

DIGITAL

THOMAS L. FLOYD

DAVID M. BUCHLA

Upper Saddle River, New Jersey
Columbus, Ohio

To our grandchildren:

Autumn, Carley, Carter, Greg, Jared, Madison, Rachel, and Taylor

Library of Congress Cataloging-in-Publication Data

Floyd, Thomas L.
The science of electronics. Digital / Thomas L. Floyd, David M. Buchla.
p. cm.
ISBN 0-13-087549-X (alk. paper)
1. Electronics–Textbooks. 2. Digital electronic–Textbooks. I. Buchla, David M. II. Title.

TK7816.F576 2005
621.381–dc22 2003069032

Editor in Chief: Stephen Helba
Acquisitions Editor: Dennis Williams
Development Editor: Kate Linsner
Production Editor: Rex Davidson
Design Coordinator: Diane Ernsberger
Cover Designer: Linda Sorrells-Smith
Cover art: Getty Images
Production Manager: Pat Tonneman
Marketing Manager: Ben Leonard

This book was set in Times Roman and Kabel by Carlisle Communications, Ltd. It was printed and bound by Courier Kendallville, Inc. The cover was printed by Phoenix Color Corp.

Pearson Education Ltd.
Pearson Education Singapore Pte. Ltd.
Pearson Education Canada, Ltd.
Pearson Education—Japan
Pearson Education Australia Pty. Limited
Pearson Education North Asia Ltd.
Pearson Educación de Mexico, S.A. de C.V.
Pearson Education Malaysia Pte. Ltd.

10 9 8 7 6 5 4 3 2 1
ISBN: 0-13-087549-X

PREFACE

Introduction to *The Science of Electronics* Series

The Science of Electronics: Digital is one of a series that also includes *The Science of Electronics: DC/AC* and *The Science of Electronics: Analog Devices.* This series presents basic electronics theory in a clear and simple, yet complete, format and shows the close relationship of electronics to other sciences. These texts are written at a level that makes them suitable as introductory texts for secondary schools as well as technical and community college programs. Pedagogical features are numerous, and they are designed to enhance the learning process and make the use of each text in the series an enjoyable experience. Inexpensive laboratory exercises for these texts are included in the accompanying lab manuals. The same author team prepared all of the texts and lab manuals in the series, providing consistency of approach and format.

The *DC/AC* text begins the series with some of the basic physics behind electronics, such as fundamental and derived units, work, energy, and energy conservation laws. Important ideas in measurement science, such as accuracy, precision, significant digits, and measurement units, are covered. In addition, the text covers passive dc and ac circuits, magnetic circuits, motors, and generators and instruments.

The *Digital* text introduces traditional topics, including number systems, Boolean algebra, combinational logic, and sequential logic, plus topics not normally found in an introductory text. The trend in industry is toward programmable devices, computers, and digital signal processing. A chapter is devoted to each of these important topics. Despite their complexity, these topics are treated with the same basic approach.

The *Analog Devices* text includes five chapters that cover diodes, transistors, and discrete amplifiers, followed by six chapters of operational amplifier coverage. Measurements are particularly important in all of the sciences, so the final chapter covers measurement and control circuits, including transducers and thyristors.

A "Science Highlight" feature opens every chapter in all textbooks. This highlight looks at scientific advances in an area related to the coverage in that chapter. Science Highlights include important related ideas in the fields of physics, chemistry, biology, computer science, and more. Electronics is a dynamic field of science, and we have tried to bring the excitement of some of the latest discoveries and advancements to the forefront for readers even as they begin their studies.

Key Features of *The Science of Electronics* Series

- A Science Highlight in each chapter looks at scientific advances in an area that is related to the coverage in that chapter.
- Easy to read and well illustrated format
- Full-color format
- "To the Student" provides an overview of the field of electronics, including careers, important safety rules, and workplace information, as well as a brief history of electronics.
- Many types of exercises reinforce knowledge and check progress, including worked-out examples, example questions, section review and chapter questions, chapter checkups, basic and basic-plus problems, and Multisim circuit simulations.
- Two-page chapter openers include a chapter outline, key objectives, list of key terms, a computer simulations directory to the appropriate figure in the chapter, and a lab experiments directory with titles of relevant exercises from the accompanying lab manual.

- Computer simulations throughout all of the books allow the student to see how specific circuits actually work.
- Safety Notes in the margins throughout all of the books continually remind students of the importance of safety.
- Historical Notes appear in margins throughout all of the books and are linked to ideas or persons mentioned in the text.
- An "On the Job" feature appears on selected chapter openers and discusses important aspects of employment.
- Key terms are indicated in red throughout the text, in addition to being defined in a key term glossary at the end of each chapter.
- A comprehensive glossary is included at the end of the book with all key terms and boldface terms from the chapters included.
- Important facts and formulas are summarized at the end of the chapters.

Introduction to *The Science of Electronics: Digital*

This text includes a wide range of topics to accommodate a variety of program requirements. Chapters 1 through 9 include fundamental topics such as number systems, Boolean algebra, logic gates, logic functions, flip-flops, timers, counters, and shift registers. Also, instruments and troubleshooting techniques are discussed. Chapters 10 through 12 cover the topics of programmable logic, computers, and digital signal processing. A variety of pedagogical features are incorporated throughout to aid in the study of this text and to enhance learning. A background in transistor circuits is not a prerequisite for this textbook.

Accompanying Student Resources

- *The Science of Electronics: Digital Lab Manual* by David M. Buchla
- *Companion Website, www.prenhall.com/SOE.* This website, created for *The Science of Electronics* series, includes:

  Computer simulation circuits designed to accompany selected examples in the text and lab manual.

  Chapter-by-chapter quizzes in true/false, fill-in-the-blank, and multiple choice formats that enable a student to check his or her understanding of the material presented in the text.
- *Prentice Hall Electronics Supersite, www.prenhall.com/electronics.* This website offers math study aids, links to career opportunities in the industry, and other useful information.

Instructor Resources

- *PowerPoint® Slides* on CD-ROM.
- *Companion Website, www.prenhall.com/SOE.* For the instructor, this website offers the ability to post your syllabus online with Syllabus Manager™. This is a great solution for classes taught online, self-paced, or in any computer-assisted manner.

- *Online Course Support.* If your program is offered in a distance-learning format, please contact your local Prentice Hall sales representative for a list of product solutions.
- *Instructor's Edition.* Includes answers to all chapter questions and worked-out solutions to all problems.
- *Lab Manual Solutions.* A separate solutions manual for all experiments in the series is available.
- *Test Item File.* A test bank of multiple choice, true/false, and fill-in-the-blank questions.
- *Prentice Hall TestGen.* This is an electronic version of the Test Item File, allowing an instructor to customize tests.
- *Prentice Hall Electronics Supersite.* Instructors can access various resources on this site, including password-protected files for the instructor supplements accompanying this text. Contact your local Prentice Hall sales representative for your user name and password.

Illustration of Chapter Features

Chapter Opener

Each chapter begins with a two-page spread, as shown in Figure P–1. The left page includes a list of the sections in the chapter and an introduction to the chapter. The right page has a list of key objectives for each section, a computer simulations directory, a laboratory experiments directory, and a list of key terms to look for in the chapter. Selected chapters contain a special feature called "On the Job."

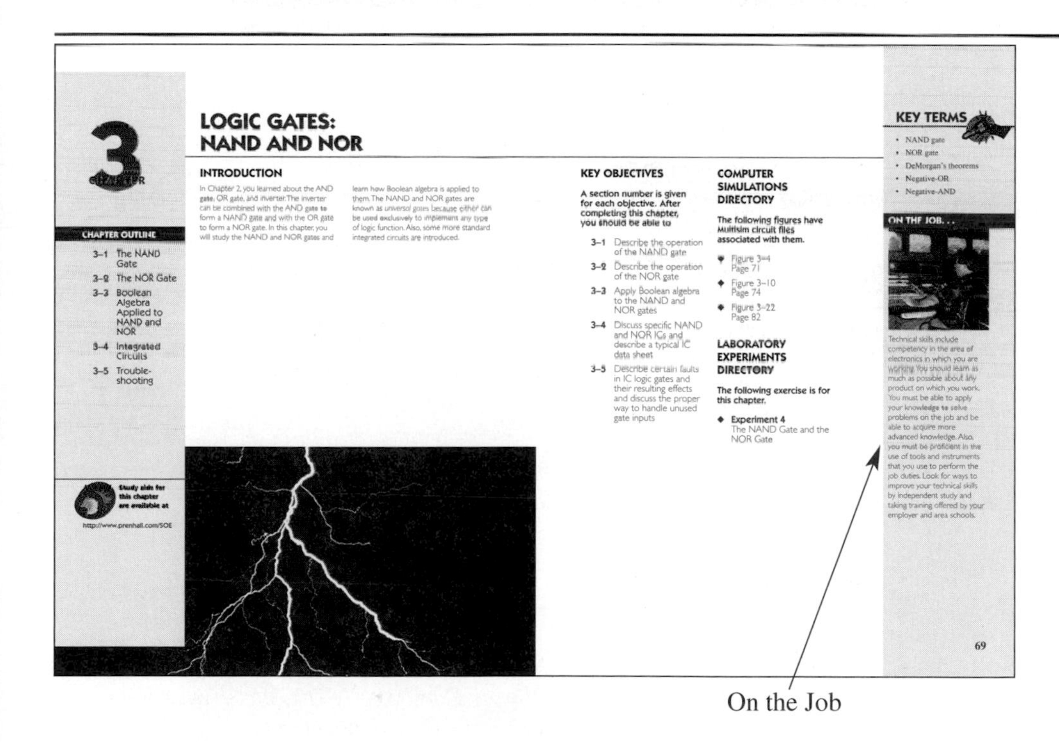

3 CHAPTER

LOGIC GATES: NAND AND NOR

INTRODUCTION

In Chapter 2, you learned about the AND gate, OR gate, and inverter. The inverter can be combined with the AND gate to form a NAND gate and with the OR gate to form a NOR gate. In this chapter, you will study the NAND and NOR gates and learn how Boolean algebra is applied to them. The NAND and NOR gates are known as *universal gates* because either can be used exclusively to implement any type of logic function. Also, some more standard integrated circuits are introduced.

CHAPTER OUTLINE

3–1 The NAND Gate
3–2 The NOR Gate
3–3 Boolean Algebra Applied to NAND and NOR
3–4 Integrated Circuits
3–5 Troubleshooting

Study aids for this chapter are available at http://www.prenhall.com/SOE

KEY OBJECTIVES

A section number is given for each objective. After completing this chapter, you should be able to

3–1 Describe the operation of the NAND gate
3–2 Describe the operation of the NOR gate
3–3 Apply Boolean algebra to the NAND and NOR gates
3–4 Discuss specific NAND and NOR ICs and describe a typical IC data sheet
3–5 Describe certain faults in IC logic gates and their resulting effects and discuss the proper way to handle unused gate inputs

COMPUTER SIMULATIONS DIRECTORY

The following figures have Multisim circuit files associated with them.

- Figure 3–4 Page 71
- Figure 3–10 Page 74
- Figure 3–22 Page 82

LABORATORY EXPERIMENTS DIRECTORY

The following exercise is for this chapter.

- Experiment 4 The NAND Gate and the NOR Gate

KEY TERMS

- NAND gate
- NOR gate
- DeMorgan's theorems
- Negative-OR
- Negative-AND

ON THE JOB...

Technical skills include competency in the area of electronics in which you are working. You should learn as much as possible about any product on which you work. You must be able to apply your knowledge to solve problems on the job and be able to acquire more advanced knowledge. Also, you must be proficient in the use of tools and instruments that you use to perform the job duties. Look for ways to improve your technical skills by independent study and taking training offered by your employer and area schools.

69

FIGURE P–1

Chapter opener.

Science Highlight

Immediately following the chapter opener is a *Sci Hi* feature that presents advanced concepts and science-related topics that tie in with the coverage of the text. A typical *Sci Hi* is shown in Figure P–2.

FIGURE P–2

Science highlight.

CHAPTER 10

The programmable logic devices covered in this chapter are complex networks of logic circuits that are joined by a matrix of interconnecting paths. One of the more exciting prospects for future electronic systems is associated with life itself. DNA (deoxyribonucleic acid) contains the "blueprint" for replicating complex molecular networks. Using DNA, scientists have constructed a DNA "computer," which has been used to solve the "traveling salesman" problem: Find the shortest route between seven cities without retracing your steps. The problem has thousands of possible solutions, but only one that is the shortest path.

To solve the problem, a unique strand of DNA was made to represent each "city." Billions of copies of the strands were mixed together, producing all possible routes between the "cities." Then with a series of biochemical reactions, the shortest strand that had each city only once in the sequence was identified. This strand represented the shortest route between the "cities."

In the future, DNA might be used to build huge networks of logic circuits.

10–1 PROGRAMMABLE LOGIC DEVICES (PLD)

The **PLD** (programmable logic device) is a programmable integrated circuit into which any digital logic design can be programmed using a PLD programming language called a hardware description language (HDL).

In this section, you will learn the basic types of PLDs and PLD arrays.

Types of PLDs

Three major types of programmable logic devices are SPLD, CPLD, and FPGA. There may be two or more categories in each type.

FIGURE 10–1

Typical SPLD package.

SPLD (Simple Programmable Logic Device)

The **SPLD** is the least complex form of PLD and was the first type available. Typically, one SPLD can replace several fixed-function SSI or MSI devices and their interconnections. A few categories of SPLD are listed here, but in this chapter we will cover only the PAL and the GAL. A typical SPLD package, such as shown in Figure 10–1, has 24 to 28 pins.

- PAL—programmable array logic
- GAL—generic array logic
- PLA—programmable logic array
- PROM—programmable read-only memory

CPLD (Complex Programmable Logic Device)

The CPLD has a much higher capacity than a SPLD so it can replace more complicated logic. Much more complex logic circuits can be programmed into them than into SPLDs. A typical **CPLD** is the equivalent of from two to sixty-four SPLDs.

298

Section Opener

Each section in a chapter begins with a brief introduction that includes a general overview. An illustration is shown in Figure P–3. This particular page also shows a computer simulation. Computer simulations are appropriately placed throughout the text.

Review Questions

Each section ends with a review consisting of five questions that emphasize the main concepts presented in the section. This feature is also shown in Figure P–3. Answers to the Section Reviews are at the end of the chapter.

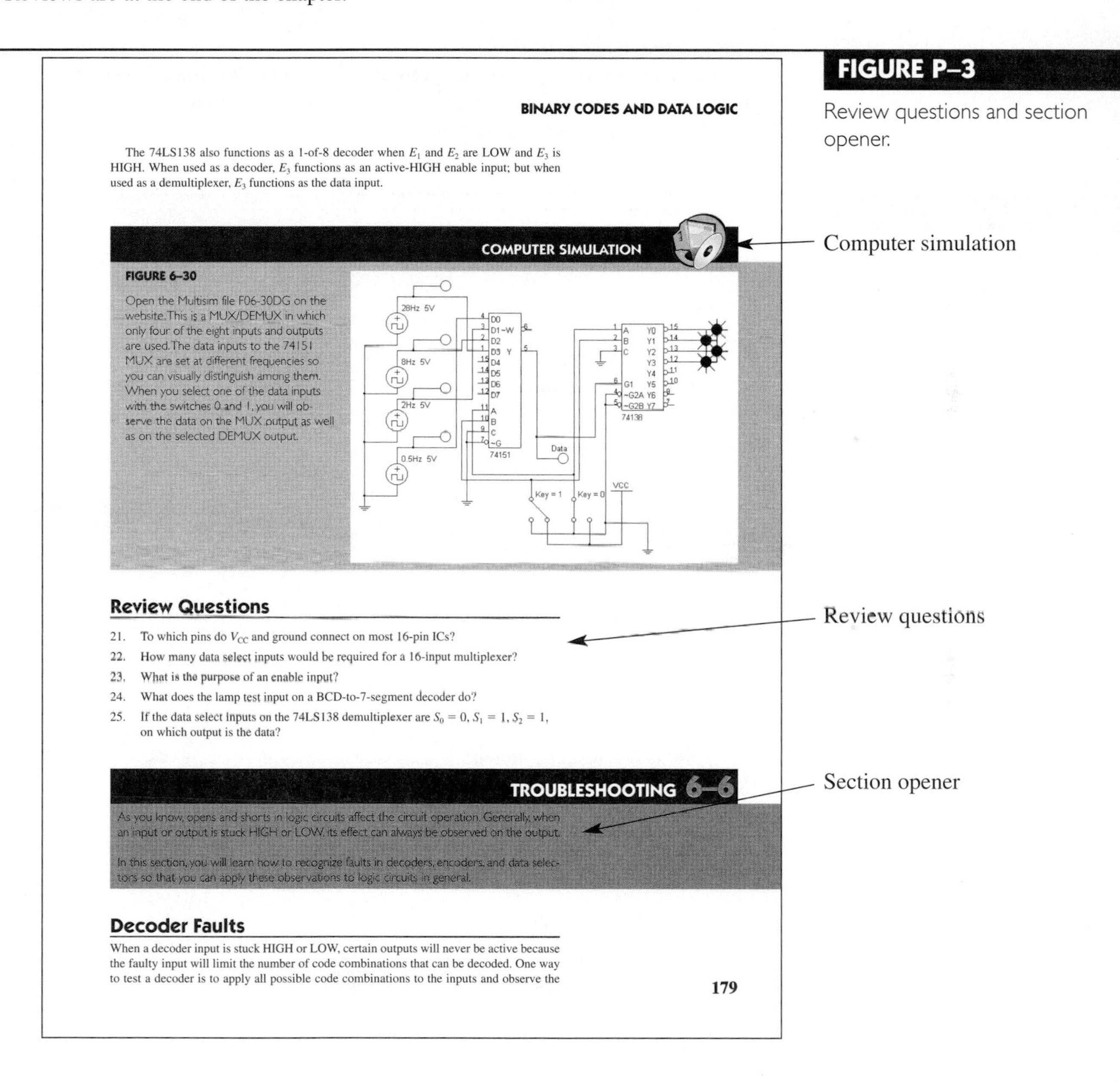

BINARY CODES AND DATA LOGIC

The 74LS138 also functions as a 1-of-8 decoder when E_1 and E_2 are LOW and E_3 is HIGH. When used as a decoder, E_3 functions as an active-HIGH enable input; but when used as a demultiplexer, E_3 functions as the data input.

COMPUTER SIMULATION

FIGURE 6–30

Open the Multisim file F06-30DG on the website. This is a MUX/DEMUX in which only four of the eight inputs and outputs are used. The data inputs to the 74151 MUX are set at different frequencies so you can visually distinguish among them. When you select one of the data inputs with the switches 0 and 1, you will observe the data on the MUX output as well as on the selected DEMUX output.

Review Questions

21. To which pins do V_{CC} and ground connect on most 16-pin ICs?
22. How many data select inputs would be required for a 16-input multiplexer?
23. What is the purpose of an enable input?
24. What does the lamp test input on a BCD-to-7-segment decoder do?
25. If the data select inputs on the 74LS138 demultiplexer are $S_0 = 0$, $S_1 = 1$, $S_2 = 1$, on which output is the data?

TROUBLESHOOTING 6–6

As you know, opens and shorts in logic circuits affect the circuit operation. Generally, when an input or output is stuck HIGH or LOW, its effect can always be observed on the output.

In this section, you will learn how to recognize faults in decoders, encoders, and data selectors so that you can apply these observations to logic circuits in general.

Decoder Faults

When a decoder input is stuck HIGH or LOW, certain outputs will never be active because the faulty input will limit the number of code combinations that can be decoded. One way to test a decoder is to apply all possible code combinations to the inputs and observe the

179

FIGURE P–3 Review questions and section opener.

Computer Simulation

Numerous circuits are provided online in Multisim. Filenames are keyed to figures within the text with the format Fxx-yyDG. The xx-yy is the figure number and DG represents a file for this text (Digital).These simulations can be used to verify the operation of selected circuits that are studied in the text. An example of a computer simulation feature is shown in Figure P–3. The Multisim circuits can be accessed on the website by going to *www.prenhall.com/SOE* and selecting this text. Choose the chapter you wish to study by clicking on that chapter, then click on the module entitled "Multisim." There you will see an introductory page with a link to the circuits for the chapter.

Worked Examples and Questions

There is an abundance of worked-out examples that help to illustrate and clarify basic concepts or specific procedures. Each example ends with a question that is related to the example. Typical examples are shown in Figure P–4.

FIGURE P–4

Worked example and related question.

LOGIC GATE COMBINATIONS

Question
Suppose another 2-input AND gate is added to the circuit in Figure 4–13. What would be the number of possible input combinations?

AND-OR Logic with Digital Waveform Inputs
Now let's look at an AND-OR circuit with digital waveform inputs. Again, we use a circuit with two 2-input AND gates, which means there will be four input waveforms. The timing relationship of these waveforms is important for determining the output waveform. When 1s are on both inputs of one or both of the AND gates, the output is 1.

EXAMPLE 4–4

Problem
Determine the output waveform *X* when the waveforms shown in Figure 4–14 are applied to the inputs of the AND-OR logic circuit. The dashed lines show how the waveforms are aligned.

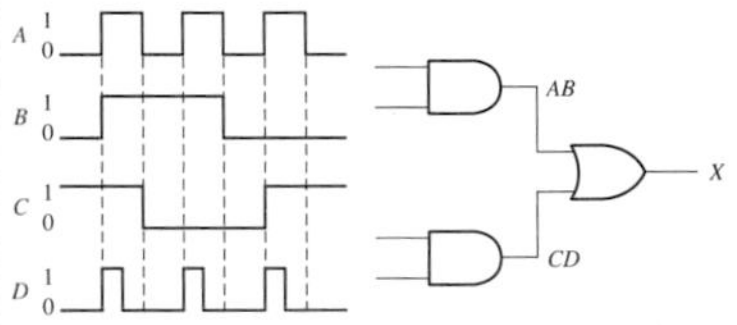

FIGURE 4–14

Solution
When input waveforms *A* and *B* are both 1 (HIGH) at the same time, the *AB* output of the top AND gate is 1. Also, when input waveforms *C* and *D* are both 1 (HIGH) at the same time, the *CD* output of the bottom AND gate is 1. Anytime waveforms *AB* or *CD* or both are 1 (HIGH), output waveform *X* is 1. The complete timing diagram is developed in Figure 4–15.

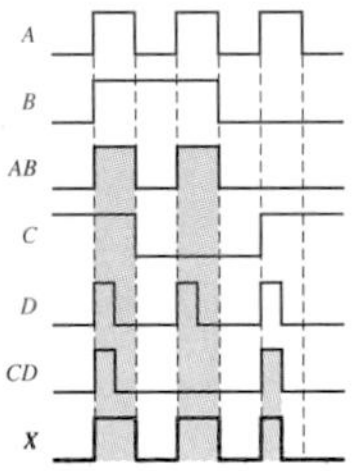

FIGURE 4–15

Question
If input waveform *B* is continuously a 1, what is output waveform *X*?

AND-OR-Invert Logic

AND-OR-Invert logic consists of AND gates with outputs that are connected to the inputs of a NOR gate. The only difference between AND-OR-Invert and AND-OR logic is that instead of an OR gate there is a NOR gate that inverts the output. Although there can be any number of AND gates with any number of inputs, there is only one NOR gate.

An AND-OR-Invert circuit with three 2-input AND gates is shown in Figure 4–16. The OR gate output is a 0 when both inputs on any or all of the AND gates are 1. The Boolean expression for the output in this case is the complemented (inverted) sum (OR) of the three AND gate outputs, as indicated by the continuous bar over the entire expression.

$$X = \overline{AB + CD + EF}$$

Remember that an AND-OR or an AND-OR-Invert circuit can have any number of AND gates with any number of inputs.

99

FIGURE P–5

Troubleshooting.

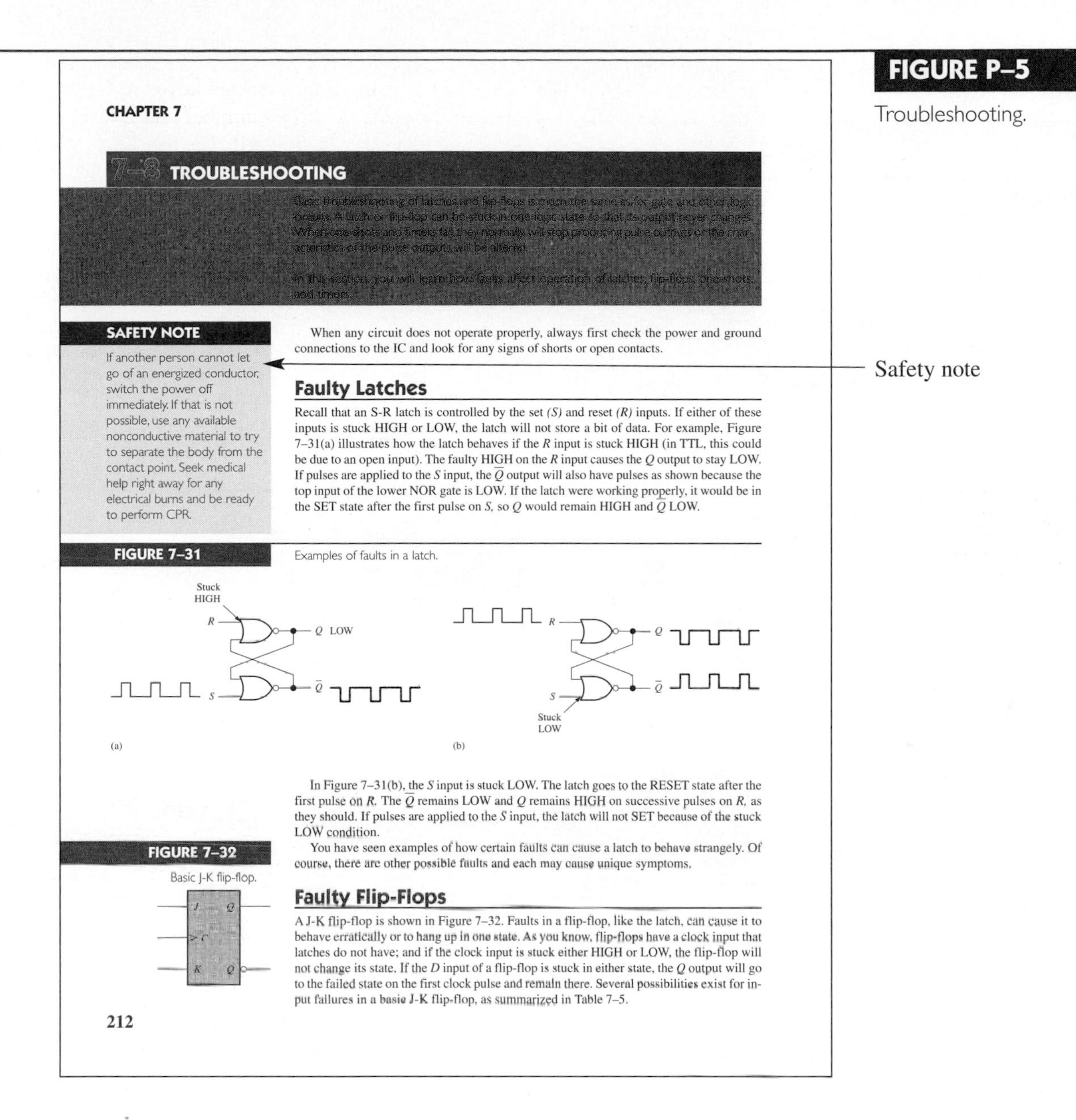

CHAPTER 7

7–8 TROUBLESHOOTING

Basic troubleshooting of latches and flip-flops is much the same as for gate and other logic circuits. A latch or flip-flop can be stuck in one logic state so that its output never changes. When one-shots and timers fail, they normally will stop producing pulse outputs or the characteristics of the pulse outputs will be altered.

In this section, you will learn how faults affect operation of latches, flip-flops, one-shots, and timers.

SAFETY NOTE

If another person cannot let go of an energized conductor, switch the power off immediately. If that is not possible, use any available nonconductive material to try to separate the body from the contact point. Seek medical help right away for any electrical burns and be ready to perform CPR.

When any circuit does not operate properly, always first check the power and ground connections to the IC and look for any signs of shorts or open contacts.

Faulty Latches

Recall that an S-R latch is controlled by the set (S) and reset (R) inputs. If either of these inputs is stuck HIGH or LOW, the latch will not store a bit of data. For example, Figure 7–31(a) illustrates how the latch behaves if the R input is stuck HIGH (in TTL, this could be due to an open input). The faulty HIGH on the R input causes the Q output to stay LOW. If pulses are applied to the S input, the $\overline{Q}$ output will also have pulses as shown because the top input of the lower NOR gate is LOW. If the latch were working properly, it would be in the SET state after the first pulse on S, so Q would remain HIGH and $\overline{Q}$ LOW.

FIGURE 7–31 Examples of faults in a latch.

Stuck HIGH
R
Q LOW
S
$\overline{Q}$
(a)
R
Q
S
$\overline{Q}$
Stuck LOW
(b)

In Figure 7–31(b), the S input is stuck LOW. The latch goes to the RESET state after the first pulse on R. The $\overline{Q}$ remains LOW and Q remains HIGH on successive pulses on R, as they should. If pulses are applied to the S input, the latch will not SET because of the stuck LOW condition.

You have seen examples of how certain faults can cause a latch to behave strangely. Of course, there are other possible faults and each may cause unique symptoms.

FIGURE 7–32 Basic J-K flip-flop.

J Q
C
K $\overline{Q}$

Faulty Flip-Flops

A J-K flip-flop is shown in Figure 7–32. Faults in a flip-flop, like the latch, can cause it to behave erratically or to hang up in one state. As you know, flip-flops have a clock input that latches do not have; and if the clock input is stuck either HIGH or LOW, the flip-flop will not change its state. If the D input of a flip-flop is stuck in either state, the Q output will go to the failed state on the first clock pulse and remain there. Several possibilities exist for input failures in a basic J-K flip-flop, as summarized in Table 7–5.

212

Troubleshooting

Many chapters include troubleshooting techniques and the use of test instruments as they relate to the topics covered. Figure P–5 shows typical troubleshooting coverage. This particular page also shows a Safety Note. Safety Notes are appropriately placed throughout the text.

Integrated Circuits

Most chapters include a section on digital integrated circuits. Typical specific devices are introduced and discussed.

FIGURE P–6 Chapter review.

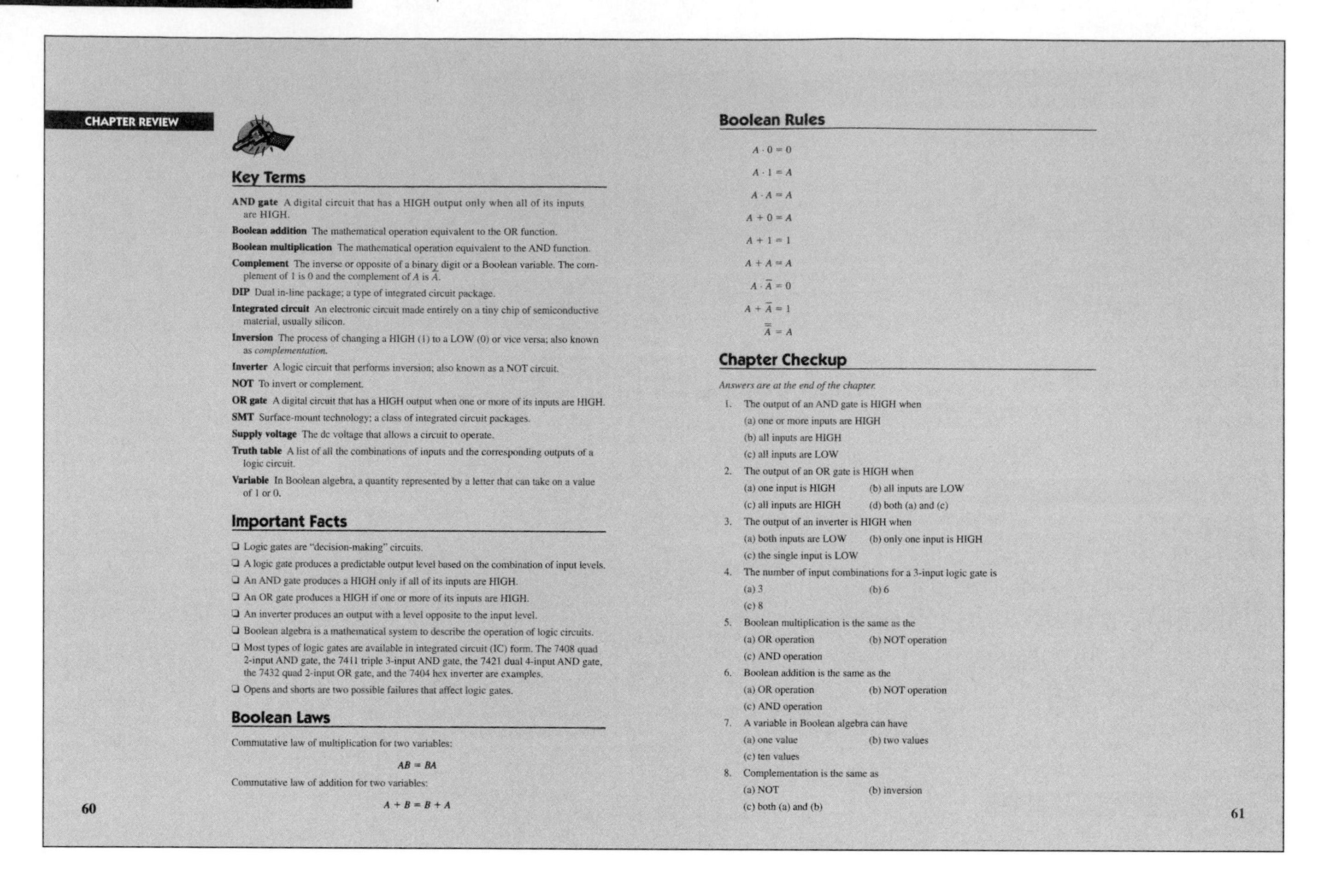

CHAPTER REVIEW

Key Terms

AND gate A digital circuit that has a HIGH output only when all of its inputs are HIGH.

Boolean addition The mathematical operation equivalent to the OR function.

Boolean multiplication The mathematical operation equivalent to the AND function.

Complement The inverse or opposite of a binary digit or a Boolean variable. The complement of 1 is 0 and the complement of A is $\overline{A}$.

DIP Dual in-line package; a type of integrated circuit package.

Integrated circuit An electronic circuit made entirely on a tiny chip of semiconductive material, usually silicon.

Inversion The process of changing a HIGH (1) to a LOW (0) or vice versa; also known as *complementation.*

Inverter A logic circuit that performs inversion; also known as a NOT circuit.

NOT To invert or complement.

OR gate A digital circuit that has a HIGH output when one or more of its inputs are HIGH.

SMT Surface-mount technology; a class of integrated circuit packages.

Supply voltage The dc voltage that allows a circuit to operate.

Truth table A list of all the combinations of inputs and the corresponding outputs of a logic circuit.

Variable In Boolean algebra, a quantity represented by a letter that can take on a value of 1 or 0.

Important Facts

- Logic gates are "decision-making" circuits.
- A logic gate produces a predictable output level based on the combination of input levels.
- An AND gate produces a HIGH only if all of its inputs are HIGH.
- An OR gate produces a HIGH if one or more of its inputs are HIGH.
- An inverter produces an output with a level opposite to the input level.
- Boolean algebra is a mathematical system to describe the operation of logic circuits.
- Most types of logic gates are available in integrated circuit (IC) form. The 7408 quad 2-input AND gate, the 7411 triple 3-input AND gate, the 7421 dual 4-input AND gate, the 7432 quad 2-input OR gate, and the 7404 hex inverter are examples.
- Opens and shorts are two possible failures that affect logic gates.

Boolean Laws

Commutative law of multiplication for two variables:

$$AB = BA$$

Commutative law of addition for two variables:

$$A + B = B + A$$

60

Boolean Rules

$$A \cdot 0 = 0$$
$$A \cdot 1 = A$$
$$A \cdot A = A$$
$$A + 0 = A$$
$$A + 1 = 1$$
$$A + A = A$$
$$A \cdot \overline{A} = 0$$
$$A + \overline{A} = 1$$
$$\overline{\overline{A}} = A$$

Chapter Checkup

Answers are at the end of the chapter.

1. The output of an AND gate is HIGH when
   (a) one or more inputs are HIGH
   (b) all inputs are HIGH
   (c) all inputs are LOW
2. The output of an OR gate is HIGH when
   (a) one input is HIGH (b) all inputs are LOW
   (c) all inputs are HIGH (d) both (a) and (c)
3. The output of an inverter is HIGH when
   (a) both inputs are LOW (b) only one input is HIGH
   (c) the single input is LOW
4. The number of input combinations for a 3-input logic gate is
   (a) 3 (b) 6
   (c) 8
5. Boolean multiplication is the same as the
   (a) OR operation (b) NOT operation
   (c) AND operation
6. Boolean addition is the same as the
   (a) OR operation (b) NOT operation
   (c) AND operation
7. A variable in Boolean algebra can have
   (a) one value (b) two values
   (c) ten values
8. Complementation is the same as
   (a) NOT (b) inversion
   (c) both (a) and (b)

61

Chapter Review

Each chapter ends in a special color section that is intended to highlight important chapter ideas. Several features are illustrated in Figure P–6. The chapter review includes:

- *Key Terms Glossary.* Terms that are in red within the chapter are defined here and in the glossary at the end of the book.
- *Important facts.* Major points from the chapter are summarized.
- *Equations and Boolean Laws/Rules.*
- *Chapter Checkup.* This is a set of multiple-choice questions with answers at the end of the chapter.
- *Questions.* This is a set of questions pertaining to the chapter. Answers to odd-numbered questions appear at the end of the book.

Problems

Pedagogical features continue with two levels of problems: Basic and Basic-Plus. In general, the Basic-Plus problems are more difficult than the Basic problems. Answers to all odd-numbered problems are at the end of the book. Also, Troubleshooting Practice problems that reference Multisim circuits containing faults are included in most chapters. The circuit files are designated with the prefix TSP.

Answers

Each chapter concludes with selected answers for questions within that chapter. These include:

- Answers to Example Questions
- Answers to Review Questions
- Answers to Chapter Checkups

End-of-Book Features

- Two Appendixes:
  Summary of Boolean Algebra
  Selected figures from the lab manual in color
- Comprehensive glossary
- Answers to odd-numbered questions
- Answers to odd-numbered problems
- Index

Acknowledgments

This text is the result of the work and the skills of many people. We think you will find this book and all the others in *The Science of Electronics* series to be valuable tools in teaching your students the basics of various areas of electronics.

Those at Prentice Hall who have, as always, contributed a great amount of time, talent, and effort to move this project through its many phases in order to produce the book as you see it, include but are not limited to Rex Davidson, Kate Linsner, and Dennis Williams. We are grateful that Lois Porter once again agreed to edit the manuscript on this project. She has done an outstanding job and we appreciate the incredible attention to detail that she has provided. Also, Jane Lopez has done a beautiful job with the graphics. Another individual who contributed significantly to this book is Doug Joksch, of Yuba College, who created all of the Multisim circuit files for the website and helped with checking the problems for accuracy. Our thanks and appreciation go to all of these people and others who were directly involved in the project.

We depend on expert input from many reviewers to create successful textbooks. We wish to express our sincere thanks to the following reviewers who submitted many valuable suggestions and provided lots of constructive criticism: Bruce Bush, Albuquerque Technical-Vocational Institute; Gary DiGiacomo, Broome Community College; Brent Donham, Richland College; J. D. Harrell, South Plains College; Benjamin Jun, Ivy Tech State College; David McKeen, Rogue Community College; Jerry Newman, Southwest Tennessee Community College; Philip W. Pursley, Amarillo College; Robert E. Magoon, Erie Institute of Technology; Dale Schaper, Lane Community College; and Arlyn L. Smith, Alfred State College.

Tom Floyd
David Buchla

CONTENTS

TO THE STUDENT

Introduction to *The Science of Electronics: Digital*

We believe that you will find *The Science of Electronics: Digital* an effective tool in your preparation for a career and should find this text useful in further studies. When you have finished this course, this book should become a valuable reference for more advanced courses or even after you have entered into the job market. We hope it provides a foundation for your continued studies in electronics.

The most complicated system in electronics can be broken down into a collection of simpler circuits. These include passive circuits (resistors, capacitors, inductors) and active circuits (integrated circuits including digital and analog devices). With a solid foundation in these topics, understanding large systems is simplified. Electronics is not an easy subject, but we have endeavored to provide help along the way to make it interesting and informative and to provide you with the preparation you need for a career in this exciting field.

Many examples in the text are worked out in detail. You should follow the steps in the examples and check your understanding with the related question. Check your understanding of each section by answering the review questions and checking your answers. At the end of each chapter are summaries, glossary terms, formulas, questions, and problems as well as many answers. If you can answer all of the questions and work the problems at the end of a chapter, you are well under way toward mastering the material presented.

Careers in Electronics

The field of electronics is diverse, and career opportunities are available in many related areas. Because electronics is currently found in so many different applications and because new technology is being developed at a fast rate, the future appears limitless. There is hardly an area of our lives that is not enhanced to some degree by electronics technology. Those who acquire a sound, basic knowledge of electrical and electronic principles and are willing to continue learning will always be in demand.

The importance of obtaining a thorough understanding of the basic principles contained in this text cannot be overemphasized. Most employers prefer to hire people who have both a thorough grounding in the basics and the ability and eagerness to grasp new concepts and techniques. If you have a good training in the basics, an employer will train you in the specifics of the job to which you are assigned.

There are many types of job classifications for which a person with training in electronics technology may qualify. Common job functions are described in the Bureau of Labor Statistics (BLS) occupational outlook handbook, which can be found on the Internet at *http://www.bls.gov/oco*. Two engineering technician's job descriptions from the BLS are as follows:

- *Electrical and electronics engineering technicians* help design, develop, test, and manufacture electrical and electronic equipment such as communication equipment, radar, industrial and medical measuring or control devices, navigational equipment, and computers. They may work in product evaluation and testing, using measuring and diagnostic devices to adjust, test, and repair equipment.
- *Broadcast and sound engineering technicians* install, test, repair, set up, and operate the electronic equipment used to record and transmit radio and television programs, cable programs, and motion pictures.

(Fluke Corporation. Reproduced with permission.)

Many other technical jobs are available in the electronics field for the properly trained person:

- *Service technicians* are involved in the repair or adjustment of both commercial and consumer electronic equipment that is returned to the dealer or manufacturer for service.
- *Industrial manufacturing technicians* are involved in the testing of electronic products at the assembly-line level or in the maintenance and troubleshooting of electronic and electromechanical systems used in the testing and manufacturing of products.
- *Laboratory technicians* are involved in testing new or modified electronic systems in research and development laboratories.
- *Field-service technicians* repair electronic equipment at the customer's site; these systems include computers, radars, automatic banking equipment, and security systems.
- *User-support technicians* are the first people called when a computer or "high-tech" electronic equipment acts up. User-support technicians must know their product inside and out and be able to troubleshoot a product over the phone. An ability to communicate well is vital.

Related jobs in electronics include technical writers, technical sales people, x-ray technicians, auto mechanics, cable installers, and many others.

Getting a Job

Once you have successfully completed a course of study in the field of electronics, the next step is to find employment. You must consider several things in the process of finding and obtaining a job.

Resources and Considerations for Locating a Job

One consideration is the job location. You must determine if you are willing to move to another town or state or if you prefer to remain near home. Depending on the economy and the current job market, you may not have much choice in location. The important thing is to get a job and gain experience. This will allow you to move up to better or more desirable jobs later. You should also try to find out if a job is one for which you are reasonably suited in terms of personality, skills, and interest.

A good resource for locating a potential employer is the classified ads in your local newspaper or in newspapers from other cities. The Internet is also an important resource for finding employment. Many large employers have a "job-line", which is a phone service dedicated to describing current openings. Another way to find a job is through an employment agency that specializes in technical jobs, but fees may be substantial with private agencies. Often, especially at the college level, employers will come to the campus to interview prospective employees. If you know someone in a technical job, contact that person to find out about job openings at his or her company.

The Resume and Application

A resume is a record of your skills, education, and job experience. Many employers request a resume before you actually apply for a job. This allows the employer to sort through many prospective applicants and narrow the field to a few of the most qualified. For this reason, your resume is very important. Don't wait until you are about to look for a job to start working on it.

The resume is the initial way in which you present yourself to a potential employer, so it is important that you create a well-organized document. There are different types of resumes, but all resumes should have certain specific information. Here are some basic guidelines:

- Your resume should be one page long unless you have significant experience or education. A shorter resume will be more likely to be read, so shorter is generally better.

- Your identification information (name, address, phone number, e-mail address) should come first.
- List your educational achievements such as diplomas, degrees, special certifications, and awards. Include the year for each item.
- List the specific courses that you have completed that relate directly to the type of job for which you are applying. Generally, you should list all the math, science, and electronics courses when applying for a job in electronics technology.
- List all of your prior job experience, especially if it is related to the job you are seeking. Most people prefer to organize it in reverse chronological order (most recent first). Show the employer, dates of employment, and a short description of your duties.
- You may include brief personal data that you feel may be to your advantage such as hobbies and interests (especially if they relate to the job).
- Do not attach any letters of recommendation, certificates, or documents to the resume when you submit it. You may indicate something like, "Letters of recommendation available on request."

If a prospective employer likes your resume, he or she will usually ask you to come in and fill out an application. As with the resume, neatness and completeness are important when completing an employment application. Often, references are required when submitting an application. Make sure you ask the person whom you want to use as reference if it's okay with him or her. Usually a reference must be from outside your family. Previous employers, teachers, school administrators, and friends are excellent choices to use as references.

The Job Interview

The interview is the most critical part of getting a job. The resume and the application are important steps because they get you to the point of having an interview. Although you might look good on paper, it is the personal contact with a prospective employer that usually determines whether or not you get the job. Two main steps in the interview process are preparing for the interview and the interview itself.

(Getty Images)

An interview helps the employer choose the best person for the job. Your goal is to prove to the employer that you are that person. Here are several guidelines for preparing for a job interview:

- Learn as much about the company as you can. Use the Internet or other local resources such as other people who work for the company to obtain information.
- Practice answering some typical questions that an interviewer may ask you.
- Make sure you know how to get to the employer's location and try to get the name of the person who will be interviewing you, if possible.
- Dress appropriately and neatly. Get your clothes and anything you plan to carry ready the night before the interview so that you will not have to rush right before leaving. The employer is looking at you as a potential representative of the company, so make a good impression with your choice of clothes.
- Bring a copy of your resume and diplomas, certificates, or other documents you believe may be of interest to the interviewer.
- Be on time. You should arrive at the interview location at least 15 minutes before the scheduled time.

During the interview itself, there are a few things to keep in mind:

- Greet the interviewer by name and introduce yourself.
- Be polite.
- Maintain good posture.

- Keep eye contact.
- Do not interrupt the interviewer.
- Answer all questions as honestly as you can. If you do not know something, say so.
- Be prepared to tell the interviewer why you believe you are the best person for the job.
- Show interest and enthusiasm. Have some questions in mind and ask the interviewer at appropriate times about the company and the job, but avoid questions about salary, raises, etc.
- When the interview is over, thank the interviewer for his/her time.

After You Get the Job

The job itself is what it's all about. It is the reason that you went through the job search, preparation, and interview. Now you have to come through for the employer because you are being paid to do a job. You must retain interest and enthusiasm for the job and apply your technical skills to the best of your ability. In addition, you will need to work as a "team player," so you must bring basic social and communication skills to the job.

Safety in the Workplace

One of the most important skills a technician brings to the job is knowledge of safe operating practice and recognition of unsafe operation, especially for shock and burn hazards. Employers expect you to work safely. It is important to learn about specific safety issues for an employer when you start a new job, as there may be special requirements that depend on the job. Workplace safety is an important issue to virtually all employers. Most employers will cover issues in an initial job orientation and/or in an employee handbook. Electricity is never to be taken lightly, and safe operation in the laboratory is vital to your own well-being and that of other employees.

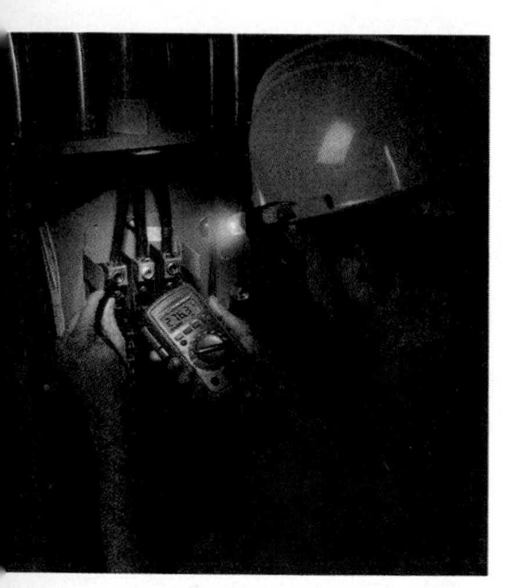

(Fluke Corporation. Reproduced with Permission.)

Companies have to comply with regulations prescribed by OSHA to ensure a safe and healthy workplace. OSHA is the Occupational Safety and Health Administration which is part of the U.S. Department of Labor. You can obtain information by accessing the OSHA website at *www.osha.gov*. In addition, you might have to be familiar with certain aspects of the National Electrical Code (NEC) and the standards issued by the American Society for Testing Materials (ASTA). Information on NEC can be obtained from the NFPA website at *www.nfpa.org* and ASTA can be accessed at *www.asta.org*. Of course, all standard electrical and electronic symbols as well as other related areas and standards are issued by IEEE (Institute of Electrical and Electronics Engineers). Their website is *www.ieee.com*.

Product Development, Marketing, and Servicing

The company you work for may be involved in developing, marketing, and servicing of products. Most electronics companies have a role in one or more of these activities. It is useful to understand the various steps in the process of bringing a new product to market.

1. Identifying the need
2. Designing
3. Prototyping and Evaluating
4. Producing
5. Marketing
6. Servicing

Identifying the Need

Before a new product is ever produced, someone (a person or group) must identify a problem or need, and suggest one or more ways a new product can solve the problem or meet the need. A company may be interested in the product but will want to do marketing analysis before expending money and time developing it. If the analysis shows a potential market for the product, design is started.

Designing

Most electronics projects can be implemented with more than one approach, so all ideas for the design of a new product are gathered together and evaluated. Usually this involves meetings with various specialists including electronic designers, test engineers, manufacturing engineers, as well as purchasing and marketing people. Selection considerations for a design must include input from these specialists. The best design is then chosen from the various ideas and a time line for completion is agreed upon. After a preliminary design is accomplished, components are selected based on cost, reliability, and availability.

Prototyping and Evaluating

Once the design of a product is complete, it must be proven that it will work properly by building and testing a prototype. Prototyping is often accomplished in two phases, preproduction and production. Usually technicians will construct a prototype of the design in a preliminary form. Test technicians will measure and evaluate the performance and report results to the design engineer. After approval and any necessary modification, the first prototype is converted to a production prototype by technicians. It is thoroughly tested and evaluated again before final approval for production.

Producing

A manufacturing engineer, familiar with the many processes that must be accomplished to produce a new product, is in charge of the production process. This person must verify the final layout and configuration and determine the sequence of all operations. The manufacturing engineer works with purchasing to determine cost of production and must ensure that the product will meet all necessary safety and reliability standards. Assemblers and technicians will produce the product and test the first production models to assure it works in accordance with the design specifications. Quality assurance technicians will test subsequent models to assure that products are ready to be shipped to customers.

Marketing

Once a product is produced, it is turned over to the marketing organization. The best product in the world is not much good if it can't be sold, so marketing is the key to success for a product. Marketing involves advertising, distribution, and followup. Technical marketing is a specialty requiring persons with both technical and marketing skills.

Servicing

Before a product is marketed, the cost of servicing it must be considered. A product may require servicing in order to assure maximum customer satisfaction and to create repeat business. Most electronics organizations maintain a service center to repair products returned for servicing. The type of servicing will depend on the cost of the product and replacement parts and the level of automation involved in the production process. Many products can be tested with automated test systems that can pinpoint a fault. Test engineers will set up the program to check the product, and service technicians will correct problems discovered as a result of testing.

The Social and Cultural Impact of Electronics Technology

Electronics technology has had a tremendous impact on our lives. The rapid advancement of technology has had both positive and negative influences on our society but by far the positives outweigh the negatives. Three areas of technological advancement that have had the most impact are the computer, communications technology, and medical technology. These areas are all interrelated because computer technology has influenced the other areas.

(Fluke Corporation. Reproduced with permission.)

The Computer and the Internet

In terms of the effect on society, the computer and the Internet have changed in a relatively short time the way we get information, communicate with friends and business associates, learn about new topics, compose letters, and pay bills. In electronics work, instruments can be connected to the computer and data sent anywhere in the world via the Internet. Computers have reduced personal contact to some degree because now we can e-mail someone instead of writing a letter or making a phone call. We tend to spend time on the computer instead of visiting with friends or neighbors, and it has made us, in a sense, more isolated. On the other hand, we can be in touch with just about anyplace in the world via the Internet and we can "talk" to people we have never met through chat rooms.

(Seth Joel/Getty Images)

Communications

Obviously, the computer is closely related to the way we communicate with each other because of the Internet. The world has "shrunk" as a result of modern communications and the instant availability of information. Television is another way in which our cultural and social values have been influenced for both good and bad. We can watch events as they unfold around the world because of television and satellite technology. Political candidates are often elected by how they present themselves on television. We learn about different cultures, people, and topics through the television media in addition to being entertained and enlightened. Unfortunately, we are also exposed to many factors that have a negative influence including significant levels of violence portrayed as "entertainment."

Another recent development that has had a major impact on our ability to communicate is the cellular telephone. Now, it is possible to contact anyone or be contacted no matter where you are. This, of course, facilitates doing business as well as providing personal benefits. Although it has added to our ability to stay in touch, it can be distracting; for example, when one is operating a motor vehicle.

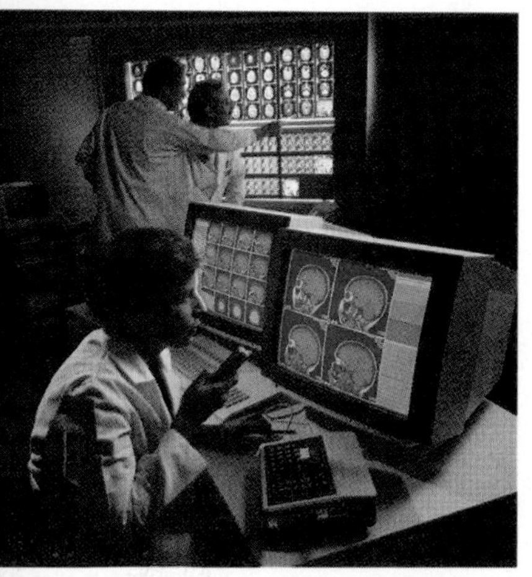

(Allan H. Shoemaker/Getty Images)

Medical Technology

Great advancements in medical technology have resulted in improvement in the quality and length of life for many people. New medical tools have provided health care professionals with the ability to diagnose illnesses, analyze test results, and determine the best course of treatment. Imaging technologies such as MRI, CATSCAN, XRAY, ultrasound, and others make effective diagnosis possible. Electronic monitoring equipment helps to supervise patients, keeping a constant watch on their condition. Operating rooms use electronic tools such as lasers and various video monitors to permit doctors to perform ever more complex surgical procedures and examinations.

Increased life span due to modern medical achievements has an impact on our society. People live longer, more productive lives and can contribute more to the improvement of the social and cultural aspects of life. On the other hand, medical advances have extended lives in some cases because of expensive life support, which may put a strain on the social and economic resources.

History of Electronics

Early experiments with electronics involved electric currents in vacuum tubes. Heinrich Geissler (1814–1879) removed most of the air from a glass tube and found that the tube glowed when there was current through it. Later, Sir William Crookes (1832–1919) found the current in vacuum tubes seemed to consist of particles. Thomas Edison (1847–1931) experimented with carbon filament bulbs with plates and discovered that there was a current from the hot filament to a positively charged plate. He patented the idea but never used it.

Other early experimenters measured the properties of the particles that flowed in vacuum tubes. Sir Joseph Thompson (1856–1940) measured properties of these particles, later called *electrons.*

Although wireless telegraphic communication dates back to 1844, electronics is basically a 20th century concept that began with the invention of the vacuum tube amplifier. An early vacuum tube that allowed current in only one direction was constructed by John A. Fleming in 1904. Called the Fleming valve, it was the forerunner of vacuum tube diodes. In 1907, Lee deForest added a grid to the vacuum tube. The new device, called the audiotron, could amplify a weak signal. By adding the control element, deForest ushered in the electronics revolution. It was with an improved version of his device that made transcontinental telephone service and radios possible. In 1912, a radio amateur in San Jose, California, was regularly broadcasting music!

In 1921, the secretary of commerce, Herbert Hoover, issued the first license to a broadcast radio station; within two years over 600 licenses were issued. By the end of the 1920s radios were in many homes. A new type of radio, the superheterodyne radio, invented by Edwin Armstrong, solved problems with high-frequency communication. In 1923, Vladimir Zworykin, an American researcher, invented the first television picture tube, and in 1927 Philo T. Farnsworth applied for a patent for a complete television system.

The 1930s saw many developments in radio, including metal tubes, automatic gain control, "midget" radios, and directional antennas. Also started in this decade was the development of the first electronic computers. In 1939, the magnetron, a microwave oscillator, was invented in Britain by Henry Boot and John Randall. In the same year, the klystron microwave tube was invented in America by Russell and Sigurd Varian.

The decade of the 1940s opened with World War II. The war spurred rapid advancements in electronics. Radar and very high-frequency communication were made possible by the magnetron and klystron. Cathode ray tubes were improved for use in radar. Computer work continued during the war. By 1946, John von Neumann had developed the first stored program computer, the Eniac, at the University of Pennsylvania. One of the most significant inventions ever occurred in 1947 with the invention of the transistor. The inventors were Walter Brattain, John Bardeen, and William Shockley. All three won Nobel prizes for their invention. PC (printed circuit) boards were also introduced in 1947. Commercial manufacturing of transistors didn't begin until 1951 in Allentown, Pennsylvania.

The most important invention of the 1950s was the integrated circuit. On September 12, 1958, Jack Kilby, at Texas Instruments, made the first integrated circuit, for which he was awarded a Nobel prize in the fall of 2000. This invention literally created the modern computer age and brought about sweeping changes in medicine, communication, manufacturing, and the entertainment industry. Many billions of "chips"—as integrated circuits came to be called—have since been manufactured.

The 1960s saw the space race begin and spurred work on miniaturization and computers. The space race was the driving force behind the rapid changes in electronics that followed. The first successful "op-amp" was designed by Bob Widlar at Fairchild Semiconductor in 1965. Called the μA709, it was very successful but suffered from "latch-up" and other problems. Later, the most popular op-amp ever, the 741, was taking shape at Fairchild. This op-amp became the industry standard and influenced design of op-amps for years to come. Precursors to the Internet began in the 1960s with remote networked computers. Systems were in place within Lawrence Livermore National Laboratory that connected over 100 terminals to a computer system (colorfully called the "Octopus system" and used by one of this text's authors). In an experiment in 1969 with very remote computers, an exchange took place between researchers at UCLA and Stanford. The UCLA group hoped to connect to a Stanford computer and began by typing the word "login" on its terminal. A separate telephone connection was set up and the following conversation occurred.

> The UCLA group asked over the phone, "Do you see the letter L?"
> "Yes, we see the L."
> The UCLA group typed an O. "Do you see the letter O?"
> "Yes, we see the O."

The UCLA group typed a G. At this point the system crashed. Such was technology, but a revolution was in the making.

By 1971, a new company that had been formed by a group from Fairchild introduced the first microprocessor. The company was Intel and the product was the 4004 chip, which had the same processing power as the Eniac computer. Later in that same year, Intel announced the first 8-bit processor, the 8008. In 1975, the first personal computer was introduced by Altair, and *Popular Science* magazine featured it on the cover of the January, 1975, issue. The 1970s also saw the introduction of the pocket calculator and new developments in optical integrated circuits.

By the 1980s, half of all U.S. homes were using cable hookups instead of television antennas. The reliability, speed, and miniaturization of electronics continued throughout the 1980s, including automated testing and calibrating of PC boards. The computer became a part of instrumentation and the virtual instrument was created. Computers became a standard tool on the workbench.

The 1990s saw a widespread application of the Internet. In 1993, there were only 130 websites; by the start of the new century (in 2001) there were over 24 million. In the 1990s, companies scrambled to establish a home page and many of the early developments of radio broadcasting had parallels with the Internet. The exchange of information and e-commerce fueled the tremendous economic growth of the 1990s. The Internet became especially important to scientists and engineers, becoming one of the most important scientific communication tools ever.

In 1995, the FCC allocated spectrum space for a new service called Digital Audio Radio Service. Digital television standards were adopted in 1996 by the FCC for the nation's next generation of broadcast television. As the 20th century drew toward a close, historians could only breathe a sigh of relief. As one person put it, "I'm all for new technologies, but I wish they'd let the old ones wear out first."

The 21st century dawned on January 1, 2001 (although most people celebrated the new century the previous year, known as "Y2K"). The major story was the continuing explosive growth of the Internet; shortly thereafter, scientists were planning a new supercomputer system that would make massive amounts of information accessible in a computer network. The new international data grid will be an even greater resource than the World Wide Web, giving people the capability to access enormous amounts of information and the resources to run simulations on a supercomputer. Research in the 21st century continues along lines of faster and smaller circuits using new technologies. One promising area of research involves carbon nanotubes, which have been found to have properties of semiconductors in certain configurations.

History of the Computer

The abacus, developed about 2000 years ago, is considered to be the beginning of computers. The abacus was a wooden rack holding rows of beads. When these beads are moved around, according to programming rules memorized by the users, all regular arithmetic problems can be done.

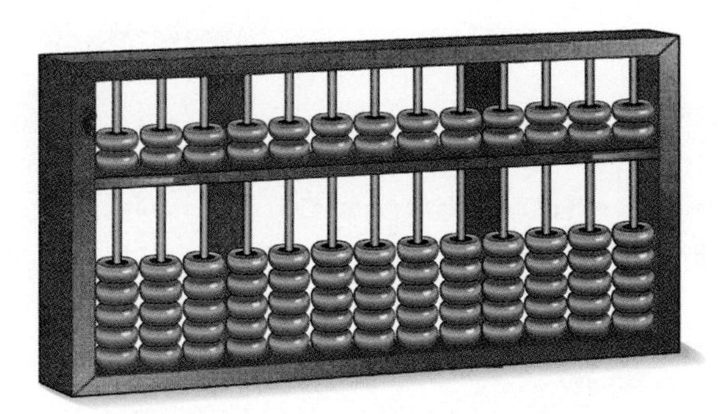

An abacus.

The first digital computer can be traced back to Pascal in 1642. It added numbers entered with dials. In 1671, Leibniz invented a computer using a stepped-gear mechanism that could add and multiply. A century later, Charles Xavier Thomas created the first successful mechanical calculator that could add, subtract, multiply, and divide. By 1890, calculating machines had been developed that could store and print results of calculations.

Babbage

An interesting series of developments in computers was started in Cambridge, England by Charles Babbage, a mathematics professor. In 1812, Babbage realized that many long calculations were really a series of repeated actions and that it was possible to do these automatically.

By 1822, he had designed an automatic mechanical calculating machine, which he called a "difference engine." It was intended to be steam powered and fully automatic, including the printing of the resulting tables, and operated by a fixed instruction program. Babbage continued to work on the difference engine for several years. His work eventually led him to develop a general-purpose, fully program-controlled, automatic mechanical digital computer. Babbage called this concept an "analytical engine," and he planned to use punched cards (similar to those used in a loom) to provide the machine with instructions to operate automatically by steam power. However, Babbage's concept of an analytical engine as a computer was never realized due to funding problems and the precision required for constructing his machine. His contributions are recognized today for their significance in the development of the stored-program computer.

Hollerith

Herman Hollerith was the first to actually use punched cards to collect and sort data after getting the idea from the weaving industry, where they were used to control looms. Hollerith developed devices that could read the information that had been punched into the cards automatically. Punched cards were used first in the 1890's census and continued to be used until the 1960s for data entry for both mechanical sorters and computers. They were also a primitive form of memory; they stored computer programs as well as data.

Electronic Digital Computers

Modern computers trace their origins to the work of John Atanasoff at Iowa State University. Beginning in 1937, he envisioned a binary machine that could do complex mathematical work. By 1939, he and graduate student Clifford Berry had constructed a binary machine called ABC, (for Atanasoff-Berry Computer) that used vacuum tubes for logic and

Example of an early punched card used for storing computer programs.

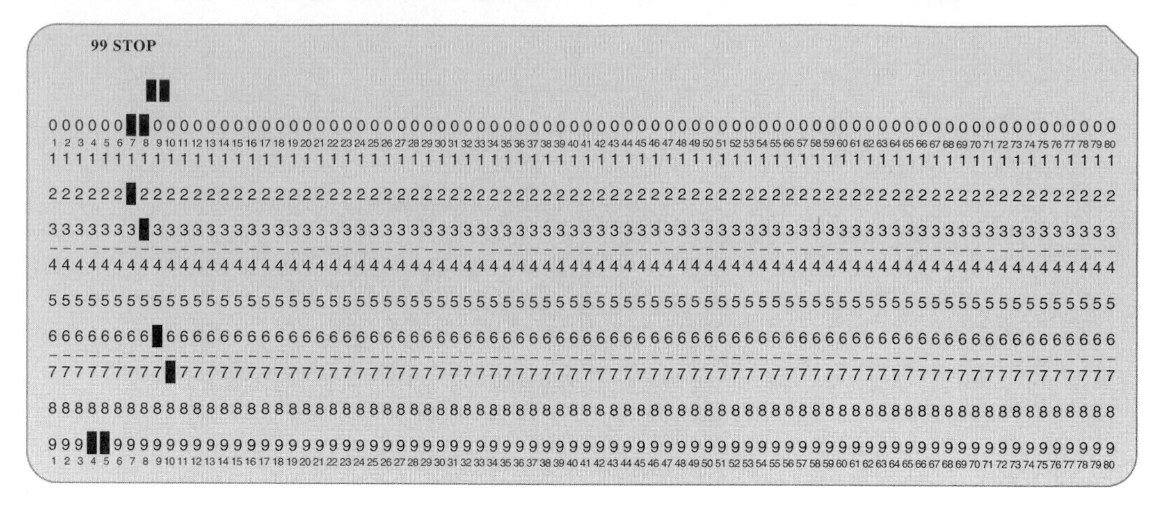

capacitors (then called condensers) for memory. Atanasoff's computer received little recognition at the time; however, in 1990 President George Bush awarded Atanasoff the National Medal of Technology for his invention.

In 1942, an electronic computer called the ENIAC (Electrical Numerical Integrator And Calculator) was developed at the Moore School of Engineering of the University of Pennsylvania. The ENIAC could multiply two numbers at a rate of 300 per second by finding the value of each product from a multiplication table stored in its memory. ENIAC was, therefore, about 1000 times faster than the early electromechanical relay computers that preceded it.

ENIAC (IBM Corporate Archives).

ENIAC used 18,000 vacuum tubes and occupied about 1800 square feet of floor space. It consumed about 180,000 watts of electrical power and used punched cards for input and output. Connections and switches had to be changed after each computation making it very inconvenient and awkward to use.

The Stored Program Computer

A major milestone in computers occurred in 1945 when mathematician John Von Neumann developed the concept of the stored-program technique upon which all modern computers are built.

The first generation of electronic computers to take advantage of the stored-program concept was built in 1947. Physically, they were much smaller than ENIAC and used only 2500 vacuum tubes. This group of computers included EDVAC and UNIVAC, the first commercially available computers.

Technological Advancements

In the 1950s, two important developments, the use of the transistor in electronics circuits and the magnetic cores in memory applications, drastically changed computer technology. These technical advances quickly found their way into new models of digital computers. Memory storage capacities and access times increased drastically. These machines were very expensive to purchase and operate. Such computers were mostly found in large computer centers operated by industry, government, and private laboratories.

During the 1960s, the major computer manufacturers began to offer a range of capabilities and prices, as well as accessories such as consoles, card readers, page printers, CRT displays, and graphing devices. In 1964, the first integrated circuit was developed. Circuits with hundreds of transistors and other components could be fabricated on a single chip of silicon. In 1965, Gordon Moore published a famous paper in *Electronics* magazine*, in which he observed an exponential growth in the number of transistors per integrated circuit and predicted that this trend would continue. All exponential growth must eventually cease, but his idea was a media hit and soon dubbed Moore's law.

The most important development in the 1970s was the introduction of the microprocessor, the complete central processor of a computer on a single integrated circuit. The 8085 was an 8-bit processor, used in a number of early computers and had 3000 transistors integrated onto a single integrated circuit "chip." Various companies, such as IBM, Apple, and Radio Shack introduced home computers during the 1970s. Advancements in software went hand-in-hand with the advancements in hardware and basic computer games became popular.

The 1980s saw the continued expansion of the capability and speed of processors. In 1987, both Motorola and Intel had 32-bit processors. The Intel 80386 was used in PCs and could operate at 20 MHz. The Intel 80486, which was the first processor to use a math coprocessor, was introduced in 1989. The coprocessor significantly increased the overall speed at which problems could be solved because it freed the main processor for other work.

The 1990s saw the introduction of first generation of the Pentium processor (which had a bug that had to be fixed at great expense to Intel!). The Pentium had 32-bit registers and a 64-bit data bus and initially had 3.1 million transistors in a single chip. At about the same time, Motorola introduced the 601 processor, which also used 32-bit internal registers and had a 64-bit data bus. In 1993, the fastest processor available ran at 80 MHz. By the end of the decade, the Pentium III was introduced, and could operate at speeds up to 1.33 GHz.

In 2000, the Pentium IV processor was introduced, which initially operated at 1.3 GHz; later speeds were doubled. In 2001, Intel introduced the Itanium™ processor, which was designed for high-end servers and workstations. In 2003, Intel introduced the Centrino mobile technology for portable computing, with built-in wireless capability. Many public locations (coffee houses, libraries, hotels) supported the new technology, allowing users to connect directly to the Internet without wires.

Amazingly, almost 40 years after being formulated, Moore's Law, which was revised to state that the number of transistors on a single chip doubles every two years, still holds true and is expected to see its 50th birthday! The number of transistors on a single chip as this is written is now in excess of 42 million! The future for computing is bright indeed!

---

* The original paper can be seen at *http://www.intel.com/research/silicon/mooreslaw.htm*

CHAPTER 1

DIGITAL QUANTITIES AND FUNCTIONS

CHAPTER OUTLINE

Study aids for this chapter available at

http://www.prenhall.com/SOE

INTRODUCTION

Digital electronics is an important and fascinating area of science. Electronics technology is changing at a rapid rate. The television industry is one example of changing electronics technology. If you know the basic concepts of digital electronics, you can always adapt to technological changes. In this chapter, the difference between analog and digital quantities is explained and illustrated. Binary numbers, which are the basis for all digital systems, are introduced. Also, hexadecimal and octal numbers are covered. The concept of logic levels and pulse waveforms are covered because they are important in any study of logic circuits and digital systems. The primary elements that make up a digital system are discussed to give you a preview of what you will learn in this book. Finally, some important instruments that are used for testing and troubleshooting logic circuits are introduced.

KEY OBJECTIVES

A section number is given for each objective. After completing this chapter, you should be able to

- **1–1** Explain the difference between analog and digital quantities
- **1–2** Evaluate binary numbers and describe their structure
- **1–3** Describe the hexadecimal and octal number systems and convert between binary numbers and hexadecimal and octal numbers
- **1–4** Use a calculator to perform number system conversion and hexadecimal arithmetic
- **1–5** Describe logic levels used in digital systems and analyze the characteristics of a pulse waveform
- **1–6** Discuss the elements that make up a digital system
- **1–7** Describe several instruments and their purpose

COMPUTER SIMULATIONS DIRECTORY

The following figure has a Multisim circuit file associated with it. To open a Multisim file, go to the website at http://www.prenhall.com/SOE, click on the cover of this book, choose this chapter, click on "Multisim", and then click on the selected file.

- ◆ Figure 1–5
  Page 18

LABORATORY EXPERIMENTS DIRECTORY

The following exercise is for this chapter. The lab manual is entitled *The Science of Electronics: Digital Lab Manual,* by David M. Buchla (ISBN-0-13-087558-9). © 2005 Prentice Hall.

- ◆ **Experiment 1**
  Constructing a Logic Probe

KEY TERMS

- Analog
- Digital
- Binary
- Bit
- Weight
- Power-of-two
- LSB
- MSB
- Byte
- Hexadecimal
- Octal
- Logic level
- Waveform
- Frequency
- Period
- Pulse width
- Duty cycle
- Amplitude
- Timing diagram

The semiconductor materials of silicon and germanium are classified as metalloids because they are elements that fall between metals and nonmetals on the periodic table. The metalloids form a diagonal band across the periodic table and include boron, arsenic, antimony, tellurium, and polonium in addition to silicon and germanium.

In 1871, Dimitri Mendeleyev, a Russian professor of chemistry, predicted the existence of an element with an atomic number of 32 (thirty-two protons in its nucleus). Mendeleyev's predictions were among the most amazing predictions in science! In 1886, a German professor of chemistry, Clemens Winkler, discovered the element predicted by Mendeleyev and called it germanium. Germanium's properties were almost identical to those predicted by Mendeleyev. This very rare metalloid was exploited after 1945 when its properties as a semiconductor were discovered.

The discovery of germanium is the genesis of the integrated circuit, which has made today's electronics technology possible. Germanium was used to create the first transistor and was used in the manufacture of the first generation of transistors. Although silicon has largely replaced germanium in transistors and is used in integrated circuits, germanium is still found in infrared detectors and in various types of lenses as well as in other applications in electronics.

1–1 ANALOG AND DIGITAL QUANTITIES

In electronics applications, measurable quantities are represented as either analog or digital.

In this section, you will learn the difference between analog and digital quantities.

Analog

Analog* quantities are those that have continuous values. Most quantities are analog when they occur in nature. Temperature is an analog quantity because it has a continuous range of values. During a 12-hour period, the temperature does not jump from one value to another but varies over a continuous range of values. For example, temperature doesn't jump from 70° to 71° instantaneously; instead, it goes through all possible values in between. A graph showing how the temperature changes with time on a particular summer day is a smooth, continuous curve, as shown in Figure 1–1. Values of temperature are on the vertical axis, and the time of day is on the horizontal axis.

FIGURE 1–1

Temperature variation during a certain summer day.

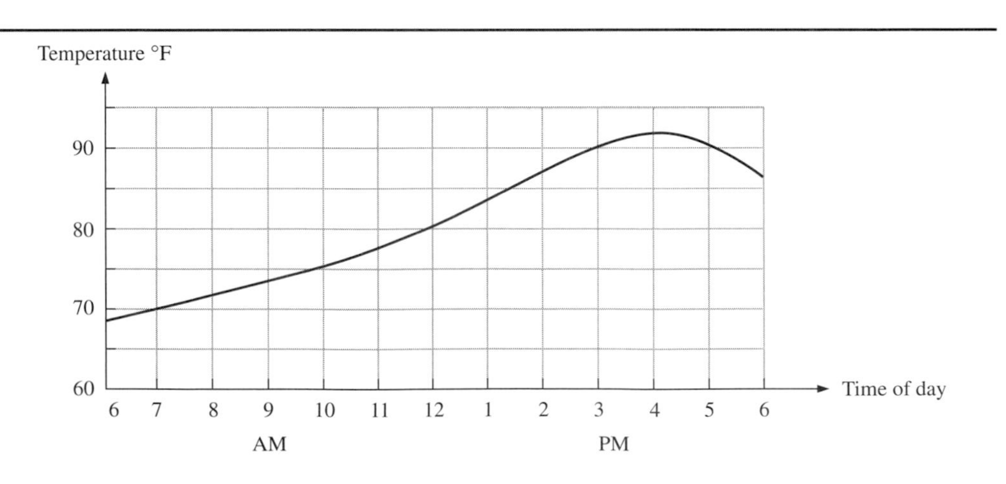

** Key terms are shown in red.*

Another example of an analog quantity is speed. When you drive down the road in your car, you are moving at a certain speed that is indicated in miles per hour (mph) on the speedometer. When you accelerate from 30 mph to 40 mph, for example, the car does not immediately jump from the lower speed to the higher speed. It actually goes through all the possible values of speed between 30 and 40 mph as the speed builds up. If you accelerate rapidly, you go from one speed to another in a very short time but never instantaneously; speed always has a continuous set of values and is therefore analog. Other examples of analog quantities are sound waves, pressure, distance, time, and just about every other natural phenomenon.

Digital

Digital quantities are those that have a series of individual values (discrete) rather than continuous values. Although most quantities are analog in nature, they can be represented in digital form. Instead of graphing the temperature on a continuous basis as shown in Figure 1–1, suppose you take a temperature reading every hour. You are actually sampling the temperature at certain points in time over the 12-hour period, as shown in Figure 1–2. There is a distinct value of temperature for each sampled point in time, but in between the sampled points you do not know the exact temperature. You can express the temperature values in degrees over the 12-hour period as a series of numbers such as 69, 70, 72, 74, 76, and so on. If you sample the temperature more frequently, you will get a more accurate representation of the temperature variation. Later, you will see how the sampled values can be represented by binary numbers as they would be in a digital system.

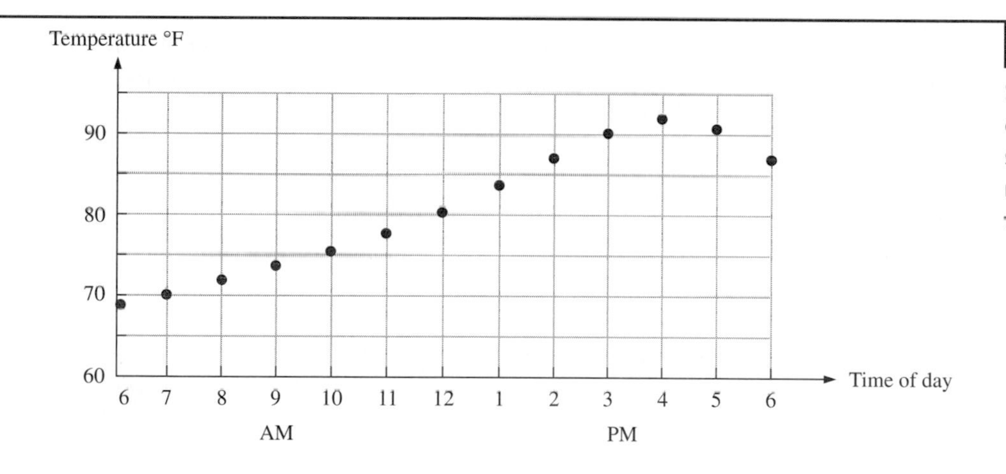

FIGURE 1–2

Sampled values of the analog quantity in Figure 1–1. Each sampled value can be represented by a number that is a series of digits.

Although speed is an analog quantity because it increases or decreases over a continuous range of values, some cars have digital speedometers that show the speed by a numerical readout. Generally, these display the speed in discrete increments of 1 mph. As you accelerate, your speedometer will jump from 50 mph to 51 mph with no values in between.

Another analog quantity that is displayed in most cars is the time of day. Some cars have an analog clock with hands that move in a smooth and continuous manner. Most cars, however, have a digital clock that jumps from one discrete value of time to another in increments of seconds.

An Analog Electronic System

A public address system is one type of an analog electronic system. A diagram of a public address system used to amplify sound, such as voice or music, so that it can be heard by an audience is shown in Figure 1–3. Sound waves, which are analog in nature, are picked up by the microphone and converted to a small analog electrical voltage called the audio signal,

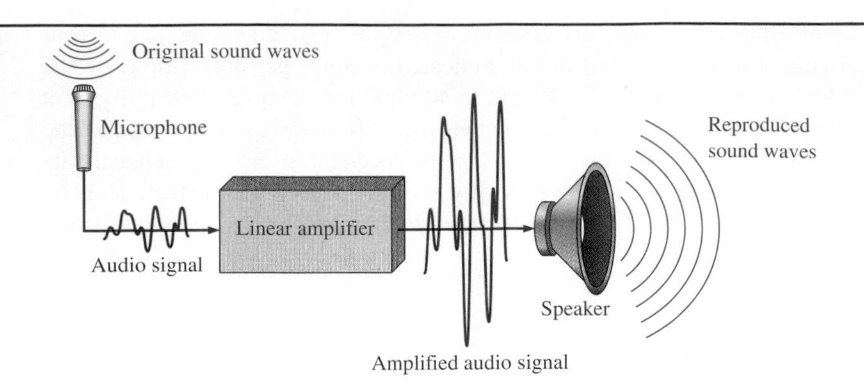

FIGURE 1–3

An analog public address system.

which varies continuously as its frequency and volume change. This small signal is increased in volume by a linear amplifier. The amplified electrical signal is then applied to the speaker, which converts it back to sound waves with much greater volume than the original sound that was picked up by the microphone.

A Combined Analog and Digital System

A compact disk (CD) player is an example of an electronic system that uses both analog and digital quantities. The diagram in Figure 1–4 shows a basic CD player. Music in digital form is stored on the CD. A laser optical system picks up the digital information from the rotating disk and transfers it to the digital-to-analog converter. There, the digital information is changed to an analog signal that is the electrical reproduction of the original music. This analog signal is amplified and sent to the speaker. When the music was originally recorded on the CD, a process known as analog-to-digital conversion was used to convert the musical sound into a digital form.

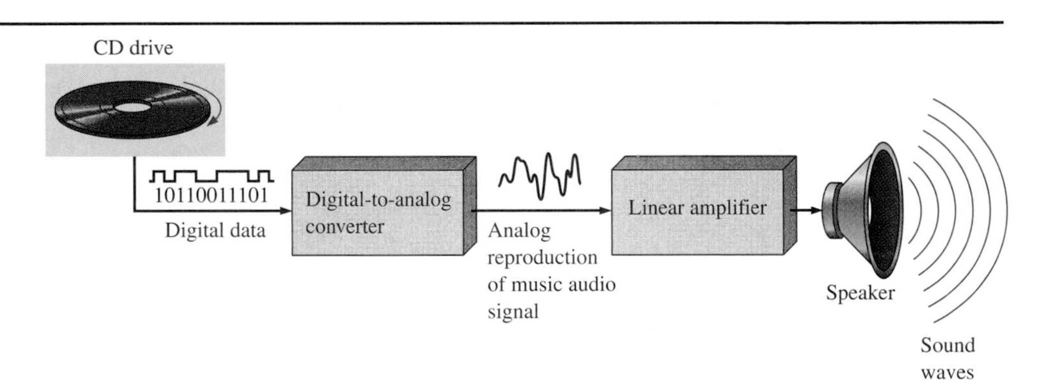

FIGURE 1–4

A basic CD player with only one channel shown.

Review Questions

Answers are at the end of the chapter.

1. What is an analog quantity?
2. What are some examples of analog quantities?
3. What is a digital quantity?
4. Can analog quantities be represented as digital quantities?
5. In a CD player, how are analog and digital used?

BINARY NUMBERS 1–2

Digital quantities are usually represented by a number system known as binary. The binary number system has two digits, a 0 and a 1, which are called bits. All computers, as well as other types of digital systems, use binary numbers.

In this section, your will learn to evaluate binary numbers and describe their structure.

The **binary** number system is a weighted number system. This means that the value of a binary digit (**bit**) depends on its position in the binary number. This concept of a weighted number system is similar to the decimal system with which you are already familiar.

The decimal number system is a weighted number system with ten digits (0 through 9). The position of each digit in a number determines its value or **weight** in the number. In decimal whole numbers, the first digit to the right has a weight of 1. The weights increase by a factor of ten for each digit position going from right to left. For example, in the decimal number 236 the digit 6 has a weight of 1 and a value of $6 \times 1 = 6$, the digit 3 has a weight of 10 and a value of $3 \times 10 = 30$, and the digit 2 has a weight of 100 and a value of $2 \times 100 = 200$. Therefore, the value of a decimal number is actually found by adding the values of each digit. For this particular number,

$$(2 \times 100) + (3 \times 10) + (6 \times 1) = 200 + 30 + 6 = 236$$

All of the weights in the decimal number system are actually powers of ten. For example, $10^0 = 1$, $10^1 = 10$, $10^2 = 100$, $10^3 = 1000$, and so on. When 10 is raised to a power, indicated by an exponent, it means that 10 is multiplied by itself the number of times indicated by the power; two exceptions are for the case of ten to the zero power, which is 1, and the case of 10 to the first power, which is 10. For example, $10^2 = 10 \times 10 = 100$. The weights are shown in Table 1–1 for decimal numbers up to seven digits. As you will see, the binary number system is based on powers of two in a similar way.

Power of ten	10^6	10^5	10^4	10^3	10^2	10^1	10^0
Weight	1,000,000	100,000	10,000	1000	100	10	1

TABLE 1–1

The decimal weights increase by 10 times for each power of ten from right to left.

The binary number system is a weighted system with only two digits (1 and 0). Like in a decimal number, the position of each binary digit (bit) determines its value in the number. In binary whole numbers, the first bit to the right has a weight of 1. The weights increase by a factor of two for each digit position going from right to left. For example, in the binary number 111 the 1 at the right has a weight of 1 and a value of $1 \times 1 = 1$, the 1 in the middle has a weight of 2 and a value of $1 \times 2 = 2$, and the 1 at the left has a weight of 4 and a value of $1 \times 4 = 4$. The decimal value of a binary number is actually found by adding the values of each bit. For this particular number,

$$(1 \times 4) + (1 \times 2) + (1 \times 1) = 4 + 2 + 1 = 7$$

The weights in the binary number system are actually powers of two. A number expressed as an exponent to the base of 2 is a **power-of-two**. For example $2^0 = 1$, $2^1 = 2$, $2^2 = 4$, $2^3 = 8$ are powers of two. When 2 is raised to a power, it means that 2 is multiplied by itself the number of times indicated by the power. For example, $2^3 = 2 \times 2 \times 2 = 8$.

In binary whole numbers, the bit at the far right has the least weight ($2^0 = 1$) and is called the **LSB** (least significant bit). The bit at the far left has the greatest weight and is called the

MSB (most significant bit). A binary number can have any number of bits, but the weights always increase by a factor of 2 (they double) for each position going from right to left. The weights are shown in Table 1–2 for binary numbers up to eight bits long. A group of eight bits is called a **byte**.

TABLE 1–2

The binary weights double for each power of two from right to left.

Power of two	2^7	2^6	2^5	2^4	2^3	2^2	2^1	2^0
Weight	128	64	32	16	8	4	2	1

EXAMPLE 1–1

Problem

Find the decimal value of each of the following binary numbers:

(a) 11 (b) 101

(c) 1101 (d) 10010

Solution

Determine the weight of each of the bits in the number and then sum the weights to find the decimal value of the number.

(a) $11 = (1 \times 2^1) + (1 \times 2^0) = (1 \times 2) + (1 \times 1) = 2 + 1 = \mathbf{3}$

(b) $101 = (1 \times 2^2) + (0 \times 2^1) + (1 \times 2^0) = (1 \times 4) + (0 \times 2) + (1 \times 1) = 4 + 1 = \mathbf{5}$

(c) $$\begin{aligned}1101 &= (1 \times 2^3) + (1 \times 2^2) + (0 \times 2^1) + (1 \times 2^0)\\ &= (1 \times 8) + (1 \times 4) + (0 \times 2) + (1 \times 1)\\ &= 8 + 4 + 1 = \mathbf{13}\end{aligned}$$

(d) $$\begin{aligned}10010 &= (1 \times 2^4) + (0 \times 2^3) + (0 \times 2^2) + (1 \times 2^1) + (0 \times 2^0)\\ &= (1 \times 16) + (0 \times 8) + (0 \times 4) + (1 \times 2) + (0 \times 1)\\ &= 16 + 2 = \mathbf{18}\end{aligned}$$

Any time you have a 0 bit, its value is always zero regardless of its weight, so you don't have to add it in to get the value of the number.

Question*

What is the decimal value of the binary number 100001?

Expressing a Decimal Number in Binary

You can express any decimal number in binary form. Place 1s in the power-of-two positions having weights that will add up to the decimal number and place 0s in the other positions. For example, to express the decimal number 25 in binary form, place 1s in the positions with weights of 16, 8, and 1 because 16 + 8 + 1 = 25. This is shown below for an 8-bit number using Table 1–2.

Power of two	2^7	2^6	2^5	2^4	2^3	2^2	2^1	2^0
Weight	128	64	32	16	8	4	2	1
Binary number for decimal 25	0	0	0	1	1	0	0	1

* *Answers are at the end of the chapter.*

The three 0s to the left are called leading zeros and can be omitted because you only need five bits for the decimal number 25. Only the bits shown in red are essential for this particular number.

For the decimal number 150, for example, all eight bits are needed because 128 + 16 + 4 + 2 = 150.

Power of two	2^7	2^6	2^5	2^4	2^3	2^2	2^1	2^0
Weight	128	64	32	16	8	4	2	1
Binary number for decimal 150	1	0	0	1	0	1	1	0

EXAMPLE 1–2

Problem

Express each decimal number in binary:

(a) 41 (b) 225

Solution

Determine the power-of-two weight values needed to add up to the decimal number. Place a 1 in each of those weight positions and a 0 in the other positions.

(a) 41 = 32 + 8 + 1. The binary number is shown in red.

32	16	**8**	4	2	**1**
1	0	1	0	0	1

(b) 225 = 128 + 64 + 32 + 1. The binary number is shown in red.

128	**64**	**32**	16	8	4	2	**1**
1	1	1	0	0	0	0	1

Question

What is the binary number for the decimal number 83?

Binary Sequences

You can use binary numbers to count in sequence, just as you do with decimal numbers. With decimal numbers, you use ten digits, 0–9. You can count to nine with one digit. With a two-digit number, you can count up to 99, and with a three-digit number you can count up to 999. Counting in the binary number system is similar except you only have two digits, called bits. Using one bit, you can only count 0 and 1. With two bits, you can count 00, 01, 10, 11 (0 to 3). With three bits, you can count from 000 to 111 (0 to 7). With four bits, you can count from 0000 to 1111 (0 to 15).

The largest number to which you can count in binary is determined by the number of bits, using the formula

$$\textbf{Largest number} = \mathbf{2}^n - \mathbf{1}$$

where n is the number of bits. For example, with eight bits you can count from zero up to $2^8 - 1 = 255$.

Review Questions

6. What are the two digits in a binary number?
7. What are the digits in a binary number called?

8. Which bit is the MSB and which is the LSB in a binary number?
9. In an 8-bit number the highest weight is 128. In a 9-bit number, what is the highest weight?
10. What is the largest number that can be represented with five bits?

1–3 HEXADECIMAL AND OCTAL NUMBERS

The hexadecimal (hex) and octal number systems are important in computer applications. These number systems are useful, particularly in relation to certain types of computer programming, because of their ease of conversion to and from the binary number system. Also, they provide an efficient way to represent large binary numbers.

In this section, you will learn about hexadecimal and octal number systems and how to convert between binary numbers and hexadecimal and octal numbers.

Hexadecimal Numbers

As you know, the decimal number system has a base of 10 and consists of ten digits and the binary number system has a base of 2 and consists of two digits. The **hexadecimal** number system has a base of 16 and, therefore, consists of sixteen digits. The first ten hexadecimal digits are the same as decimal, 0 through 9, but the remaining six digits are represented by the letters A through F. As strange as this may seem, there is no reason why a letter cannot be used to represent a numerical quantity as long as you understand what it means. Table 1–3 lists the sixteen hexadecimal digits and their decimal equivalents.

When counting in hexadecimal, you have to use a two-digit number after F, just as you do in decimal after 9. Decimal 16 is 10 in hexadecimal, and you continue with two-digit numbers to hexadecimal FF. Since many hex numbers "look" like decimal numbers, a subscript of 16 is often used to identify a hexadecimal number. For example, hexadecimal two-five is shown as 25_{16}. Another common way to identify a hexadecimal number is with an *H* following the number, such as 25H.

TABLE 1–3

Hexadecimal digits and their decimal equivalents.

Hexadecimal	Decimal
0	0
1	1
2	2
3	3
4	4
5	5
6	6
7	7
8	8
9	9
A	10
B	11
C	12
D	13
E	14
F	15

EXAMPLE 1–3

Problem
Count from 0_{16} to 30_{16} in hexadecimal.

Solution
The sequence of hexadecimal numbers is shown without the subscript 16 as follows: **0, 1, 2, 3, 4, 5, 6, 7, 8, 9, A, B, C, D, E, F, 10, 11, 12, 13, 14, 15, 16, 17, 18, 19, 1A, 1B, 1C, 1D, 1E, 1F, 20, 21, 22, 23, 24, 25, 26, 27, 28, 29, 2A, 2B, 2C, 2D, 2E, 2F, 30**.

Question
What are the next five hexadecimal numbers after 30_{16}?

Conversion Between Hexadecimal and Binary

It is easy to go from hex to binary or from binary to hex. The largest decimal value represented by a hex digit is 15, (F in hex). The binary system requires four bits to represent 15, ($2^4 - 1 = 15$). Therefore, each hexadecimal digit represents four binary bits, as shown in Table 1–4.

TABLE 1–4

Hexadecimal	Binary
0	0000
1	0001
2	0010
3	0011
4	0100
5	0101
6	0110
7	0111
8	1000
9	1001
A	1010
B	1011
C	1100
D	1101
E	1110
F	1111

Conversion from Binary to Hexadecimal

Binary to hex conversion is a straightforward process. Simply break the binary number into 4-bit groups starting at the right-most bit (LSB) and replace each 4-bit group with the equivalent hexadecimal digit. If needed, you should add leading zeros to complete the left-most 4-bit group.

EXAMPLE 1–4

Problem
Convert the following binary numbers to hexadecimal:

(a) 1100010111100001
(b) 111011000001001111

Solution

Use Table 1–4 to determine the equivalent hexadecimal digit for each 4-bit group.

(a)	1100	0101	1110	0001	Binary
	C	**5**	**E**	**1**	Hexadecimal

(b) Add the 0s shown in red to complete the left-most 4-bit group.

0011	1011	0000	0100	1111	Binary
3	**B**	**0**	**4**	**F**	Hexadecimal

Question

What is the hexadecimal number for the binary number 1000111001110100?

Conversion from Hexadecimal to Binary

Hex to binary conversion is also straightforward. Simply write the 4-bit binary number for each hexadecimal digit. You should add leading zeros to complete the left-most 4-bit group if necessary.

EXAMPLE 1–5

Problem

Convert the following hexadecimal numbers to binary:

(a) $942A_{16}$ (b) $3DA71_{16}$

Solution

Use Table 1–4 to determine the equivalent 4-bit binary number for each hex digit.

(a)	9	4	2	A		Hexadecimal
	1001	**0100**	**0010**	**1010**		Binary
(b)	3	D	A	7	1	Hexadecimal
	0011	**1101**	**1010**	**0111**	**0001**	Binary

Question

What is the binary number for the hexadecimal number 5C08?

You can use a calculator to do arithmetic operations in hexadecimal. This topic will be discussed in Section 1–4.

Octal Numbers

Although the octal number system is not as commonly used as the hexadecimal number system, you should be familiar with it. The **octal** number system has a base of 8 and, therefore, consists of eight digits. The eight octal digits are 0 through 7.

When counting in octal, you have to use a two-digit number after 7, just as you do in decimal after 9. Decimal 8 is 10 in octal, and you continue with two-digit numbers to 77. Since the octal numbers "look" like decimal numbers, a subscript of 8 is often used to identify an octal number. For example, octal two-five is shown as 25_8. Another common way to identify an octal number is with a *Q* following the number, such as 25Q.

EXAMPLE 1–6

Problem
Count from 0_8 to 30_8 in octal.

Solution
The sequence of octal numbers is shown without the subscript 8 as follows:
0, 1, 2, 3, 4, 5, 6, 7, 10, 11, 12, 13, 14, 15, 16, 17, 20, 21, 22, 23, 24, 25, 26, 27, 30.

Question
What are the next five octal numbers after 30_8?

Conversion Between Octal and Binary

Just like in hexadecimal, octal provides for easy conversion to and from binary and is an efficient way to represent large binary numbers. The largest decimal value that can be represented by an octal digit is 7. The binary system requires three bits to represent the number 7, ($2^3 - 1 = 7$). Therefore, each octal digit represents three binary bits, as shown in Table 1–5.

TABLE 1–5

Octal	Binary
0	000
1	001
2	010
3	011
4	100
5	101
6	110
7	111

Conversion from Binary to Octal

To convert a binary number to octal, break the binary number into 3-bit groups starting at the right-most bit (LSB) and replace each 3-bit group with the equivalent octal digit. If needed, you should add leading zeros to complete the left-most 3-bit group.

EXAMPLE 1–7

Problem
Convert the following binary numbers to octal:

(a) 110001011110 (b) 1100110000011

Solution
Determine the octal digit for each 3-bit group.

(a)

110	001	011	110	Binary
6	**1**	**3**	**6**	Octal

(b) Add the 0s shown in red to complete the left-most 3-bit group.

001	100	110	000	011	Binary
1	**4**	**6**	**0**	**3**	Octal

Question
What is the octal number for the binary number 1000111001110100?

Conversion from Octal to Binary

To convert an octal number to binary, write the 3-bit binary number for each octal digit. You should add leading zeros to complete the left-most 3-bit group if necessary.

EXAMPLE 1–8

Problem

Convert the following octal numbers to binary:

(a) 375_8 (b) 6024_8

Solution

Determine the equivalent 3-bit binary number for each octal digit.

(a)	3	7	5		Octal
	011	**111**	**101**		Binary
(b)	6	0	2	4	Octal
	110	**000**	**010**	**100**	Binary

Question

What is the binary number for the octal number 1027_8?

You can use your calculator to do arithmetic operations with octal numbers as described in the next section.

Review Questions

11. How many digits does the hexadecimal number system have?
12. When converting between hex and binary, how many bits does it take to represent a hex digit?
13. How many digits does the octal number system have?
14. When converting between octal and binary, how many bits does it take to represent an octal digit?
15. What is the decimal equivalent of 5_{16}?

1–4 NUMBER SYSTEMS ON THE CALCULATOR

You can use a calculator to convert between decimal, hexadecimal, and octal numbers. Also, you can evaluate powers of two on a calculator and do arithmetic operations with hexadecimal and octal numbers. Scientific calculators differ in terms of the various key functions, display format, and capability. We use the TI-36X to show calculator operations. If you have a different calculator, refer to the owner's manual.

In this section, you will learn to do number system conversions and hexadecimal arithmetic on a calculator.

Powers of Two

Only four keystrokes are necessary to evaluate a power of two on the TI-36X: [2], followed by [y^x], followed by the power of two, followed by [=].

EXAMPLE 1–9

Problem
Use a calculator to find each of the following powers of two:

(a) 2^3 (b) 2^6 (c) 2^8 (d) 2^{10}

Solution

(a) 2 yˣ 3 = The display shows **8**
(b) 2 yˣ 6 = The display shows **64**
(c) 2 yˣ 8 = The display shows **256**
(d) 2 yˣ 10 = The display shows **1024**

Question
What is the value of 2^{15}?

Converting from Decimal to Hexadecimal

To put the TI-36X calculator in the decimal mode, press 3rd then EE; DEC is a tertiary function of the EE key. In this mode, the numbers that you enter are all decimal numbers.

To convert a decimal number to hexadecimal, first enter the decimal number; then press 3rd and ((HEX is a tertiary function of the (key) to display the equivalent hexadecimal number. The calculator is now in the hex mode and will show a small HEX on the screen.

EXAMPLE 1–10

Problem
Use a calculator to convert each of the following decimal numbers to hexadecimal:

(a) 12 (b) 97 (c) 653

Solution

(a) 3rd EE (DEC) 1 2 3rd ((HEX) The display shows **C**
(b) 3rd EE (DEC) 9 7 3rd ((HEX) The display shows **61**
(c) 3rd EE (DEC) 6 5 3 3rd ((HEX) The display shows **28d**

Question
What is the hexadecimal number for the decimal number 215?

Converting from Hexadecimal to Decimal

To put the TI-36X calculator in the hexadecimal mode, press 3rd then (. In this mode, the numbers that you enter are all hexadecimal numbers. To enter hexadecimal digits A through F, you must first press the 3rd key before you enter the digit.

To convert a hexadecimal number to decimal, first enter the hexadecimal number; then press 3rd and EE to display the equivalent decimal number. The calculator is now in the decimal mode.

EXAMPLE 1–11

Problem

Use a calculator to convert each of the following hexadecimal numbers to decimal:

(a) $39A_{16}$ (b) $DF50_{16}$

Solution

(a) 3rd ([HEX] 3 9 3rd 1/x [A] 3rd EE [DEC] The display shows **922**

(b) 3rd ([HEX] 3rd SIN [D] 3rd TAN [F] 5 0 3rd EE [DEC] The display shows **57168**

Question

What is the decimal number for the hexadecimal number $81B_{16}$?

Hexadecimal Arithmetic

You can add, subtract, multiply, and divide hexadecimal numbers on your calculator. The arithmetic operations are done just as you would for decimal numbers except you must have the calculator in the hex mode.

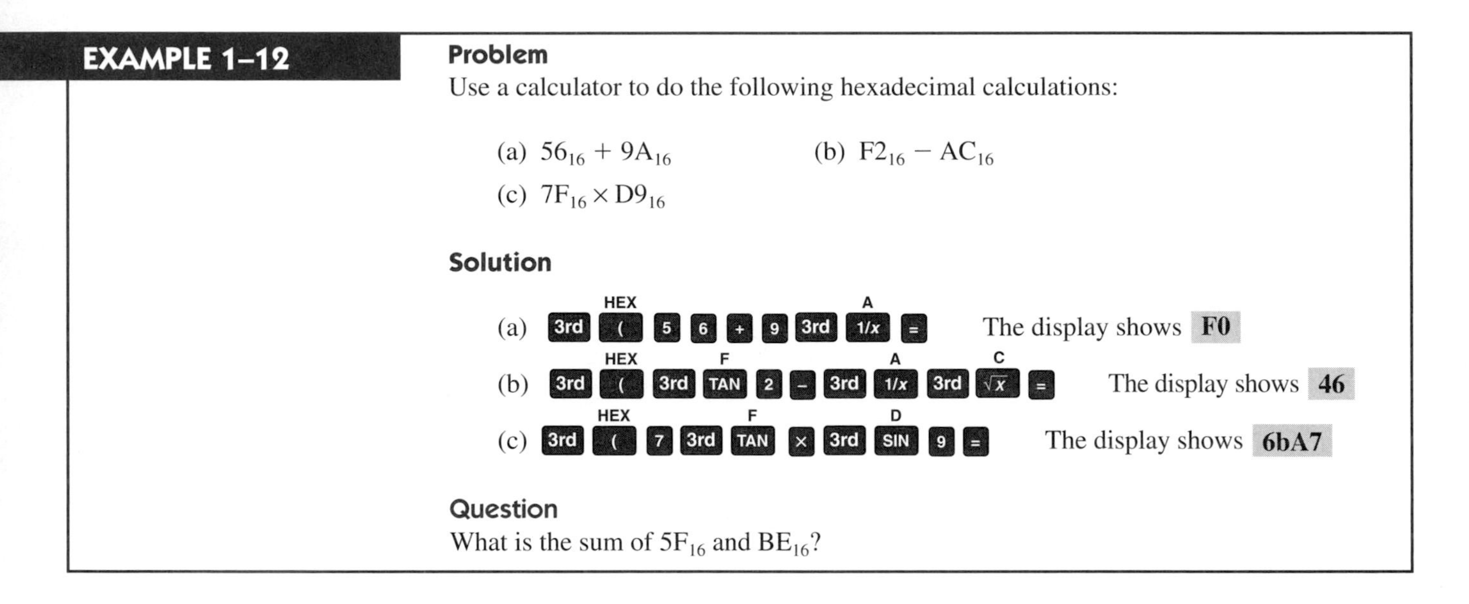

EXAMPLE 1–12

Problem

Use a calculator to do the following hexadecimal calculations:

(a) $56_{16} + 9A_{16}$ (b) $F2_{16} - AC_{16}$

(c) $7F_{16} \times D9_{16}$

Solution

(a) 3rd ([HEX] 5 6 + 9 3rd 1/x [A] = The display shows **F0**

(b) 3rd ([HEX] 3rd TAN [F] 2 − 3rd 1/x [A] 3rd √x [C] = The display shows **46**

(c) 3rd ([HEX] 7 3rd TAN [F] × 3rd SIN [D] 9 = The display shows **6bA7**

Question

What is the sum of $5F_{16}$ and BE_{16}?

Decimal/Octal Conversions

Conversions from decimal to octal and octal to decimal are the same as the hexadecimal conversions except that you use the octal mode. OCT on the calculator is the tertiary function of the) key.

EXAMPLE 1–13

Problem

(a) Convert the decimal number 92 to octal.

(b) Convert the octal number 47_8 to decimal.

Solution

(a) [3rd] [EE] (DEC) [9] [2] [3rd] [)] (OCT) The display shows **134**

(b) [3rd] [)] (OCT) [4] [7] [3rd] [EE] (DEC) The display shows **39**

Question

What is the octal number for the decimal number 289? What is the decimal number for 614_8?

Decimal/Binary Conversions

Conversions from decimal to binary and binary to decimal are the same as the hexadecimal and octal conversions except that you use the binary mode. BIN on the calculator is the tertiary function of the [×] key.

EXAMPLE 1–14

Problem

(a) Convert the decimal number 35 to binary.

(b) Convert the binary number 101011 to decimal.

Solution

(a) [3rd] [EE] (DEC) [3] [5] [3rd] [X] (BIN) The display shows **100011**

(b) [3rd] [X] (BIN) [1] [0] [1] [0] [1] [1] [3rd] [EE] (DEC) The display shows **43**

Question

What is the binary number for the decimal number 289? What is the decimal number for the binary number 11101100?

Review Questions

16. How do you put the TI-36X calculator in the decimal mode?
17. How do you put the calculator in the hexadecimal mode?
18. How do you enter one of the hexadecimal digits A through F?
19. How do you put the calculator in the octal mode?
20. How do you put the calculator in the binary mode?

LOGIC LEVELS AND DIGITAL WAVEFORMS 1–5

Digital electronic circuits use two voltages called **logic levels**. The two logic levels in a digital circuit represent the binary digits 1 and 0. In most cases, a HIGH logic level represents a 1, and a LOW logic level represents a 0.

In this section, you will learn to describe the logic levels in digital systems and analyze the characteristics of a pulse waveform.

Logic Levels

The two logic levels used in digital circuits can be illustrated with a simple circuit that produces two voltage levels using a battery and switch, as shown in Figure 1–5 in the computer simulation feature. The battery is connected to one contact on the switch, and ground is connected to the other. The positive terminal of the battery is at some voltage, in this case 5 V, and ground is 0 V. When the switch contact is in the upper position, it is connected to 5 V; and when it is in the lower position, it is connected to 0 V. Point *A* is either at the higher voltage (5 V) or at the lower voltage (0 V), depending on the switch position. If a probe light is connected to the switch as shown, it will indicate the voltage level.

COMPUTER SIMULATION

FIGURE 1–5

Open the Multisim file F01-05DG on the website. To make the switch move back and forth, press the space bar. The high voltage level is represented by the battery, and the low voltage is represented by the ground symbol. The probe light is red when point *A* is HIGH.

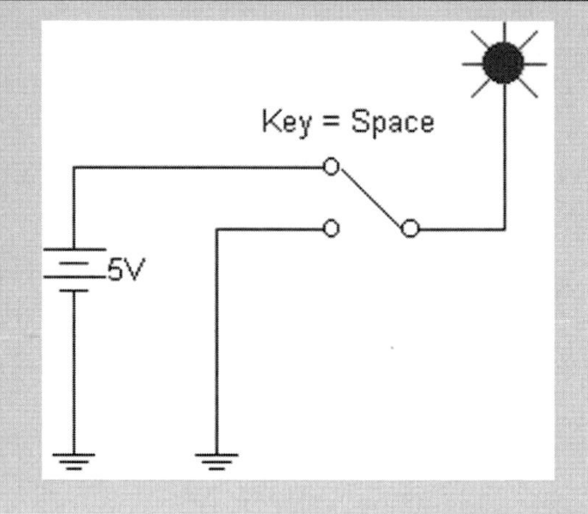

Digital Waveforms

A **waveform** is the pattern followed when voltage changes with time. When the voltage level changes back and forth between a higher voltage and a lower voltage, it creates a digital waveform. Suppose that you change the switch in Figure 1–5 back and forth every second. The voltage at point *A* remains at 5 V for one second and then goes to 0 V for one second, as indicated by the probe light in the Multisim circuit. This pattern continues to repeat as the switch is moved back and forth. If the voltage at point *A* is graphed with respect to time, you have the waveform diagram shown in Figure 1–6. This waveform diagram

FIGURE 1–6

A digital waveform diagram.

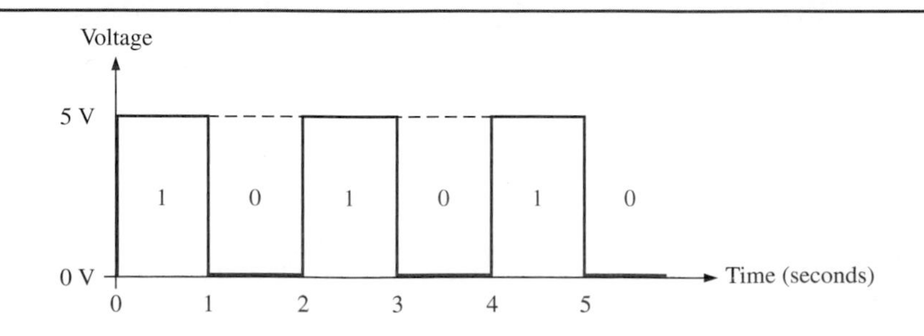

shows how the voltage changes with time and can be used to represent a sequence of binary digits 1, 0, 1, 0, 1, 0, as indicated in red in the figure. When the waveform is at a high level (HIGH), a 1 is represented; and when it is at a low level (LOW), a 0 is represented. This method of representation, called positive logic, is the most commonly used method.

Digital waveforms are made up of a series of pulses. In Figure 1–6, three pulses are shown. A positive-going pulse occurs when the voltage makes a transition from the lower level to the upper level, remains at the upper level for a specific time, and then makes a transition back from the higher level to the lower level. Two types of digital waveforms are timing waveforms and data waveforms.

Timing Waveforms

Timing waveforms provide precise time intervals upon which all operations in a digital system are based. A timing waveform is sometimes called the **clock** because it is the basic timing signal. For example, the statement that a certain microprocessor runs at 2 GHz refers to the maximum clock frequency at which the device will operate. A timing waveform or clock, such as shown in Figure 1–7, has five important characteristics: *frequency*, *period*, *pulse width*, *duty cycle*, and *amplitude*.

- The **frequency** of a digital waveform is the number of pulses that repetitively occur in one second, expressed in the unit of hertz (Hz). The symbol for frequency is f.
- The **period** is the time interval measured from one pulse to the same point on the next pulse, expressed in the unit of seconds (s). The symbol for period is T. Frequency is equal to the reciprocal of the period, $f = 1/T$.
- The **pulse width** is the time duration of each pulse, expressed in the unit of seconds. The symbol for pulse width is t_W.
- The **duty cycle** is the ratio of the pulse width to the period and has no units. It may be expressed as a percentage.
- The **amplitude** is the "height" of the pulses measured from the baseline, usually expressed in volts, V.

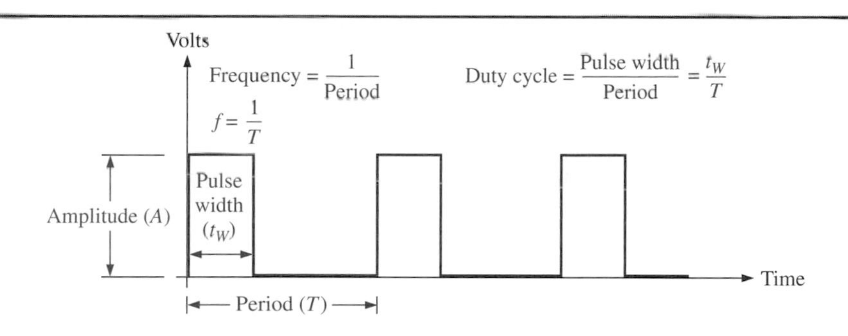

FIGURE 1–7

A timing waveform.

Data Waveforms

Unlike timing waveforms, **data waveforms** contain binary information; they represent sequences of bits with a series of high and low levels. Data waveforms do not have a definable pulse width or period because these values can vary depending on the sequence of bits. For example, there can be several consecutive HIGHS or several consecutive LOWS. The frequency of a data waveform is usually defined by the clock frequency to which it is **synchronized**.

When a data waveform is synchronized to a timing waveform (clock), the data waveform changes only when the clock changes, as illustrated in Figure 1–8. This particular data waveform contains the bit sequence 110100010. Notice that this data waveform does not change on every clock pulse but when it does, the change occurs only at a precise point (front or leading edge) on a clock pulse. When two or more waveforms are shown in the proper time relationship to each other, it is called a **timing diagram**.

FIGURE 1–8

A timing diagram showing the relationship of a timing waveform and a data waveform.

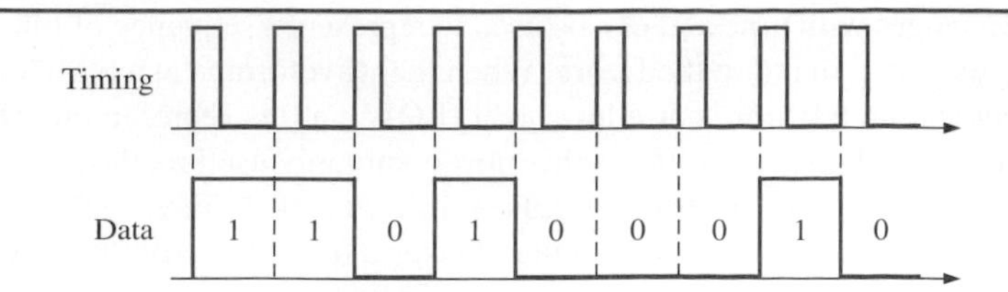

EXAMPLE 1–15

Problem

(a) Find the following values for the timing waveform in Figure 1–9(a): period; frequency; duty cycle; and amplitude.

(b) What is the bit sequence in Figure 1–9(b)?

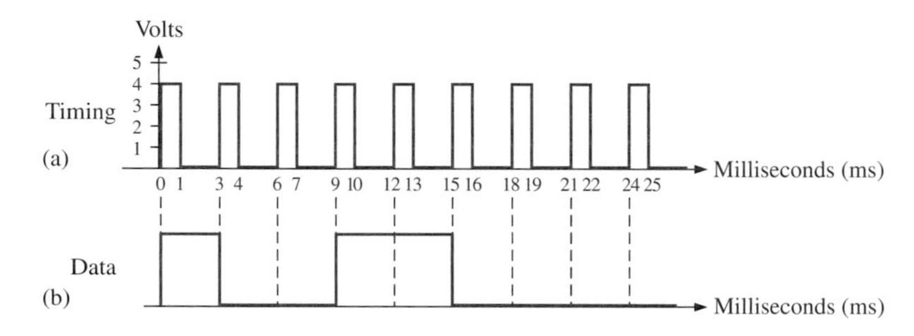

FIGURE 1–9

Solution

(a) *Period:* The time scale (on the horizontal axis) of the timing waveform is in milliseconds. The pulses occur at 3 ms intervals. The period of the timing waveform is the time from the beginning of one pulse to the beginning of the next pulse and is equal to 3 milliseconds.

$$T = \mathbf{3\ ms}$$

Frequency: The frequency is the reciprocal of the period.

$$f = \frac{1}{T} = \frac{1}{3\text{ ms}} = \frac{1}{3 \times 10^{-3}\text{ s}} = \mathbf{333\ Hz}$$

Duty cycle: The duty cycle is the ratio of the pulse width to the period. The pulse width is 1 ms.

$$\text{Duty cycle} = \frac{t_W}{T} = \frac{1\text{ ms}}{3\text{ ms}} = \mathbf{0.333}$$

Expressed as a percentage, the duty cycle is 33.3%.

Amplitude: The amplitude is the "height" of the pulses in volts and is read on the vertical axis.

$$\text{Amplitude} = \mathbf{4\ V}$$

(b) The bit sequence of the data waveform is **100110000**.

Question
What is the frequency of a waveform if the period is 10 μs?

Review Questions

21. How are the two binary digits represented by logic levels?
22. What is the difference between a timing waveform and a data waveform?
23. How is the period of a timing waveform determined?
24. If the period of a certain waveform is 1 s, what is the frequency?
25. What bit sequence is represented by a waveform with the following sequence of logic levels: HIGH, HIGH, LOW, HIGH, LOW, LOW, LOW, LOW, HIGH, HIGH, LOW?

ELEMENTS OF A DIGITAL SYSTEM 1–6

All digital systems, from a digital watch to a complex digital computer, are made up of basic "building block" circuits called logic gates. Most logic circuits are made on tiny chips of silicon and encased in a small package with leads or pins for connections. These are called integrated circuits.

In this section, you will learn about the elements that make up a digital system.

Digital Logic Circuits

All digital circuits produce, operate on, or store binary digits in the form of logic levels. These circuits vary from very simple to very complex. For example, a logic gate is a relatively simple circuit and a microprocessor is a very complex device. All of the circuits discussed here are used to create digital systems by interconnecting them in such a way that they perform a specified result.

Logic circuits that you will learn about in this book are logic gates, arithmetic logic, decoders, encoders, data selectors, timers, latches, flip-flops, counters, registers, programmable logic, memories, and data converters.

Logic Gates

Logic gates are the "building blocks" of digital systems. Most of the other logic circuits that you will study in this text are made from logic gates connected together in various ways. The type of logic gate determines how it responds to various combinations of logic levels that are applied to it.

In the next three chapters, you will learn about the characteristics and operation of the AND gate, OR gate, Inverter, NAND gate, NOR gate, Exclusive-Or gate, and Exclusive-NOR gate.

Arithmetic Logic

This type of logic circuit performs mathematical and logical operations on binary numbers, including addition, subtraction, multiplication, and division. The part of a microprocessor that does these operations is called the Arithmetic Logic Unit or ALU.

Decoders and Encoders

Decoder circuits are a combination of logic gates. Decoders can detect when certain binary numbers or other codes are applied and produce an output that indicates the value of the binary input. Encoders are a combination of logic gates and can produce a binary number on the output corresponding to an applied decimal number on the input.

Data Selectors

These logic circuits are also known as multiplexers and demultiplexers depending on their operation. A multiplexer is used to select data from several lines and connect the data to a single line in a prescribed time sequence. A demultiplexer reverses the multiplexer operation. It takes the data from a single line and connects the data to several lines in a prescribed time sequence.

Timers

Timers are digital circuits that are used to produce timing waveforms and precise time intervals. One type of timer is the one-shot, a type of circuit that has only one stable state (monostable) that can produce a single pulse for a precise duration of time.

Another type of timer is a circuit that has no stable states (astable) and is used primarily to produce timing or clock waveforms. An astable circuit is also known as an *oscillator.*

Latches and Flip-Flops

These are not things that go on doors or on your feet! A latch is a digital circuit for storing a single bit of data and is made up of logic gates. The latch is a bistable device having two stable states. The flip-flop is also a bistable device that is a close relative of the latch. The basic difference between a latch and a flip-flop is that the flip-flop is synchronized with other circuits with a timing waveform or clock.

Counters

A counter is a group of flip-flops connected in a certain way so that it exhibits a prescribed sequence of binary states when a timing or clock waveform is applied. The counter is used to count the number of electrical events or produce a prescribed sequence of binary numbers. Counters are important in all digital systems, including computers and microprocessors.

Shift Registers

A register is basically a group of flip-flops used to store more than one bit at a time. Registers are an important part of computers and microprocessors, and most microprocessors use many registers for different purposes.

Programmable Logic

The area of programmable logic devices (PLDs) is a relatively new technology and is being used more and more in digital systems. All of the types of logic circuits that have been mentioned are generally available in individual integrated circuit packages. However, the trend is toward implementing many types of logic circuits in a device that can be programmed to be just about any type of logic circuit that you want it to be. A PLD can be programmed as logic gates, flip-flops, decoders, encoders, data selectors, registers, counters, or combinations of these.

Memory

Storing large amounts of data is an important function in many types of systems, especially the computer. The part of a system that is used to store data is called the memory. Some types of memories use latches as the basic storage element, while others utilize capacitors, magnetic material, or other methods.

Conversion and Interfacing

When two or more components of a system are connected together, they must be properly interfaced in order to work. *Interfacing* is the process of making two or more circuits or systems electronically compatible so they will operate properly when connected to each other to perform a desired function. Many digital systems require the conversion of an analog quantity to digital and vice versa. One important application is in digital signal processors (DSPs).

Review Questions

26. What type of logic circuit can be considered as a "building block" for digital systems?
27. What is a latch used for?
28. What is a PLD?
29. What is interfacing?
30. What is the purpose of a memory?

INTRODUCTION TO INSTRUMENTS 1–7

In order to troubleshoot digital circuits, you must know how to use certain instruments. You may already be familiar with the basic instruments from your dc/ac circuits studies. Therefore, this discussion is limited to a basic review and applications to digital measurements.

In this section, you will learn about common instruments and how to use them in the lab.

The General-Purpose Digital Oscilloscope

The oscilloscope (scope) is used to display waveforms and measure their parameters such as amplitude, frequency, period, and duty cycle. Two types of oscilloscopes are analog and digital. The digital scope is the most widely used type. The front panel of a typical digital oscilloscope is shown in Figure 1–10. A special type of digital oscilloscope is the digital storage oscilloscope (DSO), which allows you to store digital waveforms for later analysis and viewing.

FIGURE 1–10 Digital oscilloscope with displayed sample waveforms.

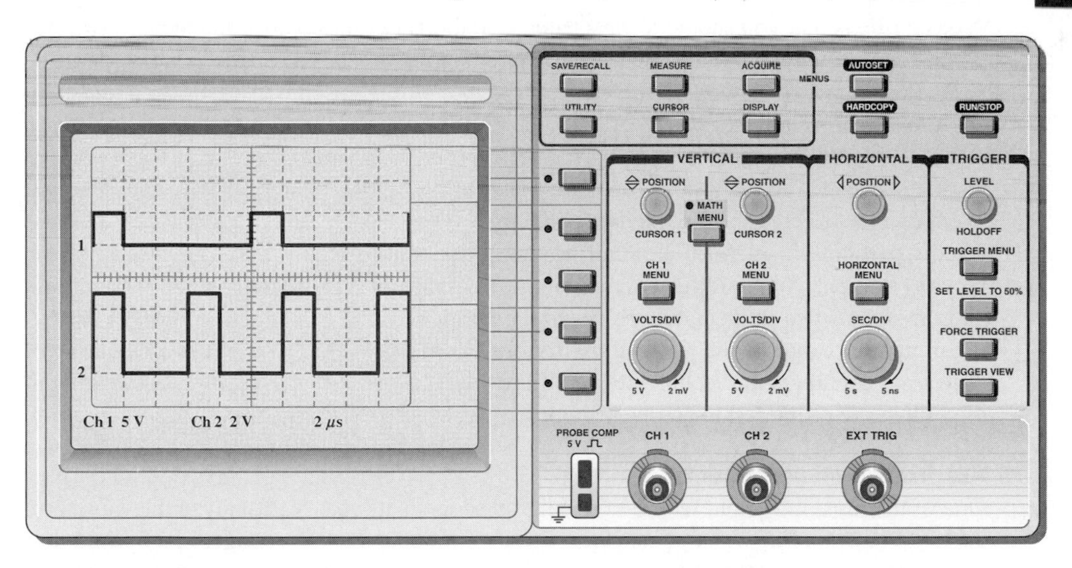

The Oscilloscope Screen

In addition to displaying waveforms, the oscilloscope screen has many features that provide information about the waveform and the control settings. The screen is divided into eight major vertical divisions and ten major horizontal divisions. Voltage is measured on the vertical scale where the Volts/Div control sets the volts for each division. There are two Volts/Div controls, one for each channel. Time is measured on the horizontal scale where

SAFETY NOTE

Proper grounding is important when you set up to take measurements or work on a circuit because proper grounding protects you from shock and assures accurate measurements. Grounding the oscilloscope simply means to connect it to earth ground by plugging the three-prong power cord plug into a grounded outlet. Also, for accurate measurements, make sure that the ground in the circuit you are testing is the same as the scope ground. This can be done by connecting the ground lead on the scope probe to a known ground point in the circuit, such as the metal chassis or a ground contact on the circuit board.

the Sec/Div control sets the seconds for each division. There is only one Sec/Div control for both channels.

DC Coupling

You should always use dc coupling when measuring digital waveforms so that you can see the waveform with respect to a ground reference. DC coupling is selected from the Vertical Menu for each channel. Each waveform shown in Figure 1–10 has a base line at zero volts (ground) because dc coupling is used. The on-screen numbered markers show the ground reference for each waveform. The "1" is for the channel 1 waveform, and the "2" is for the channel 2 waveform.

Measuring the Amplitude

The Volts/Div settings for both channels are displayed along the bottom left side of the screen. You can measure the amplitude of a waveform by multiplying the number of vertical divisions from the baseline to the peak by the Volts/Div setting displayed for that channel. For example, the channel 1 waveform in Figure 1–10 is one division high and the displayed setting is 5 V, so its amplitude is 1×5 V = 5 V. The channel 2 waveform is 2.5 divisions high and the displayed setting is 2 V, so its amplitude is $2.5 \times 2 = 5$ V.

Measuring the Period

The Sec/Div setting for the time base is displayed at the bottom of the screen, as shown in Figure 1–10. You can measure the period of a repeating waveform by counting the number of horizontal divisions covered by one cycle and multiplying that number by the Sec/Div setting. For example, the channel 1 waveform in Figure 1–10 shows five divisions from one pulse to the next and the displayed setting is 2 μs, so its period is $5 \times 2\ \mu s = 10\ \mu s$. The channel 2 waveform shows three divisions from one pulse to the next, so its period is $3 \times 2\ \mu s = 6\ \mu s$. Remember, both channels have the same time base (Sec/Div setting). Once you know the period, you can also determine the frequency using the formula $f = 1/T$.

Triggering

A basic rule is that you should always trigger on the slowest (lowest frequency) waveform when comparing two digital signals. The trigger source is selected by the Trigger Menu, as shown in Figure 1–10. One useful feature of digital storage oscilloscopes is their ability to capture waveforms either *before* or *after* the trigger event. Any segment of the waveform can be captured for analysis. With pretrigger capture the acquisition of data occurs before a trigger event. This is possible because the data is digitized continuously, and a trigger event can be selected to stop the data collection at some point within the sample window. This is particularly useful if a circuit has an occasional problem such as a random noise spike or "glitch." With pretrigger capture, the scope can be triggered on the fault condition, and the signals that immediately preceded the fault condition can be observed. By employing pretrigger capture, one can analyze trouble leading to the fault. A similar application of pretrigger capture is in certain types of failure analysis studies where the events leading to the failure are of interest to an investigator but the failure itself causes the scope triggering.

Automated Measurements

One of the most important features on a digital scope is its ability to analyze the waveform and show certain important parameters such as frequency or rms voltage. In addition, digital scopes can measure and display the time or voltage between a set of two user-controlled markers (called *cursors*). In this way, the user can read the time or voltage directly between the cursors and avoid the possibility of misreading a dial or making a calculational mistake. Results are generally more accurate than reading the number of divisions and multiplying by a dial setting. The voltage of a given wave can be displayed in various forms also. For example, a sine wave can be analyzed for its peak-to-peak or its rms voltage without the user having to read the settings of the controls.

Because of the large number of functions that can be accomplished by even basic DSOs, manufacturers have largely replaced the large number of controls with menu options, similar to computer menus. The settings of controls can be easily viewed on the display and options can be selected, depending on the control. CRTs have been replaced by liquid crystal displays, similar to those on laptop computers.

Scope Probe

A probe, such as the one shown in Figure 1–11, can be attached to each of the channel connectors and then connected to the point in a circuit at which you wish to see the waveform. The probe has a connector that connects to the scope, a tip which is touched to the point in a circuit where a waveform is to be measured, and a ground lead that should be connected to circuit ground.

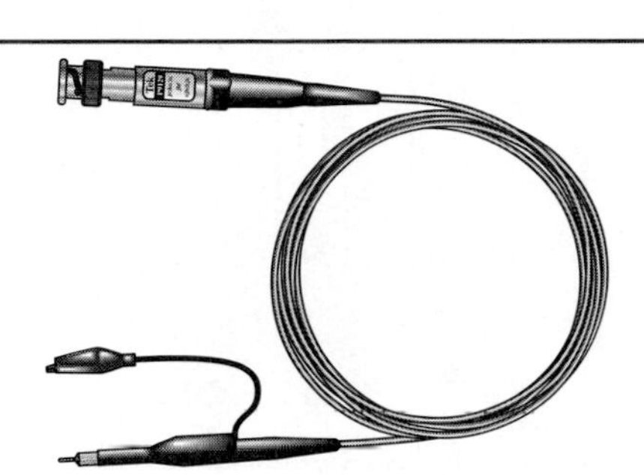

FIGURE 1–11

A typical oscilloscope voltage probe.

Probe Attenuation

Probes generally are either classified as ×1 (times 1) or ×10 (times 10) based on the attenuation or reduction of the measured voltage to which they are connected. A ×1 probe does not reduce the measured voltage. A ×10 probe reduces the measured voltage by 10 but results in less of a loading effect on the circuit than a ×1 probe. The ×1 probe is sometimes used to measure very small voltages, but the ×10 probe is the one that is most commonly used. Digital oscilloscopes automatically offset the attenuation of the ×10 probe so that exact voltage readings are provided. Some analog oscilloscopes require a manual selection for either a ×1 or ×10 so that the Volts/Div reading takes into account the attenuation.

Probe Compensation

Voltage probes generally require compensation to adjust for any mismatch between the probe and the oscilloscope input. If a probe is not properly compensated, a digital waveform displayed on the scope screen may show a certain amount of distortion, as illustrated in Figure 1–12. The figure shows how an ideal square wave being measured by a probe would appear on the scope screen for three different probe compensation conditions. These conditions are properly compensated, undercompensated, and overcompensated.

Oscilloscope waveforms for three possible conditions of probe compensation.

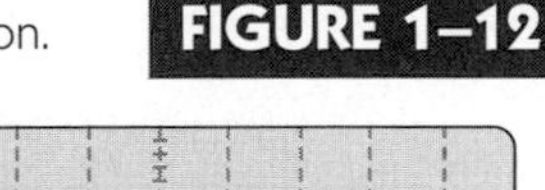

FIGURE 1–12

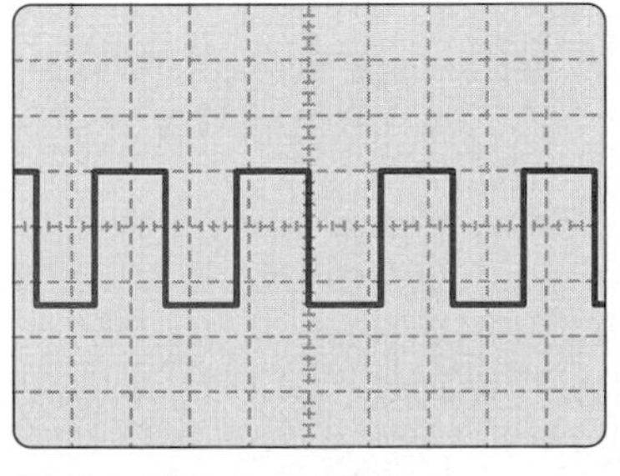

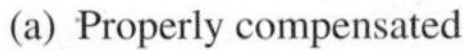

(a) Properly compensated

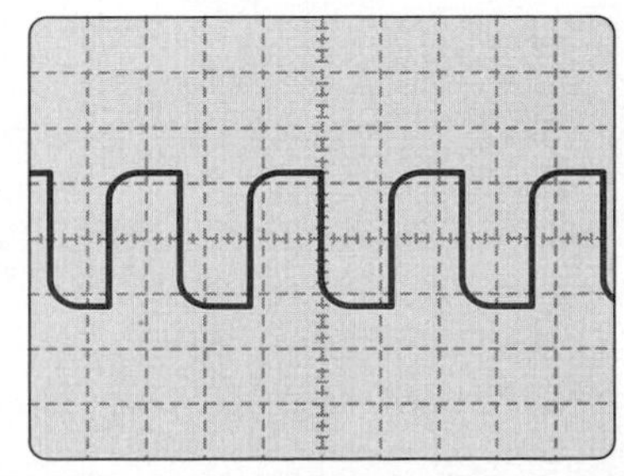

(b) Undercompensated

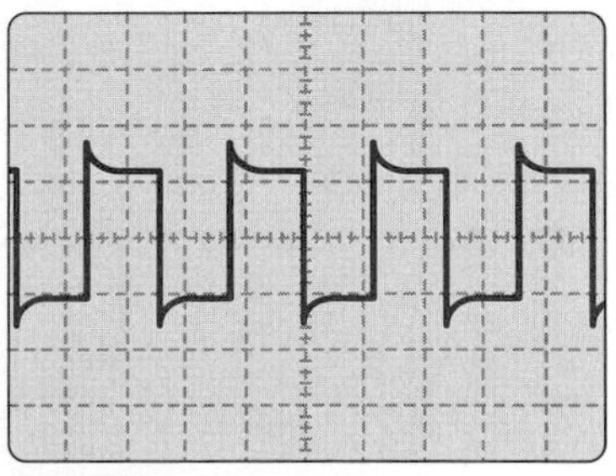

(c) Overcompensated

Most oscilloscopes provide a compensating waveform from a front panel connector, as shown in Figure 1–13. The probe to be compensated is attached to the compensation output and a small adjustment screw is "tweaked" until the waveform on the screen appears to be an exact square wave. It is good practice to check the compensation of your probe periodically, especially if you are using a different scope.

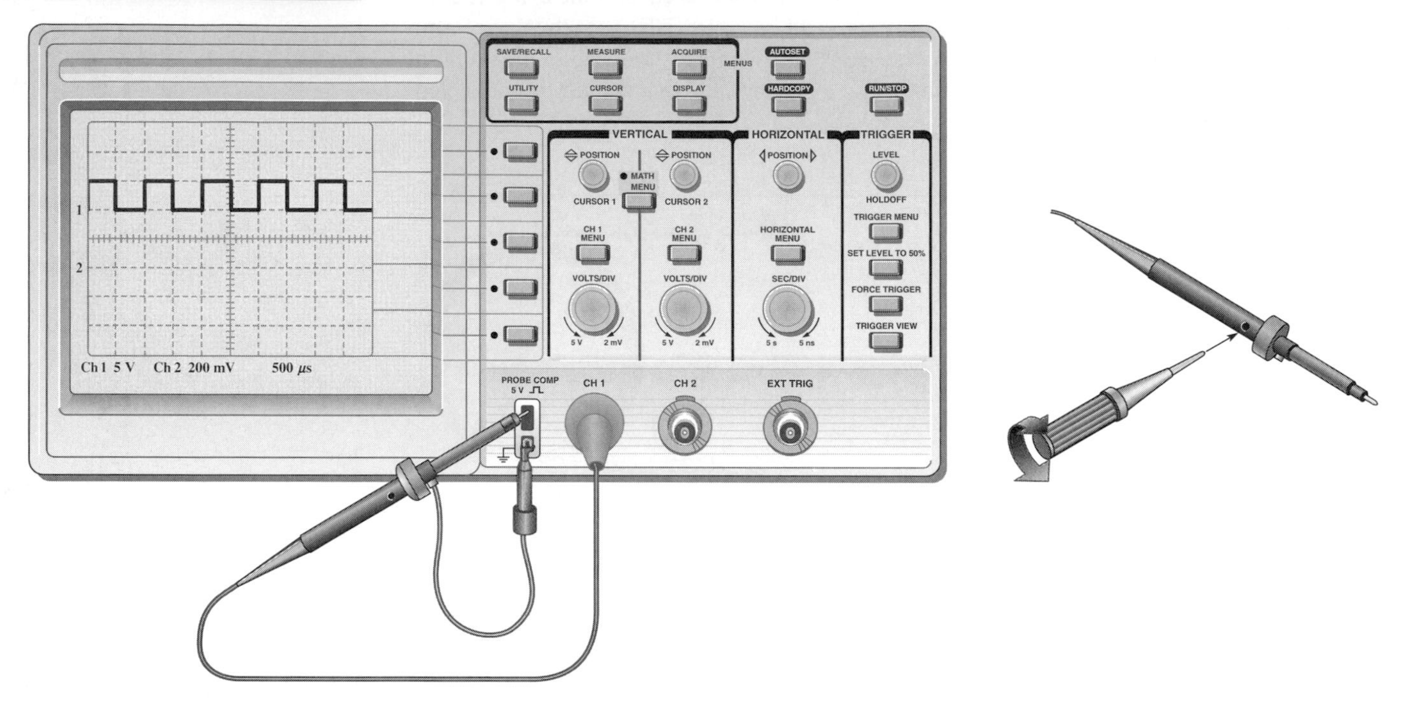

FIGURE 1–13 Compensation of an oscilloscope probe.

The Logic Analyzer

The logic analyzer is an instrument that displays digital data in several formats. Three basic formats are oscilloscope, timing diagram, and state table. In the oscilloscope format, the logic analyzer can be used to display single or dual waveforms on the screen so that characteristics of individual pulses or waveform parameters can be measured. In the timing diagram format, the logic analyzer can display typically up to sixteen waveforms in proper time relationship so that you can analyze sets of waveforms and determine how they change in time with respect to each other. In the state table format, the logic analyzer can display binary data in tabular form. For example, various memory locations in a microprocessor-based system can be examined to determine the contents. The data can be displayed in a variety of number systems and codes such as binary, hexadecimal, octal, and other codes. A typical logic analyzer is shown in Figure 1–14.

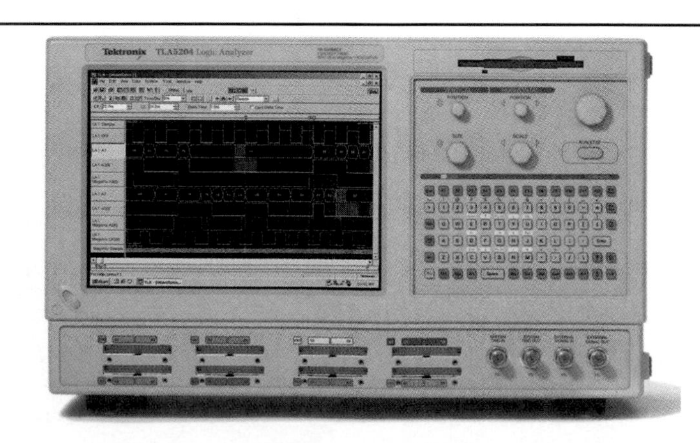

FIGURE 1–14 A typical logic analyzer. (Used with permission from Tektronix, Inc.)

The Digital Multimeter (DMM)

The digital multimeter is widely used for electrical measurements. The basic DMM measures voltage, current, and resistance. Some DMMs measure other quantities such as decibels and perform a diode test.

Although you can use the oscilloscope to measure dc voltage, the DMM is often more convenient and provides a quicker way to check a dc voltage, such as the supply voltage or a logic level, in a digital circuit.

One situation in which you must use a DMM is in resistance and continuity measurements. You can check for continuity to detect a short or an open between two points or the resistance between two points using the ohmmeter function of the DMM. Resistance and continuity measurements should be done with the power off and with no input voltage to the circuit.

An autoranging DMM is shown in Figure 1–15. Autoranging means that you do not have to manually select the voltage, current, or resistance range with a switch. When you measure dc voltage, the lead (usually red) connected to the V-Ω jack must connect to the positive voltage and the lead (usually black) connected to the COM jack must connect to ground.

FIGURE 1–15

An autoranging digital multimeter (DMM). (Courtesy of B+K Precision)

The Logic Probe and Logic Pulser

The logic probe is a convenient, handheld instrument that can be used to troubleshoot a digital circuit by sensing voltage conditions at a point in the circuit. The logic probe detects high-level voltage, low-level voltage, single pulses, repetitive pulses, and opens. You can't use the logic probe to determine measured values because it only displays the presence or absence of a voltage or pulse with a light-emitting diode. Figure 1–16 shows a typical logic probe and indications. Logic probes from various manufacturers may differ.

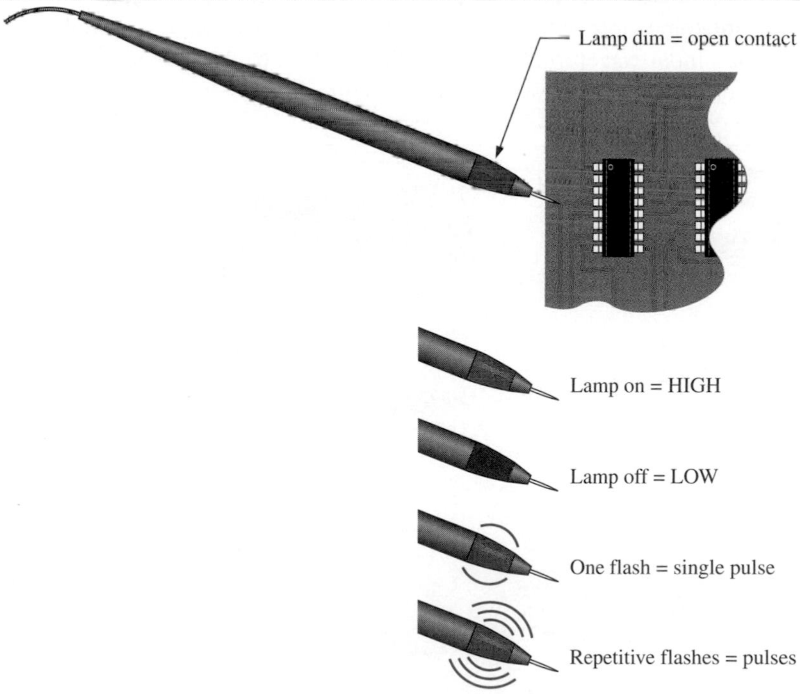

Lamp dim = open contact

Lamp on = HIGH

Lamp off = LOW

One flash = single pulse

Repetitive flashes = pulses

FIGURE 1–16

A typical logic probe and the indications for the various conditions.

The logic pulser produces repetitive pulses that can be applied to any point in a digital circuit. You can apply pulses at one point in a circuit with the pulser and check for resulting pulses at another point with a logic probe.

The DC Power Supply

The dc power supply is an indispensable instrument on any test bench. The power supply converts the ac voltage from your wall outlet at its input into a regulated dc voltage at its output. The power that can be provided depends on the amount of current the power supply can produce for the available dc output voltage. Any type of electronic circuit requires a dc supply voltage in order to operate. For example, many integrated logic circuits operate from a +5 V supply voltage. You will use a power supply when you are testing a newly constructed circuit or when you troubleshoot a pc board that is no longer operating from the internal system power supply. A typical power supply is shown in Figure 1–17.

FIGURE 1–17

A typical dc power supply that is used on the test bench. (Courtesy of B+K Precision)

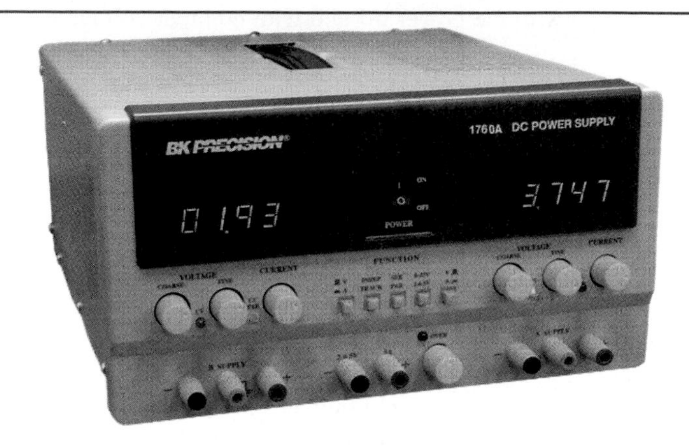

The Function Generator

The function generator is a source for electrical signals and provides pulse waveforms, sinusoidal waveforms, and triangular waveforms. Many function generators have outputs that are compatible with logic levels required by various IC technologies and are used to provide input waveforms to digital circuits for checking operation or troubleshooting. A typical function generator is shown in Figure 1–18.

FIGURE 1–18

A typical function generator that is used on the test bench. (Courtesy of B+K Precision)

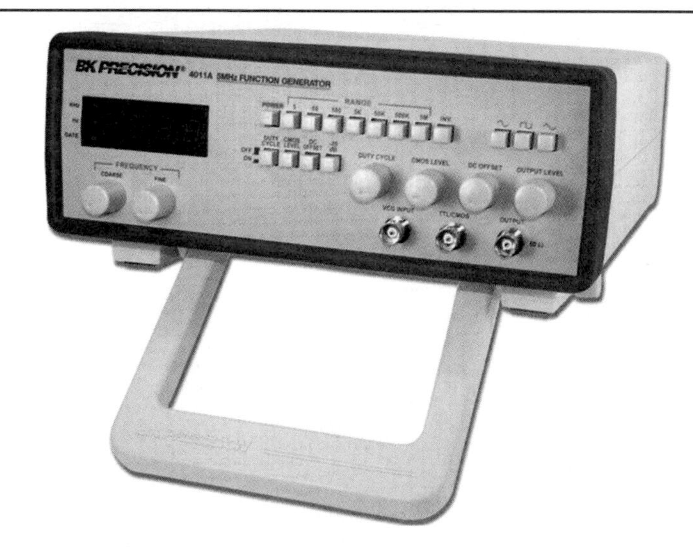

Review Questions

31. What is the purpose of an oscilloscope?
32. What type of coupling should be used when you measure a digital waveform with the oscilloscope?

33. What is the purpose of an oscilloscope voltage probe?
34. What are two types of probes in terms of attenuation?
35. What are three common instruments for making measurements.

CHAPTER REVIEW

Key Terms

Amplitude The "height" of a pulse measured from the baseline, expressed in volts.

Analog Having a continuous set of values.

Binary A number system with a base of 2 that has two digits, 1 and 0.

Bit A binary digit.

Byte A group of eight bits.

Digital Having a series of individual values (discrete).

Duty cycle The ratio of the pulse width to the period of a digital waveform, usually expressed as a percentage.

Frequency For a digital waveform, the number of pulses that occur in one second expressed in hertz (Hz).

Hexadecimal A number system with a base of 16 that has sixteen digits (0–9 and A–F).

Logic level A voltage level that represents a binary digit.

LSB Least significant bit.

MSB Most significant bit.

Octal A number system with a base of 8 that has eight digits (0–7).

Period The time interval between pulses expressed in units of seconds.

Power of two A number expressed as an exponent to the base of 2.

Pulse width The time duration of a pulse expressed in units of seconds.

Timing diagram A graph showing the time relationship of two or more waveforms.

Waveform A pattern of the changes in voltage (or current) with time.

Weight The value of a digit in a number based on its position in the number.

Important Facts

- ❑ Most natural quantities are analog.
- ❑ A digital quantity can be reproduced accurately and stored easily.
- ❑ Binary numbers have two digits, 1 and 0.
- ❑ The binary number system is a weighted system in which the weights are powers of two.
- ❑ The hexadecimal number system uses sixteen digits, 0 through 9 and A through F.
- ❑ The octal number system uses eight digits, 0 through 7.

- A timing waveform is a type of digital waveform in which the pulses occur at regular intervals.
- A data waveform is a type of digital waveform that represents sequences of binary digits that contain information.
- Logic gates are the "building blocks" of digital systems.
- Digital waveforms can be observed and measured on an oscilloscope.

Chapter Checkup

Answers are at the end of the chapter.

1. A quantity that has a continuous set of values is
   (a) a data quantity (b) an analog quantity
   (c) a digital quantity
2. A quantity that has a discrete set of values is
   (a) an analog quantity (b) a binary quantity
   (c) a digital quantity
3. The weights of a decimal number are
   (a) powers of ten (b) powers of two
   (c) equal to the digit in the number
4. The weights of a binary number are
   (a) powers of ten (b) powers of two
   (c) 1 or 0, depending on the position
5. The binary number system consists of
   (a) one digit (b) no digits
   (c) two digits
6. A digit in the binary system is called a
   (a) byte (b) bit
   (c) power of two
7. When 2 is raised to the fifth power, the 2 is actually
   (a) added to itself 5 times (b) multiplied by itself 5 times
   (c) multiplied by 5
8. In a binary whole number, the right-most bit has a weight of
   (a) zero (b) one
   (c) two
9. The least significant bit (LSB) in a whole binary number is always
   (a) at the far right (b) at the far left
   (c) depends on the number
10. A binary number can have
    (a) only four bits (b) only two bits
    (c) any number of bits
11. MSB means
    (a) most significant base (b) major size bit
    (c) most significant bit

12. In the hexadecimal number system, there are
    (a) six digits (b) sixteen digits
    (c) ten digits
13. In the octal number system, there are
    (a) eight digits (b) sixteen digits
    (c) ten digits
14. A digital waveform is
    (a) a series of 1s and 0s (b) a series of pulses
    (c) a series of time intervals
15. The frequency, in hertz, of a repetitive digital waveform is
    (a) the number of pulses in 1 minute
    (b) the duration of each pulse
    (c) the number of pulses in 1 second
16. Frequency is expressed in the unit of
    (a) seconds (b) hertz
    (c) time
17. The "building blocks" of digital systems are
    (a) logic gates (b) timers
    (c) registers

Questions

Answers to odd-numbered questions are at the end of the book.

1. How does an analog quantity differ from a digital quantity?
2. What is a weighted number system?
3. On what are the weights in the decimal system based?
4. How many digits are in the decimal number system?
5. What does it mean to raise 10 to a certain power?
6. How many digits are in the binary number system?
7. What are the binary digits?
8. What is a bit? What is a byte?
9. What are the weights in the binary system?
10. What does it mean to raise 2 to a certain power?
11. What is an LSB? What is an MSB?
12. What is the resulting number for each of the following conversions?
    (a) Convert decimal 7 to binary.
    (b) Convert binary 1101 to decimal.
    (c) Convert binary 10111 to decimal.
    (d) Convert octal 6 to binary.
    (e) Convert hexadecimal 5B to binary.
    (f) Convert binary 101001 to octal.
    (g) Convert binary 101011010011 to hexadecimal.

13. What is the highest decimal number that you can express with eight bits? With ten bits?
14. What are the digits in the hexadecimal number system?
15. What are the digits in the octal number system?
16. On the TI-36X calculator, how do you find the value of 2^{16}?
17. What is the amplitude of a digital waveform?
18. If a certain digital waveform has 2500 pulses per second, how do you express the frequency?
19. If a pulse occurs every 10 millisconds (ms), what is the period of the digital waveform? A millisecond is 10^{-3} second (s).
20. If the pulses in a certain digital waveform are 1 microsecond (μs) wide and the period is 4 μs, what is the duty cycle? A microsecond is 10^{-6} second.
21. What is the bit sequence represented by the following series of logic levels: HIGH, LOW, LOW, LOW, HIGH, LOW, HIGH, HIGH, HIGH, HIGH, LOW, LOW, HIGH, LOW?
22. What is a timing diagram?

PROBLEMS

Basic Problems

Answers to odd-numbered problems are at the end of the book.

1. Find the decimal value of each of the following binary numbers:
   (a) 10 (b) 110
   (c) 1010 (d) 11011
2. Find the decimal value of each of the following binary numbers:
   (a) 11 (b) 101
   (c) 1110 (d) 10101
3. Express each decimal number in binary:
   (a) 25 (b) 76
   (c) 139 (d) 245
4. Express each decimal number in binary:
   (a) 18 (b) 99
   (c) 325 (d) 560
5. Determine the highest number to which you can count with each of the following numbers of bits?
   (a) 3 (b) 5
   (c) 7 (d) 9
   (e) 12
6. Determine the highest number to which you can count with each of the following numbers of bits?
   (a) 4 (b) 6
   (c) 8 (d) 10
   (e) 20

7. Express each hexadecimal number as a binary number:
   (a) 73_{16} (b) $A50_{16}$
   (c) $8B2C_{16}$
8. Express each binary number as a hexadecimal number:
   (a) 101011001000 (b) 0010111100001001
   (c) 1101110100111000
9. Express each octal number as a binary number:
   (a) 67_8 (b) 146_8
   (c) 3052_8
10. Use a calculator to find each of the following powers of two:
    (a) 2^2 (b) 2^5
    (c) 2^9 (d) 2^{14}
11. Use a calculator to find each of the following powers of two:
    (a) 2^3 (b) 2^8
    (c) 2^{16} (d) 2^{32}
12. Use a calculator to convert each of the following decimal numbers to hexadecimal:
    (a) 14 (b) 56
    (c) 435 (d) 841
13. Use a calculator to convert each of the following decimal numbers to octal:
    (a) 18 (b) 72
    (c) 555 (d) 1075
14. Use a calculator to convert each of the following hexadecimal numbers to decimal:
    (a) FFE_{16} (b) $56A_{16}$
    (c) $C082_{16}$
15. Use a calculator to convert each of the following octal numbers to decimal:
    (a) 27_8 (b) 670_8
    (c) 5331_8
16. Find the following values for the timing waveform in Figure 1–19.
    (a) period (b) frequency
    (c) duty cycle (d) amplitude

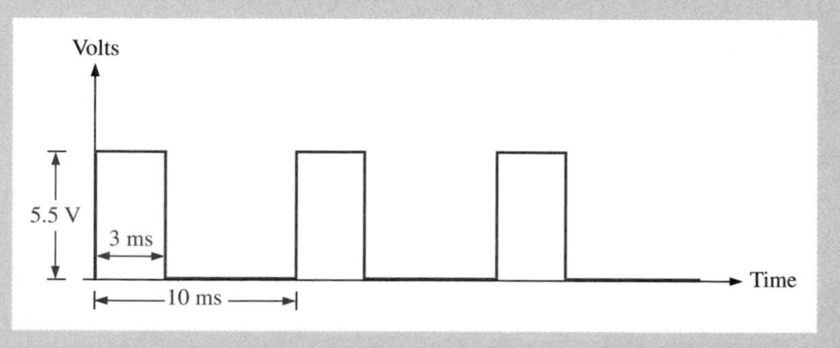

FIGURE 1–19

17. Find the following values for the timing waveform in Figure 1–20.

(a) period (b) frequency

(c) duty cycle (d) amplitude

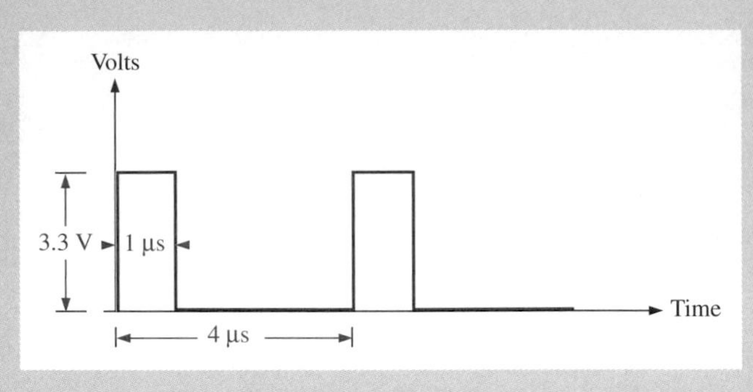

FIGURE 1–20

Basic-Plus Problems

18. Express each of the decimal numbers as a sum of weights times the appropriate digits:

(a) 248 (b) 688

(c) 1250 (d) 2495

19. Express each of the decimal numbers as a sum of weights times the appropriate digits:

(a) 295 (b) 850

(c) 1801 (d) 39,472

20. Evaluate the following powers of ten using a series of multiplications:

(a) 10^2 (b) 10^4

(c) 10^6

21. Evaluate the following powers of ten using a series of multiplications:

(a) 10^3 (b) 10^5

(c) 10^8

22. Express each of the binary numbers as a sum of weights times the appropriate digits:

(a) 111 (b) 11011

(c) 1010111 (d) 1100011100

23. Express each of the binary numbers as a sum of weights times the appropriate digits:

(a) 110 (b) 10010

(c) 1110100 (d) 1111000100

24. Show the sequence of binary numbers for counting from 1 to 16.

25. Show the sequence of binary numbers for counting from 17 to 31.

26. Convert the following decimal numbers to binary with the aid of your calculator:

(a) 84 (b) 831

(c) 2,597

27. Convert the following binary numbers to decimal with the aid of your calculator.
   (a) 100101011110
   (b) 11000011010111
   (c) 101000101111010
28. Perform the following binary operations with the aid of your calculator.
   (a) 101000110011 + 100001111101
   (b) 111001011110 − 101111111111
29. If the frequency of a certain timing waveform is 10 kHz and pulses have a width of 25 μs, what is the duty cycle?
30. If the frequency of a certain timing waveform is 500 MHz and pulses have a width of 1 nanosecond (ns), what is the duty cycle? A nanosecond is 10^{-9} second.
31. The duty cycle of a waveform is 0.35 and t_W = 5 ns. Determine the frequency.
32. The duty cycle of a waveform is 0.12 and t_W = 1 μs. Determine the frequency.
33. Draw a timing diagram for a 10 MHz timing waveform with a duty cycle of 50% and a synchronized data waveform representing the bit pattern 1011100101011110.
34. Draw a timing diagram for a 500 kHz timing waveform with a duty cycle of 20% and a synchronized data waveform representing the bit pattern 1000100010000100001.
35. Draw a data waveform that represents the binary number for decimal 1680.
36. Draw a data waveform that represents the binary number for decimal 4739.

ANSWERS

Example Questions

1–1: 100001 = 33

1–2: 83 = 1010011

1–3: 31_{16}, 32_{16}, 33_{16}, 34_{16}, 35_{16}

1–4: 1000111001110100 = $8E74_{16}$

1–5: $5C08_{16}$ = 0101110000001000

1–6: 31_8, 32_8, 33_8, 34_8, 35_8

1–7: 1000111001110100 = 107164_8

1–8: 1027_8 = 001000010111

1–9: 2^{15} = 32,768

1–10: 215 = $D7_{16}$

1–11: $81B_{16}$ = 2075

1–12: $5F_{16}$ + BE_{16} = $11D_{16}$

1–13: 289 = 441_8; 614_8 = 396

1–14: 100100001; 236

1–15: f = 100 kHz

Review Questions

1. An analog quantity is one having a continuous set of values.
2. Examples of analog quantities are time, temperature, pressure, and sound.
3. A digital quantity is one having a series of individual values.

LOGIC GATES: AND, OR, AND INVERTER

CHAPTER OUTLINE

Study aids for this chapter are available at

http://www.prenhall.com/SOE

INTRODUCTION

All logic circuits and functions are made from AND gates, OR gates, and inverters. Think of these logic gates as bricks in a structure. Individual bricks can be arranged to form various types of buildings, and bricks can be used to build fireplaces, steps, walkways, walls, and floors. Likewise, individual logic gates are arranged and interconnected to form various functions in a digital system. Regardless of the type of function, all logic circuits are all made from the same basic component, the logic gate. There are several types of logic gates, just as there may be several shapes or sizes of bricks in a structure.

Boolean algebra is a special form of math that lets you express the operation of a logic circuit with an equation. Boolean algebra is essentially a concise way to indicate how a logic circuit works instead of having to describe the operation in words. Also, the Boolean expression for a logic circuit allows you to analyze exactly what the circuit will do under various conditions.

All logic circuits used today are in the form of integrated circuits (ICs) that fall into one of two classes: fixed-function logic or programmable logic. Although many digital systems are made of integrated circuits containing large numbers of gates, in this chapter you will be introduced to fixed-function ICs that contain only a few gates.

KEY OBJECTIVES

A section number is given for each objective. After completing this chapter, you should be able to

- **2–1** Describe the AND, OR, and NOT logic functions and their switch equivalents
- **2–2** Describe the operation of an AND gate
- **2–3** Describe the operation of an OR gate
- **2–4** Describe the operation of an inverter
- **2–5** Apply Boolean algebra to an AND gate, OR gate, and inverter
- **2–6** Discuss several integrated circuits that provide AND, OR, and inverter functions
- **2–7** Describe certain faults in IC logic gates and their resulting effects

COMPUTER SIMULATIONS DIRECTORY

The following figures have Multisim circuit files associated with them.

- Figure 2–6 Page 44
- Figure 2–12 Page 47
- Figure 2–18 Page 50
- Figure 2–31 Page 57

LABORATORY EXPERIMENTS DIRECTORY

The following exercises are for this chapter.

- **Experiment 2** The AND Gate and the OR Gate
- **Experiment 3** The Inverter

KEY TERMS

- AND gate
- Truth table
- OR gate
- Inversion
- Inverter
- NOT
- Asserted state
- Variable
- Boolean multiplication
- Boolean addition
- Complement
- Integrated circuit
- DIP
- SMT
- Supply voltage

Today's microprocessors have wires and logic gates hundreds of times smaller than the width of a human hair. In the future, digital circuit components will be even smaller and inevitably reach a point where logic gates are so small that they are made from only a handful of atoms. The rules that determine the properties of logic gates are very different on the atomic scale. So, if computers are to become smaller and faster in the future, new quantum technology will replace the current technology. One research area for computers of the future (perhaps within 50 years) is the quantum computer, which will use certain effects that occur only at the atomic level to perform computations at speeds far beyond today's computer.

2–1 BASIC LOGIC FUNCTIONS

Three basic logic functions are AND, OR, and NOT. To help understand the AND and OR functions, they can be implemented with switches, which are the mechanical equivalents of the actual electronic logic gates. The logic function NOT is implemented by a circuit called an inverter.

In this section, you will learn to describe the AND, OR, and NOT logic functions and their switch equivalents.

The AND Function

All digital systems operate with two states or levels that represent the two binary digits. These two states can be defined in terms of binary digits as 1 and 0 or in terms of electrical voltages as HIGH and LOW. Additionally, the states can also be thought of as the electrical states *on* and *off* or *open* and *closed.*

The AND function indicates when several conditions are all true. In terms of logical thinking, the AND function can be described as a conditional statement using the two conditions, *true* and *false.*

A conditional AND statement is true only if *all* of its conditions are true.

An example of a conditional AND statement is *The lamp is on only if the bulb works AND the switch is closed.* One of the conditions is *the bulb works.* The other condition is *the switch is closed.* If both conditions are true, the conditional statement is true. If either or both of the conditions are not true (false), the conditional statement is not true. Generally, a conditional statement can have any number of conditions.

Switch Equivalent of the AND Function

Although AND gates are made with transistors that act as switches, the AND function can be illustrated using an arrangement of mechanical switches in a basic circuit, as shown in Figure 2–1. Each of the two switches represents a condition, and each can be open (*on*) or closed (*off*). Assume that a fixed HIGH-level voltage (battery) is connected to point *A*. In order for point *B* to be HIGH and the lamp to turn *on*, both switches must be closed, as shown in Figure 2–1(a). That is, switch 1 (SW1) must be closed *and* switch 2 (SW2) must be closed. When one or both switches are open, the lamp is *off,* as illustrated in parts (b), (c), and (d).

The OR Function

The OR function indicates when any one or more conditions are true. A conditional statement to describe the OR function is

A conditional OR statement is true only if *one or more* of its conditions are true.

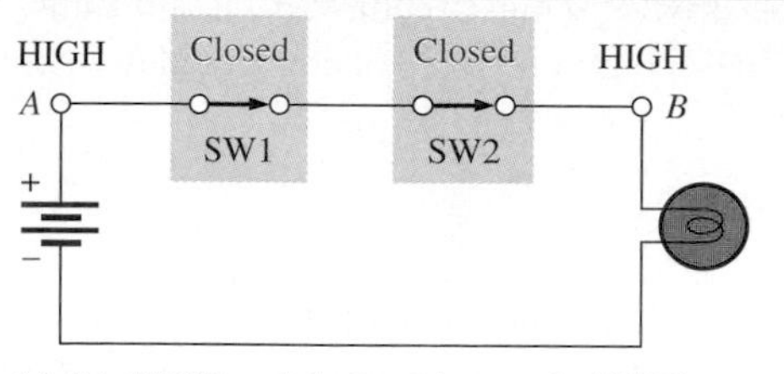

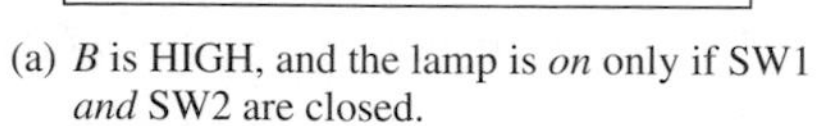
(a) B is HIGH, and the lamp is *on* only if SW1 *and* SW2 are closed.

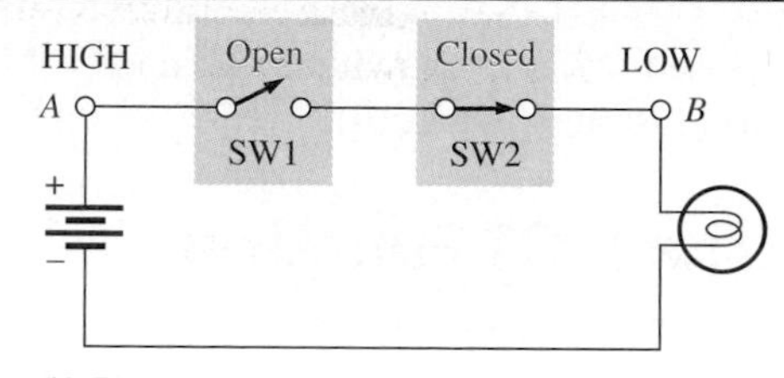

(b) The lamp is *off* when switch SW1 is open.

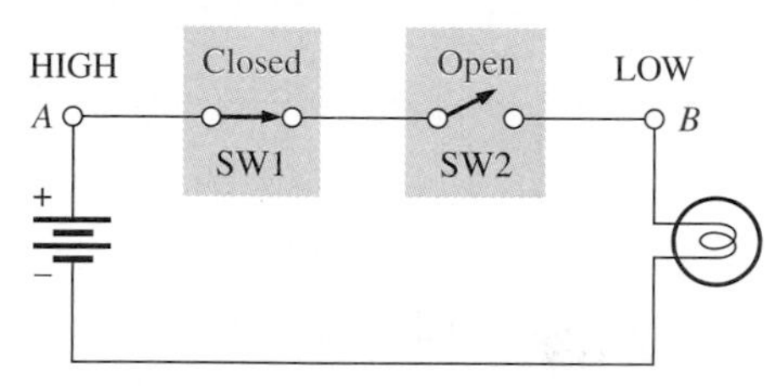

(c) The lamp is *off* when switch SW2 is open.

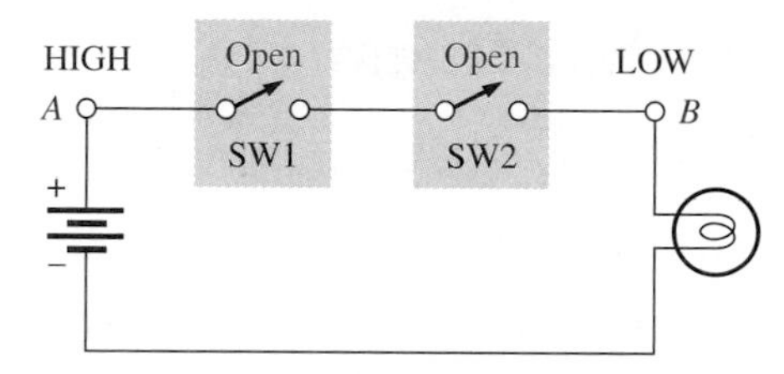

(d) The lamp is *off* when both switches are open.

FIGURE 2–1

A switch representation of a two-condition AND function. The lamp is *on* only if SW1 and SW2 are closed.

An example of a conditional OR statement is *The portable radio will work only if it is plugged into an electrical outlet OR its batteries are good.* One of the conditions is *the radio is plugged in.* The other condition is *the batteries are good.* If one or both of the conditions in this example are true, the conditional statement is true. If both of the conditions are not true (false), the conditional statement is not true.

Switch Equivalent of the OR Function

Although OR gates are made with transistors that act as switches, the OR function can be illustrated using an arrangement of mechanical switches in a basic circuit, as shown in Figure 2–2. Each of the two switches represents a condition, and each can be open or closed. A fixed HIGH-level voltage (battery) is connected to point A. In order for the lamp to turn *on,* either or both switches must be closed, as shown in parts (a), (b), and (c). That

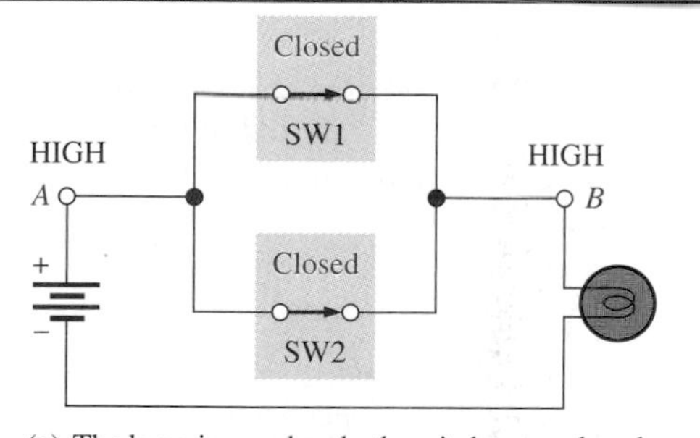

(a) The lamp is *on* when both switches are closed.

(b) The lamp is *on* when switch SW1 is closed.

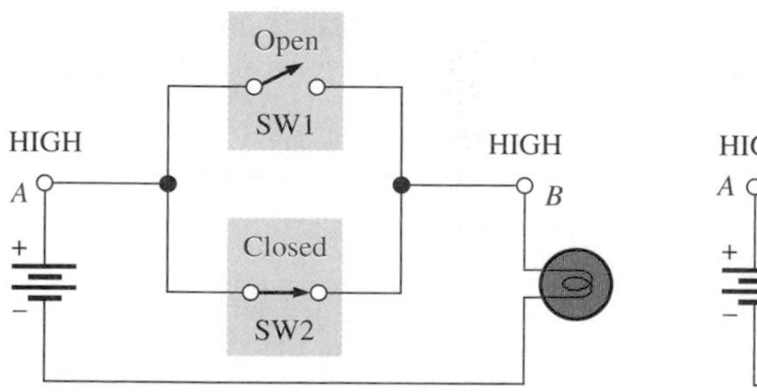

(c) The lamp is *on* when switch SW2 is closed.

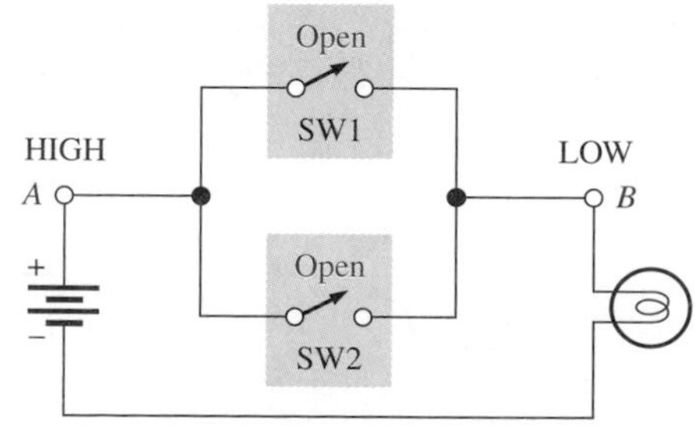

(d) The lamp is *off* when both switches are open.

FIGURE 2–2

A switch representation of a two-condition OR function. The lamp is *on* only if SW1 or SW2 or both are closed.

is, switch 1 (SW1) must be closed *or* switch 2 (SW2) must be closed *or* both must be closed. When both switches are open, the HIGH-level voltage is not connected to point *B*, as illustrated in part (d).

The NOT Function

The NOT function indicates an opposite condition. Examples of a NOT statement in terms of a switch equivalent are *The light is on if the switch is not open* and *The light is off if the switch is not closed.*

Review Questions

Answers are at the end of the chapter.

1. What is an AND function?
2. How many conditions must there be in an AND conditional statement?
3. What is an OR function?
4. How many conditions must there be in an OR conditional statement?
5. What is the NOT function?

2–2 THE AND GATE

The AND gate is one of the basic "building blocks" of digital systems. When AND gates are used in combination with other types of basic logic gates, any type of logic function can be implemented.

In this section, you will learn how an AND gate operates.

An **AND gate** is an electronic digital circuit in which the output is at a high logic level (HIGH) only if all of its inputs are at a high logic level. The AND gate, is a "decision-making" circuit. It can "decide" when all specified conditions are met (true) and produce an indication of this by a high logic level on its output.

An AND gate has only one output and at least two inputs. The AND gate output is analogous to point *B* shown in the switch arrangement for the AND function in Figure 2–1. Each AND gate input is analogous to a switch in the switch arrangement. A high logic level (HIGH) input compares to a closed switch and a low logic level (LOW) input compares to an open switch in the switch representation of the AND function.

AND Gate Symbols

Two standard AND gate logic symbols are used in electronic schematics for digital circuits and systems. One symbol is the distinctive shape symbol and is shown in Figure 2–3(a) for an AND gate with two inputs. Inputs are on the flat side of the symbol, and the output is on the curved side. An AND gate can have any number of inputs. Figure 2–3(b) shows a 4-input

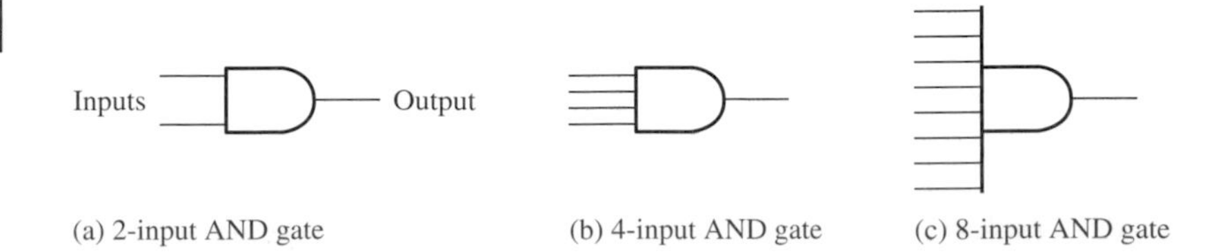

FIGURE 2–3

Distinctive shape AND gate symbol with various numbers of inputs.

AND gate symbol. When there are more than a few inputs, extensions are added to the gate symbol to accommodate the additional lines, as shown in Figure 2–3(c). The distinctive shape symbol is used in this book.

The second type of standard AND gate symbol is the rectangular outline symbol. Examples are shown in Figure 2–4. You will see this type of symbol in various publications and documents as well as in some schematics. This symbol was established by the ANSI/IEEE standard 91-1984 created jointly by the American National Standards Institute and the Institute of Electrical and Electronic Engineers. The rectangular outline symbol consists of a box with an internal label (&) to indicate the AND logic function.

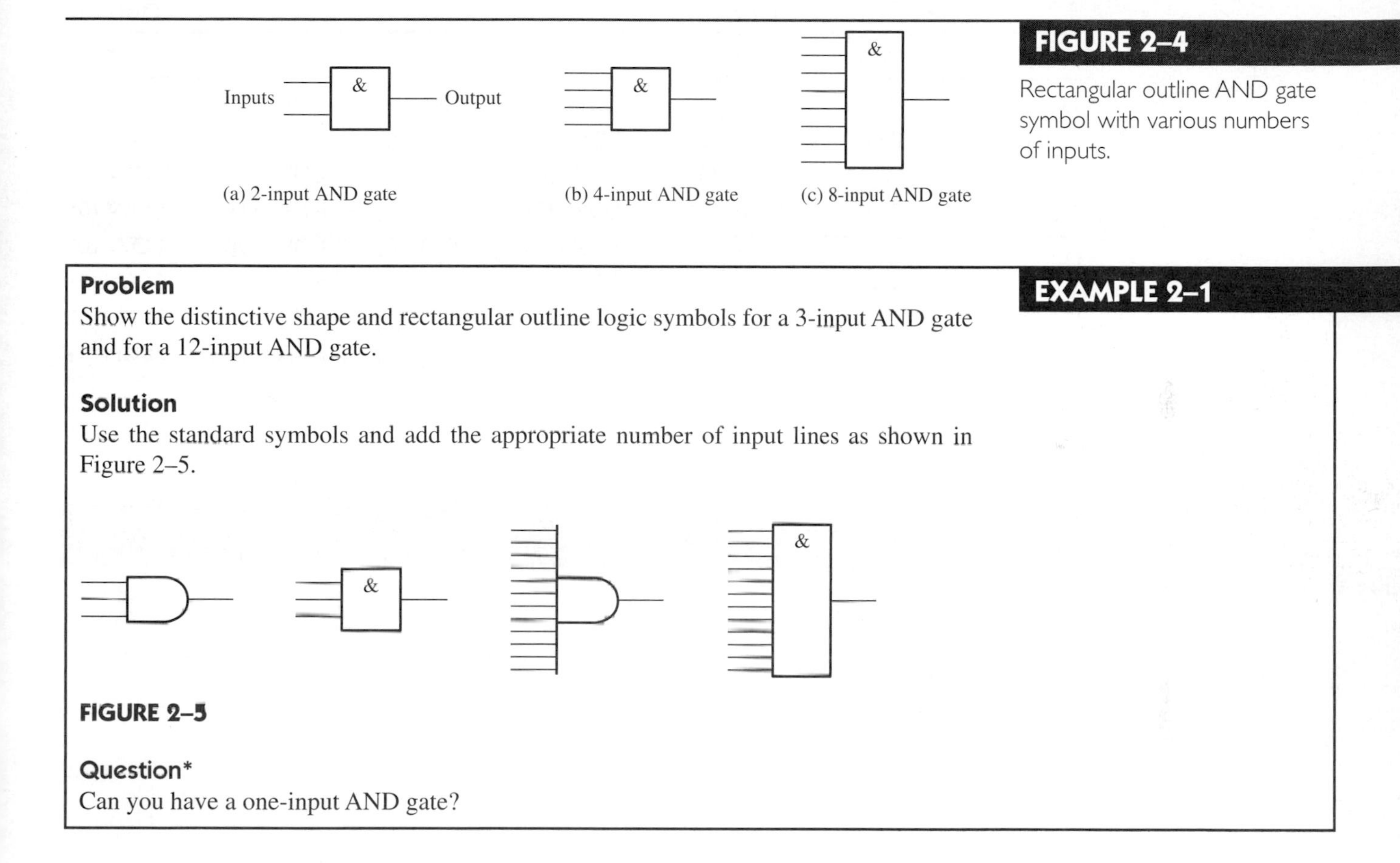

(a) 2-input AND gate (b) 4-input AND gate (c) 8-input AND gate

FIGURE 2–4

Rectangular outline AND gate symbol with various numbers of inputs.

EXAMPLE 2–1

Problem

Show the distinctive shape and rectangular outline logic symbols for a 3-input AND gate and for a 12-input AND gate.

Solution

Use the standard symbols and add the appropriate number of input lines as shown in Figure 2–5.

FIGURE 2–5

Question*

Can you have a one-input AND gate?

AND Gate Operation

The number of gate inputs determines the number of possible input state combinations, which is defined by a power of two.

$$\text{Number of input state combinations} = 2^n$$

where n is the number of inputs to a gate. For example, a 2-input gate has $2^2 = 4$ input state combinations and a 4-input gate has $2^4 = 16$ input state combinations.

The logic operation of an AND gate is often described by a truth table. A **truth table** lists all the possible input state combinations and the corresponding output state for each input combination. Either *HIGH* and *LOW* or *1* and *0* can be used on a truth table to represent the input and output states.

** Answers are at the end of the chapter.*

Truth tables for a 2-input AND gate and a 3-input AND gate are given in Table 2–1. A 2-input gate has $2^2 = 4$ input combinations, and a 3-input gate has $2^3 = 8$ input combinations. For reference purposes, the gate inputs are labeled *A, B,* and *C,* and the output is labeled *X.* As you can see, the output of an AND gate is HIGH (1) only if all of the inputs are HIGH (1); otherwise the output is LOW (0).

TABLE 2–1 Truth tables for (a) 2-input AND gate and (b) 3-input AND gate.

Inputs		Output
A	*B*	*X*
LOW (0)	LOW (0)	LOW (0)
LOW (0)	HIGH (1)	LOW (0)
HIGH (1)	LOW (0)	LOW (0)
HIGH (1)	HIGH (1)	HIGH (1)

(a)

Inputs			Output
A	*B*	*C*	*X*
LOW (0)	LOW (0)	LOW (0)	LOW (0)
LOW (0)	LOW (0)	HIGH (1)	LOW (0)
LOW (0)	HIGH (1)	LOW (0)	LOW (0)
LOW (0)	HIGH (1)	HIGH (1)	LOW (0)
HIGH (1)	LOW (0)	LOW (0)	LOW (0)
HIGH (1)	LOW (0)	HIGH (1)	LOW (0)
HIGH (1)	HIGH (1)	LOW (0)	LOW (0)
HIGH (1)	HIGH (1)	HIGH (1)	HIGH (1)

(b)

COMPUTER SIMULATION

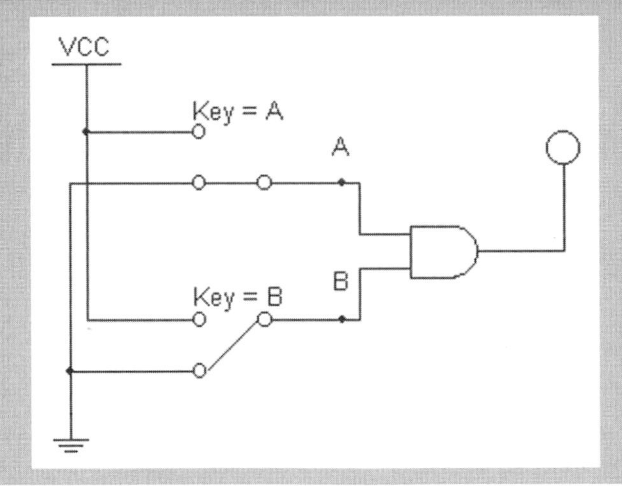

FIGURE 2–6

Open the Multisim file F02-06DG on the website. Check the 2-input AND gate truth table by applying LOWs and HIGHs with the switches. The ground line is the LOW and the $+V_{CC}$ line is the HIGH. For the simulation, operate the top switch with the *A* key and the bottom switch with the *B* key. When the probe light is white, it indicates a LOW; when it is red, it indicates a HIGH.

AND Gates with Digital Waveform Inputs

In an actual digital system, the AND gate inputs are usually not at a constant level all of the time. Most of the time you will find that gate inputs are digital waveforms that transition between high and low levels. In terms of their truth table operation, AND gates with digital waveform inputs work the same as they do with constant input levels because, at any time, the output is determined by the inputs. The output is HIGH only when all inputs are HIGH; otherwise, the output is LOW. Example 2–2 illustrates this with a 2-input AND gate.

EXAMPLE 2–2

Problem

Determine the output waveform X when the waveforms shown in Figure 2–7 are applied to the 2-input AND gate.

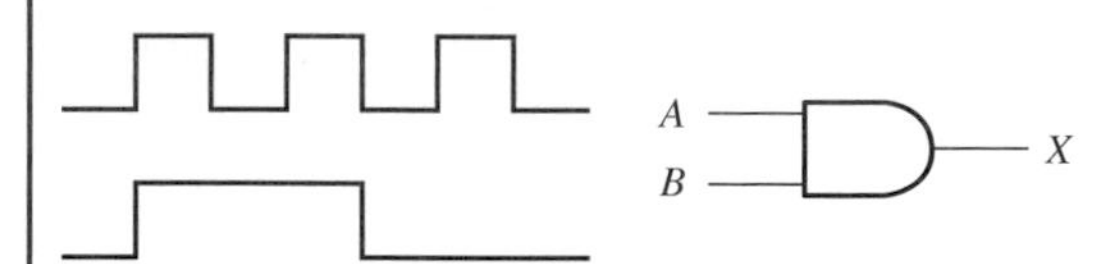

FIGURE 2–7

Solution

The times that the two input waveforms A and B are HIGH at the same time are indicated by the blue-shaded areas in Figure 2–8. The output waveform X with the corresponding HIGHs is shown in the figure.

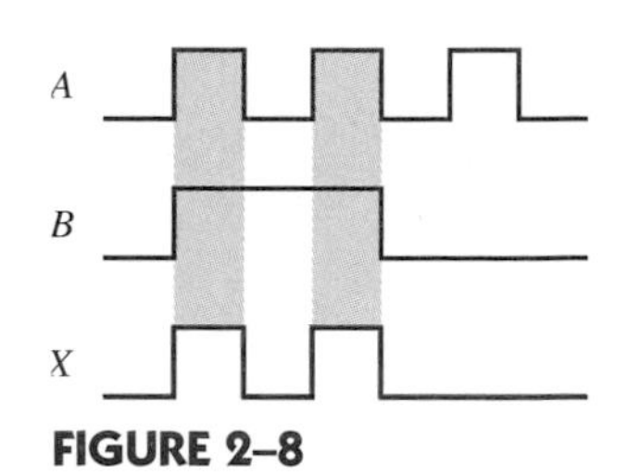

FIGURE 2–8

Question

If input waveform A stays the same as shown in Figure 2–7 but input waveform B is always HIGH, what would the output waveform X look like?

Review Questions

6. When is the output of an AND gate at the high logic level (HIGH)?
7. How many possible combinations of input states are there for a 4-input AND gate?
8. How many rows and how many columns does the truth table for a 5-input AND gate contain?
9. If the inputs to an AND gate are digital waveforms, when will the output be HIGH?
10. What is the standard label in the rectangular outline symbol for an AND gate?

THE OR GATE 2–3

The OR gate, like the AND gate, is one of the basic "building blocks" of digital systems. When OR gates are used in combination with other types of basic logic gates, any type of logic function can be implemented.

In this section, you will learn how an OR gate operates.

An **OR gate** is an electronic digital circuit in which the output is at a high logic level (HIGH) only if one or more of its inputs are at a high logic level. The OR gate is also a "decision-making" circuit. It can "decide" when one or more of several input conditions are met (true) and produce an indication of this by a high logic level on its output.

An OR gate has only one output and at least two inputs. The OR gate output is analogous to point B shown in the switch arrangement for the OR function in Figure 2–2 in Section 2–1. Each OR gate input is analogous to a switch in the switch arrangement that was described. A high logic level (HIGH) input compares to a closed switch and a low logic level (LOW) input compares to an open switch in the switch representation of the OR function.

OR Gate Symbols

Two standard OR gate logic symbols are used in electronics schematics for digital circuits and systems. One symbol is the distinctive shape symbol and is shown in Figure 2–9(a) for

FIGURE 2–9

Distinctive shape OR gate symbol with various numbers of inputs.

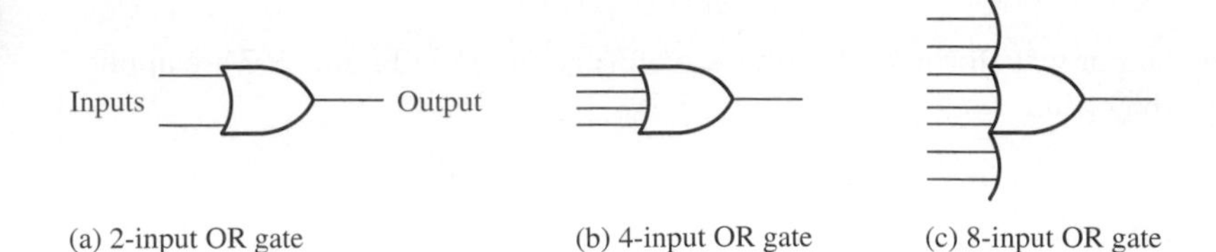

(a) 2-input OR gate (b) 4-input OR gate (c) 8-input OR gate

an OR gate with two inputs. Inputs are on the curved side of the symbol, and the output is on the pointed side. An OR gate can have any number of inputs. Figure 2–9(b) shows a 4-input OR gate symbol. When there are more than a few inputs, extensions are added to the gate symbol to accommodate the additional lines, as shown in Figure 2–9(c). The distinctive shape symbol is used in this book.

The second type of standard OR gate symbol is the rectangular outline symbol. Examples are shown in Figure 2–10. The rectangular outline symbol consists of a box, just as for the AND gate symbol, but with a different internal label (≥1) to indicate the OR function.

FIGURE 2–10

Rectangular outline OR gate symbol with various numbers of inputs.

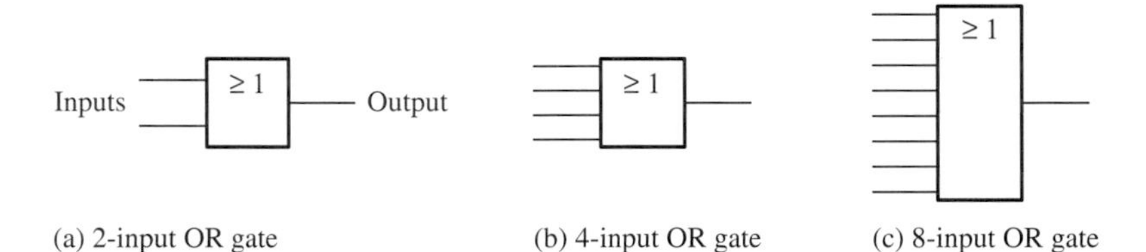

(a) 2-input OR gate (b) 4-input OR gate (c) 8-input OR gate

EXAMPLE 2–3

Problem

Show the distinctive shape and rectangular outline logic symbols for a 3-input OR gate and for a 12-input OR gate.

Solution

Use the standard symbols and add the appropriate number of input lines as shown in Figure 2–11.

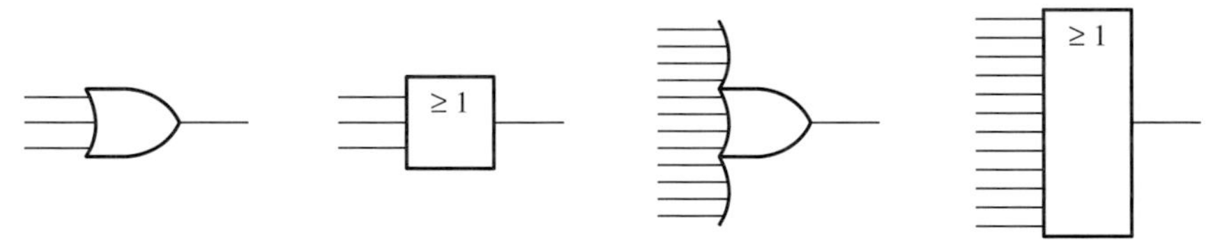

FIGURE 2–11

Question

Why do you think the ≥1 label is used for the OR gate in the rectangular outline symbol?

OR Gate Operation

Just as with the AND gate, the number of gate inputs determines the number of possible input state combinations, which is defined by a power of two.

$$\text{Number of input state combinations} = 2^n$$

where n is the number of inputs to a gate.

The logic operation of an OR gate can be described by a truth table. As you have learned, a truth table lists all the possible input state combinations and the corresponding output state for each input combination. Again, we will use HIGH (1) and LOW (0) for the input and output states to represent high logic levels and low logic levels, respectively.

Truth tables for a 2-input OR gate and a 3-input OR gate are given in Table 2–2. A 2-input gate has $2^2 = 4$ input combinations and a 3-input gate has $2^3 = 8$ input combinations. The gate inputs are labeled *A, B,* and *C,* and the output is labeled *X.* As you can see, the output of an OR gate is HIGH (1) when one or more inputs are HIGH (1).

TABLE 2–2

Truth tables for (a) 2-input OR gate and (b) 3-input OR gate.

Inputs		Output
A	*B*	*X*
LOW (0)	LOW (0)	LOW (0)
LOW (0)	HIGH (1)	HIGH (1)
HIGH (1)	LOW (0)	HIGH (1)
HIGH (1)	HIGH (1)	HIGH (1)

(a)

Inputs			Output
A	*B*	*C*	*X*
LOW (0)	LOW (0)	LOW (0)	LOW (0)
LOW (0)	LOW (0)	HIGH (1)	HIGH (1)
LOW (0)	HIGH (1)	LOW (0)	HIGH (1)
LOW (0)	HIGH (1)	HIGH (1)	HIGH (1)
HIGH (1)	LOW (0)	LOW (0)	HIGH (1)
HIGH (1)	LOW (0)	HIGH (1)	HIGH (1)
HIGH (1)	HIGH (1)	LOW (0)	HIGH (1)
HIGH (1)	HIGH (1)	HIGH (1)	HIGH (1)

(b)

COMPUTER SIMULATION

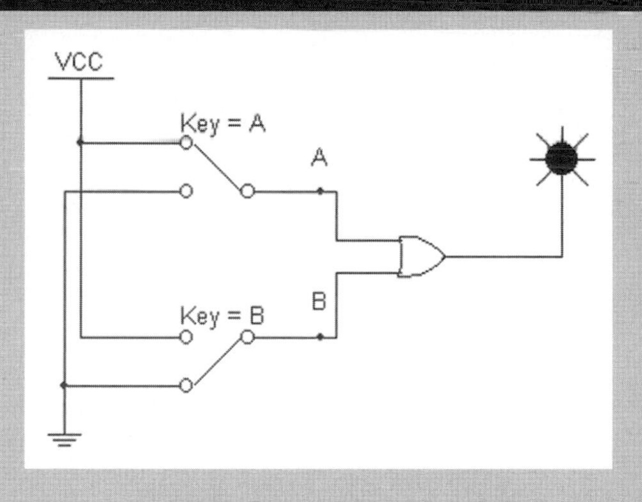

FIGURE 2–12
Open the Multisim file F02-12DG on the website. Check the 2-input OR gate truth table by applying LOWs and HIGHs with the switches. The ground line is the LOW and the $+V_{CC}$ line is the HIGH. For the simulation, operate the top switch with the *A* key and the bottom switch with the *B* key. When the probe light is red, it indicates a HIGH; when it is white, it indicates a LOW.

OR Gates with Digital Waveform Inputs

In an actual digital system, the OR gate inputs are usually not at a constant level all of the time. Most of the time you will find that gate inputs are digital waveforms. In terms of their truth table operation, OR gates with digital waveform inputs work the same as they do with constant input levels. The output is HIGH when one or more inputs are HIGH, and it is LOW when all the inputs are LOW.

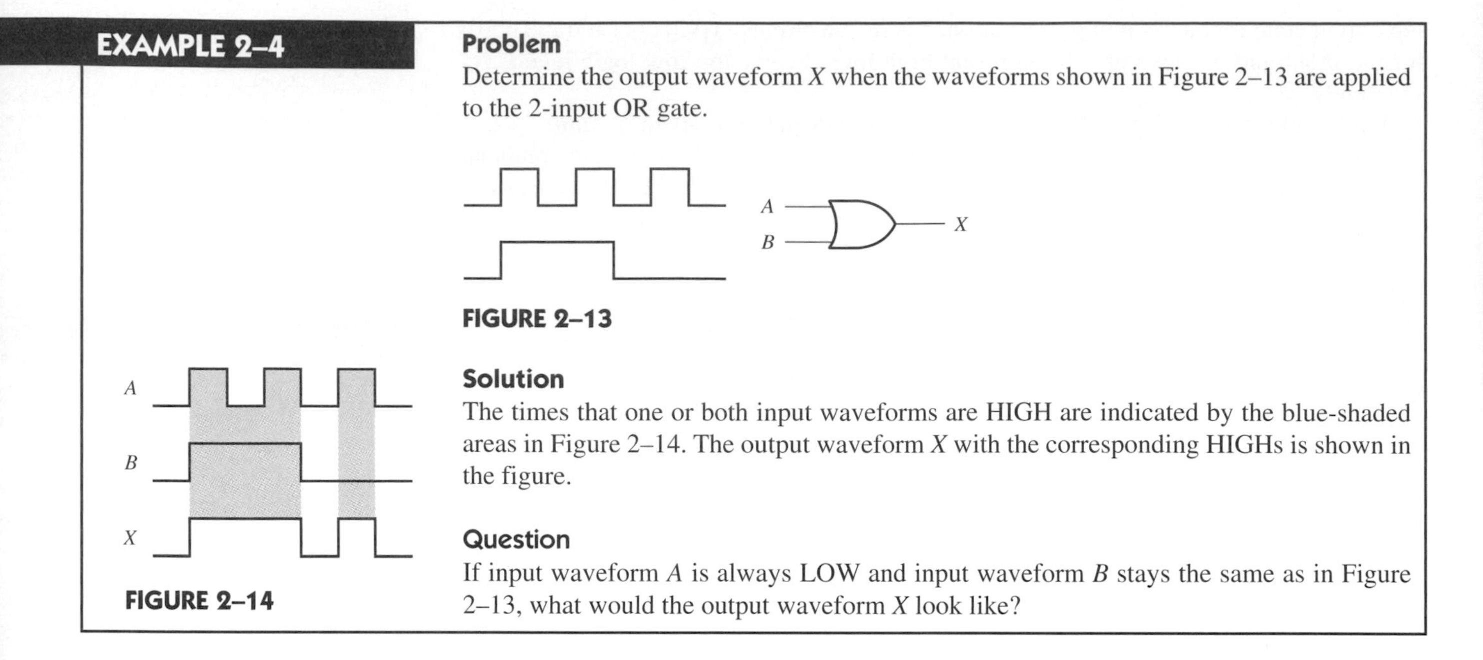

EXAMPLE 2–4

Problem

Determine the output waveform *X* when the waveforms shown in Figure 2–13 are applied to the 2-input OR gate.

FIGURE 2–13

Solution

The times that one or both input waveforms are HIGH are indicated by the blue-shaded areas in Figure 2–14. The output waveform *X* with the corresponding HIGHs is shown in the figure.

FIGURE 2–14

Question

If input waveform *A* is always LOW and input waveform *B* stays the same as in Figure 2–13, what would the output waveform *X* look like?

Review Questions

11. When is the output of an OR gate at the high logic level (HIGH)?
12. How many possible combinations of input states are there for a 4-input OR gate?
13. How many rows and how many columns does the truth table for a 5-input OR gate contain?
14. If the inputs to an OR gate are digital waveforms, when will the output be HIGH?
15. What is the standard label in the rectangular outline symbol for the OR gate?

2–4 THE INVERTER

An inverter produces a logic level on its output that is the opposite of the logic level on the input. The inverter is also known as a NOT circuit.

In this section, you will learn how an inverter operates.

Inversion is the process of changing a logic level to the opposite logic level. An **inverter** is a logic circuit that has only one input and one output. The inverter changes a LOW to a HIGH or a HIGH to a LOW. In terms of binary digits, the inverter changes a 0 to a 1 or a 1 to a 0. The inverter is called a **NOT** circuit because the output is not the same state as the input.

Inverter Symbols

Two standard logic symbols for the inverter are used in schematics for digital circuits and systems. One symbol is the distinctive shape symbol and is shown in Figure 2–15(a). The input is on the flat side of the symbol and the output is on the pointed side. The little bubble, called the *negation indicator,* is a symbol for inversion and can be placed on either the input side or the output side of the inverter symbol, as shown in the figure. The term *negation* means the same as inversion. The placement of the bubble does not change the logic.

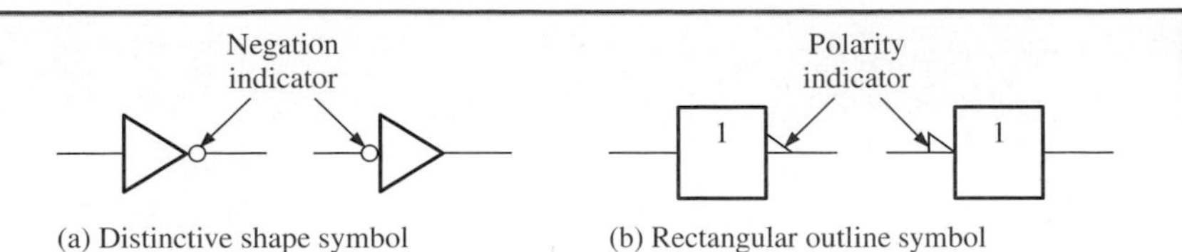

FIGURE 2–15

Standard symbols for an inverter (NOT circuit).

The second type of standard inverter symbol is the rectangular outline symbol. An example is shown in Figure 2–15(b). You will see this type of symbol in various publications and documents as well as in some schematics. The rectangular outline symbol for the NOT function consists of a box with an internal label (1). The little "right triangle," called the *polarity* or *level indicator,* is a symbol for inversion used with the rectangular outline symbols and is equivalent to the bubble used in the distinctive shape symbols. The polarity indicator can appear on either the input or the output to indicate the asserted LOW state, which is explained next.

Inverter Operation

The term *asserted state* is also known as *active state.* An **asserted state** is the logic state or level (LOW or HIGH) that brings about the action that a logic gate or inverter is intended to produce. For example, suppose that the output of an inverter drives an LED (light-emitting diode) and in order to light the LED, a LOW output is required. Therefore, a LOW is the asserted output state of the inverter and, because a HIGH is necessary on the input to produce a LOW output, a HIGH is the asserted input state. In this case, the inverter would be drawn with the bubble on the output. Of course, when the input is LOW and output is HIGH, the states are *nonasserted* and the LED is off.

The logic operation of an inverter can be described by a simple truth table, as given in Table 2–3. The inverter input is labeled *A* and the output is labeled *X*.

TABLE 2–3

Truth table for an inverter (NOT circuit).

Input	Output
A	*X*
LOW (0)	HIGH (1)
HIGH (1)	LOW (0)

Inverter with a Digital Waveform Input

In an actual digital system, the inverter input is usually not at a constant level all of the time. Most of the time, inputs are digital waveforms. In terms of their truth table operation, inverters work the same whether or not the input is always constant or is changing. The output is always inverted so when the input is LOW, the output is HIGH, and when the input is HIGH, the output is LOW.

EXAMPLE 2–5

Problem

Determine the output waveform *X* when the waveform shown in Figure 2–16 is applied to the inverter.

FIGURE 2–16

Solution

When the input waveform is HIGH, the output is LOW and vice versa, as shown in Figure 2–17.

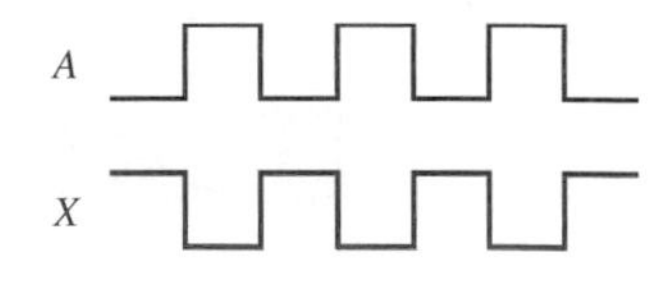

FIGURE 2–17

Question

What is the asserted state of the output as indicated by the placement of the bubble in Figure 2–16?

COMPUTER SIMULATION

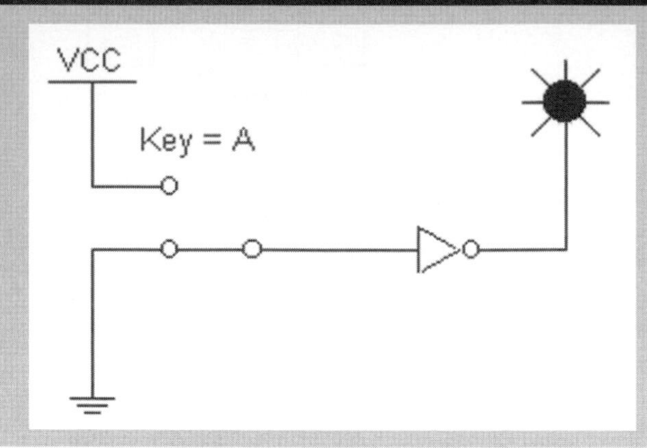

FIGURE 2–18

Open the Multisim file F02-18DG on the website. Verify the inverter truth table by applying a LOW and a HIGH with the switch. The ground line is the LOW and the $+V_{CC}$ line is the HIGH. For the simulation, operate the switch with the *A* key. When the probe light is red, it indicates a HIGH; when it is white, it indicates a LOW.

Review Questions

16. How does an inverter perform the NOT function?
17. How many inputs does an inverter have?
18. What is an asserted level and how is it indicated?
19. What do the terms *inversion, negation,* and *NOT* have in common?
20. What is the standard label in a rectangular outline symbol for an inverter?

2–5 BOOLEAN ALGEBRA APPLIED TO AND, OR, AND NOT

Boolean algebra is the mathematics of logic circuits. The operation of all types of gates can be described using symbolic notation similar to regular algebra. Boolean laws, rules, and theorems are summarized in Appendix A.

In this section, you will learn how to use Boolean algebra to express the operation of the AND gate, OR gate, and inverter.

In Boolean algebra, a **variable** is a quantity represented by a letter that can take on either of two values, a 1 or a 0, which corresponds to the bits of 1 and 0 in the binary number system. Up to this point, low and high logic levels (LOWs and HIGHs) have been used to represent "physical" voltage levels that can be observed and measured. When you evaluate logic circuits using Boolean algebra, substitute 1s and 0s for the variables instead of HIGHs and LOWs. In this text, we will use the positive logic convention.

$$1 = \text{HIGH} \qquad 0 = \text{LOW}$$

Generally, input variables are labeled *A, B, C,* and *D* and output variables are labeled *X, Y,* or *Z.* Also, a variable can be designated as a letter with a subscript, such as A_1, A_2, or A_3. Figure 2–19 shows examples of typical labeling of gate inputs and outputs with variable designations.

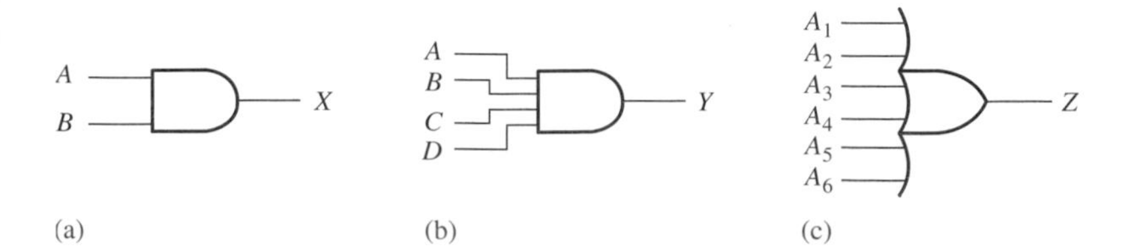

FIGURE 2–19

Gates with variables representing inputs and outputs.

Table 2–4 shows the truth table in terms of 1s and 0s for a 2-input AND gate. The truth table for any gate can be expressed in terms of 1s and 0s as well as HIGHs and LOWs.

TABLE 2–4

Truth table for a 2-input AND gate.

A	B	X
0	0	0
0	1	0
1	0	0
1	1	1

Boolean Multiplication

In terms of Boolean algebra, the AND gate is a multiplier. This is the same as regular decimal multiplication because when you multiply 0 times 1 you get 0. When you multiply 1 times 1, you get 1, just as the AND gate truth table indicates.

Boolean multiplication (ANDing) of two variables is indicated as AB, which is called a **product term**. For three variables, Boolean multiplication is stated as ABC. Letters continue for any number of variables. Of course, other letters can be used instead of A, B, and C. This is the same way that multiplication of variables is usually stated in regular algebra. Sometimes, in an algebraic expression a dot (·) between variables is used to indicate multiplication, such as $A \cdot B$, and sometimes parentheses are used, such as $(A)(B)$, especially when there are multiple functions in an expression.

HISTORICAL NOTE

George Boole (1815–1864), the son of an English shoemaker, developed an early interest in mathematics. In 1844 he published a paper on the calculus of operators for which he was widely recognized. Later he published *The Mathematical Analysis of Logic*, which introduced his ideas on symbolic logic. In 1854, *An Investigation of the Laws of Thought, on Which are Founded the Mathematical Theories of Logic and Probabilities* was published. This expanded on his earlier work and contained concepts that are now known collectively as Boolean algebra.

Boolean Expression for the AND Gate

The expression for the operation of a 2-input AND gate in Boolean algebra is

$$X = AB$$

This Boolean expression states that the *output X equals the product (AND) of the inputs A and B* and is read "$X = A$ and B."

To evaluate $X = AB$ and see how it actually describes an AND gate operation, simply substitute a 1 or a 0 for each variable. For two variables, there are $2^2 = 4$ combinations of input states.

For $A = 0, B = 0$: $X = AB = 0 \cdot 0 = 0$
For $A = 0, B = 1$: $X = AB = 0 \cdot 1 = 0$
For $A = 1, B = 0$: $X = AB = 1 \cdot 0 = 0$
For $A = 1, B = 1$: $X = AB = 1 \cdot 1 = 1$

EXAMPLE 2–6

Problem

Write the Boolean expression for the 3-input AND gate in Figure 2–20. Evaluate the expression for all possible combinations of input states, and express the results in the form of a truth table.

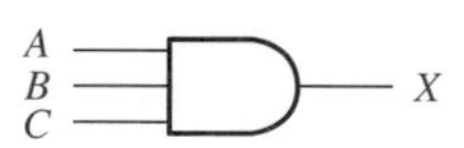

FIGURE 2–20

Solution

The Boolean expression for the 3-input AND gate is

$$X = ABC$$

To evaluate the expression, substitute the binary number values in all possible combinations ($2^3 = 8$) of input states. The corresponding truth table is shown in Table 2–5.

Table 2–5

A	B	C	X
0	0	0	0
0	0	1	0
0	1	0	0
0	1	1	0
1	0	0	0
1	0	1	0
1	1	0	0
1	1	1	1

For $A = 0, B = 0, C = 0$: $X = ABC = 0 \cdot 0 \cdot 0 = 0$
For $A = 0, B = 0, C = 1$: $X = ABC = 0 \cdot 0 \cdot 1 = 0$
For $A = 0, B = 1, C = 0$: $X = ABC = 0 \cdot 1 \cdot 0 = 0$
For $A = 0, B = 1, C = 1$: $X = ABC = 0 \cdot 1 \cdot 1 = 0$
For $A = 1, B = 0, C = 0$: $X = ABC = 1 \cdot 0 \cdot 0 = 0$
For $A = 1, B = 0, C = 1$: $X = ABC = 1 \cdot 0 \cdot 1 = 0$
For $A = 1, B = 1, C = 0$: $X = ABC = 1 \cdot 1 \cdot 0 = 0$
For $A = 1, B = 1, C = 1$: $X = ABC = 1 \cdot 1 \cdot 1 = 1$

Question
If an AND gate has four inputs, how many rows would there be in the truth table?

TABLE 2–6

Truth table for a 2-input OR gate.

A	B	X
0	0	0
0	1	1
1	0	1
1	1	1

Boolean Addition

The OR gate performs Boolean addition, which is indicated by the + operator. This differs from regular decimal addition because, in Boolean addition, when you add 1 + 1 you get 1. Table 2–6 shows the truth table for a 2-input OR gate.

Boolean addition (ORing) of two variables is indicated as $A + B$, which is called a **sum term**. For three variables, Boolean addition is stated as $A + B + C$. This is the same way that addition of variables is stated in regular algebra.

Boolean Expression for the OR Gate

The expression for the operation of a 2-input OR gate in Boolean algebra is

$$X = A + B$$

This Boolean expression states that *the output X is equal to the Boolean sum (OR) of the inputs A and B* and is read "X equals A or B."

To evaluate $X = A + B$ and see how it actually describes an OR gate operation, simply substitute a 1 or a 0 for each variable. As before, there are $2^2 = 4$ possible combinations of input states.

For $A = 0, B = 0$: $X = A + B = 0 + 0 = 0$
For $A = 0, B = 1$: $X = A + B = 0 + 1 = 1$
For $A = 1, B = 0$: $X = A + B = 1 + 0 = 1$
For $A = 1, B = 1$: $X = A + B = 1 + 1 = 1$

EXAMPLE 2–7

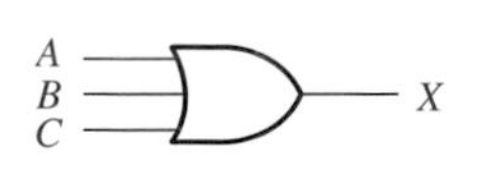

FIGURE 2–21

Problem
Write the Boolean expression for the 3-input OR gate in Figure 2–21. Evaluate the expression for all possible combinations of input states, and express the results in the form of a truth table.

Solution
The Boolean expression for the 3-input OR gate is

$$X = A + B + C$$

To evaluate the expression, substitute the binary number values in all possible combinations ($2^3 = 8$) of input states. The corresponding truth table is shown in Table 2–7.

For $A = 0, B = 0, C = 0$: $X = A + B + C = 0 + 0 + 0 = 0$
For $A = 0, B = 0, C = 1$: $X = A + B + C = 0 + 0 + 1 = 1$
For $A = 0, B = 1, C = 0$: $X = A + B + C = 0 + 1 + 0 = 1$
For $A = 0, B = 1, C = 1$: $X = A + B + C = 0 + 1 + 1 = 1$
For $A = 1, B = 0, C = 0$: $X = A + B + C = 1 + 0 + 0 = 1$
For $A = 1, B = 0, C = 1$: $X = A + B + C = 1 + 0 + 1 = 1$
For $A = 1, B = 1, C = 0$: $X = A + B + C = 1 + 1 + 0 = 1$
For $A = 1, B = 1, C = 1$: $X = A + B + C = 1 + 1 + 1 = 1$

Table 2–7

A	B	C	X
0	0	0	0
0	0	1	1
0	1	0	1
0	1	1	1
1	0	0	1
1	0	1	1
1	1	0	1
1	1	1	1

Question
For an OR gate with five inputs, how many rows would there be in the truth table?

The Commutative Law

Three important laws of Boolean algebra—the *commutative law,* the *associative law,* and the *distributive law*—are related to AND and OR gates. The commutative law is discussed here, and the other two laws will be introduced in Chapter 4, Section 4–4, when you study combinations of gates.

The **commutative law of multiplication** applies to the AND gate and is stated as follows:

The result of an AND operation does not depend on the order in which the variables are ANDed.

In terms of a 2-input AND gate, the commutative law is expressed as

$$AB = BA$$

Figure 2–22 illustrates that it doesn't matter to which input each variable is assigned. This law applies to AND gates with any number of inputs.

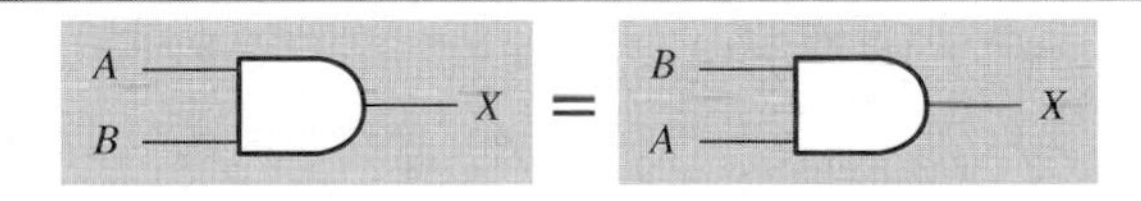

FIGURE 2–22
Illustration of the commutative law of Boolean multiplication.

The **commutative law of addition** applies to the OR gate and is stated as follows:

The result of an OR operation does not depend on the order in which variables are ORed.

In terms of a 2-input OR gate, the commutative law is expressed as

$$A + B = B + A$$

Figure 2–23 illustrates that it doesn't matter to which input each variable is assigned. This law applies to OR gates with any number of inputs.

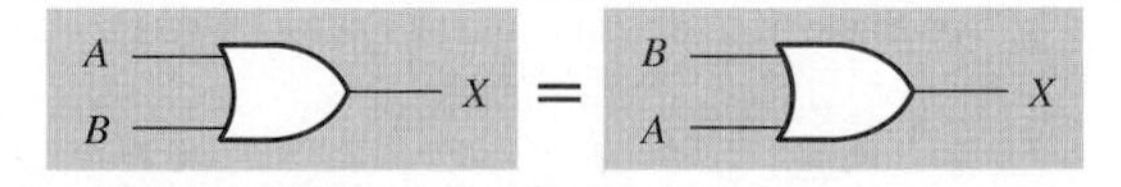

FIGURE 2–23
Illustration of the commutative law of Boolean addition.

Boolean Complementation

In Boolean algebra, the inverter performs complementation. The terms *complementation, inversion, negation,* and *NOT* all mean the same thing.

Boolean complementation is usually indicated by a bar over the complemented variable. For example, the **complement**, or opposite, of the Boolean variable A is shown as $\overline{A}$. It is read as "not A" or "A bar". Occasionally, you will see a prime symbol instead of an overbar, such as A'. The overbar is used in this book.

The complement of 0 is 1 and the complement of 1 is 0.

Rules of Boolean Algebra

Boolean algebra includes several rules that are helpful in working with Boolean expressions. Three Boolean rules apply to AND gates. They are illustrated in Figure 2–24.

$A \cdot 0 = 0$ When a variable is ANDed with 0, the output is always 0.
$A \cdot 1 = A$ When a variable is ANDed with 1, the output equals the variable.
$A \cdot A = A$ When a variable is ANDed with itself, the output equals the variable.

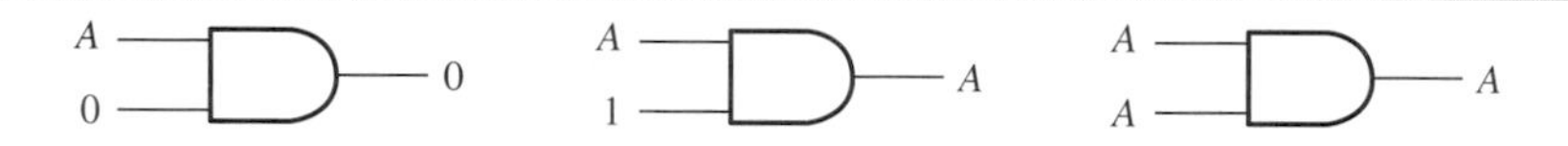

FIGURE 2–24
Illustration of Boolean rules for AND gates.

Three other Boolean rules apply to OR gates. They are illustrated in Figure 2–25.

$A + 0 = A$ When a variable is ORed with 0, the output equals the variable.
$A + 1 = 1$ When a variable is ORed with 1, the output is always 1.
$A + A = A$ When a variable is ORed with itself, the output equals the variable.

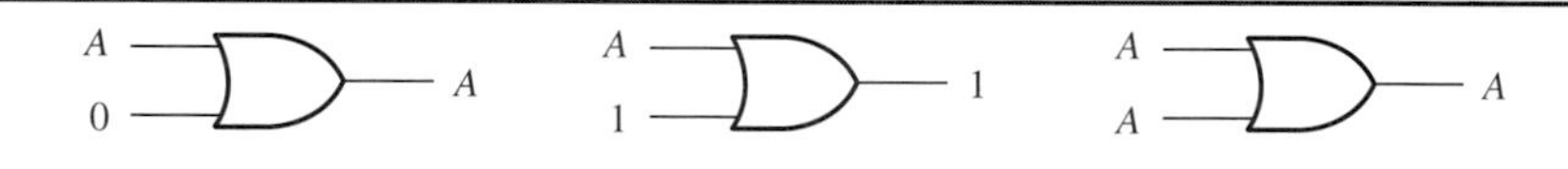

FIGURE 2–25
Illustration of Boolean rules for OR gates.

Three additional Boolean rules apply to complementation (inversion). They are illustrated in Figure 2–26.

$A \cdot \overline{A} = 0$ When a variable is ANDed with its complement, the output is always 0.
$A + \overline{A} = 1$ When a variable is ORed with its complement, the output is always 1.
$\overline{\overline{A}} = A$ When a variable is complemented (NOTed) twice, the output equals the variable.

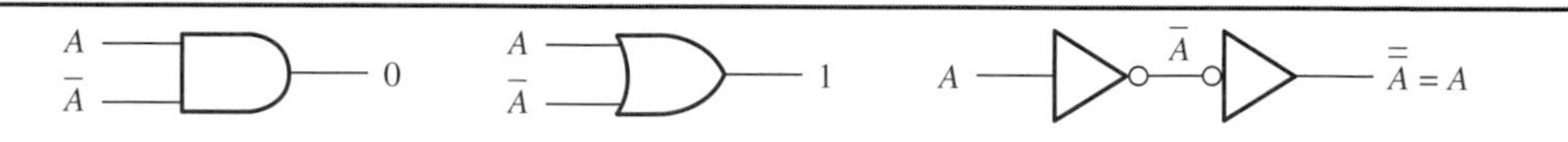

FIGURE 2–26
Illustration of Boolean rules involving complementation.

Review Questions

21. What is a Boolean variable and what values can it have?
22. What Boolean operation does the AND gate perform?
23. What Boolean operation does the OR gate perform?
24. What does the commutative law of addition mean?
25. What Boolean operation does the inverter perform?

INTEGRATED CIRCUITS 2–6

All of the logic gates that you will use are in integrated circuit (IC) form. An individual gate is made from transistors, resistors, and other components. Digital ICs range from configurations that contain one to several logic gates up to configurations that contain hundreds of thousands or even millions of logic gates. Digital ICs are available in circuit technologies that differ from each other by the type of transistor used. The two main types are CMOS (complementary metal-oxide semiconductors) and TTL (transistor-transistor logic).

The trend in technology is toward programmable logic devices (PLDs) into which any logic function can be programmed. PLDs are introduced in Chapter 10. The fixed-function devices, such as the ones introduced here and in other chapters are useful as teaching tools to help you better understand how digital logic works. A study of certain IC devices in this text will familiarize you with the ones that you will use in the lab.

In this section, you will learn about integrated circuits and specifically those devices that provide the AND, OR, and inverter functions.

An **integrated circuit (IC)** is an electronic circuit that is constructed entirely on a tiny chip of semiconductive material, usually silicon. A cutaway view of one type of IC package is shown in Figure 2–27. The tiny silicon chip is encased in a plastic or ceramic housing with input and output pins that allow the package to be connected to an external circuit. All of the logic circuits are integrated on the chip that is inside the package and connects to the pins by very small wires that are bonded to the chip and to the pins.

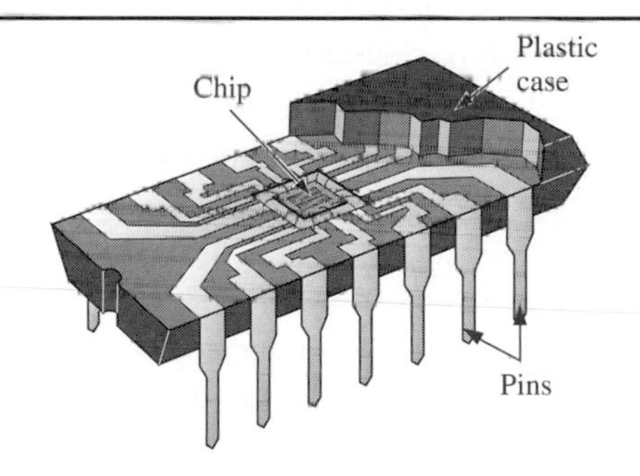

FIGURE 2–27

Cutaway view of one type of IC package (called the dual in-line package or DIP). The tiny chip is mounted inside with connections to input and output pins.

Types of IC Packages

Numerous types and configurations of packages are available for digital integrated circuits. The **DIP** (dual in-line package), which was illustrated in a 14-pin configuration in Figure 2–27 and is shown again in Figure 2–28(a) in a 16-pin configuration, is used

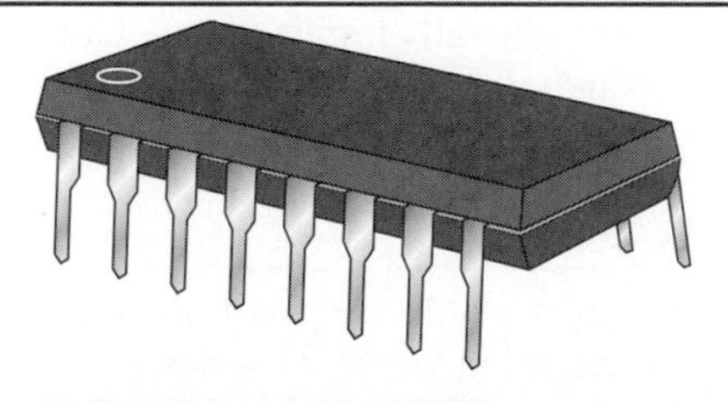

(a) Dual in-line package (DIP)

(b) Small-outline IC (SOIC)

FIGURE 2–28

Two examples of digital IC packages with 16 pins. The DIP is larger than an SOIC with the same number of pins. The SOIC is about half the size of the DIP.

for through-hole mounting on printed circuit (PC) boards. The pins project downward so that they can be inserted into holes in the board to which connections to other circuits are made on both the top and bottom side of the board. Like other IC packages, the DIP comes in several varieties with different numbers of pins depending on the complexity of the circuit inside. You will use DIPs for connecting and testing circuits on a solderless breadboard in the lab.

Another category of IC package is the **SMT** (surface-mount technology) package. There are many types of SMT packages and one type, known as the SOIC (small-outline IC), is shown in Figure 2–28(b) as an example. SMT packages are used for mounting on the surface of a PC board, where higher density (more ICs in a given area) is a requirement.

FIGURE 2–29

Example of how pins are numbered on a DIP or SOIC.

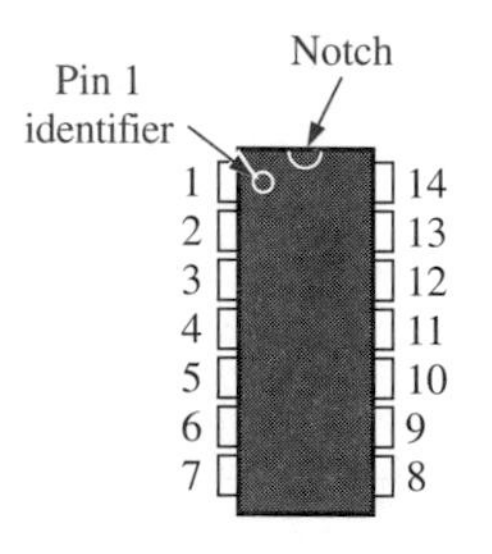

Pin Numbering on an IC

The top view of a 14-pin IC (either DIP or SOIC) is shown in Figure 2–29. The pins on these types of ICs are numbered beginning with pin 1 in the upper left, which can be located by a dot, notch, or beveled edge on the package. The numbers increase down along the left side and then over to the bottom right. From there the pin numbers increase up the right side. The highest pin number is always the upper right pin.

Complexity Classifications for Fixed-Function ICs

Fixed-function digital ICs are classified according to their complexity. They are listed here from the least complex to the most complex. The complexity figures stated here for SSI, MSI, LSI, VLSI, and ULSI are generally accepted, but definitions may vary from one source to another.

- **Small-scale integration (SSI)** describes fixed-function ICs that have up to twelve equivalent gate circuits on a single chip, and they include basic gates and flip-flops.
- **Medium-scale integration (MSI)** describes integrated circuits that have from 12 to 99 equivalent gates on a chip. They include logic functions such as encoders, decoders, counters, registers, multiplexers, arithmetic circuits, and others.
- **Large-scale integration (LSI)** is a classification of ICs with complexities of 100 to 9999 equivalent gates per chip, including small memories.
- **Very large-scale integration (VLSI)** describes integrated circuits with complexities of 10,000 to 99,999 equivalent gates per chip.
- **Ultra large-scale integration (ULSI)** describes very large memories, larger microprocessors, and larger single-chip computers. Complexities of 100,000 equivalent gates and greater are classified as ULSI.

HISTORICAL NOTE

Jack Kilby was born in Missouri in 1923. He earned degrees in electrical engineering from the University of Illinois and the University of Wisconsin. From 1947 to 1958, he worked at the Centralab Division of Globe Union, Inc. in Milwaukee. In 1958, he joined Texas Instruments in Dallas. Within a year after joining TI he invented the monolithic integrated circuit, which has revolutionized electronics. Kilby left TI in 1970 and received the Nobel prize in Physics in 2000.

Integrated Circuit Technologies

The types of transistors with which all integrated circuits are implemented are either MOSFETs (metal-oxide semiconductor field-effect transistors) or bipolar junction transistors. The major circuit technologies that use MOSFETs are CMOS (complementary MOS) and NMOS (*n*-channel MOS). The type of digital circuit technology that uses bipolar junction transistors is TTL (transistor-transistor logic).

All gates and other functions can be implemented with either type of circuit technology. SSI and MSI circuits are generally available in both CMOS and TTL although CMOS is more common. LSI, VLSI, and ULSI are generally implemented with CMOS or NMOS because they require less area on a chip and consume less power.

AND, OR, and Inverter ICs

AND gates, OR gates, and inverters are available in fixed-function IC form. Standard devices in both CMOS logic and TTL logic are in the 54/74 series. All device numbers in this series begin with the prefix 54 or 74. The prefix 54 is for the military grade and 74 is for the commercial grade. The prefix is followed by a letter designation for the particular logic family, such as LS (low-power Schottky), and then a device number. For example, 74LS08 is the number for a particular TTL IC that has four 2-input AND gates in the package (quad 2-input AND gate). Regardless of the particular IC family, the 08 designation always indicates a quad 2-input AND gate IC. The various IC families differ only in circuit technology, not in the logic functions themselves.

Figure 2–30 shows pin diagrams for common AND, OR, and inverter ICs. The dc **supply voltage** (V_{CC}) is connected to pin 14 and ground is connected to pin 7 on this particular 14-pin IC package, which is usually the case, but not always. Although they are not shown in the diagram, V_{CC} and ground connections internally go to each gate. For illustration, the LS designation is shown on the devices in this text. Other families such as ALS, HC, and AHC are logically equivalent.

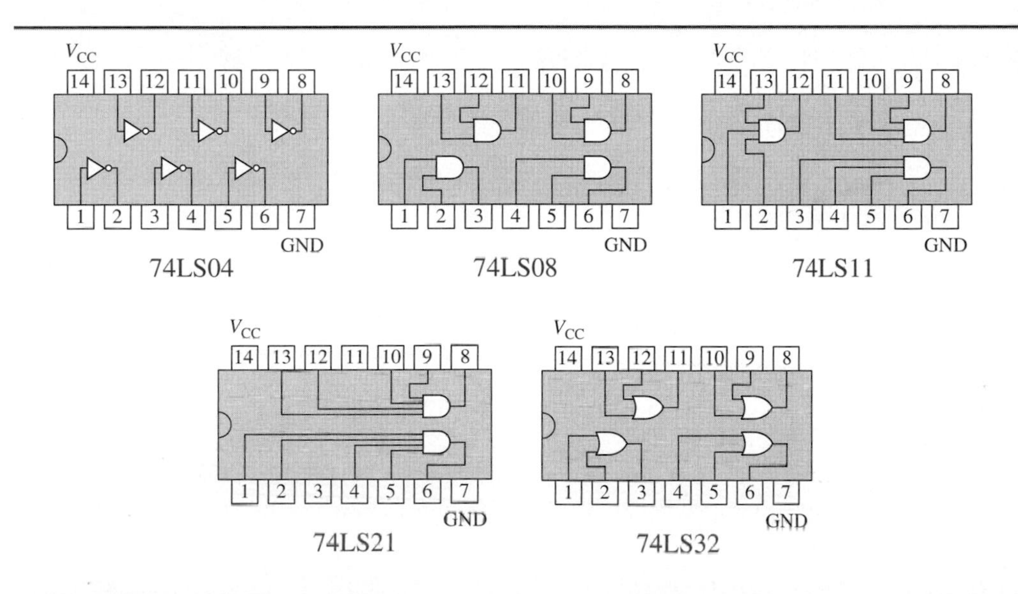

FIGURE 2–30

AND, OR, and inverter ICs. Two or three letters may be shown after the 74 designation to indicate a specific logic family, such as LS, ALS, HC, AHC, and others within a circuit technology (CMOS or TTL).

COMPUTER SIMULATION

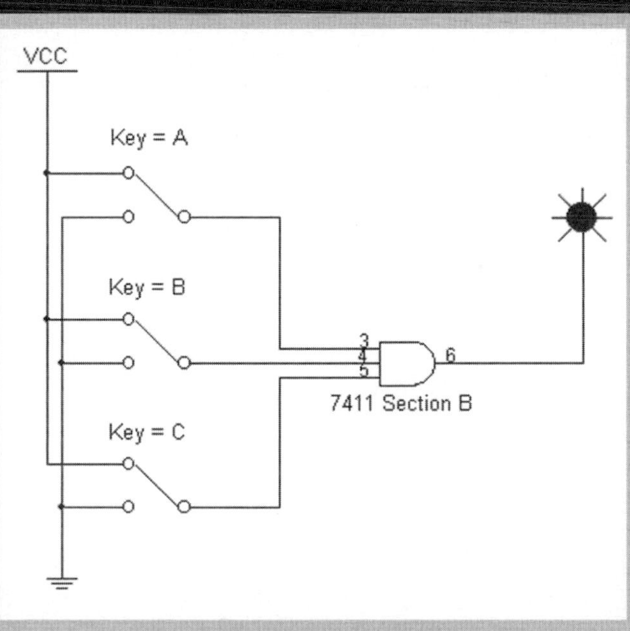

FIGURE 2–31

Open the Multisim file F02-31DG on the website. Check the 7411 3-input AND gate truth table operation by applying LOWs and HIGHs with the switches. The 7411 contains three AND gates, but only one is shown for simulation. For the simulation, operate the top switch with the *A* key, the middle switch with the *B* key, and the bottom key with the *C* key. When the probe light is red, it indicates a HIGH; when it is white, it indicates a LOW. Multisim does not distinguish between IC families such as LS, ALS, AHC, and others.

Review Questions

26. What is an IC?
27. What does VLSI stand for?
28. What does the term *quad 2-input AND gate* mean?
29. What two things are connected to all gates in an IC?
30. What does DIP stand for?

2–7 TROUBLESHOOTING

Troubleshooting is the process of identifying and isolating a fault in a circuit so that the fault can be repaired. Troubleshooting begins when a circuit or system fails to operate properly. In order to troubleshoot digital circuits, you must understand how they work and be able to determine when they are not working properly.

Although the basic troubleshooting skills that you will learn in this text are useful, faults in low-cost modules such as ICs and printed circuit boards are often diagnosed using computers. In many cases, the faulty unit is discarded and replaced by a new one. For high-cost units or when replacement units are not available, you will have to resort to your basic troubleshooting skills.

In this section, you will learn how to determine certain faults in logic gates and recognize their resulting effects.

Faults in AND and OR Gates

In general, IC logic gates are very reliable and seldom fail. However, when they do fail, there are certain types of failures that are more common than others. Open inputs, outputs, supply voltage, or ground lines are the most common types of faults. Shorts are less likely to occur internally to the IC but can occur externally in associated circuitry.

Open Input

An open input on an AND or OR gate effectively disconnects the gate from one of its inputs. An open input has the effect of a high level in TTL gates. In a CMOS gate you cannot reliably predict how the gate will behave with an open input.

When an open AND gate input acts as a HIGH, the output will depend on the signals on the remaining inputs. For example, if a 2-input AND gate has an open input that appears HIGH and there is a digital waveform on the other input, the output looks like the digital waveform, as illustrated in Figure 2–32(a).

FIGURE 2–32

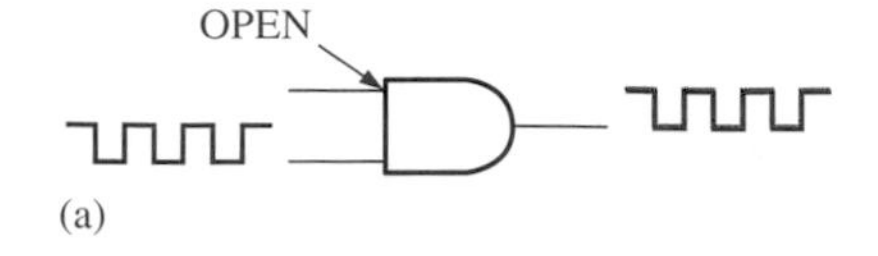

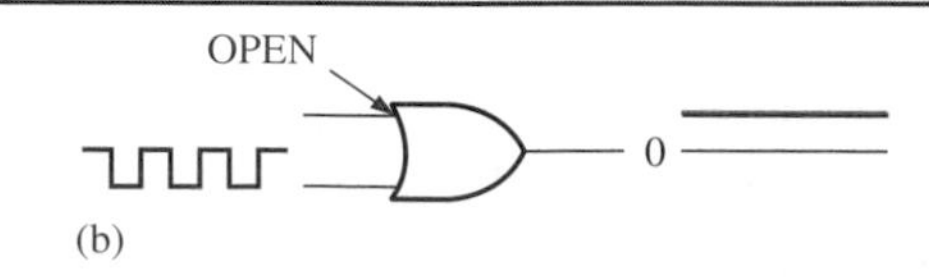

The effects on the output of an open input that appears as a HIGH.

When an open OR gate input acts as a HIGH, the output will be continuously high (stuck HIGH) regardless of the other inputs. For example, if a 2-input OR gate has an open input that appears HIGH and there is a digital waveform on the other input, the output is always HIGH, as shown in Figure 2–32(b).

SAFETY NOTE

To avoid electrical shock, never touch a circuit while it is connected to a voltage source. If you need to handle a circuit, remove a component, or change a component, first make sure the voltage source is disconnected.

Open Output

An open output on an AND or OR gate effectively disconnects the gate from its output pin. If a gate has an open output, there will be no definite output level no matter what the inputs are. An open output produces the same effect for both AND and OR gates in either CMOS or TTL technologies. An open output pin is just "hanging in mid air" and is said to be "floating." This can produce an erratic or "noisy" line at 0 Vdc on your oscilloscope.

Shorted Input

An internal short in an IC is much less likely than an internal open. However, shorts may occur between pins or printed circuit board contacts and therefore affect gate operation. Shorted inputs may be caused by "solder bridges," small wire clippings, or imperfections in the printed circuit board. A short can occur between an input or output pin and ground, V_{CC}, or another gate input or output.

When a gate input is shorted to ground, it has the same effect as a LOW on all types of gates. When an input of an AND gate is shorted to ground, the output is always low (stuck LOW) no matter what the other inputs are. For example, if one input of a 2-input AND gate is shorted to ground, and there is a digital waveform on the other input, the output is constantly LOW, as shown in Figure 2–33(a).

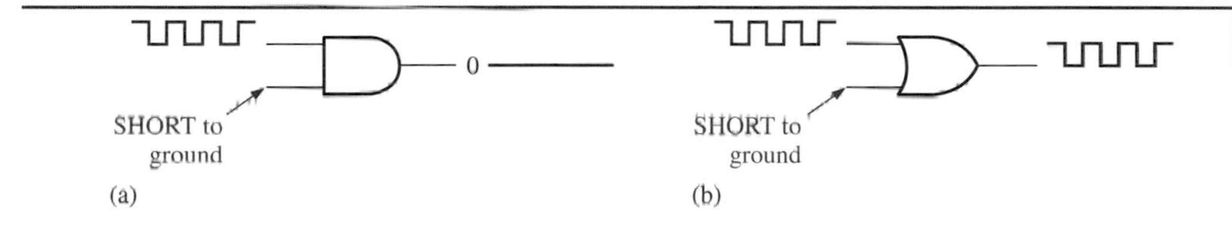

FIGURE 2–33

The effects on the output of an input that is shorted to ground.

When an input of an OR gate is shorted to ground, the output depends on the other inputs. For example, if one input of a 2-input OR gate is shorted to ground and there is a digital waveform on the other input, the output looks like the digital waveform, as illustrated in Figure 2–33(b).

A short to ground is not necessarily the most common type of short, but its effects on gate operation are predictable. Other shorted conditions, such as an input shorted to another signal line or an input and output shorted, are not always so predictable in terms of the effects on the gate operation. In fact, many types of failures can cause very "bizarre" behavior in the way the gate operates.

Review Questions

31. What are two types of faults that are associated with logic gates?
32. Which type of internal fault is most common in an IC?
33. What are two possible causes of a short on a printed circuit board?
34. If one input of a 2-input AND gate is a pulse waveform and the other input is shorted to ground, what will the output be?
35. If one input to a 2-input OR gate is shorted to ground and a HIGH is applied to the other input, what will the output be?

LOGIC GATES: NAND AND NOR

CHAPTER OUTLINE

3–1 The NAND Gate

3–2 The NOR Gate

3–3 Boolean Algebra Applied to NAND and NOR

3–4 Integrated Circuits

3–5 Trouble-shooting

INTRODUCTION

In Chapter 2, you learned about the AND gate, OR gate, and inverter. The inverter can be combined with the AND gate to form a NAND gate and with the OR gate to form a NOR gate. In this chapter, you will study the NAND and NOR gates and learn how Boolean algebra is applied to them. The NAND and NOR gates are known as *universal gates* because either can be used exclusively to implement any type of logic function. Also, some more standard integrated circuits are introduced.

Study aids for this chapter are available at

http://www.prenhall.com/SOE

KEY TERMS

- NAND gate
- NOR gate
- DeMorgan's theorems
- Negative-OR
- Negative-AND

KEY OBJECTIVES

A section number is given for each objective. After completing this chapter, you should be able to

3–1 Describe the operation of the NAND gate

3–2 Describe the operation of the NOR gate

3–3 Apply Boolean algebra to the NAND and NOR gates

3–4 Discuss specific NAND and NOR ICs and describe a typical IC data sheet

3–5 Describe certain faults in IC logic gates and their resulting effects and discuss the proper way to handle unused gate inputs

COMPUTER SIMULATIONS DIRECTORY

The following figures have Multisim circuit files associated with them.

◆ Figure 3–4
Page 71

◆ Figure 3–10
Page 74

◆ Figure 3–22
Page 82

LABORATORY EXPERIMENTS DIRECTORY

The following exercise is for this chapter.

◆ **Experiment 4**
The NAND Gate and the NOR Gate

ON THE JOB. . .

Technical skills include competency in the area of electronics in which you are working. You should learn as much as possible about any product on which you work. You must be able to apply your knowledge to solve problems on the job and be able to acquire more advanced knowledge. Also, you must be proficient in the use of tools and instruments that you use to perform the job duties. Look for ways to improve your technical skills by independent study and taking training offered by your employer and area schools.

Physical barriers tend to place a limit on how much faster computers can run using conventional technology because the speed of computers is limited by the speed of electron motion. Manufacturers of integrated circuits are exploring some new, more innovative methods that hold a great deal of promise.

One approach takes advantage of the steadily shrinking trace size on microchips. Smaller traces mean that as many as 100 million transistors can now be fabricated on a single silicon chip. Increasing transistor densities allows for more and more functions to be integrated onto a single chip much closer together so that the electrons have extremely short distances to travel. High-density chips also now allow data to be processed 64 bits at a time instead of 32 bits or less.

3–1 THE NAND GATE

The NAND gate is an AND gate followed by an inverter. When NAND gates are used in combination, any type of logic function can be implemented.

In this section, you will learn how the NAND gate operates.

The **NAND gate** is a logic circuit in which the output is at a low logic level (LOW) only if all of its inputs are at a high logic level (HIGH). A NAND gate has only one output and at least two inputs. The NAND gate, like the AND and OR gates, is a "decision-making" circuit. It can "decide" when all of a specified number of conditions are met (true) and produce an indication of this by a low logic level on its output. The NAND gate is formed with an AND gate followed by an inverter (NOT circuit), as shown in Figure 3–1. The term *NAND* is a contraction of *NOT-AND.*

FIGURE 3–1

An AND gate followed by an inverter produces the NOT-AND or NAND function.

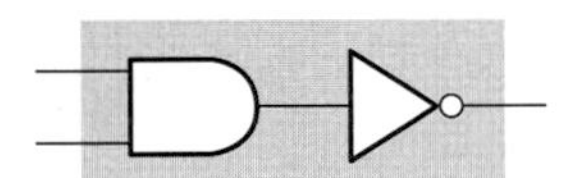

NAND Gate Symbols

Two standard NAND gate logic symbols are the distinctive shape symbol and the rectangular outline symbol. The distinctive shape symbol is shown in Figure 3–2(a) for a NAND gate with two inputs. The bubble represents inversion. The bubble also indicates that a LOW is the asserted output level. As with the AND gate, a NAND gate can have any number of inputs. Figure 3–2(b) shows a 4-input NAND gate symbol. When there are more than a few inputs, extensions are added to the gate symbol to accommodate the additional lines, as shown in Figure 3–2(c). The distinctive shape symbol is used in this book.

FIGURE 3–2

Distinctive shape NAND gate symbols with various numbers of inputs.

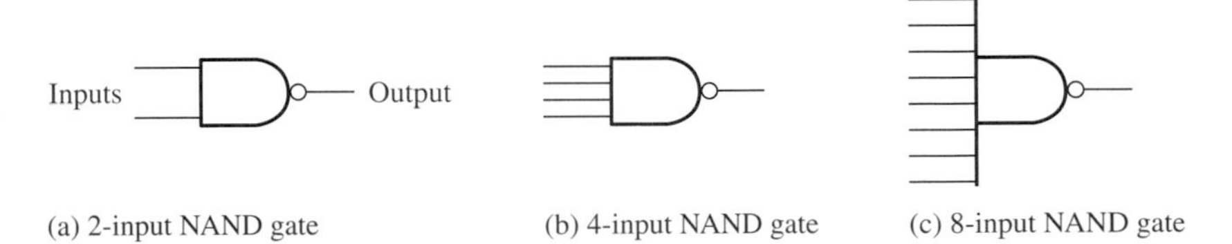

(a) 2-input NAND gate (b) 4-input NAND gate (c) 8-input NAND gate

Figure 3–3 shows examples of the rectangular outline NAND gate symbol. The rectangular outline symbol consists of a box with an internal label (&) to indicate the AND function and the addition of a polarity indicator (small right triangle) on the output to indicate inversion.

FIGURE 3–3

Rectangular outline NAND gate symbol with various numbers of inputs.

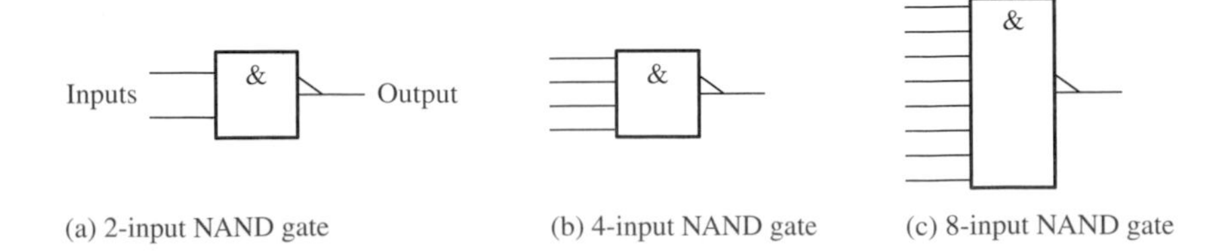

(a) 2-input NAND gate (b) 4-input NAND gate (c) 8-input NAND gate

NAND Gate Operation

The number of gate inputs determines the number of possible input state combinations, which is defined by a power of two.

$$\text{Number of input state combinations} = 2^n$$

where n is the number of inputs to a gate. For example, a 2-input gate has $2^2 = 4$ input state combinations and a 4-input gate has $2^4 = 16$ input state combinations.

The logic operation of a NAND gate can be described by a truth table, which lists all the possible input state combinations and the corresponding output state for each input combination. As before, we will use HIGH and LOW for the input and output states to represent high logic levels and low logic levels, respectively. Remember that with positive logic a HIGH and a 1 can be used interchangeably and that a LOW and 0 can be used interchangeably.

Truth tables for a 2-input NAND gate and a 3-input NAND gate are given in Table 3–1. A 2-input gate has $2^2 = 4$ input combinations and a 3-input gate has $2^3 = 8$ input combinations. The gate inputs are labeled *A, B,* and *C,* and the output is labeled *X*. As you can see, the output of a NAND gate is LOW (0) only if all of the inputs are HIGH (1).

TABLE 3–1

Truth tables for (a) 2-input NAND gate and (b) 3-input NAND gate.

Inputs		Output
A	*B*	*X*
LOW	LOW	HIGH
LOW	HIGH	HIGH
HIGH	LOW	HIGH
HIGH	HIGH	LOW

(a)

Inputs			Output
A	*B*	*C*	*X*
LOW	LOW	LOW	HIGH
LOW	LOW	HIGH	HIGH
LOW	HIGH	LOW	HIGH
LOW	HIGH	HIGH	HIGH
HIGH	LOW	LOW	HIGH
HIGH	LOW	HIGH	HIGH
HIGH	HIGH	LOW	HIGH
HIGH	HIGH	HIGH	LOW

(b)

COMPUTER SIMULATION

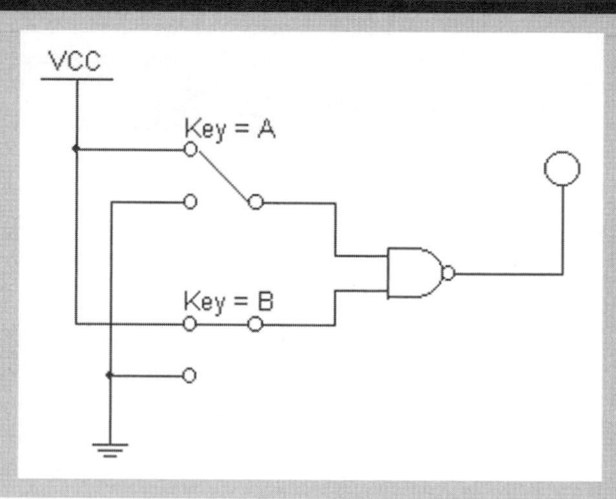

FIGURE 3–4

Open the Multisim file F03-04DG on the website. Verify the 2-input NAND gate truth table by applying LOWs and HIGHs with the switches. The ground line is the LOW and the $+V_{CC}$ line is the HIGH. For the simulation, operate the top switch with the *A* key and the bottom switch with the *B* key. When the probe light is white, it indicates a LOW.

NAND Gates with Digital Waveform Inputs

Gate inputs are usually not at a constant level all of the time. Most of the time you will find that gate inputs are digital waveforms. In terms of their truth table operation, NAND gates work the same whether or not the inputs are always constant or are changing. The output is LOW only when all inputs are HIGH; otherwise, the output remains HIGH. Example 3–1 illustrates this with a 2-input NAND gate.

EXAMPLE 3–1

Problem

Determine the output waveform X when the waveforms shown in Figure 3–5 are applied to the 2-input NAND gate.

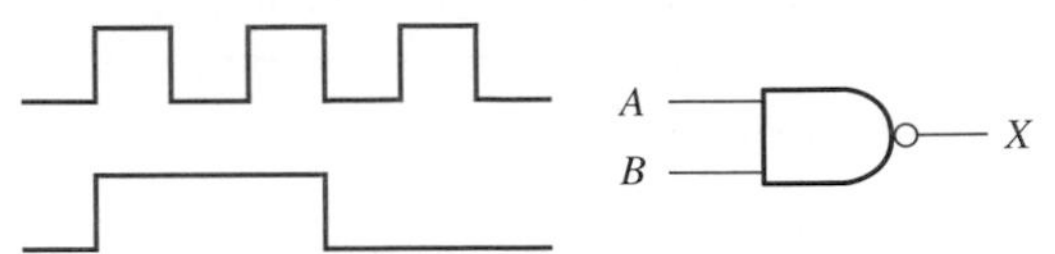

FIGURE 3–5

Solution

The times that the two input waveforms A and B are HIGH at the same time are indicated by the blue-shaded areas in Figure 3–6. The output waveform X is LOW as shown. The diagram in Figure 3–6 is called a timing diagram.

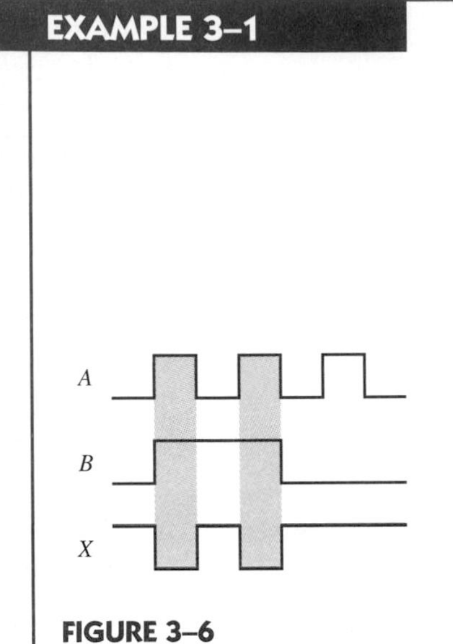

FIGURE 3–6

Question*

If input waveform A stays the same as shown and input waveform B is always HIGH, what would the output waveform X look like?

Review Questions

Answers are at the end of the chapter.

1. When is the output of a NAND gate at the low logic level (LOW)?
2. How many possible combinations of input states are there for a 5-input NAND gate?
3. How many rows and how many columns are in the truth table for a 4-input NAND gate?
4. If the inputs to a NAND gate are digital waveforms, when will the output be LOW?
5. If all possible input state combinations are applied to an 8-input NAND gate, for how many of these combinations is the output LOW?

3–2 THE NOR GATE

A NOR gate is an OR gate followed by an inverter. When NOR gates are used in combination, any type of logic function can be implemented.

In this section, you will learn how a NOR gate operates.

The **NOR gate** is a logic circuit in which the output is at a low logic level (LOW) only if one or more of its inputs are at a high logic level (HIGH). A NOR gate has only one output and at least two inputs. The NOR gate (like the NAND, AND, and OR gates) is a "decision-making" circuit. The NOR gate can "decide" when one or more conditions are met (true) and produce an indication of this by a low logic level on its output. The NOR gate is formed with an OR gate followed by an inverter (NOT circuit), as shown in Figure 3–7. The term *NOR* is a contraction of *NOT-OR*.

FIGURE 3–7

An OR gate followed by an inverter produces the NOT-OR or NOR function.

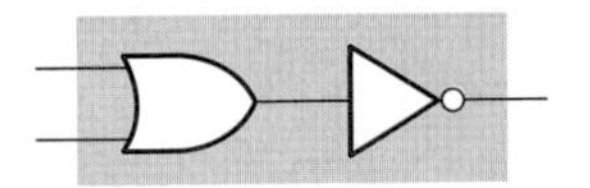

* *Answers are at the end of the chapter.*

NOR Gate Symbols

The distinctive shape symbol is shown in Figure 3–8(a) for a NOR gate with two inputs. Figure 3–8(b) shows a 4-input NOR gate symbol. The bubble represents inversion. The bubble also indicates that a LOW is the asserted output level. When there are more than a few inputs, extensions are added to the gate symbol to accommodate the additional lines, as shown in Figure 3–8(c). The distinctive shape symbol is the one used in this book.

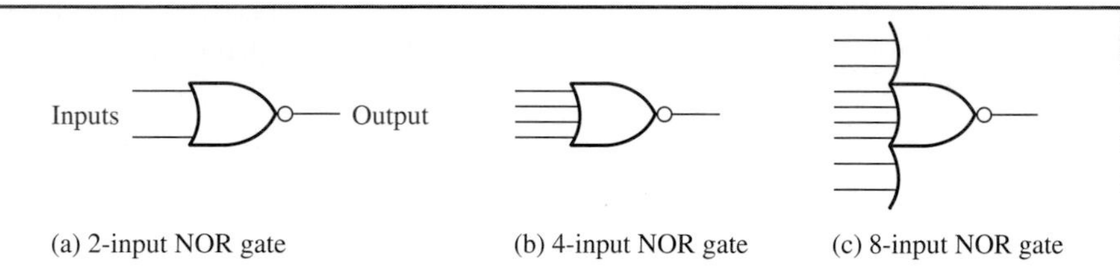

FIGURE 3–8

Distinctive shape NOR gate symbol with various numbers of inputs.

Figure 3–9 shows examples of the rectangular outline NOR gate symbol. The rectangular outline symbol consists of a box with an internal label (≥1) to indicate the OR function and the addition of a small right triangle to indicate inversion.

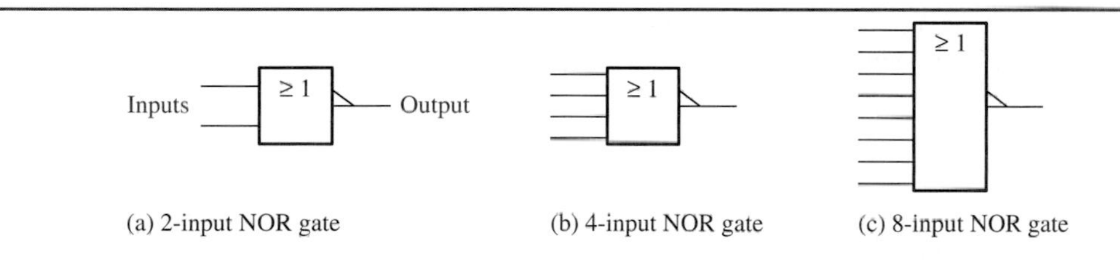

FIGURE 3–9

Rectangular outline NOR gate symbol with various numbers of inputs.

NOR Gate Operation

The logic operation of a NOR gate can also be described by a truth table, which lists all the possible input state combinations and the corresponding output state for each input combination. Again, we will use HIGH and LOW for the input and output states to represent high logic levels and low logic levels, respectively.

Truth tables for a 2-input NOR gate and a 3-input NOR gate are given in Table 3–2. A 2-input gate has $2^2 = 4$ input combinations and a 3-input gate has $2^3 = 8$ input combinations. The gate inputs are labeled *A, B,* and *C,* and the output is labeled *X.* As you can see, the output of a NOR gate is LOW (0) when one or more inputs are HIGH (1).

Inputs		Output
A	B	X
LOW	LOW	HIGH
LOW	HIGH	LOW
HIGH	LOW	LOW
HIGH	HIGH	LOW

(a)

Inputs			Output
A	B	C	X
LOW	LOW	LOW	HIGH
LOW	LOW	HIGH	LOW
LOW	HIGH	LOW	LOW
LOW	HIGH	HIGH	LOW
HIGH	LOW	LOW	LOW
HIGH	LOW	HIGH	LOW
HIGH	HIGH	LOW	LOW
HIGH	HIGH	HIGH	LOW

(b)

TABLE 3–2

Truth tables for (a) 2-input NOR gate and (b) 3-input NOR gate.

COMPUTER SIMULATION

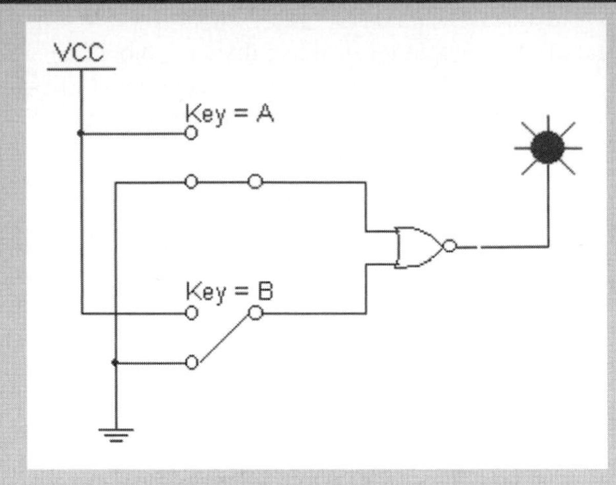

FIGURE 3–10

Open the Multisim File F03-10DG on the website. Verify the 2-input NOR gate truth table by applying LOWs and HIGHs with the switches. The ground line is the LOW and the $+V_{CC}$ line is the HIGH. For the simulation, operate the top switch with the *A* key and the bottom switch with the *B* key.

NOR Gates with Digital Waveform Inputs

As with the other gates, NOR gates work the same in terms of their truth table operation whether or not the inputs are always constant or are changing. The output is LOW when one or more inputs are HIGH, and it is HIGH when all the inputs are LOW.

EXAMPLE 3–2

Problem

Determine the output waveform *X* when the waveforms shown in Figure 3–11 are applied to the 2-input NOR gate.

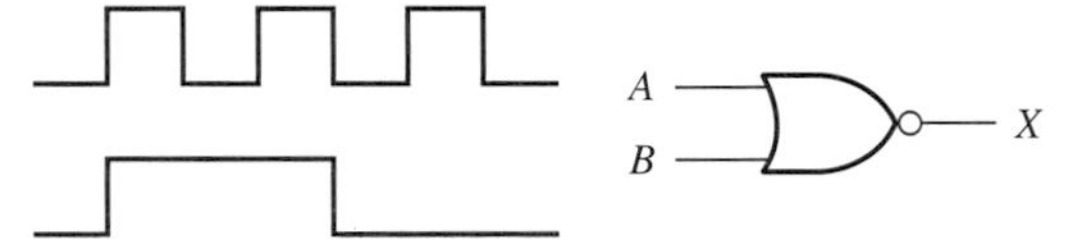

FIGURE 3–11

Solution

The times that one or both input waveforms are HIGH are indicated by the blue-shaded areas in the timing diagram of Figure 3–12. The output waveform *X* with the corresponding LOWs is shown in the figure.

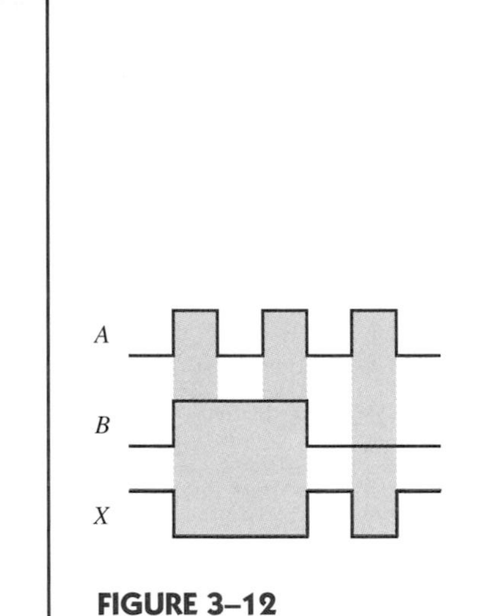

FIGURE 3–12

Question

If input waveform *A* is always LOW and input waveform *B* stays the same as shown, what would the output waveform *X* look like?

Review Questions

6. When is the output of a NOR gate at the LOW logic level?
7. How many possible combinations of input states are there for a 4-input NOR gate?
8. How many rows and how many columns does the truth table for a 5-input NOR gate contain?
9. If the inputs to a NOR gate are digital waveforms, when will the output be LOW?
10. How do you make a NOR gate using an OR gate and an inverter?

BOOLEAN ALGEBRA APPLIED TO NAND AND NOR 3–3

The operation of the AND gate, the OR gate and the inverter can be described using Boolean algebra. Boolean algebra can also be used to express the operation of the NAND and NOR gates.

In this section, you will learn to apply Boolean algebra to the NAND and NOR gates.

Recall that in Boolean algebra a variable can have only two values, a 1 or a 0, which corresponds to the bits of 1 and 0 in the binary number system. When you evaluate logic circuits using Boolean algebra, 1s and 0s are used for the values of variables instead of HIGHs and LOWs.

Also recall that a variable in Boolean algebra can be designated with a letter. Generally, input variables will be labeled *A, B, C,* and *D* and output variables will be labeled *X, Y,* or *Z*. Also, a variable can be designated as a letter with a subscript, such as A_1, A_2, or A_3.

Table 3–3 shows the truth table in terms of 1s and 0s for a 2-input NAND gate. The truth table for any logic circuit can be expressed in terms of 1s and 0s or HIGHs and LOWs.

TABLE 3–3

Truth table for a 2-input NAND gate.

A	B	X
0	0	1
0	1	1
1	0	1
1	1	0

Boolean Expression for the NAND Gate

To express the operation of a 2-input NAND gate in Boolean algebra, place a bar over the AND term.

$$X = \overline{AB}$$

This Boolean expression states that *the output X equals the complement of the product (AND) of the inputs A and B.*

To evaluate $X = \overline{AB}$ and see how it actually describes the operation of a NAND gate, simply substitute a 1 or a 0 for each variable. The AND operation must be done before the complement operation. For two inputs, there are $2^2 = 4$ possible combinations of input states.

For $A = 0, B = 0$: $X = \overline{AB} = \overline{0 \cdot 0} = \overline{0} = 1$
For $A = 0, B = 1$: $X = \overline{AB} = \overline{0 \cdot 1} = \overline{0} = 1$
For $A = 1, B = 0$: $X = \overline{AB} = \overline{1 \cdot 0} = \overline{0} = 1$
For $A = 1, B = 1$: $X = \overline{AB} = \overline{1 \cdot 1} = \overline{1} = 0$

EXAMPLE 3–3

Problem

Write the Boolean expression for the 3-input NAND gate in Figure 3–13. Evaluate the expression for all possible combinations of input states, and express the results in the form of a truth table.

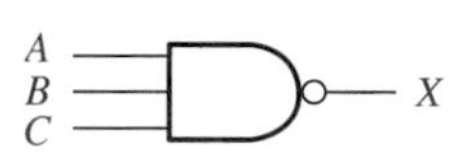

FIGURE 3–13

Solution

The Boolean expression for the 3-input NAND gate is

$$X = \overline{ABC}$$

To evaluate the expression, substitute 1s and 0s for the variables to create all possible combinations ($2^3 = 8$) of input states. The corresponding truth table is shown in Table 3–4.

For $A = 0, B = 0, C = 0$: $X = \overline{ABC} = \overline{0 \cdot 0 \cdot 0} = \overline{0} = 1$
For $A = 0, B = 0, C = 1$: $X = \overline{ABC} = \overline{0 \cdot 0 \cdot 1} = \overline{0} = 1$
For $A = 0, B = 1, C = 0$: $X = \overline{ABC} = \overline{0 \cdot 1 \cdot 0} = \overline{0} = 1$
For $A = 0, B = 1, C = 1$: $X = \overline{ABC} = \overline{0 \cdot 1 \cdot 1} = \overline{0} = 1$

Table 3–4

A	B	C	X
0	0	0	1
0	0	1	1
0	1	0	1
0	1	1	1
1	0	0	1
1	0	1	1
1	1	0	1
1	1	1	0

For $A = 1, B = 0, C = 0$: $X = \overline{ABC} = \overline{1 \cdot 0 \cdot 0} = \overline{0} = 1$
For $A = 1, B = 0, C = 1$: $X = \overline{ABC} = \overline{1 \cdot 0 \cdot 1} = \overline{0} = 1$
For $A = 1, B = 1, C = 0$: $X = \overline{ABC} = \overline{1 \cdot 1 \cdot 0} = \overline{0} = 1$
For $A = 1, B = 1, C = 1$: $X = \overline{ABC} = \overline{1 \cdot 1 \cdot 1} = \overline{1} = 0$

Question
If a NAND gate has six inputs, how many possible input combinations are there?

Boolean Expression for the NOR Gate

To express the operation of a 2-input NOR gate in Boolean algebra, place a bar over the OR term.

$$X = \overline{A + B}$$

This Boolean expression states that the *output X equals the complement of the Boolean sum (OR) of the inputs A and B.*

To evaluate $X = \overline{A + B}$ and see how it actually describes the operation of a NOR gate, simply substitute a 1 or a 0 for each variable. The OR operation must be done before the complement operation. As before, there are $2^2 = 4$ possible combinations of input states.

For $A = 0, B = 0$: $X = \overline{A + B} = \overline{0 + 0} = \overline{0} = 1$
For $A = 0, B = 1$: $X = \overline{A + B} = \overline{0 + 1} = \overline{1} = 0$
For $A = 1, B = 0$: $X = \overline{A + B} = \overline{1 + 0} = \overline{1} = 0$
For $A = 1, B = 1$: $X = \overline{A + B} = \overline{1 + 1} = \overline{1} = 0$

EXAMPLE 3–4

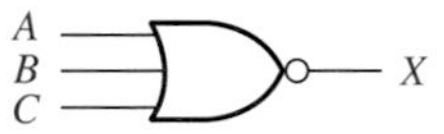

A, B, C → NOR → X

FIGURE 3–14

Problem
Write the Boolean expression for the 3-input NOR gate in Figure 3–14. Evaluate the expression for all possible combinations of input states, and express the results in the form of a truth table.

Solution
The Boolean expression for the 3-input NOR gate is

$$X = \overline{A + B + C}$$

To evaluate the expression, substitute 1s and 0s for the variables to create all possible combinations ($2^3 = 8$) of input states. The corresponding truth table is shown in Table 3–5.

For $A = 0, B = 0, C = 0$: $X = \overline{A + B + C} = \overline{0 + 0 + 0} = \overline{0} = 1$
For $A = 0, B = 0, C = 1$: $X = \overline{A + B + C} = \overline{0 + 0 + 1} = \overline{1} = 0$
For $A = 0, B = 1, C = 0$: $X = \overline{A + B + C} = \overline{0 + 1 + 0} = \overline{1} = 0$
For $A = 0, B = 1, C = 1$: $X = \overline{A + B + C} = \overline{0 + 1 + 1} = \overline{1} = 0$
For $A = 1, B = 0, C = 0$: $X = \overline{A + B + C} = \overline{1 + 0 + 0} = \overline{1} = 0$
For $A = 1, B = 0, C = 1$: $X = \overline{A + B + C} = \overline{1 + 0 + 1} = \overline{1} = 0$
For $A = 1, B = 1, C = 0$: $X = \overline{A + B + C} = \overline{1 + 1 + 0} = \overline{1} = 0$
For $A = 1, B = 1, C = 1$: $X = \overline{A + B + C} = \overline{1 + 1 + 1} = \overline{1} = 0$

Table 3–5

A	B	C	X
0	0	0	1
0	0	1	0
0	1	0	0
0	1	1	0
1	0	0	0
1	0	1	0
1	1	0	0
1	1	1	0

Question
If an OR gate has eight inputs, how many possible input combinations are there?

The Commutative Law

You learned about the commutative laws of multiplication and addition in Chapter 2 and now we apply the commutative laws to NAND and NOR gates. Two more Boolean laws, the distributive law and the associative law, will be introduced in Chapter 4 when you study combinations of gates.

The commutative law of multiplication applies to the NAND gate as follows:

In terms of the result, the order in which variables are NANDed makes no difference.

In terms of a 2-input NAND gate, the commutative law is expressed as

$$\overline{AB} = \overline{BA}$$

The commutative law of addition applies to the NOR gate and is stated as follows:

In terms of the result, the order in which variables are NORed makes no difference.

In terms of a 2-input NOR gate, the commutative law is expressed as

$$\overline{A + B} = \overline{B + A}$$

DeMorgan's Theorems

Two theorems that are important in Boolean algebra because of their application to NAND and NOR gates for logic simplification are **DeMorgan's theorems**. DeMorgan's first theorem is stated as follows:

The complement of a product (AND) of variables is equal to the sum (OR) of the complemented variables.

This theorem applies to any number of variables. For two variables, it is expressed as

$$\overline{AB} = \overline{A} + \overline{B}$$

In practice this means that a NAND gate ($\overline{AB}$) is equivalent to an OR gate with its input variables complemented ($\overline{A} + \overline{B}$), as illustrated in Figure 3–15. The OR gate with inputs complemented, as indicated by the bubbles, is known as a *negative-OR gate.*

As illustrated in Figure 3–15, when a NAND gate is used to detect all HIGHs on its inputs, it is performing the NAND function. In this case, the asserted level of the inputs is HIGH and the asserted level of its output is LOW. When a NAND gate is used to detect one or more LOWs on its inputs, it is performing the **negative-OR** operation. In this case, the asserted level of the inputs is LOW and the asserted level of its output is HIGH.

HISTORICAL NOTE

Augustus DeMorgan (1807–1871) was born in India while his father was stationed there with the military. Shortly afterwards, the family returned to London. DeMorgan was blind in one eye from an early age, and, as a result, he had difficulty in school. However, he graduated from Trinity College in Cambridge and eventually became the first professor of mathematics at the University College in London.

DeMorgan published several books and corresponded with Charles Babbage, the man who is credited with the development of the first programmable computing machine. However, there is no indication that DeMorgan knew George Boole.

DeMorgan recognized the purely symbolic nature of algebra and was aware of algebras other than ordinary algebra. This indicates that he most likely was familiar with Boole's work. He is best known for the two laws that we know today as DeMorgan's theorems.

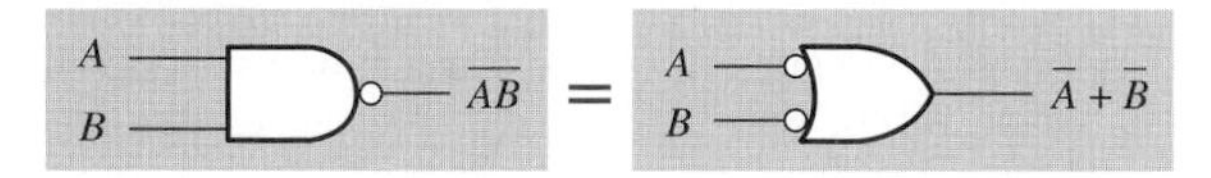

FIGURE 3–15

DeMorgan's first theorem shows that a NAND gate is equivalent to an OR gate with complemented inputs (negative-OR).

DeMorgan's second theorem is stated as follows:

The complement of a sum (OR) of variables is equal to the product (AND) of the complemented variables.

This theorem also applies to any number of variables. For two variables, it is expressed as

$$\overline{A + B} = \overline{A}\,\overline{B}$$

In practice this means that a NOR gate ($\overline{A + B}$) is equivalent to an AND gate with its input variables complemented ($\overline{A}\,\overline{B}$), as illustrated in Figure 3–16. The AND gate with inputs complemented, as indicated by the bubbles, is known as a *negative-AND gate.*

As illustrated in Figure 3–16, when a NOR gate is used to detect one or more HIGHs on its inputs, it is performing the NOR function. In this case, the asserted level of the inputs is HIGH and the asserted level of its output is LOW. When a NOR gate is used to detect all LOWs on its inputs, it is performing the **negative-AND** operation. In this case, the asserted level of the inputs is LOW and the asserted level of its output is HIGH.

FIGURE 3–16

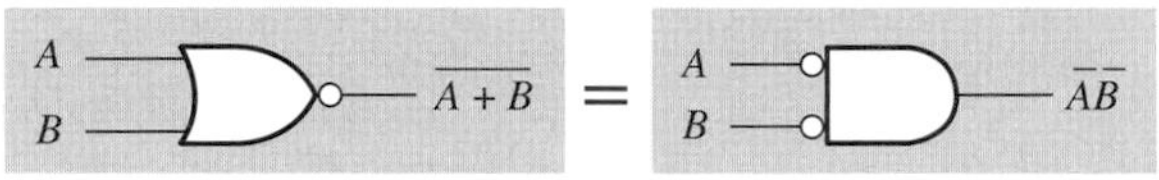

DeMorgan's second theorem shows that a NOR gate is equivalent to an AND gate with complemented inputs (negative-AND).

Review Questions

11. What is the Boolean expression for a 2-input NAND gate?
12. What is the Boolean expression for a 2-input NOR gate?
13. What Boolean operations does the NAND gate perform?
14. What Boolean operations does the NOR gate perform?
15. What do the terms *negative-OR* and *negative-AND* mean?

3–4 INTEGRATED CIRCUITS

NAND and NOR gate ICs are available in both CMOS (Complementary Metal-Oxide Semiconductor) and TTL (Transistor-Transistor Logic) circuit technologies. The CMOS devices are more widely used, but they must be carefully handled because they are susceptible to damage by static electrical charge (electrostatic discharge).

In this section, you will learn about specific NAND and NOR ICs and how to handle unused gate inputs. Also, you will see a typical IC data sheet.

NAND Gate and NOR Gate ICs

Standard NAND and NOR gate fixed-function ICs in both CMOS and TTL logic are available in the 54/74 series. All device numbers in this series begin with the prefix 54 or 74. The prefix is followed by a letter designation for the particular logic family and then a device number. For example, 74AHC00 is the number for a particular CMOS IC that has four 2-input NAND gates in the package (quad 2-input NAND gate). Regardless of the particular IC family, the 00 designation always indicates a quad 2-input NAND gate IC.

Figure 3–17 shows pin diagrams for common NAND and NOR gate ICs. The typical input and output 14-pin connections are shown for each gate. The dc supply voltage (V_{CC}) is connected to pin 14 and ground is connected to pin 7 for these ICs, but you should always check the manufacturer's data sheet if you aren't sure. Although they are not shown in the diagram, the V_{CC} and ground connections go to each gate within the IC.

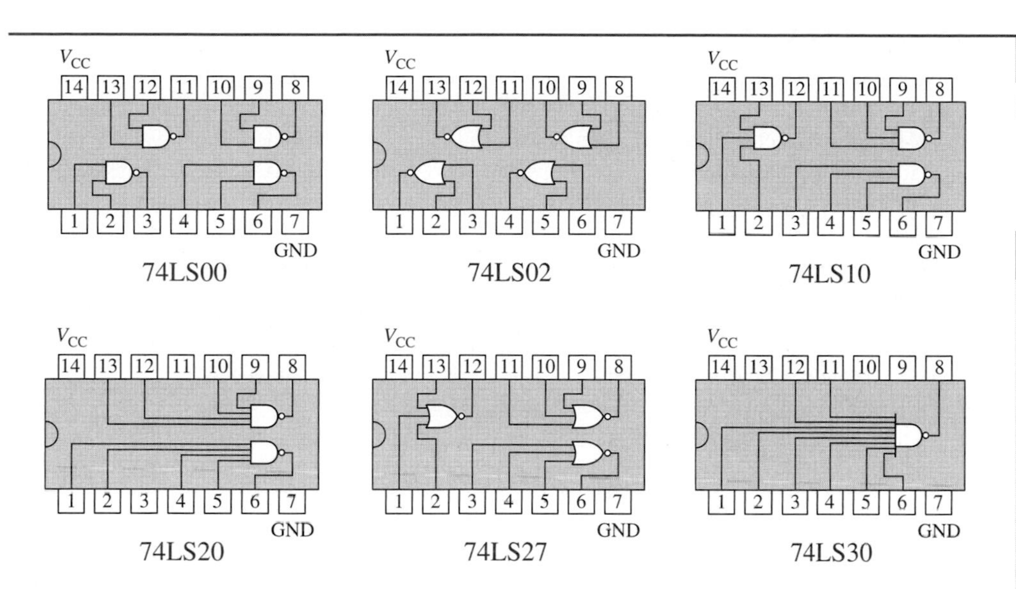

FIGURE 3–17

NAND and NOR gate ICs.

Unused Gate Inputs

Sometimes all of the available inputs on a logic gate are not needed. Any excess inputs in a given gate application become unused inputs and must be properly connected to prevent potentially incorrect gate operation due to noise. For example, suppose that only three inputs of a 4-input gate are required for a certain application. The fourth input of this gate is unused and must be properly connected, so the gate effectively operates as a 3-input gate. Proper connection of an unused gate input means that it is connected to either a high level or a low level, depending on the type of gate. An unused input can also be connected to another input on the gate that is used.

All unused inputs on a NAND gate or an AND gate should be connected to the dc supply voltage (V_{CC}) through a "pull-up" resistor, as shown in Figure 3–18. As an example, part (a) shows one unused input of a 3-input NAND gate, and part (b) shows two unused inputs of a 4-input NAND gate. Another option is to connect the unused input to a used input, as shown in part (c).

HISTORICAL NOTE

Robert Noyce (1927–1990) was born in Iowa and earned a Ph.D. in physics from MIT in 1953. He was a pioneer in semiconductor development and cofounded Fairchild Semiconductor in 1957. Noyce, along with Gordon Moore, founded the Intel Corporation in 1968.

Noyce, independent of Jack Kilby's work, came up with his own ideas for an integrated circuit. His invention, however, was about six months behind Kilby's. Noyce improved on Kilby's invention by replacing wire connections with metal connections integrated into the chip, making it more suitable for mass production. He is generally recognized as the coinventor of the integrated circuit.

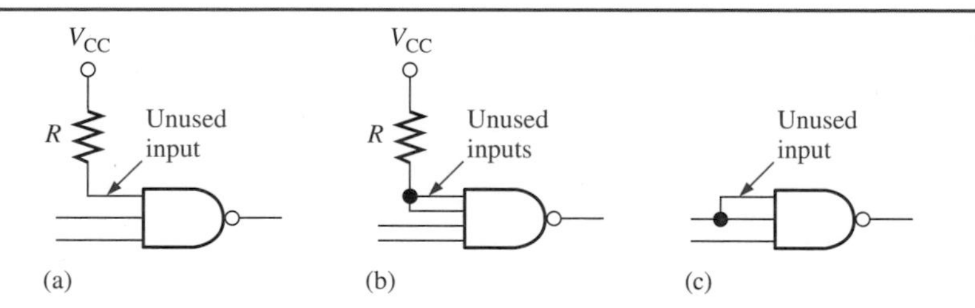

FIGURE 3–18

Connection of unused inputs for NAND gates. The same applies to AND gates.

All unused inputs on a NOR gate or an OR gate should be connected directly to ground, as shown in Figure 3–19. As an example, part (a) shows one unused input of a 3-input NOR gate, and part (b) shows two unused inputs of a 4-input NOR gate. Another option is to connect the unused input to a used input, as shown in part (c).

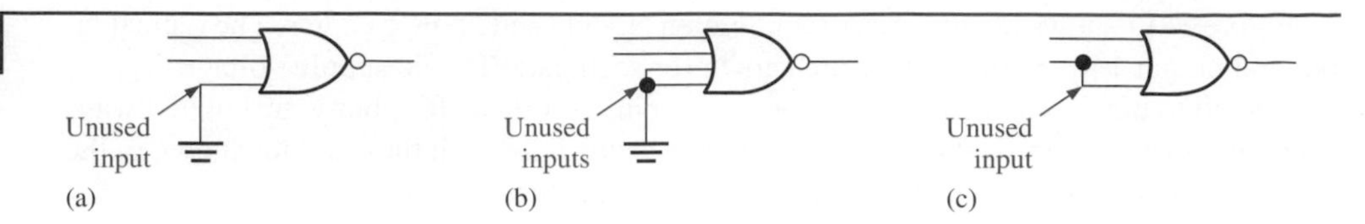

FIGURE 3–19 Connection of unused inputs for NOR gates. The same applies to OR gates.

Data Sheets

A typical data sheet consists of an information page that shows, among other things, the logic diagram and packages, the recommended operating conditions, the electrical characteristics, and the switching characteristics. Partial data sheets for a TTL 74LS00 quad 2-input NAND gate and a CMOS 74HC00A quad 2-input NAND gate are shown in Figures 3–20 and 3–21, respectively. The length of data sheets vary and some have much more information than others.

FIGURE 3–20 The partial data sheet for a TTL 74LS00.

QUAD 2-INPUT NAND GATE

• ESD > 3500 Volts

SN54/74LS00

QUAD 2-INPUT NAND GATE
LOW POWER SCHOTTKY

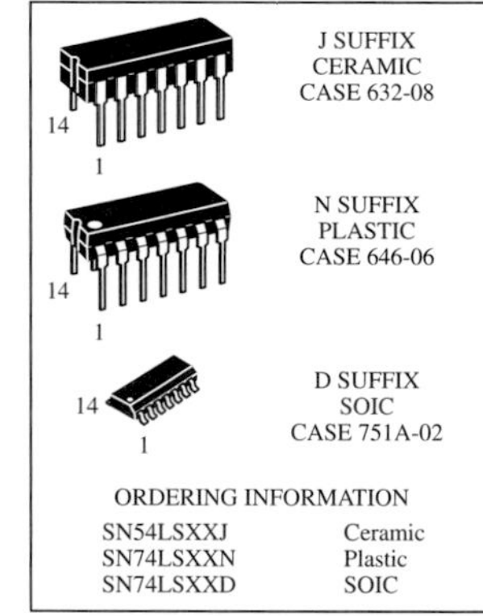

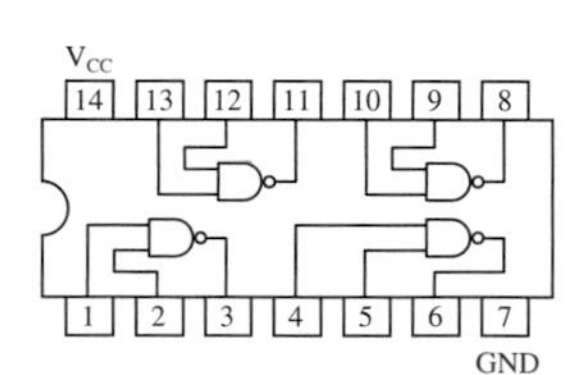

SN54/74LS00

DC CHARACTERISTICS OVER OPERATING TEMPERATURE RANGE (unless otherwise specified)

Symbol	Parameter		Limits Min	Typ	Max	Unit	Test Conditions
V_{IH}	Input HIGH Voltage		2.0			V	Guaranteed Input HIGH Voltage for All Inputs
V_{IL}	Input LOW Voltage	54			0.7	V	Guaranteed Input LOW Voltage for All Inputs
		74			0.8		
V_{IK}	Input Clamp Diode Voltage			−0.65	−1.5	V	V_{CC} = MIN, I_{IN} = −18 mA
V_{OH}	Ouput HIGH Voltage	54	2.5	3.5		V	V_{CC} = MIN, I_{OH} = MAX, V_{IN} = V_{IH} or V_{IL} per Truth Table
		74	2.7	3.5		V	
V_{OL}	Ouput LOW Voltage	54, 74		0.25	0.4	V	I_{OL} = 4.0 mA; V_{CC} = V_{CC} MIN, V_{IN} = V_{IL} or V_{IH} per Truth Table
		74		0.35	0.5	V	I_{OL} = 8.0 mA
I_{IH}	Input HIGH Current				20	µA	V_{CC} = MAX, V_{IN} = 2.7 V
					0.1	mA	V_{CC} = MAX, V_{IN} = 7.0 V
I_{IL}	Input LOW Current				−0.4	mA	V_{CC} = MAX, I_N = 0.4 V
I_{OS}	Short Circuit Current (Note 1)		−20		−100	mA	V_{CC} = MAX
I_{CC}	Power Supply Current Total, Output HIGH				1.6	mA	V_{CC} = MAX
	Total, Output LOW				4.4		

NOTE 1: Not more than one output should be shorted at a time, nor for more than 1 second.

AC CHARACTERISTICS (T_A = 25°C)

Symbol	Parameter	Limits Min	Typ	Max	Unit	Test Conditions
t_{PLH}	Turn-Off Delay, Input to Output		9.0	15	ns	V_{CC} = 5.0 V C_L = 15 pF
t_{PHL}	Turn-On Delay, Input to Output		10	15	ns	

GUARANTEED OPERATING RANGES

Symbol	Parameter		Min	Typ	Max	Unit
V_{CC}	Supply Voltage	54 74	4.5 4.75	5.0 5.0	5.5 5.25	V
T_A	Operating Ambient Temperature Range	54 74	−55 0	25 25	125 70	°C
I_{OH}	Output Current — High	54, 74			−0.4	mA
I_{OL}	Output Current — Low	54 74			4.0 8.0	mA

The partial data sheet for a CMOS 74HC00A. **FIGURE 3–21**

Quad 2-Input NAND Gate **High-Performance Silicon–Gate CMOS**

The MC54/74HC00A is identical in pinout to the LS00. The device inputs are compatible with Standard CMOS outputs; with pullup resistors, they are compatible with LSTTL outputs.

- Output Drive Capability: 10 LSTTL Loads
- Outputs Directly Interface to CMOS, NMOS and TTL
- Operating Voltage Range: 2 to 6 V
- Low Input Current: 1 μA
- High Noise Immunity Characteristic of CMOS Devices
- In Compliance With the JEDEC Standard No. 7A Requirements
- Chip Complexity: 32 FETs or 8 Equivalent Gates

LOGIC DIAGRAM

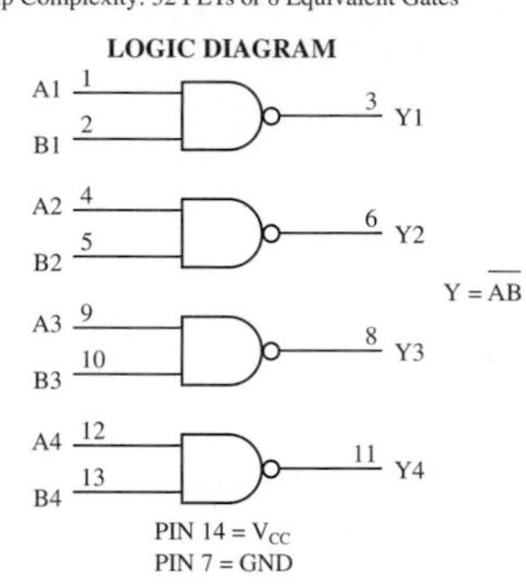

PIN 14 = V_{CC}
PIN 7 = GND

Pinout: 14–Load Packages (Top View)

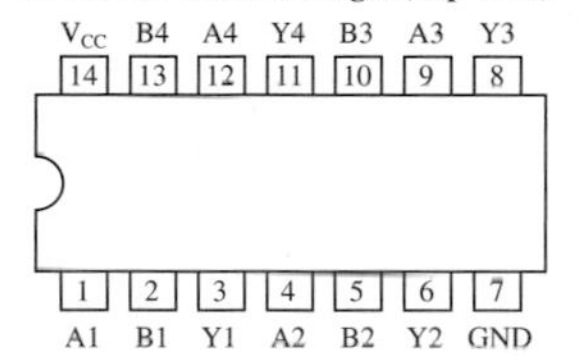

MC54/74HC00A

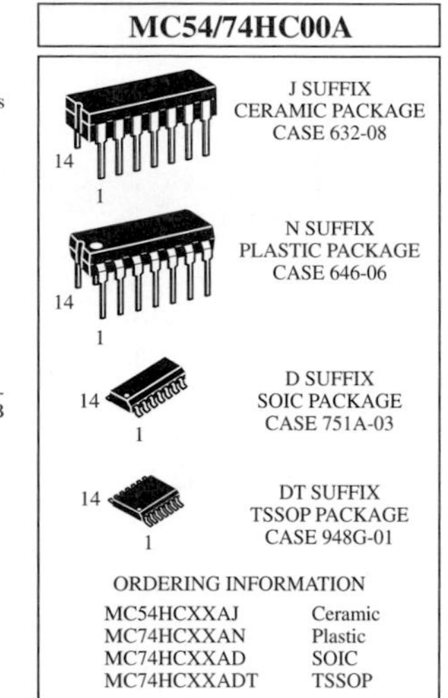

ORDERING INFORMATION

MC54HCXXAJ	Ceramic
MC74HCXXAN	Plastic
MC74HCXXAD	SOIC
MC74HCXXADT	TSSOP

FUNCTION TABLE

Inputs		Output
A	B	Y
L	L	H
L	H	H
H	L	H
H	H	L

MAXIMUM RATINGS*

Symbol	Parameter	Value	Unit
V_{CC}	DC Supply Voltage (Referenced to GND)	–0.5 to + 7.0	V
V_{in}	DC Input Voltage (Referenced to GND)	–0.5 to V_{CC} + 0.5	V
V_{out}	DC Output Voltage (Referenced to GND)	–0.5 to V_{CC} + 0.5	V
I_{in}	DC Input Current, per Pin	± 20	mA
I_{out}	DC Output Current, per Pin	± 25	mA
I_{CC}	DC Supply Current, V_{CC} and GND Pins	± 50	mA
P_D	Power Dissipation in Still Air, Plastic or Ceramic DIP†	750	mW
	SOIC Package†	500	
	TSSOP Package†	450	
T_{stg}	Storage Temperature	–65 to + 150	°C
T_L	Lead Temperature, 1 mm from Case for 10 Seconds		°C
	Plastic DIP, SOIC or TSSOP Package	260	
	Ceramic DIP	300	

* Maximum Ratings are those values beyond which damage to the device may occur. Functional operation should be restricted to the Recommended Operating Conditions.

† Derating — Plastic DIP: – 10 mW/°C from 65° to 125° C
Ceramic DIP: – 10 mW/°C from 100° to 125° C
SOIC Package: – 7 mW/°C from 65° to 125° C
TSSOP Package: – 6.1 mW/°C from 65° to 125° C

RECOMMENDED OPERATING CONDITIONS

Symbol	Parameter		Min	Max	Unit
V_{CC}	DC Supply Voltage (Referenced to GND)		2.0	6.0	V
V_{in}, V_{out}	DC Input Voltage, Output Voltage (Referenced to GND)		0	V_{CC}	V
T_A	Operating Temperature, All Package Types		–55	+125	°C
t_r, t_f	Input Rise and Fall Time	V_{CC} = 2.0 V	0	1000	ns
		V_{CC} = 4.5 V	0	500	
		V_{CC} = 6.0 V	0	400	

DC CHARACTERISTICS (Voltages Referenced to GND) MC54/74HC00A

Symbol	Parameter	Condition		V_{CC} V	Guaranteed Limit –55 to 25°C	≤85°C	≤125°C	Unit
V_{IH}	Minimum High-Level Input Voltage	V_{out} = 0.1V or V_{CC} – 0.1V		2.0	1.50	1.50	1.50	V
		$\|I_{out}\| \le 20\mu A$		3.0	2.10	2.10	2.10	
				4.5	3.15	3.15	3.15	
				6.0	4.20	4.20	4.20	
V_{IL}	Maximum Low-Level Input Voltage	V_{out} = 0.1V or V_{CC} – 0.1V		2.0	0.50	0.50	0.50	V
		$\|I_{out}\| \le 20\mu A$		3.0	0.90	0.90	0.90	
				4.5	1.35	1.35	1.35	
				6.0	1.80	1.80	1.80	
V_{OH}	Minimum High-Level Output Voltage	V_{in} = V_{IH} or V_{IL}		2.0	1.9	1.9	1.9	V
		$\|I_{out}\| \le 20\mu A$		4.5	4.4	4.4	4.4	
				6.0	5.9	5.9	5.9	
		V_{in} = V_{IH} or V_{IL}	$\|I_{out}\| \le 2.4mA$	3.0	2.48	2.34	2.20	
			$\|I_{out}\| \le 4.0mA$	4.5	3.98	3.84	3.70	
			$\|I_{out}\| \le 5.2mA$	6.0	5.48	5.34	5.20	
V_{OL}	Maximum Low-Level Output Voltage	V_{in} = V_{IH} or V_{IL}		2.0	0.1	0.1	0.1	V
		$\|I_{out}\| \le 20\mu A$		4.5	0.1	0.1	0.1	
				6.0	0.1	0.1	0.1	
		V_{in} = V_{IH} or V_{IL}	$\|I_{out}\| \le 2.4mA$	3.0	0.26	0.33	0.40	
			$\|I_{out}\| \le 4.0mA$	4.5	0.26	0.33	0.40	
			$\|I_{out}\| \le 5.2mA$	6.0	0.26	0.33	0.40	
I_{in}	Maximum Input Leakage Current	V_{in} = V_{CC} or GND		6.0	±0.1	±1.0	±1.0	μA
I_{CC}	Maximum Quiescent Supply Current (per Package)	V_{in} = V_{CC} or GND, I_{out} = 0μA		6.0	1.0	10	40	μA

AC CHARACTERISTICS (C_L = 50 pF, Input $t_r = t_f$ = 6 ns)

Symbol	Parameter	V_{CC} V	Guaranteed Limit –55 to 25°C	≤85°C	≤125°C	Unit
t_{PLH}, t_{PHL}	Maximum Propagation Delay, Input A or B to Output Y	2.0	75	95	110	ns
		3.0	30	40	55	
		4.5	15	19	22	
		6.0	13	16	19	
t_{TLH}, t_{THL}	Maximum Output Transition Time, Any Output	2.0	75	95	110	ns
		3.0	27	32	36	
		4.5	15	19	22	
		6.0	13	16	19	
C_{in}	Maximum Input Capacitance		10	10	10	pF

		Typical @ 25°C, V_{CC} = 5.0 V, V_{EE} = 0 V	
C_{PD}	Power Dissipation Capacitance (Per Buffer)	22	pF

COMPUTER SIMULATION

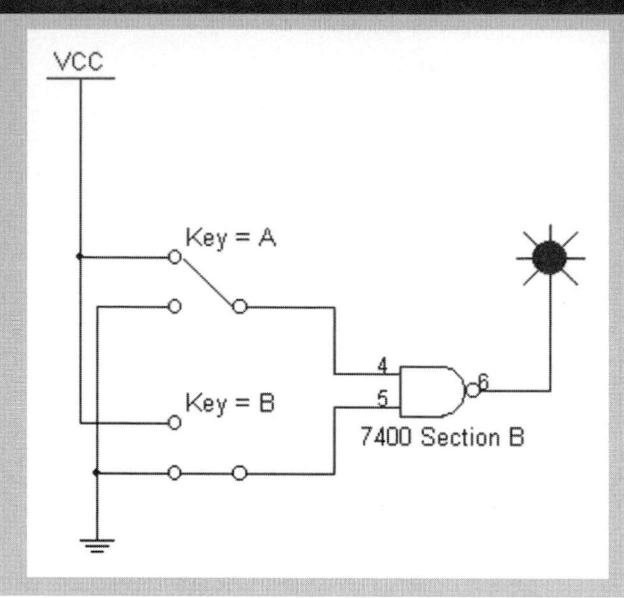

FIGURE 3–22

Open the Multisim File F03-22DG on the website. The 7400 contains four NAND gates, but only one is shown for simulation. Verify the 7400 2-input NAND gate truth table operation by applying LOWs and HIGHs with the switches. For the simulation, operate the top switch with the *A* key and the bottom switch with the *B* key. Multisim does not show logic family designations.

Review Questions

16. What does the term *dual 4-input NAND gate* mean?
17. On most 14-pin ICs, to which pins are V_{CC} and ground connected?
18. How do you properly connect an unused NAND gate input?
19. How do you properly connect an unused NOR gate input?
20. What is a "pull-up" resistor?

3–5 TROUBLESHOOTING

Opens and shorts can affect the operation of NAND and NOR gates in the same way as they affected the operation of AND and OR gates except that their outputs are inverted.

In this section, you will review basic faults and their occurrence in NAND and NOR gates.

Faults in NAND and NOR Logic Gates

Open Input

An open input on a NAND or NOR gate effectively disconnects the gate from one of its inputs. An open input has the effect of a high level in TTL gates. In a CMOS gate you cannot reliably predict how the gate will behave with an open input.

When an open NAND gate input acts as a HIGH, the output will depend on the signals on the remaining inputs. For example, if a 2-input NAND gate has an open input that appears HIGH and there is a digital waveform on the other input, the output is the inverted digital waveform, as illustrated in Figure 3–23(a).

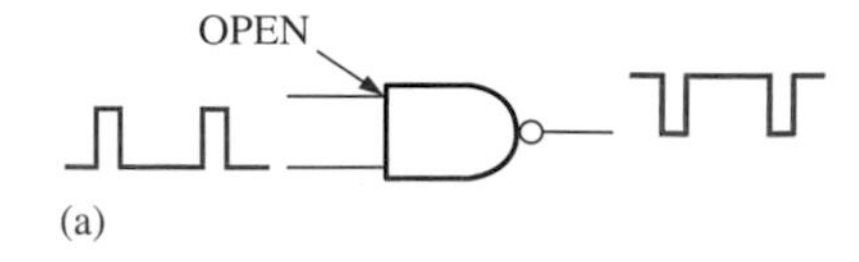

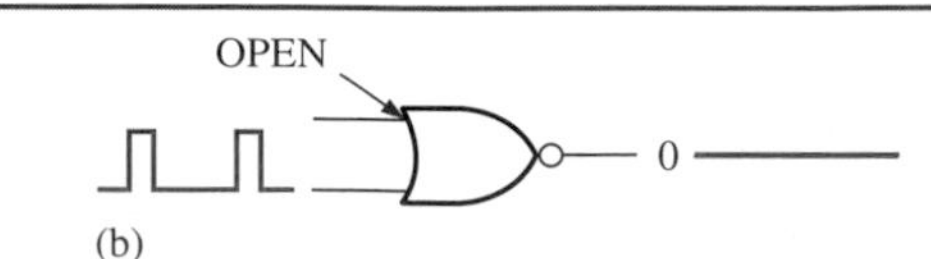

FIGURE 3–23

The effects on the output of an open input that appears as a HIGH.

When an open NOR gate input acts as a HIGH, the output will be continuously low (stuck LOW) regardless of the other inputs. For example, if a 2-input NOR gate has an open input that appears HIGH and there is a digital waveform on the other input, the output is always LOW, as shown in Figure 3–23(b).

Open Output

An open output on a NAND or NOR gate effectively disconnects the gate from its output pin. If a gate has an open output, there will be no definite output level no matter what the inputs are. An open output produces the same effect for both NAND and NOR gates in either CMOS or TTL technologies. Recall that an open output is said to be "floating." This can produce an erratic or "noisy" line near 0 V on your oscilloscope.

Shorted Input

An internal short in an IC is less likely than an internal open. However, shorted inputs may be caused by "solder bridges," small wire clippings, or imperfections in the printed circuit board. A short can occur between an input or output pin and ground, V_{CC}, or another gate input or output.

When a gate input is shorted to ground, it has the same effect as a LOW on all types of gates. When an input of a NAND gate is shorted to ground, the output is always high stuck HIGH no matter what the other inputs are. For example, if one input of a 2-input NAND gate is shorted to ground and there is a digital waveform on the other input, the output is constantly HIGH, as shown in Figure 3–24(a).

SAFETY NOTE

You need to have eye protection with approved safety glass whenever you are doing electrical or electronic hardware repair or using tools that can produce flying cuttings such as drills, saws, hammers, air hoses, or grinders. Flying cuttings can also be produced whenever small wires are clipped from a circuit board. In addition to wearing eye protection, you should cut in a manner that directs the cuttings downward, making sure that they do not go onto the circuit board or the lab area.

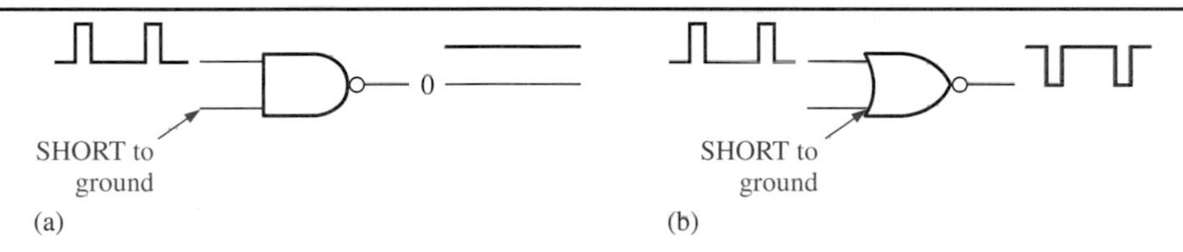

FIGURE 3–24

The effects on the output of an input that is shorted to ground.

When an input of a NOR gate is shorted to ground, the output depends on the other inputs. For example, if one input of a 2-input NOR gate is shorted to ground and there is a digital waveform on the other input, the output is the inverted digital waveform, as illustrated in Figure 3–24(b).

Review Questions

21. Which is more common in an IC gate, an open or a short?
22. Is the result of an open TTL NOR gate input the same as for an OR gate?
23. How does an input that is shorted to ground affect a NAND gate in terms of its output?
24. How does an input that is shorted to ground affect a NOR gate in terms of its output?
25. If a NOR gate input is shorted to V_{CC}, how is its output affected?

CHAPTER REVIEW

Key Terms

DeMorgan's theorems Two Boolean theorems that relate to the complementation of products and sums.

PROBLEMS

Basic Problems

Answers to odd-numbered problems are at the end of the book.

1. For a 2-input NAND gate, determine the output for each of the following combinations of input levels:

   (a) LOW, LOW (b) LOW, HIGH

   (c) HIGH, LOW (d) HIGH, HIGH

2. For a 4-input NAND gate, determine the output (0 or 1) for the following combinations of inputs:

   (a) 1110 (b) 1010

   (c) 1111 (d) 0000

   (e) 0011

3. Identify each of the gate symbols in Figure 3–25.

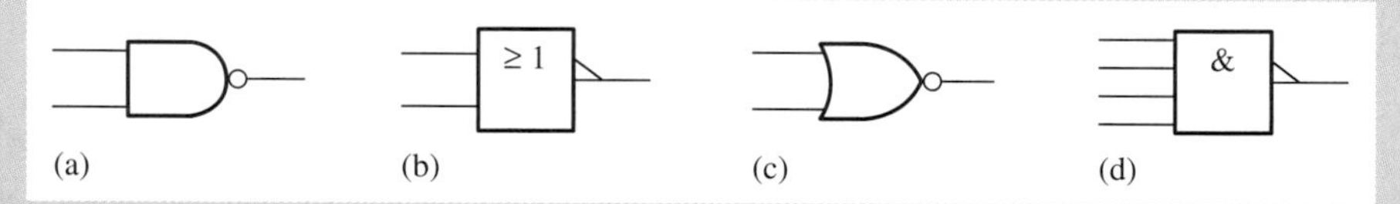

FIGURE 3–25

4. Show the sequence of binary digits that represents each of the following sequences of HIGHs and LOWs. Then show the complement of each binary sequence.

   (a) HIGH, LOW, LOW, HIGH

   (b) LOW, HIGH, LOW, LOW, HIGH, HIGH

   (c) HIGH, HIGH, LOW, LOW, HIGH, LOW, LOW, LOW

5. Name the Boolean operation that is represented by each of the following expressions and specify the number of variables in each case.

   (a) $\overline{AB}$ (b) $\overline{A + B + C}$

   (c) $\overline{ABCD}$

6. What type of gate does each of the expressions in Problem 5 represent?

7. Write the Boolean expression for a NAND gate with each of the following groups of input variables:

   (a) A, B (b) A, B, C

   (c) A, B, C, D

8. Write the Boolean expression for a NOR gate with each of the following groups of input variables:

   (a) A, B (b) A, B, C

   (c) A, B, C, D

9. If a NAND gate has two inputs, A and B, determine the output (1 or 0) for each of the following combinations of input states:

   (a) $A = 1, B = 0$ (b) $A = 1, B = 1$

   (c) $A = 0, B = 0$ (d) $A = 0, B = 1$

10. If a NOR gate has three inputs, A, B, and C, determine the output (1 or 0) for each of the following combinations of input states:

    (a) $A = 0, B = 0, C = 0$

    (b) $A = 0, B = 1, C = 0$

    (c) $A = 1, B = 1, C = 1$

11. Define the Boolean law represented by each of the expressions:

    (a) $\overline{AB} = \overline{BA}$

    (b) $\overline{A + B} = \overline{B + A}$

Basic-Plus Problems

12. Show a truth table for each of the following NAND gates using 1s and 0s:

    (a) 2-input (b) 3-input

    (c) 4-input

13. Show a truth table for each of the following NOR gates using 1s and 0s:

    (a) 2-input (b) 3-input

    (c) 4-input

14. Connect the output of one 2-input NOR gate to the input of a second 2-input NOR gate and show the truth table for the resulting logic circuit with inputs *A*, *B*, and *C*.

15. Complete each timing diagram in Figure 3–26 by drawing the output waveform of a 2-input NAND gate having the input waveforms shown.

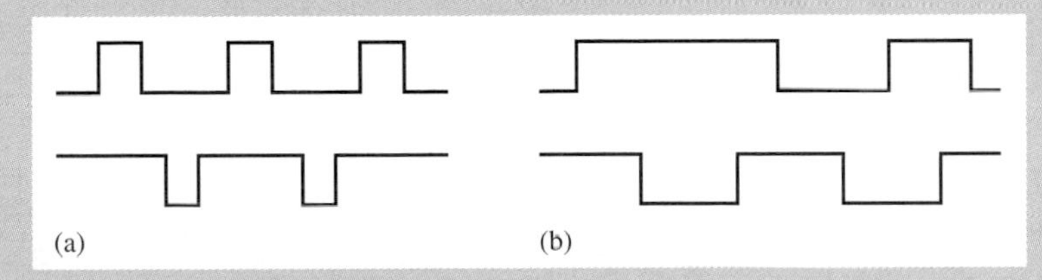

FIGURE 3–26

16. Complete each timing diagram in Figure 3–26 by drawing the output waveform of a 2-input NOR gate having the input waveforms shown.

17. Determine the amplitude, period, and frequency of each waveform displayed on the oscilloscope screen in Figure 3–27.

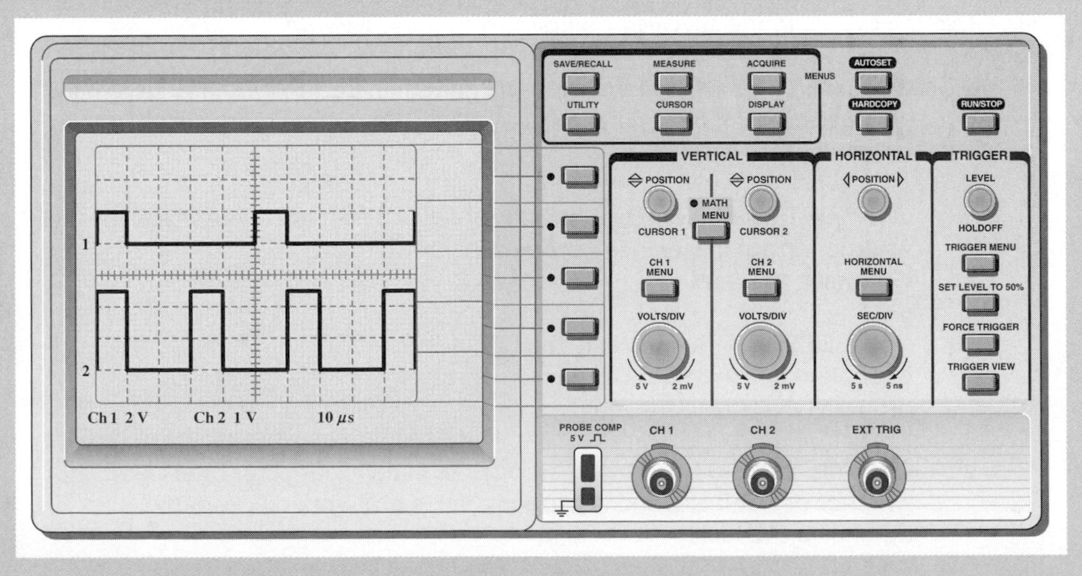

FIGURE 3–27

18. Prove DeMorgan's first theorem by showing that the truth table for a 2-input NAND gate is the same as the truth table when a NAND gate is operated as a negative-OR.
19. Prove DeMorgan's second theorem by showing that the truth table for a 2-input NOR gate is the same as the truth table when a NOR gate is operated as a negative-AND.
20. Find the most likely fault, if any, in each NAND gate in Figure 3–28.

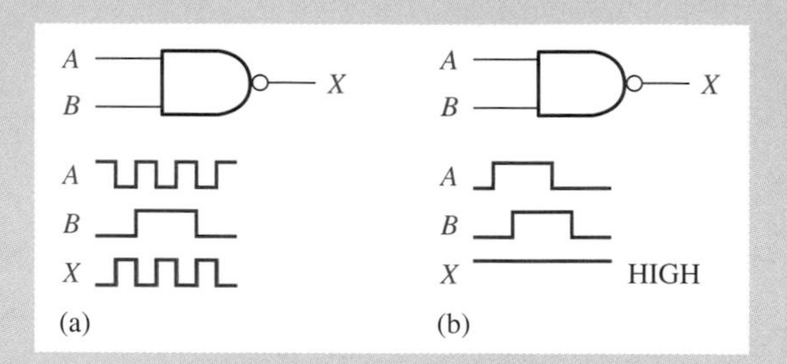

FIGURE 3–28

21. Find the most likely fault, if any, in each NOR gate in Figure 3–29.

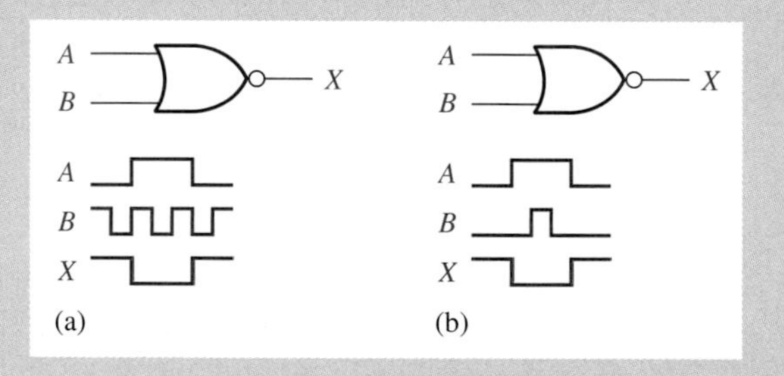

FIGURE 3–29

TROUBLESHOOTING PRACTICE

Use the circuits in the specified Multisim files on the website.

22. Open file TSP03-01 and troubleshoot the simulated NAND gate. Determine whether or not it is working properly. If it is not working properly, determine the fault.
23. Open file TSP03-02 and troubleshoot the simulated NOR gate. Determine whether or not it is working properly. If it is not working properly, determine the fault.
24. Open file TSP03-03 and verify the operation of the two circuits is equivalent according to DeMorgan's first theorem. If they are not working properly, determine the fault.
25. Open file TSP03-04 and troubleshoot the multi-input gate. Determine whether or not it is working properly. If it is not working properly, determine the fault.

Example Questions

ANSWERS

3–1: Waveform A inverted

3–2: Waveform B inverted

3–3: 64

3–4: 256

Review Questions

1. The output of a NAND is LOW when all inputs are HIGH.
2. $2^5 = 32$ combinations
3. $2^4 = 16$ rows and 5 columns (4 inputs and 1 output)
4. The output is LOW when all inputs are HIGH at the same time.
5. One
6. The output of a NOR is LOW when one or more inputs are HIGH.
7. $2^4 = 16$ combinations
8. $2^5 = 32$ rows and 6 columns (5 inputs and 1 output)
9. The output is LOW when any of the inputs are HIGH.
10. Connect the OR gate output to the inverter input.
11. $X = \overline{AB}$
12. $X = \overline{A + B}$
13. The NAND performs multiplication and complementation.
14. The NOR performs addition and complementation.
15. Negative-OR means that a 0 on one or more inputs of a NAND produces a 1 on the output. Negative-AND means that a 0 on all inputs of a NOR produces a 1 on the output.
16. Two 4-input NAND gates in an IC package.
17. V_{CC} is pin 14 and ground in pin 7.
18. Connect an unused NAND gate input to V_{CC} through a pull-up resistor.
19. Connect an unused NOR gate input to ground.
20. A pull-up resistor is a resistor connected from a point to V_{CC} to create a HIGH at the point.
21. Opens are more common.
22. No
23. A shorted to ground NAND input keeps the output HIGH.
24. The output is determined by the other input(s).
25. The output is always LOW.

Chapter Checkup

1. (b)	2. (d)	3. (d)	4. (b)	5. (c)
6. (c)	7. (a)			

CHAPTER

LOGIC GATE COMBINATIONS

CHAPTER OUTLINE

Study aids for this chapter are available at

http://www.prenhall.com/SOE

INTRODUCTION

You have studied all the basic logic gates on an individual basis; now let's look at how they can be used in combination. Many types of logic functions such as decoders, encoders, and multiplexers are combinations of various logic gates as you will learn in Chapter 6. When logic gates are connected together to perform some specific function in which the output level is at all times dependent on the combination of input levels, it is usually referred to as **combinational logic**. Sometimes it is called *combinatorial logic*.

In this chapter, you will learn how exclusive-OR (XOR) and exclusive-NOR (XNOR) gates are made by combining AND gates, OR gates, and inverters. Also, simple combinational logic functions called AND/OR and AND-OR-Invert are covered. You will also learn how any combinational logic function can be implemented in either of two standard forms: *sum-of-products* and *product-of-sums*. You may want to review the Boolean rules, laws, and theorems in Appendix A.

Finally, some fixed-function integrated circuits and basic troubleshooting are covered.

KEY OBJECTIVES

A section number is given for each objective. After completing this chapter, you should be able to

- **4–1** Describe how the exclusive-OR (XOR) gate operates
- **4–2** Describe how the exclusive-NOR (XNOR) gate operates
- **4–3** Describe the AND-OR and AND-OR-Invert logic functions
- **4–4** Apply Boolean algebra to combinational logic functions
- **4–5** Describe how any combinational logic can be implemented in sum-of-products form or in product-of-sums form
- **4–6** Use a Karnaugh map to simplify a Boolean expression
- **4–7** Discuss the basic combinational logic functions available in IC form and define some important IC parameters
- **4–8** Discuss some troubleshooting concepts and methods

COMPUTER SIMULATIONS DIRECTORY

The following figures have Multisim circuit files associated with them.

LABORATORY EXPERIMENTS DIRECTORY

The following exercises are for this chapter.

- **Experiment 5**
  The XOR Gate
- **Experiment 6**
  Logic Simplification

KEY TERMS

- Combinational logic
- XOR
- XNOR
- AND-OR
- AND-OR-Invert
- Sum-of-products
- Product-of-sums
- Karnaugh map
- Power dissipation
- Propagation delay
- Fan-out
- Unit load
- Node
- Signal tracing

A well-known axiom in technology is called Moore's law. It states that the number of transistors that can be put on a silicon integrated circuit chip is doubling every 18 to 24 months.

After four decades, solid-state electronics has advanced to the point at which 100 million transistors can be put on a piece of silicon with an area of just a few square centimeters. However, these transistors are still much larger than molecular-scale devices that are in the research stage. For a perspective, if a conventional transistor were scaled up to the size of this page you are reading, a molecular transistor would be the size of the period at the end of this sentence.

No one thinks that conventional silicon-based microelectronics will continue to follow Moore's law indefinitely. At some point, a physical limitation will be reached; and when that happens, molecular electronics may allow Moore's law to extend further into the future.

4–1 EXCLUSIVE-OR (XOR) GATE

An exclusive-OR (XOR) gate is a composite of two AND gates, one OR gate, and two inverters. Although the XOR is not a basic gate, it is considered an individual logic gate with its own unique symbol.

In this section, you will learn how the exclusive-OR (XOR) gate operates.

An **XOR** gate is a logic gate that produces a 1 (HIGH) if one and only one of its inputs is 1 (HIGH). Like the other logic gates you have studied, two standard XOR gate logic symbols are the the distinctive shape symbol and the rectangular outline symbol. The distinctive shape symbol is shown in Figure 4–1(a). Inputs are on the "curved" side of the symbol, and the output is on the "pointed" side. Notice that it is similar to the OR gate symbol but it has an extra curved line on the input side. The rectangular outline symbol is shown in Figure 4–1(b). As you can see, the rectangular outline symbol consists of a box, just as for the AND and OR gate symbols, but with a different internal label (=1) to indicate that the exclusive-OR logic gate requires exactly one input equal to 1 for the output to equal 1.

FIGURE 4–1

Symbols for a 2-input exclusive-OR (XOR) gate.

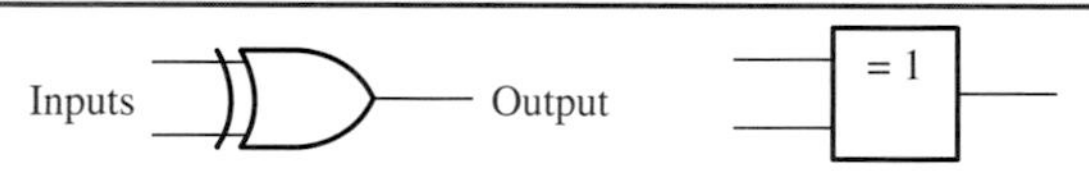

(a) Distinctive shape symbol

(b) Rectangular outline symbol

TABLE 4–1

Truth table for an XOR gate.

Inputs		Output
A	*B*	*X*
0	0	0
0	1	1
1	0	1
1	1	0

XOR Gate Operation

Although AND and OR gates can have any number of inputs, the XOR gate has only two inputs. Table 4–1 is a truth table that shows the logic operation for a 2-input XOR gate. Notice that binary numbers 1 and 0 are used for the input and output states to represent HIGH and LOW, respectively. There are four input combinations of the variables A and B, and the output is labeled X. As you can see, the output is 1 only if exactly one input is a 1. The operation of the XOR gate is similar to the OR gate except that the XOR produces a 0 (LOW) output when both inputs are 1.

COMPUTER SIMULATION

FIGURE 4–2

Open the Multisim file F04-02DG on the website. Check the 2-input XOR gate truth table by applying 0s and 1s with the switches. The ground line is the 0 and the $+V_{CC}$ line is the 1. For the simulation, operate the top switch with the *A* key and the bottom switch with the *B* key. When the probe light is red, it indicates a 1.

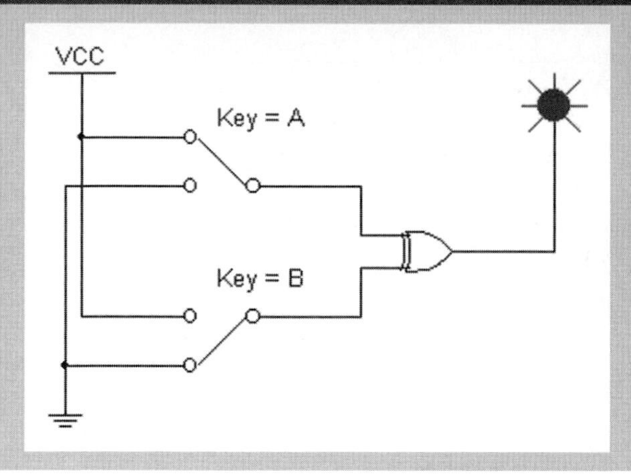

XOR gates with digital waveform inputs work the same as they do with constant input levels. The output will be 1 when one and only one input is 1.

EXAMPLE 4–1

Problem

Determine the output waveform *X* when the waveforms shown in Figure 4–3 are applied to the XOR gate.

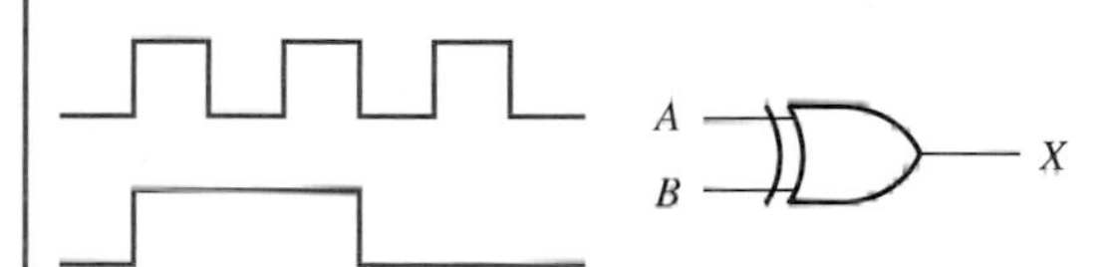

FIGURE 4–3

Solution

The times that one and only one input waveform is 1 (HIGH) is indicated by the blue-shaded areas in Figure 4–4. The output waveform *X* with the corresponding 1s (HIGHs) is shown in the figure.

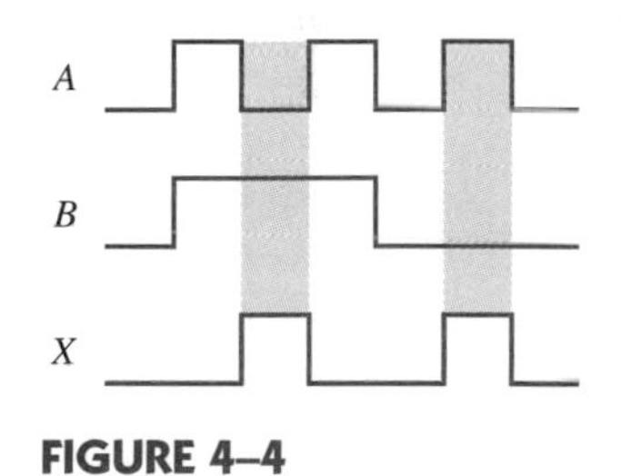

FIGURE 4–4

Question*

If input waveform *A* is always 1 (HIGH) and input waveform *B* stays the same as in Figure 4–3, what would the output waveform *X* look like?

XOR Logic

The Boolean expression for the XOR is

$$X = A\overline{B} + \overline{A}B$$

where *X* is the output and *A* and *B* are the inputs. The first term $A\overline{B}$ is for an AND gate with the *B* input inverted (complemented). The second term $\overline{A}B$ is for an AND gate with the *A* input inverted. The + indicates that these two AND terms are ORed.

* *Answers are at the end of the chapter.*

The logic diagram for the XOR is shown in Figure 4–5(a). As you can see, the upper inverter and AND gate produce the term $A\overline{B}$. The lower inverter and AND gate produce the term $\overline{A}B$. The OR gate produces the final output $A\overline{B} + \overline{A}B$ by ORing the outputs of the two AND gates.

FIGURE 4–5

XOR logic.

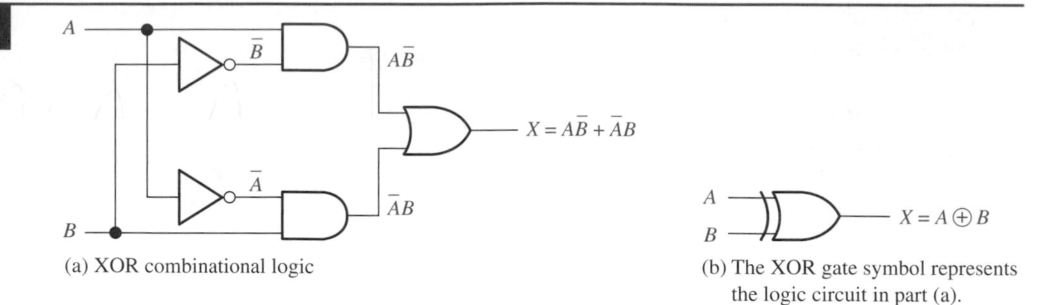

(a) XOR combinational logic

(b) The XOR gate symbol represents the logic circuit in part (a).

When you substitute all four possible combinations of 1s and 0s for A and B, you see that X is a 1 only when either A is a 1 or B is a 1. These substitutions are as follows:

$$\text{For } A = 0, B = 0: \quad X = A\overline{B} + \overline{A}B = 0\cdot\overline{0} + \overline{0}\cdot 0 = 0\cdot 1 + 1\cdot 0 = 0 + 0 = 0$$
$$\text{For } A = 0, B = 1: \quad X = A\overline{B} + \overline{A}B = 0\cdot\overline{1} + \overline{0}\cdot 1 = 0\cdot 0 + 1\cdot 1 = 0 + 1 = 1$$
$$\text{For } A = 1, B = 0: \quad X = A\overline{B} + \overline{A}B = 1\cdot\overline{0} + \overline{1}\cdot 0 = 1\cdot 1 + 0\cdot 0 = 1 + 0 = 1$$
$$\text{For } A = 1, B = 1: \quad X = A\overline{B} + \overline{A}B = 1\cdot\overline{1} + \overline{1}\cdot 1 = 1\cdot 0 + 0\cdot 1 = 0 + 0 = 0$$

The combinational logic shown in Figure 4–5(a) is represented by the standard XOR symbol, shown in Figure 4–5(b). A shorthand expression for the exclusive-OR function is

$$X = A \oplus B$$

where the Boolean operator $\oplus$ stands for XOR.

COMPUTER SIMULATION

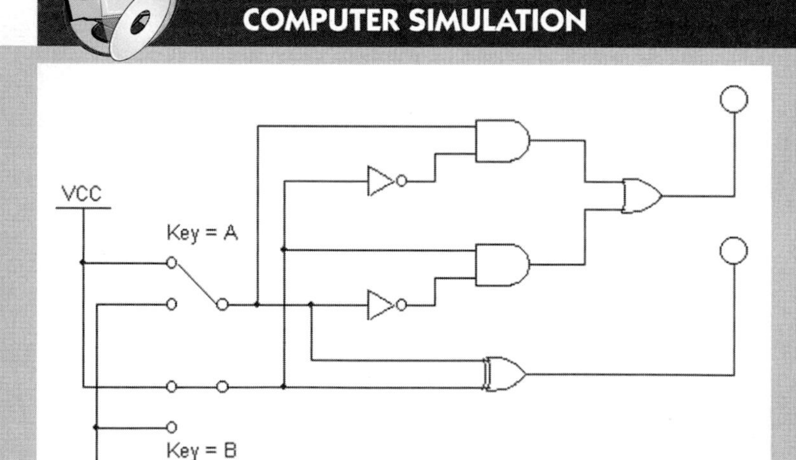

FIGURE 4–6

Open the Multisim file F04-06DG on the website. Observe that the combinational logic and the XOR gate always have the same output by applying 0s and 1s to the inputs with the switches. The ground line is the 0 and the $+V_{CC}$ line is the 1. Operate the top switch with the A key and the bottom switch with the B key. Both probe lights should always be the same to indicate that the two circuits are equivalent.

Review Questions

Answers are at the end of the chapter.

1. When is the output of an XOR gate 1 (HIGH)?
2. How does the XOR gate differ from the OR gate in terms of its operation?

3. Explain the significance of the =1 label in the rectangular outline XOR symbol.
4. Can the XOR gate be used to uniquely detect when both of its inputs are 1 (HIGH)?
5. How can an XOR gate be used as an inverter?

EXCLUSIVE-NOR (XNOR) GATE 4–2

The exclusive-NOR (XNOR) gate is made up of two AND gates, one OR gate, and three inverters. Like the XOR, it is considered as a single gate with its own unique symbol.

In this section, you will learn how the exclusive-NOR (XNOR) gate operates.

An **XNOR** gate is a logic gate that produces a 0 (LOW) if one and only one of its inputs is 1 (HIGH). The distinctive shape symbol is shown in Figure 4–7(a.). The XNOR gate symbol has a bubble on the output, indicating that the asserted output state is LOW (0). The rectangular outline symbol is shown in Figure 4–7(b). It has a small triangle on the output.

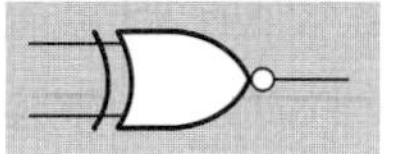

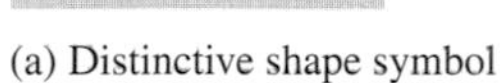

(a) Distinctive shape symbol

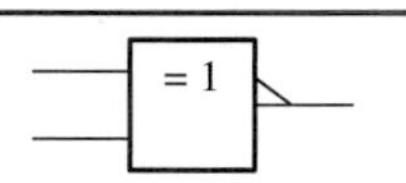

(b) Rectangular outline symbol

FIGURE 4–7

Symbols for a 2-input exclusive-NOR gate (XNOR).

TABLE 4–2

Truth table for a 2-input XNOR gate.

Inputs		Output
A	*B*	*X*
0	0	1
0	1	0
1	0	0
1	1	1

XNOR Gate Operation

Table 4–2 is a truth table that shows the logic operation for a 2-input XNOR gate. There are four input combinations of the variables *A* and *B*, and the output is labeled *X*. As you can see, the output is 0 only if exactly one input is 1. The operation of the XNOR gate is similar to the NOR gate except the XNOR produces a 1 output when both inputs are 1.

COMPUTER SIMULATION

FIGURE 4–8

Open the Multisim file F04-08DG on the website. Verify the 2-input XNOR gate truth table by applying 0s and 1s with the switches. The ground line is the 0 and the $+V_{CC}$ line is the 1. For the simulation, operate the top switch with the *A* key and the bottom switch with the *B* key.

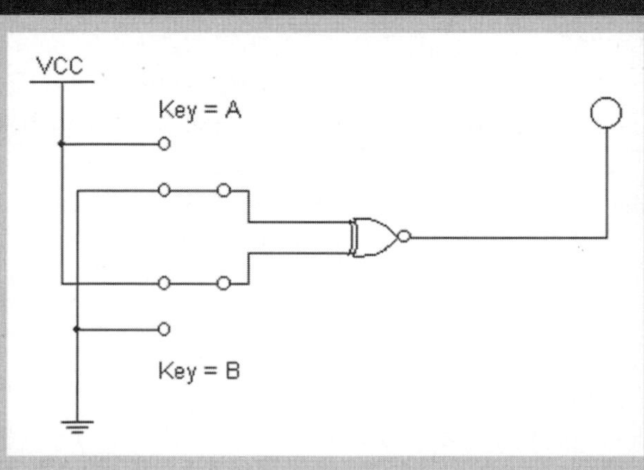

XNOR gates work the same in terms of their truth table operation whether or not the inputs are always constant or are changing. The output will be 0 when one and only one input is 1.

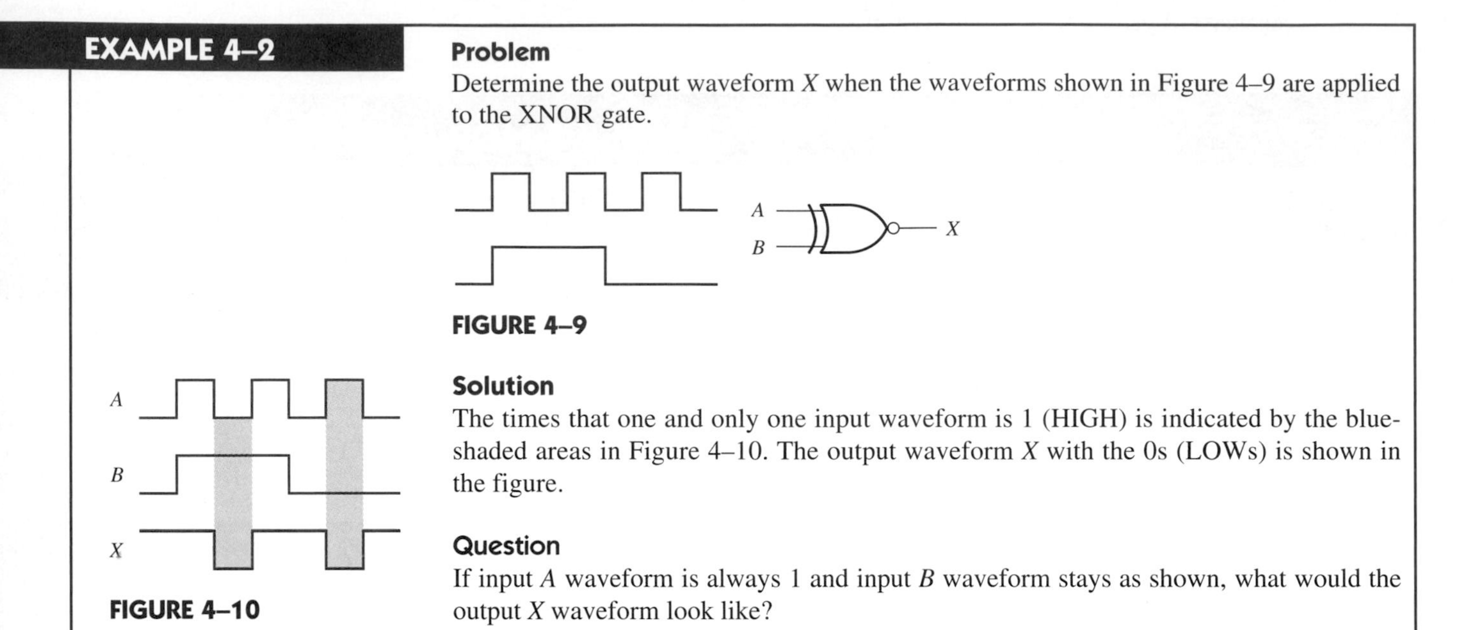

EXAMPLE 4–2

Problem

Determine the output waveform X when the waveforms shown in Figure 4–9 are applied to the XNOR gate.

FIGURE 4–9

Solution

The times that one and only one input waveform is 1 (HIGH) is indicated by the blue-shaded areas in Figure 4–10. The output waveform X with the 0s (LOWs) is shown in the figure.

FIGURE 4–10

Question

If input A waveform is always 1 and input B waveform stays as shown, what would the output X waveform look like?

XNOR Logic

The XNOR logic is the same as XOR logic except that there is an inverter on the output, as shown in Figure 4–11. The Boolean expression for the XNOR is

$$X = \overline{A\overline{B} + \overline{A}B}$$

The bar over both terms indicates the inversion.

FIGURE 4–11

XNOR logic.

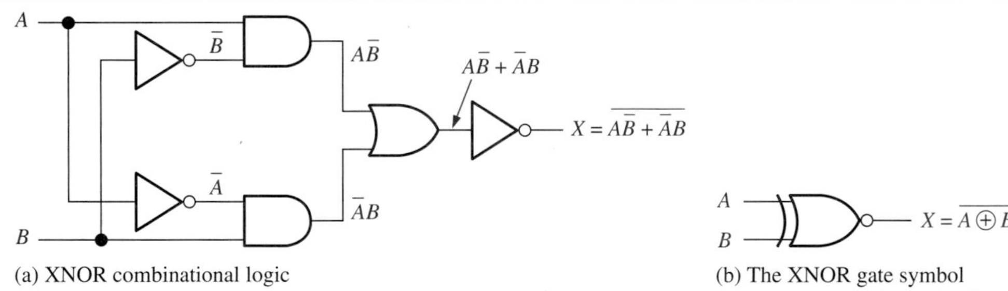

(a) XNOR combinational logic

(b) The XNOR gate symbol represents the logic circuit in part (a).

As you can see in the logic diagram in Figure 4–11(a), the upper inverter and AND gate produce the term $A\overline{B}$. The lower inverter and AND gate produce the term $\overline{A}B$. The OR gate produces the output $A\overline{B} + \overline{A}B$ by ORing the outputs of the two AND gates. The inverter at the output produces the final expression: $\overline{A\overline{B} + \overline{A}B}$.

When you substitute all four possible combinations of 1s and 0s for A and B, you see that X is a 0 only when either A is a 1 or B is a 1.

For $A = 0, B = 0$: $X = \overline{A\overline{B} + \overline{A}B} = \overline{0\cdot\overline{0} + \overline{0}\cdot 0} = \overline{0\cdot 1 + 1\cdot 0} = \overline{0 + 0} = 1$

For $A = 0, B = 1$: $X = \overline{A\overline{B} + \overline{A}B} = \overline{0\cdot\overline{1} + \overline{0}\cdot 1} = \overline{0\cdot 0 + 1\cdot 1} = \overline{0 + 1} = 0$

For $A = 1, B = 0$: $X = \overline{A\overline{B} + \overline{A}B} = \overline{1\cdot\overline{0} + \overline{1}\cdot 0} = \overline{1\cdot 1 + 0\cdot 0} = \overline{1 + 0} = 0$

For $A = 1, B = 1$: $X = \overline{A\overline{B} + \overline{A}B} = \overline{1\cdot\overline{1} + \overline{1}\cdot 1} = \overline{1\cdot 0 + 0\cdot 1} = \overline{0 + 0} = 1$

The combinational logic shown in Figure 4–11(a) is represented by the standard XNOR symbol, shown in Figure 4–11(b). A shorthand expression for the exclusive-NOR function is

$$X = \overline{A \oplus B}$$

where the Boolean operator $\oplus$ stands for XOR.

COMPUTER SIMULATION

FIGURE 4–12

Open the Multisim file F04-12DG on the website. Observe that the combinational logic and the XNOR gate always have the same output by applying 0s and 1s to the inputs with the switches. The ground line is the 0 and the $+V_{CC}$ line is the 1. Operate the top switch with the *A* key and the bottom switch with the *B* key. The probe lights should always be the same, indicating that the circuits are equivalent.

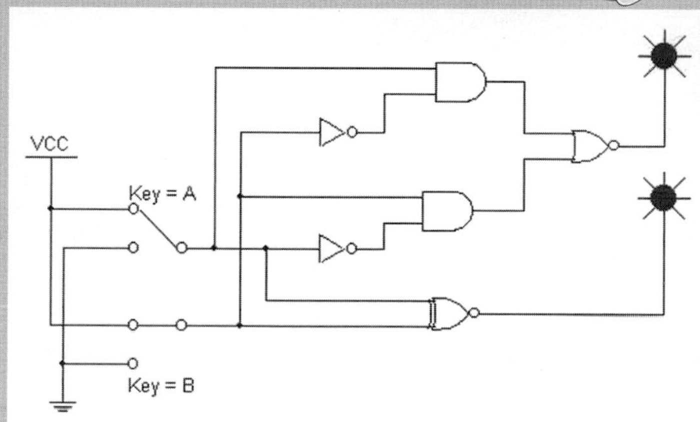

Review Questions

6. When is the output of an XNOR gate LOW?
7. How does the XNOR gate differ from the NOR gate in terms of its operation?
8. What is the significance of the ⊾ on the rectangular outline symbol?
9. Can the XNOR gate be used to uniquely detect when both of its inputs are HIGH?
10. How can an XNOR gate be used as an inverter?

AND-OR/AND-OR-INVERT LOGIC 4–3

AND-OR logic and AND-OR-Invert logic are examples of combinational logic functions. AND-OR logic uses AND gates and an OR gate. AND-OR-Invert logic uses AND gates and a NOR gate. The XOR and XNOR gates are actually types of AND-OR and AND-OR-Invert circuits with only two variables and their complements as inputs.

In this section, you will learn to describe the AND-OR and AND-OR-Invert logic functions.

AND-OR Logic

AND-OR logic consists of AND gates with outputs that are connected to the inputs of an OR gate. The OR gate output is a 1 only when all the inputs of at least one AND gate are all 1s. Although there can be any number of AND gates with any number of inputs, there is only one OR gate.

The Boolean expression for AND-OR logic with just two 2-input AND gates is

$$X = AB + CD$$

where X is the output and A, B, C, and D are the inputs. The first term AB is for one of the AND gates and the second term CD is for the other AND gate. The + indicates that these two AND terms are ORed.

The logic diagram for the AND-OR logic circuit is shown in Figure 4–13. As you can see, the upper AND gate produces the term AB. The lower AND gate produces the term CD. The OR gate produces the final output $AB + CD$ by ORing the outputs of the two AND gates.

FIGURE 4–13

AND-OR logic with two 2-input AND gates.

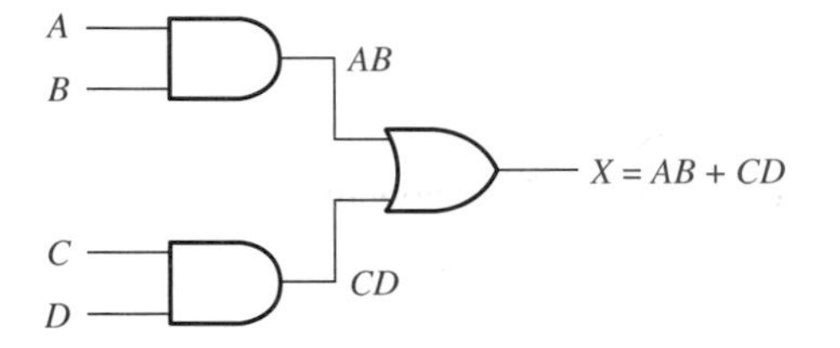

Because there are four inputs, there are $2^4 = 16$ combinations of the input variables. However, the output X is 1 only in those combinations where $AB = 1$ and/or $CD = 1$. The output X is 0 for any of the other input variable combinations.

EXAMPLE 4–3

Problem

Show that the AND-OR logic circuit in Figure 4–13 produces an output of 1 only when $A = 1$ and $B = 1$ or $C = 1$ and $D = 1$.

Solution

There are four input variables and sixteen possible combinations of those variables. By substituting all of these 1 and 0 combinations into the Boolean expression for the AND-OR function and applying the Boolean rules you have learned, you can show that X is a 1 only when one or both AND terms (AB, CD) are 1. Table 4–3 summarizes the results.

Table 4–3

A	B	C	D	AB	CD	X
0	0	0	0	0	0	0
0	0	0	1	0	0	0
0	0	1	0	0	0	0
0	0	1	1	0	1	1
0	1	0	0	0	0	0
0	1	0	1	0	0	0
0	1	1	0	0	0	0
0	1	1	1	0	1	1
1	0	0	0	0	0	0
1	0	0	1	0	0	0
1	0	1	0	0	0	0
1	0	1	1	0	1	1
1	1	0	0	1	0	1
1	1	0	1	1	0	1
1	1	1	0	1	0	1
1	1	1	1	1	1	1

Question

Suppose another 2-input AND gate is added to the circuit in Figure 4–13. What would be the number of possible input combinations?

AND-OR Logic with Digital Waveform Inputs

Now let's look at an AND-OR circuit with digital waveform inputs. Again, we use a circuit with two 2-input AND gates, which means there will be four input waveforms. The timing relationship of these waveforms is important for determining the output waveform. When 1s are on both inputs of one or both of the AND gates, the output is 1.

EXAMPLE 4–4

Problem

Determine the output waveform X when the waveforms shown in Figure 4–14 are applied to the inputs of the AND-OR logic circuit. The dashed lines show how the waveforms are aligned.

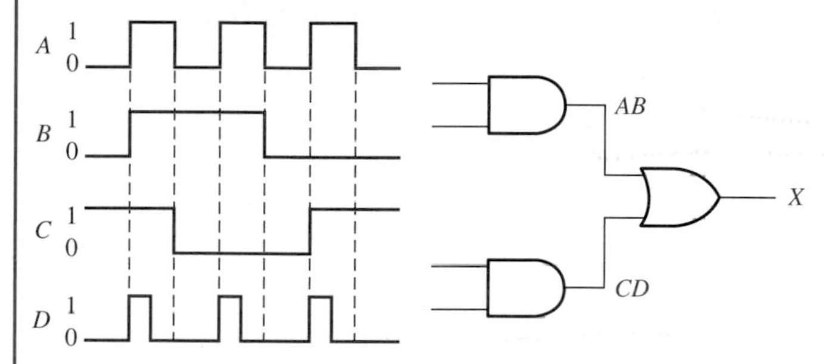

FIGURE 4–14

Solution

When input waveforms A and B are both 1 (HIGH) at the same time, the AB output of the top AND gate is 1. Also, when input waveforms C and D are both 1 (HIGH) at the same time, the CD output of the bottom AND gate is 1. Anytime waveforms AB or CD or both are 1 (HIGH), output waveform X is 1. The complete timing diagram is developed in Figure 4–15.

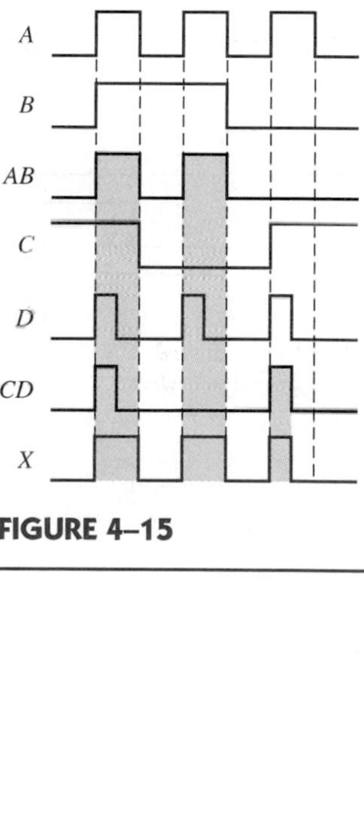

A
B
AB
C
D
CD
X

FIGURE 4–15

Question

If input waveform B is continuously a 1, what is output waveform X?

AND-OR-Invert Logic

AND-OR-Invert logic can consist of AND gates with outputs that are connected to the inputs of a NOR gate. The only difference between AND-OR-Invert and AND-OR logic is that instead of an OR gate there is a NOR gate that inverts the output. Although there can be any number of AND gates with any number of inputs, there is only one NOR gate.

An AND-OR-Invert circuit with three 2-input AND gates is shown in Figure 4–16. The OR gate output is a 0 when both inputs on any or all of the AND gates are 1. The Boolean expression for the output in this case is the complemented (inverted) sum (OR) of the three AND gate outputs, as indicated by the continuous bar over the entire expression.

$$X = \overline{AB + CD + EF}$$

Remember that an AND-OR or an AND-OR-Invert circuit can have any number of AND gates with any number of inputs.

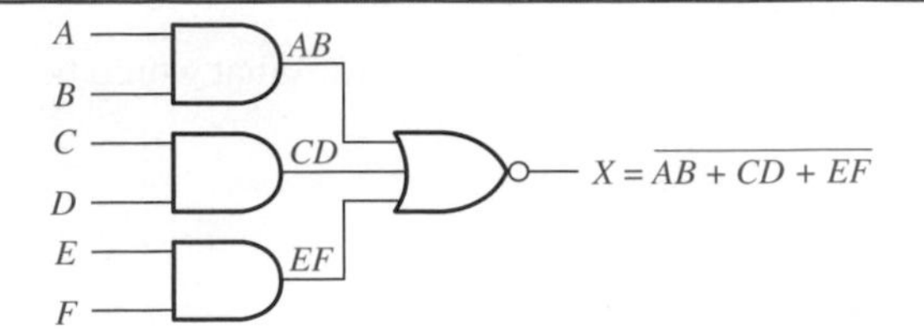

FIGURE 4–16

AND-OR-Invert logic with three 2-input AND gates.

AND-OR-Invert Logic with Digital Waveform Inputs

The output waveform is 0 any time both input waveforms on any or all of the AND gates are 1. Example 4–5 uses the circuit in Figure 4–16 to illustrate AND-OR-Invert logic. With this circuit, you will have six input waveforms.

EXAMPLE 4–5

Problem

Determine the output waveform X when the waveforms shown in Figure 4–17 are applied to the inputs of the AND-OR-Invert logic circuit. The dashed lines show how the waveforms are aligned.

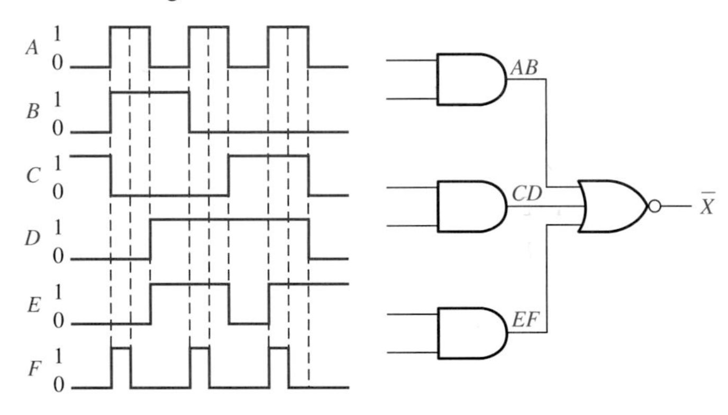

FIGURE 4–17

Solution

When input waveforms A and B are both 1 at the same time, the AB output of the top AND gate is 1. When input waveforms C and D are both 1 at the same time, the CD output of the middle AND gate is 1. When input waveforms E and F are both 1 at the same time, the EF output of the bottom AND gate is 1. When at least one AND gate output is 1, output $\overline{X}$ is 0. The complete timing diagram is developed in Figure 4–18.

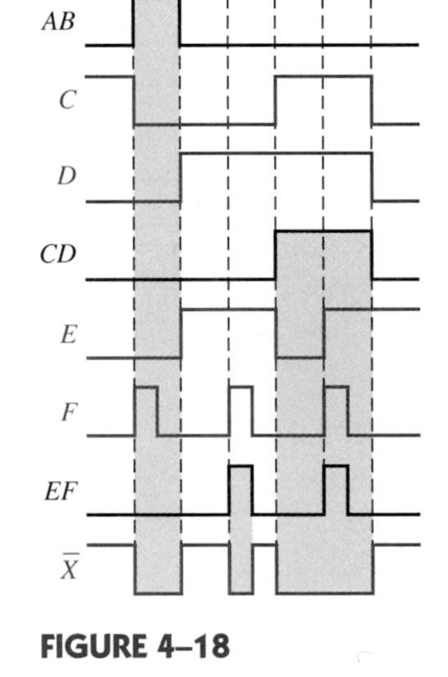

FIGURE 4–18

Question

If the output of the top AND gate fails so that its output is continuously stuck in the 1 state, what is the final output of the AND-OR-Invert circuit?

COMPUTER SIMULATION

Open the Multisim file F04-19DG on the website. Observe that the operation of the AND-OR logic verifies the truth table in Table 4–3 by applying 0s and 1s to the inputs with the switches. Operate the top switch with the *A* key, the second switch with the *B* key, the third switch with the *C* key, and the bottom switch with the *D* key.

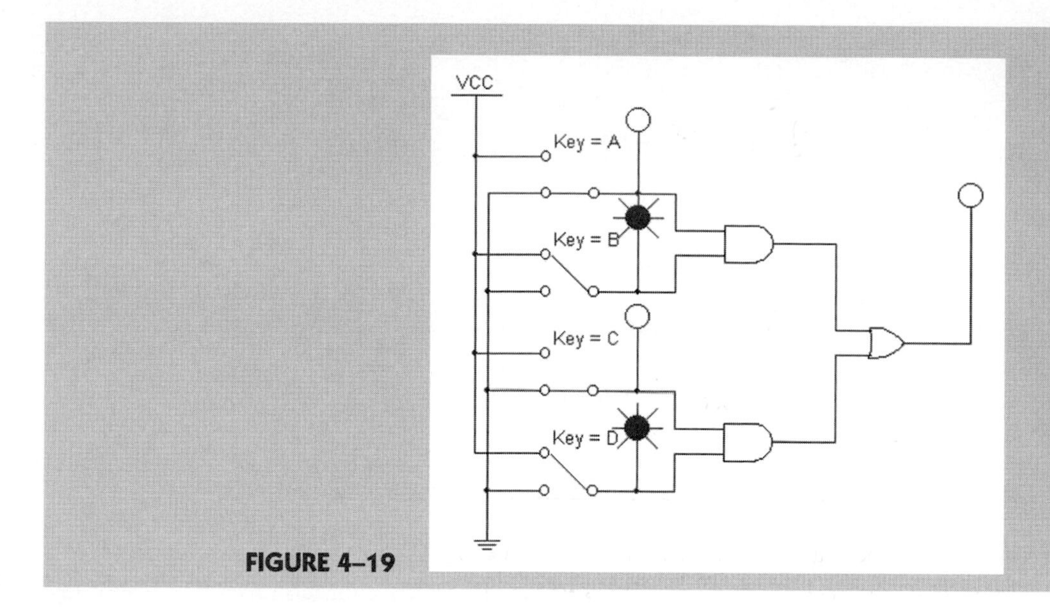

FIGURE 4–19

Review Questions

11. How many OR gates can an AND-OR circuit have?
12. How many AND gates can an AND-OR circuit have?
13. What is the Boolean expression for AND-OR logic with two 2-input AND gates?
14. What is the Boolean expression for AND-OR-Invert logic with three 2-input AND gates?
15. The AND-OR logic with two 2-input AND gates has the following inputs: $A = 1$, $B = 0$, $C = 1$, $D = 1$. What is the output?

BOOLEAN ALGEBRA APPLIED TO GATE COMBINATIONS 4–4

Certain laws and rules of Boolean algebra apply to combinational logic. The associative law and the distributive law are used to analyze these types of logic circuits.

In this section, you will learn to apply Boolean algebra to combinational logic functions.

Laws of Boolean Algebra

The Associative Law of Multiplication

When two or more AND gates are combined in a logic circuit, the **associative law** of multiplication applies. Recall that the AND gate performs Boolean multiplication. For three variables,

$$A(BC) = (AB)C$$

This expression for the associative law of multiplication shows that it makes no difference how the variables are grouped when ANDing more than two variables.

The associative law of multiplication is illustrated by the two 2-input AND gates in Figure 4–20, which are used to AND three variables. Keep in mind that you can also produce the term ABC with one 3-input AND gate.

FIGURE 4–20

Illustration of the associative law of Boolean multiplication. Both of these ways of ANDing the variables are equivalent.

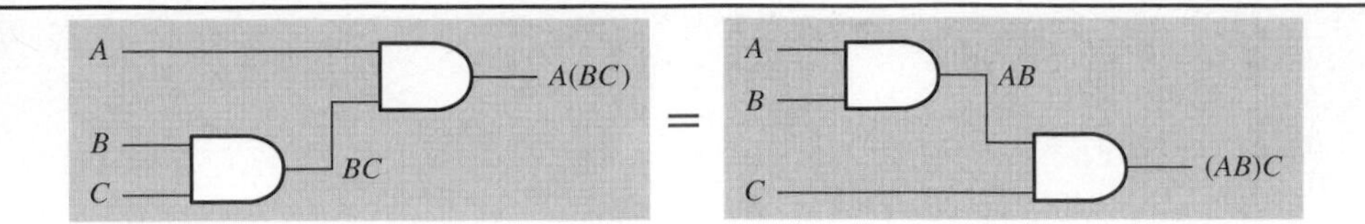

The Associative Law of Addition

When two or more OR gates are combined in a logic circuit, the associative law of addition applies. Recall that the OR gate performs Boolean addition. For three variables,

$$A + (B + C) = (A + B) + C$$

This expression for the associative law of addition shows that in terms of the result, the order in which the variables are grouped makes no difference when ORing more than two variables.

The associative law of addition is illustrated by the two 2-input OR gates in Figure 4–21, which are used to OR three variables. You can also produce the term $A + B + C$ with one 3-input OR gate.

FIGURE 4–21

Illustration of the associative law of Boolean addition. Both of these ways of ORing the variables are equivalent.

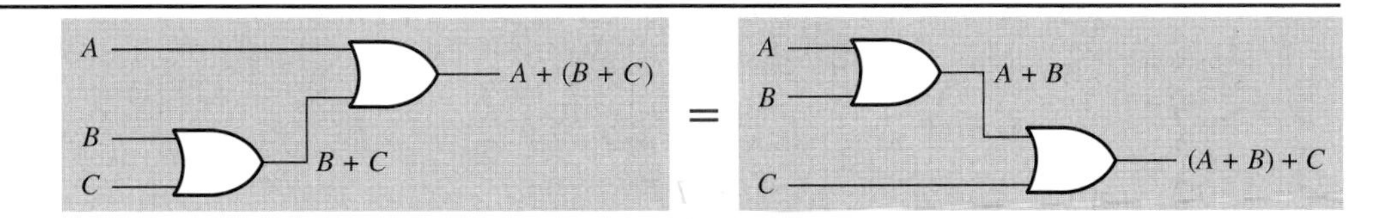

The Distributive Law

The **distributive law** applies to AND and OR gates in combination. For three variables,

$$A(B + C) = AB + AC$$

In terms of combinational logic, the distributive law states that ORing two or more variables and then ANDing the result with a single variable is equivalent to ANDing the single variable with each of the other variables. The distributive law is illustrated for three variables in Figure 4–22.

FIGURE 4–22

Illustration of the distributive law.

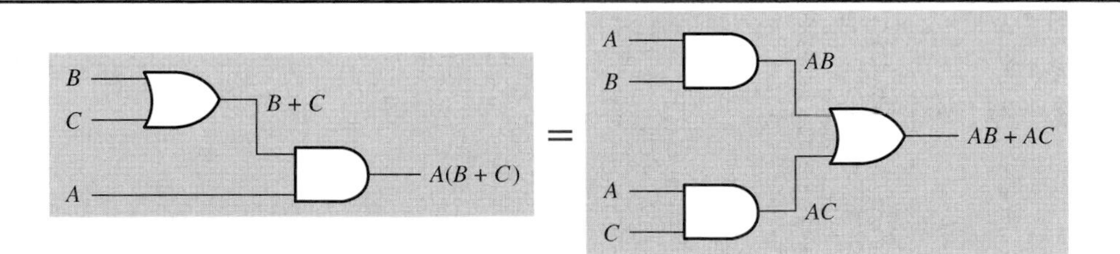

Rules of Boolean Algebra

Several Boolean rules were covered in Chapter 2. Three additional rules can be applied to simplify gate combinations.

$$A + AB = A$$
$$A + \overline{A}B = A + B$$
$$(A + B)(A + C) = A + BC$$

The first rule, $A + AB = A$, states that a logic gate combination with a Boolean expression of the form $X = A + AB$ can be simplified to $X = A$. The output X equals the input A, which is a direct connection with no gates. The original logic consisting of an AND gate and an OR gate can be replaced by a single wire, as illustrated in Figure 4–23.

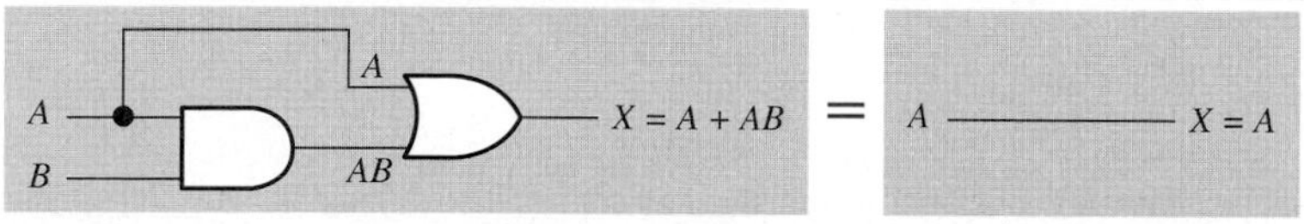

FIGURE 4–23

Illustration of the Boolean rule $A + AB = A$.

The second rule, $A + \overline{A}B = A + B$, states that a logic gate combination with a Boolean expression of the form $X = A + \overline{A}B$ can be simplified to $X = A + B$. The original logic consisting of an AND gate, OR gate, and inverter can be replaced by a single OR gate, as illustrated in Figure 4–24.

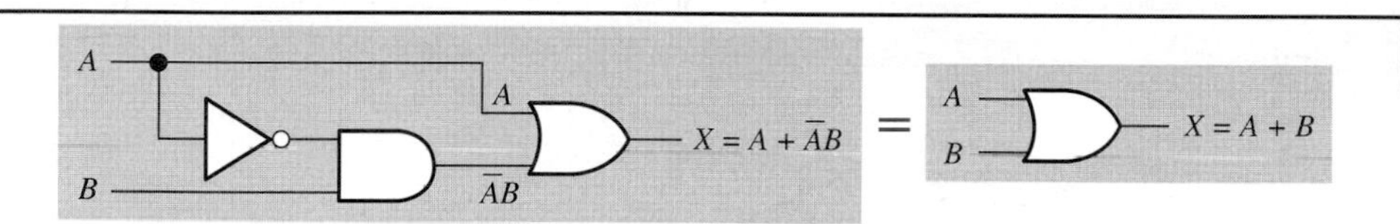

FIGURE 4–24

Illustration of the Boolean rule $A + \overline{A}B = A + B$.

The third rule, $(A + B)(A + C) = A + BC$, states that a logic gate combination with a Boolean expression of the form $X = (A + B)(A + C)$ can be simplified to $X = A + BC$. The original logic consisting of two OR gates and an AND gate can be replaced by one OR gate and one AND gate, as illustrated in Figure 4–25.

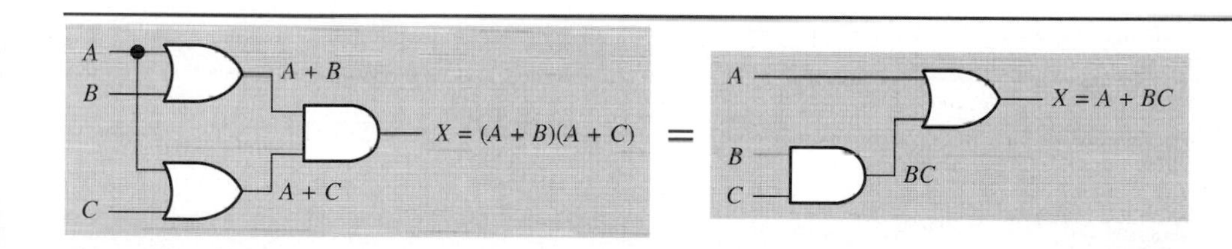

FIGURE 4–25

Illustration of the Boolean rule $(A + B)(A + C) = A + BC$.

EXAMPLE 4–6

Problem

Simplify the following expression using applicable Boolean laws and rules.

$$X = ABC + (\overline{AB})C + AC$$

Solution

$$\begin{aligned} X &= ABC + (\overline{AB})C + AC \\ &= C(AB + \overline{AB} + A) \\ &= C(1 + A) \\ &= C(1) \\ &= \mathbf{C} \end{aligned}$$

Question

Which Boolean law was applied in the second line of the solution?

COMPUTER SIMULATION

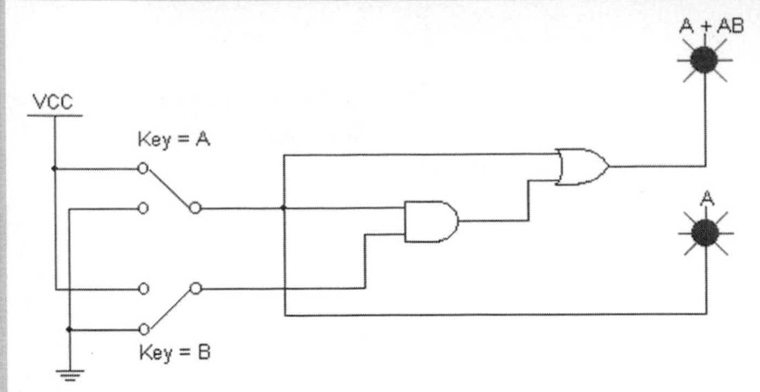

FIGURE 4–26

Open the Multisim file F04-26DG on the website. Observe that the operation verifies the Boolean rule $A + AB = A$ by applying 0s and 1s to the inputs with the switches and observing the outputs. Operate the top switch with the *A* key and the bottom switch with the *B* key. Notice that the probe lights are always the same no matter what the inputs are. This shows that the gates can be replaced by a wire connected to the *A* switch.

COMPUTER SIMULATION

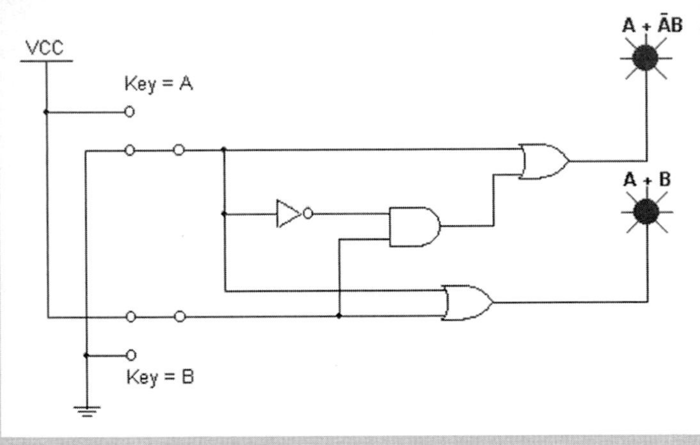

FIGURE 4–27

Open the Multisim file F04-27DG on the website. Observe that the operation verifies the Boolean rule $A + \overline{A}B = A + B$ by applying 0s and 1s to the inputs with the switches and observing the outputs. Operate the top switch with the *A* key and the bottom switch with the *B* key. Notice that the probe lights are always the same no matter what the inputs are. This shows the upper logic is equivalent to an OR gate.

COMPUTER SIMULATION

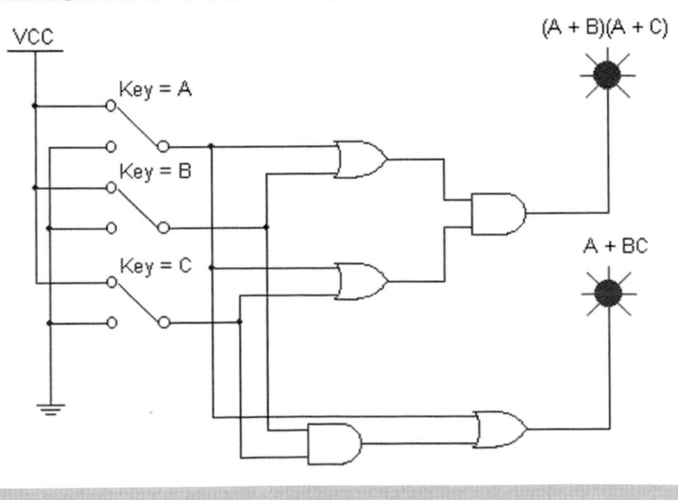

FIGURE 4–28

Open the Multisim file F04-28DG on the website. Observe that the operation verifies the Boolean rule $(A + B)(A + C) = A + BC$ by applying 0s and 1s to the inputs with the switches and observing the outputs. Operate the top switch with the *A* key and the bottom switch with the *B* key. Notice that the probe lights are always the same no matter what the inputs are. This shows that the upper logic is equivalent to the lower logic.

Review Questions

16. What does the associative law of multiplication state?
17. What does the associative law of addition state?
18. What does the distributive law state?
19. How can you tell when two logic circuits are equivalent?
20. What is the Boolean expression $A(B + C + D)$ equivalent to according to the distributive law?

STANDARD FORMS OF COMBINATIONAL LOGIC 4–5

All combinational logic circuits, no matter what their form, can be converted into a standard form. This standard form is the sum-of-products. You know that in Boolean algebra, as in mathematics, a multiplication (AND) results in a product and an addition (OR) results in a sum. In addition to the sum-of-products form, there is another standard form, called product-of-sums, which is less frequently used.

In this section, you will learn how to implement logic in a standard sum-of-products (SOP) form and in a standard product-of-sums (POS) form.

Sum-of-Products (SOP) Logic

A logic circuit in the **sum-of-products** form is composed of AND gates with their outputs going into a single OR gate. The AND-OR circuit that you learned about in Section 4–3 is an example of SOP logic. As Figure 4–29 shows, the AND gates form the product terms, which are then ORed together to form the sum. Also, the XOR gate is a type of SOP logic because its output can be expressed as $\overline{A}B + A\overline{B}$.

An SOP expression is equal to 1 when one or more of the product terms are equal to 1.

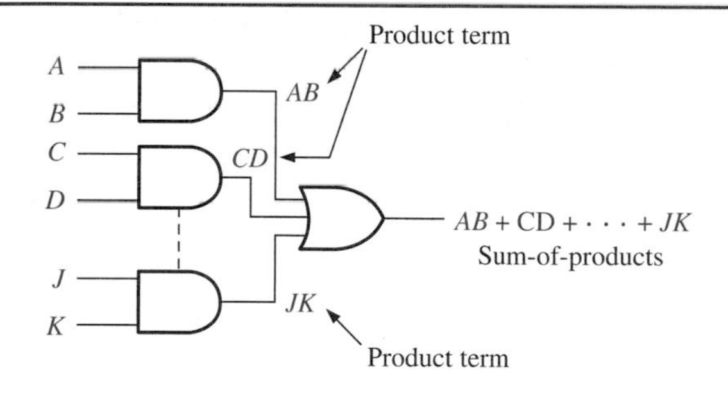

FIGURE 4–29

Basic SOP logic. The dashed line indicates that there can be any number of AND gates and product terms. Each AND gate output goes to an OR gate input.

Sum-of-Products Logic Using NAND Gates

Recall that NAND gates can be used in combination to form any logic function. SOP logic can be directly implemented with only NAND gates, as shown in Figure 4–30 where the NAND gate at the output functions as a negative-OR. As you know, the bubbles indicate inversion. Since each NAND gate output bubble is connected to a negative-OR gate input bubble, there is a double inversion. Recall from Boolean rules that if a variable is complemented twice, there is effectively no inversion ($\overline{\overline{A}} = A$). This means that the NAND implementation of an SOP function is the same as an AND-OR implementation.

You can convert any Boolean expression to an SOP expression by applying the laws and rules of Boolean algebra that you have learned to each specific expression.

FIGURE 4–30

SOP logic using NAND gates. The dashed line indicates that there can be any number of NAND gates.

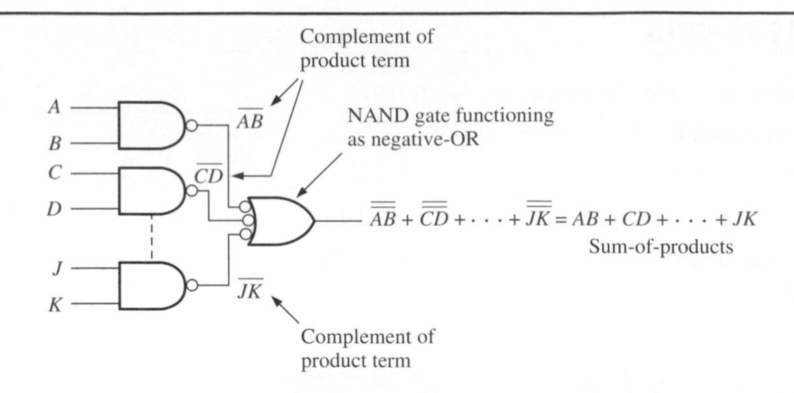

EXAMPLE 4–7

Problem

Convert the expression $A(B + C + DE)$ to an SOP form and show the logic circuits for both.

Solution

This expression $A(B + C + DE)$ is converted to SOP by applying the distributive law.

$$A(B + C + DE) = \overbrace{AB + AC + ADE}^{\text{SOP}}$$

Figure 4–31(a) shows the logic circuit for the original expression. Part (b) shows the SOP form of the circuit. Both circuits are equivalent.

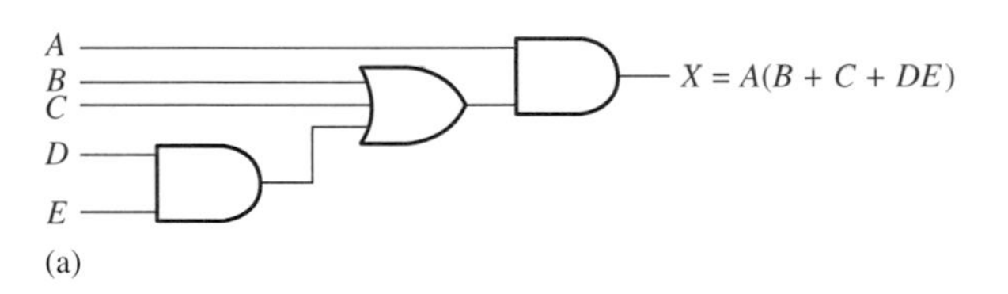

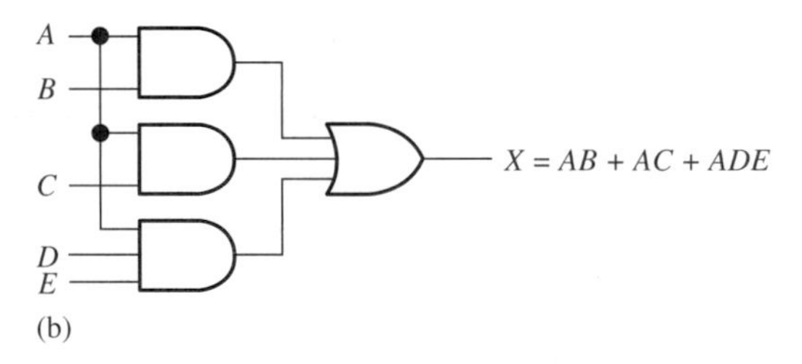

FIGURE 4–31

Question

How can the logic diagram in Figure 4–31(b) be modified to show all NAND gates?

COMPUTER SIMULATION

Open the Multisim file F04-32DG on the website. Observe that the operation verifies that the two logic circuits in Figure 4–31 are equivalent because the probes always indicate the same state regardless of what the inputs are. Operate the top switch with the *A* key, the second switch with the *B* key, the third switch with the *C* key, the fourth switch with the *D* key, and the bottom switch with the *E* key.

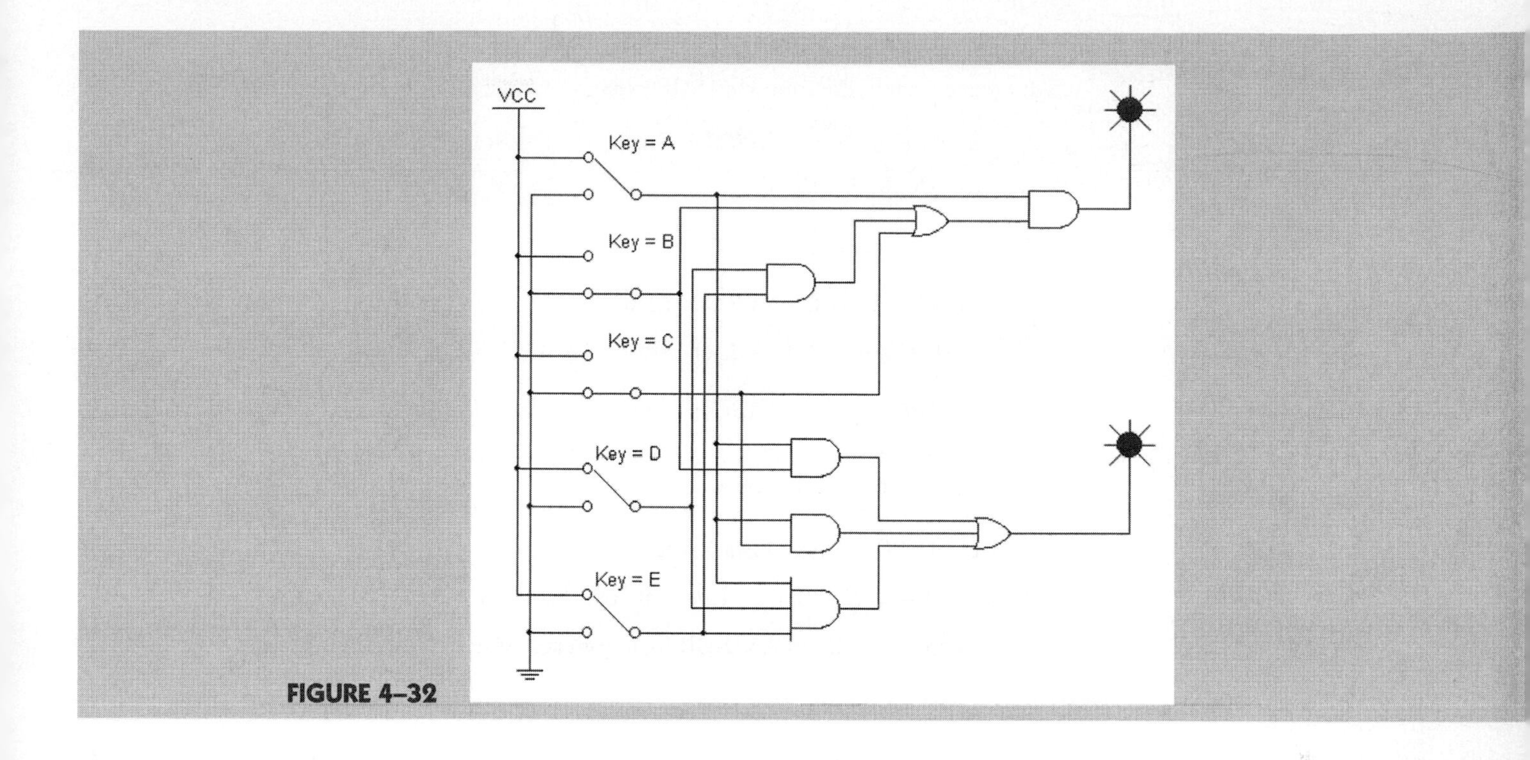

FIGURE 4–32

Evaluation of an SOP Expression

To evaluate a product term generally means to determine the values of the variables that make the term a 1. As previously indicated, a product term can contain both variables and complements of variables. For example, the product term $A\overline{B}C$ contains two uncomplemented variables, A and C, and one complemented variable, $\overline{B}$. In order for a term to be equal to 1, all the variables and complements of variables in the term must be 1. This means that $A = 1$, $\overline{B} = 1$, and $C = 1$. For $\overline{B}$ to be a 1, the variable B must be a 0, since the complement of 0 is 1. Therefore, the evaluation of the product term $A\overline{B}C$ is found by substituting either a 1 or a 0 for each variable in order to make the term a 1.

$$A\overline{B}C = 1 \cdot \overline{0} \cdot 1 = 1 \cdot 1 \cdot 1 = 1$$

This evaluation of the product term shows that the bit combination $1\overline{0}1$ makes the term equal to 1.

To evaluate an SOP expression, you must determine the combination of 1s and 0s that make each product term a 1. For example, let's evaluate the SOP expression $\overline{A}B\overline{C} + AB\overline{C}$, which contains two product terms. The SOP expression is equal to 1 when either or both terms are equal to 1.

Recall that for a term to equal 1, a complemented variable must be 0 and an uncomplemented variable must be 1. The bit combination 010 makes the term $\overline{A}B\overline{C}$ a 1 and the combination 110 makes the term $AB\overline{C}$ a 1.

$$\overline{A}B\overline{C} = \overline{0} \cdot 1 \cdot \overline{0} = 1 \cdot 1 \cdot 1 = 1$$

$$AB\overline{C} = 1 \cdot 1 \cdot \overline{0} = 1 \cdot 1 \cdot 1 = 1$$

The SOP expression is, therefore, equal to 1 when $\overline{A}B\overline{C} = 010$ or when $AB\overline{C} = 110$.

EXAMPLE 4–8

Problem

Evaluate each of the following SOP expressions and show a logic circuit for each:

(a) $ABC + \overline{A}BC + \overline{A}\,\overline{B}\,\overline{C}$ (b) $AB\overline{C} + \overline{A}BC + A\overline{B}C + \overline{A}\,\overline{B}C$

(c) $A\overline{B}C\overline{D} + ABCD$

Solution

(a) $ABC + \overline{A}BC + \overline{A}\,\overline{B}\,\overline{C}$ is a 1 when

$A = \mathbf{1}, B = \mathbf{1}, C = \mathbf{1}$ or $A = \mathbf{0}, B = \mathbf{1}, C = \mathbf{1}$ or $A = \mathbf{0}, B = \mathbf{0}, C = \mathbf{0}$

(b) $AB\overline{C} + \overline{A}BC + A\overline{B}C + \overline{A}\,\overline{B}C$ is a 1 when

$A = \mathbf{1}, B = \mathbf{1}, C = \mathbf{0}$ or $A = \mathbf{0}, B = \mathbf{1}, C = \mathbf{1}$ or $A = \mathbf{1}, B = \mathbf{0}, C = \mathbf{1}$ or $A = \mathbf{0}, B = \mathbf{0}, C = \mathbf{1}$

(c) $A\overline{B}C\overline{D} + ABCD$ is a 1 when

$A = \mathbf{1}, B = \mathbf{0}, C = \mathbf{1}, D = \mathbf{0}$ or $A = \mathbf{1}, B = \mathbf{1}, C = \mathbf{1}, D = \mathbf{1}$

Figure 4–33 shows AND-OR logic for each SOP expression.

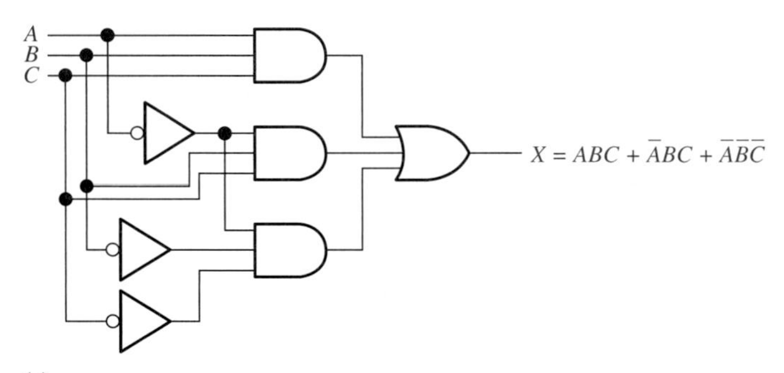

(a)

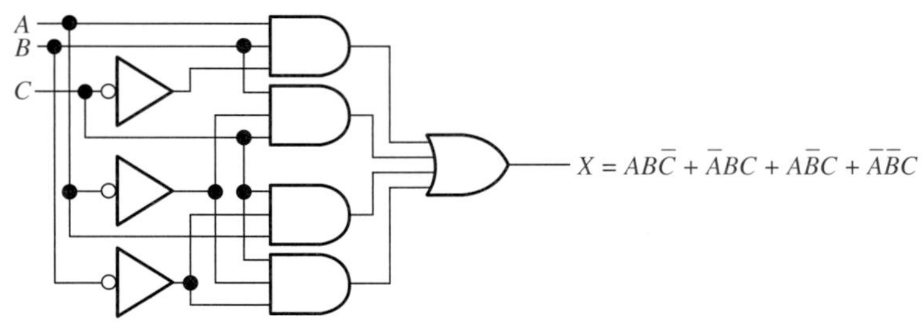

(b)

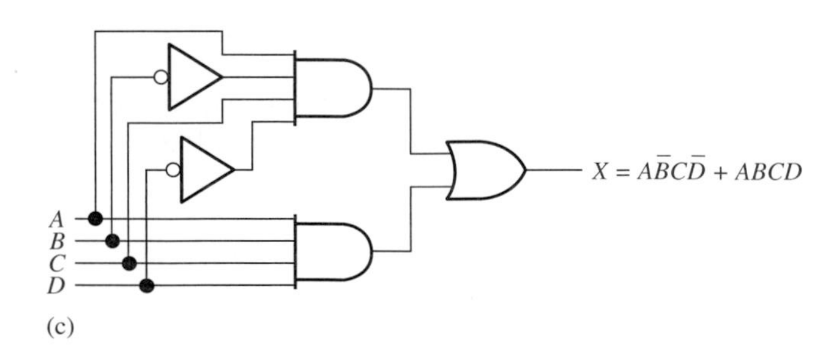

(c)

FIGURE 4–33

Question
When is the SOP expression $\overline{A}BC + AB\overline{C}$ equal to 1?

Product-of-Sums (POS) Logic

Another standard form of Boolean expression is the product-of-sums form. The SOP form is more widely implemented, so this is a brief introduction to POS.

A logic circuit in the **product-of-sums** form is composed of OR gates with their outputs going into a single AND gate. As Figure 4–34 shows, the OR gates form the sum terms that are then ANDed together to form the product.

A POS expression is equal to 1 only when all of the sum terms are equal to 1.

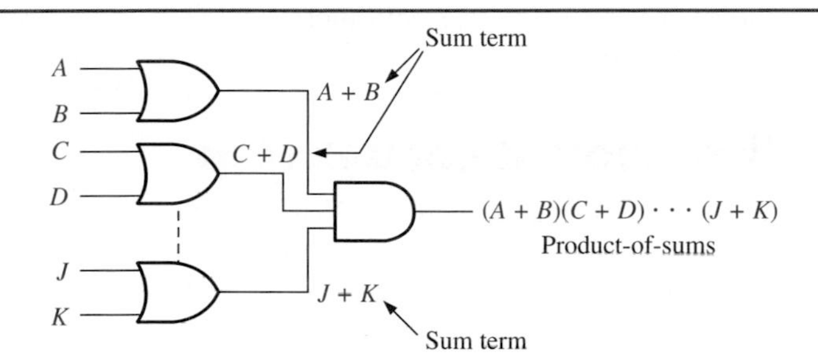

FIGURE 4–34

Basic POS logic. The dashed line indicates that there can be any number of OR gates and sum terms.

Product-of-Sums Logic Using NOR Gates

POS logic can be directly implemented with only NOR gates, as shown in Figure 4–35 where the NOR gate at the output functions as a negative-AND. As you know, the bubbles indicate inversion. Since each NOR gate bubble is connected to a negative-AND gate bubble, there is a double inversion, which is the same as no inversion.

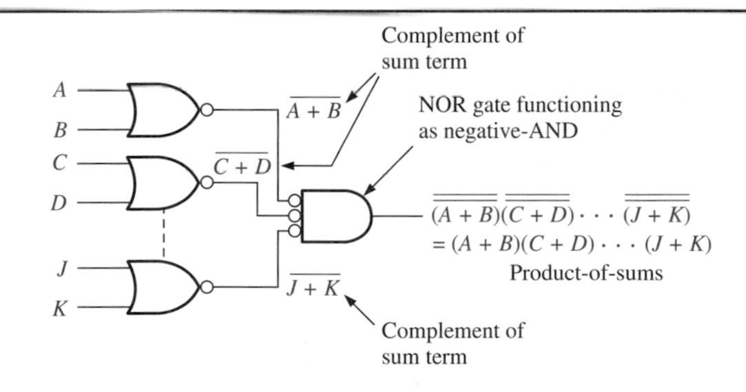

FIGURE 4–35

POS logic using NOR gates.

Review Questions

21. To what logic function does a Boolean product term correspond?
22. To what logic function does a Boolean sum correspond?
23. When is a product term equal to 1?
24. What does SOP mean?
25. What type of logic implements an SOP expression?

4–6 KARNAUGH MAPS

Simplification of certain Boolean expressions is sometimes useful in order to reduce the amount of logic (number of gates) required to implement the expression. The Karnaugh (pronounced "car no") map method provides a "cookbook" step-by-step approach to simplifying expressions with four or less variables. For more than four variables, the Karnaugh map method becomes more difficult and is not usually applied.

In this section, you will learn to use a Karnaugh map to simplify a Boolean expression.

A **Karnaugh map** is a graphical tool for simplifying Boolean expressions. It is divided into cells, each representing a specific combination of variable values. For a given Karnaugh map, the total number of cells equals the total possible combinations of the variables in an expression. For example, the Karnaugh map for an SOP expression with three variables has $2^3 = 8$ cells and the Karnaugh map for a four-variable expression has $2^4 = 16$ cells.

3-Variable Karnaugh Map

The Karnaugh map in Figure 4–36 has eight cells and each cell is for one combination of three variables, as shown in part (a). Cells with a common border differ by only one variable and are classified as *adjacent* cells as indicated by the blue arrows. Notice that the two cells at the top and the two cells at the bottom are adjacent due to "wraparound." The variable labels are usually placed outside the map, as shown in part (b). The variable combination for each cell is the variables associated with the row and column in which the cell is located, as the two cells in part (b) indicate.

FIGURE 4–36

A 3-variable Karnaugh map. Blue arrows in part (a) show adjacent cells.

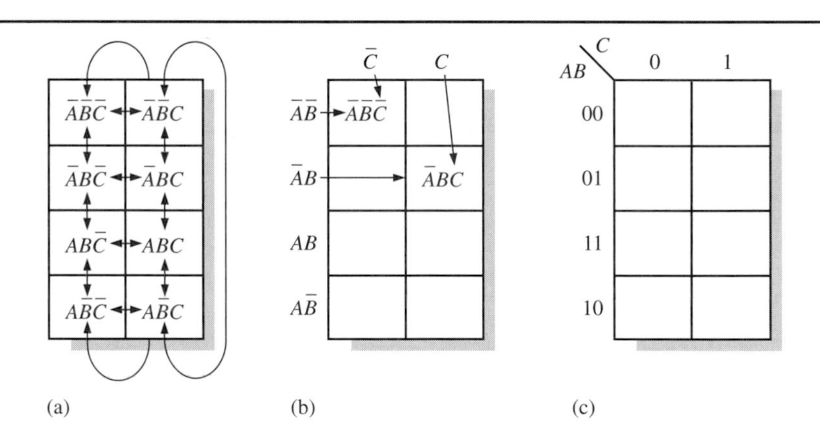

HISTORICAL NOTE

Maurice Karnaugh, a telecommunications engineer at Bell Labs, invented the Karnaugh map method for reducing logic circuits described by Boolean expressions. His paper, *The Map Method for Synthesis of Combinational Logic Circuits*, was published by the AIEE in November, 1953.

An alternate way to label a Karnaugh map is to write the variable names at the top of the map and place 0s and 1s along the side and top, as shown in Figure 4–36(c). A 1 is read as the true form of the variable and a 0 is read as its complement. To familiarize you with it, the lab manual will show all maps with this alternate method of labelling.

Simplifying an SOP Expression

Not all SOP expressions can be simplified, but many can. Simplification with a Karnaugh map is a fairly straightforward five-step process. Each term in the expression must contain all the variables. Such a term containing all the variables is called a *minterm*.

Step 1: Place a 1 in each cell corresponding to a minterm in the SOP expression.

Step 2: Group all 1s that are in adjacent cells into groupings of the maximum number of cells. A group can *only* contain a number of 1s equal to a power of two and groups can overlap.

Step 3: For each group, determine the variables that are the same. A complemented and an uncomplemented variable are not the same and cancel each other.

Step 4: Write the resulting product terms for each group, using only the variables that are the same within the group.

Step 5: Combine the resulting product terms into an SOP expression which will be equivalent to the original SOP expression.

EXAMPLE 4–9

Problem

Use a Karnaugh map to simplify the following 3-variable minterm SOP expression:

$$A\overline{B}C + \overline{A}BC + \overline{A}B\overline{C} + ABC + AB\overline{C}$$

Solution

Step 1: Place a 1 in the Karnaugh map in Figure 4–37 for each of the product terms in the SOP expression.

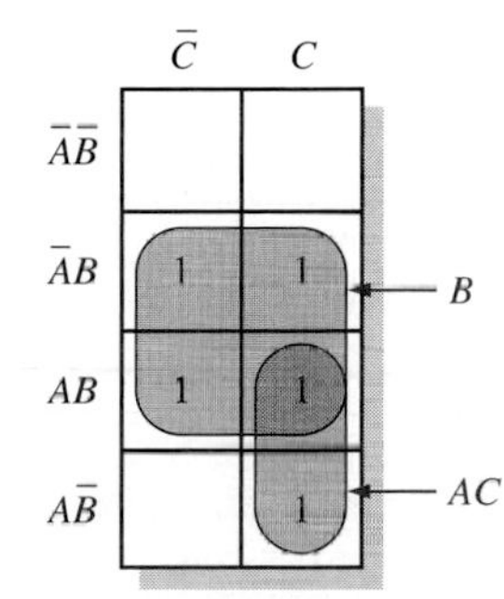

FIGURE 4–37

Step 2: Group the 1s into two groups as indicated. Remember, only 1s in adjacent cells can be placed in a group and each group must have 1, 2, 4, or 8 cells. Groups can overlap and do in this case.

Step 3: For each group determine the variables that are the same. In the group of four 1s, only the variable *B* is the same in all four cells. Eliminate *A* and *C* because both appear complemented and uncomplemented. This means that the four product terms represented by the 1s in the four cells can be reduced to the single-variable term *B*. In the group of two 1s, the variables *A* and *C* are the same in both cells. The variable *B* appears both complemented and uncomplemented, so eliminate it. This means that the two product terms represented by the 1s in the two cells can be reduced to the term *AC*.

Step 4: The product terms are *B* and *AC*.

Step 5: The SOP expression that is equivalent to the original SOP expression is

$$A\overline{B}C + \overline{A}BC + \overline{A}B\overline{C} + ABC + AB\overline{C} = \mathbf{B + AC}$$

The Computer Simulation on the next page shows that the logic circuits for both of these expressions produce the same output states for any combination of inputs. This proves that they are equivalent.

Question

What is the maximum grouping of cells possible in a 3-variable Karnaugh map?

COMPUTER SIMULATION

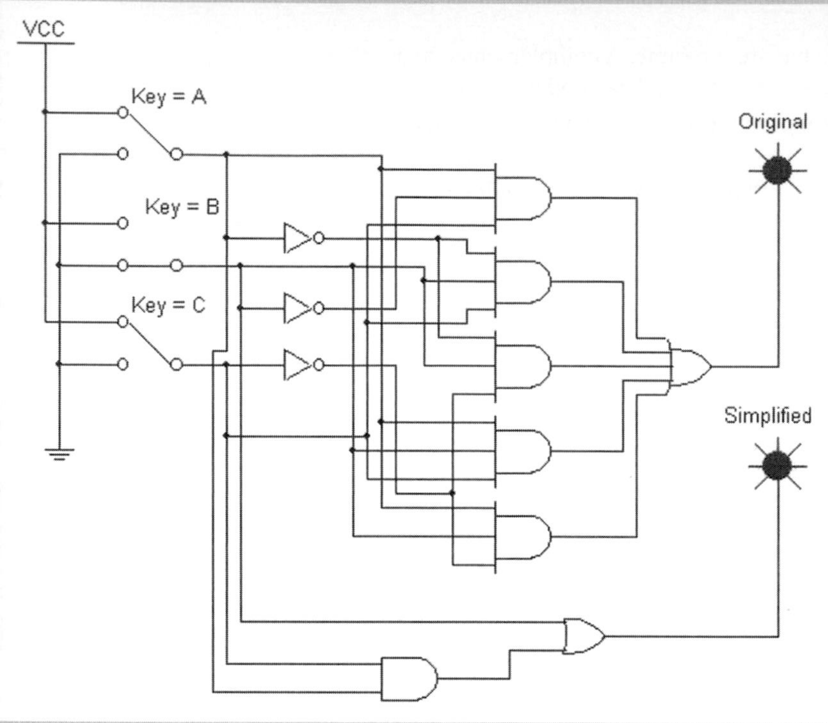

FIGURE 4–38

Open the Multisim file F04-38DG on the website. Observe that the operation verifies that the two logic expressions in Example 4–9 are equivalent because the probe lights always indicate the same state regardless of what the inputs are. Operate the top switch with the *A* key, the middle switch with the *B* key, and the bottom switch with the *C* key.

4-Variable Karnaugh Map

A 4-variable Karnaugh map is shown in Figure 4–39. It has $2^4 = 16$ cells and, as with the 3-variable map, the product terms for each adjacent cell differ by only one variable. Adjacent cells are indicated by the blue arrows and the steps listed before apply except now you can have a 16-cell grouping in addition to 1, 2, 4, and 8 cell groupings.

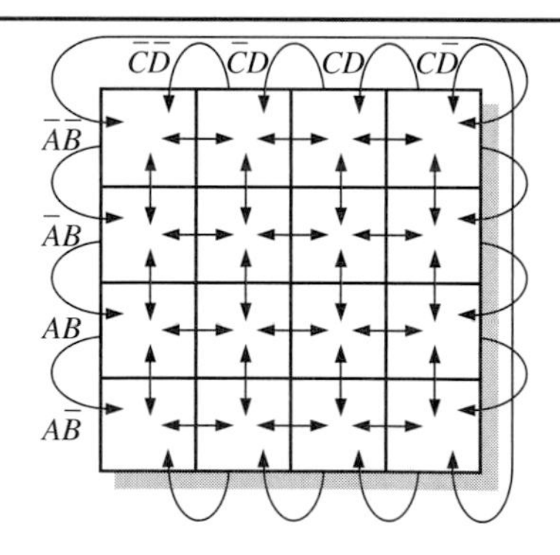

FIGURE 4–39

The 4-variable Karnaugh map showing cell adjacencies.

EXAMPLE 4–10

Problem

Apply the Karnaugh map method to simplify, if possible, the following 4-variable SOP minterm expression:

$$\overline{A}\,\overline{B}\,\overline{C}\,\overline{D} + \overline{A}B\overline{C}\,\overline{D} + AB\overline{C}\,\overline{D} + A\overline{B}\,\overline{C}\,\overline{D} + \overline{A}\,\overline{B}\,\overline{C}D + \overline{A}BCD + ABC\overline{D}$$

Solution

Step 1: Place a 1 in the Karnaugh map in Figure 4–40 for each of the product terms (minterms) in the SOP expression.

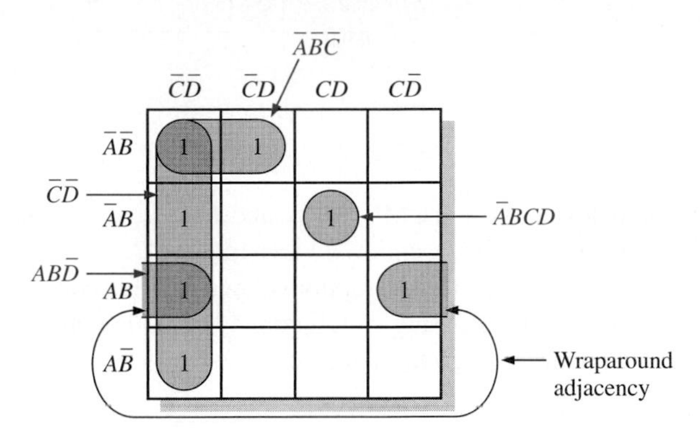

FIGURE 4–40

Step 2: Group the 1s as indicated. Remember, only 1s in adjacent cells can be placed in a group and each group must have either 1, 2, 4, 8, or 16 cells. Groups can and do overlap in this case.

Step 3: Cancel each variable that appears both complemented and uncomplemented in a group.

Step 4: Write the resulting simplified product term for each group. In this case, you cannot simplify the product term for the single 1 because the 1 cannot be grouped with any other 1s.

Step 5: This Karnaugh map process simplifies the original SOP expression to the equivalent expression

$$\overline{C}\overline{D} + AB\overline{D} + \overline{A}\overline{B}\overline{C} + \overline{A}BCD$$

Question

How many variables in a 4-variable Karnaugh map would be eliminated by a grouping of eight 1s?

Review Questions

26. What is the purpose of a Karnaugh map?
27. How many cells are in a 3-variable Karnaugh map?
28. How many cells are in a 4-variable Karnaugh map?
29. How do you simplify an SOP minterm expression with a Karnaugh map?
30. Are the terms $ABCD$ and $\overline{A}BCD$ represented by adjacent cells in a Karnaugh map?

4–7 INTEGRATED CIRCUITS

XOR and AND-OR-Invert fixed-function ICs are available. Power dissipation, propagation delay, and fan-out are three important IC logic gate parameters.

In this section, you will learn about specific combinational logic ICs and three IC parameters.

As you have learned, standard devices in both CMOS logic and TTL logic are usually available in the 54/74 series. Figure 4–41 shows XOR and AND-OR-Invert ICs. The input and output pin connections are shown for each gate. The dc supply voltage (V_{CC}) is connected to pin 14, and ground is connected to pin 7 for the ICs shown in Figure 4–41. Although not shown in the diagram, the V_{CC} and ground connections go to each gate in the IC.

FIGURE 4–41 XOR and AND-OR-Invert integrated circuits.

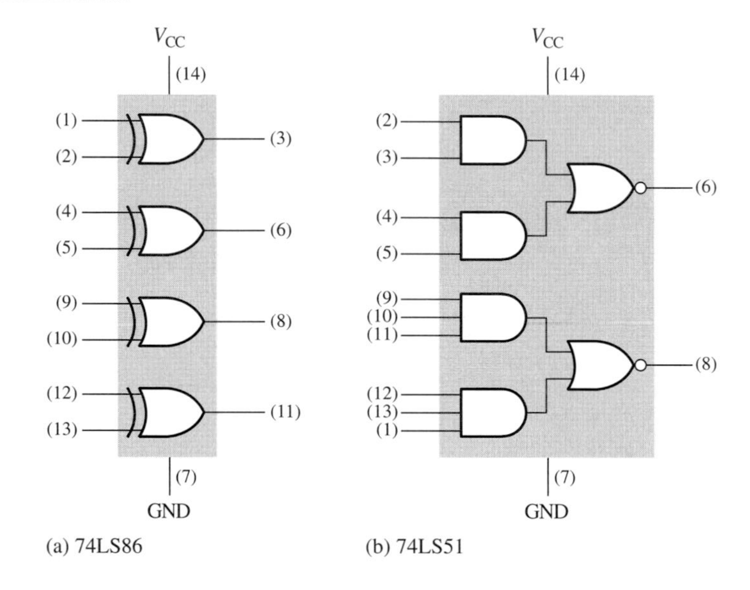

COMPUTER SIMULATION

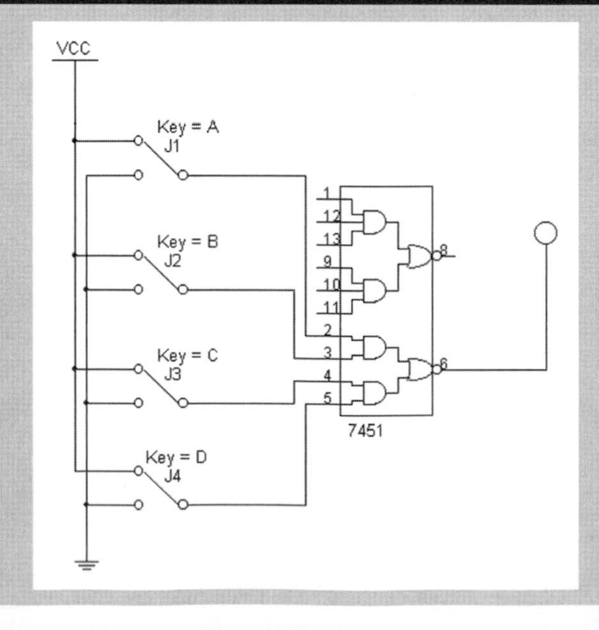

FIGURE 4–42

Open the Multisim file F04-42DG on the website. Only the AND-OR-Invert logic with the two 2-input AND gates is connected. By now you should know how to operate the switches to simulate 1 (HIGH) and 0 (LOW) inputs. The output is 0 (white probe light) when at least one AND gate has both inputs equal to 1.

Parameters of IC Logic

Certain aspects of gate operation depend on the internal components and circuits in the IC chip. These elements determine important parameters that apply to all IC gates and that are usually found on IC data sheets. These parameters include power dissipation, propagation delay, and fan-out. Data sheets show other parameters, but these are the most important.

Power Dissipation

All logic gates draw current from the dc supply voltage source (V_{CC}). When a gate output is in the HIGH state, it draws a different amount of current than when the output is in the LOW state. Supply current in the HIGH state is designated I_{CCH}, and current in the LOW state is designated I_{CCL}. The average supply current (I_{CC}) is specified as the average of I_{CCH} and I_{CCL} when the gate is driven by a digital waveform with a 50% duty cycle so that the gate is in each state half the time. The average supply current is

$$I_{CC} = \frac{I_{CCH} + I_{CCL}}{2}$$

The average **power dissipation** of a gate is

$$P_{AVG} = V_{CC}\, I_{CC}$$

EXAMPLE 4–11

Problem

A particular gate draws 2.0 μA when its output is in the HIGH state and 3.6 μA when its output is in the LOW state. What is the average power dissipation of the gate if V_{CC} is 5 V?

Solution

The average current is

$$I_{CC} = \frac{I_{CCH} + I_{CCL}}{2} = \frac{2.0\ \mu\text{A} + 3.6\ \mu\text{A}}{2} = 2.8\ \mu\text{A}$$

The average power dissipation is

$$P_{AVG} = V_{CC}I_{CC} = (5\ \text{V})(2.8\ \mu\text{A}) = \mathbf{14\ \mu W}$$

Question

What is the power dissipation when the gate output remains in the HIGH state?

Propagation Delay

When there is a change or transition in the input level, it takes a short time, called the **propagation delay**, for a corresponding change in output level. Two propagation delay times are associated with logic gates:

- t_{PHL} The time between an input transition and the corresponding HIGH-to-LOW output transition.
- t_{PLH} The time between an input transition and the corresponding LOW-to-HIGH output transition.

Propagation delay times vary depending on the logic family and generally range from a fraction of a nanosecond (ns) to several nanoseconds. A nanosecond is one-billionth of a second. The propagation delay time determines how fast a gate can switch from one state to the other, which affects the maximum operating frequency.

Fan-Out

The **fan-out** of a logic gate is the maximum number of inputs of other gates to which a gate output can be connected. Fan-out is specified in terms of unit loads. A **unit load** for a logic gate equals one input to a similar logic gate. If the fan-out is exceeded, the output voltage of the gate may not stay within specified limits.

Review Questions

31. What is the logic in the 74LS51?
32. On what does the power dissipation of a logic gate depend?
33. What is the propagation delay of a logic gate?
34. What is fan-out?
35. What is a unit load?

4–8 TROUBLESHOOTING

When gates are connected together as they are in combinational logic, the failure of one gate can affect the operation of others.

In this section, you will learn about the circuit node and the troubleshooting methods of signal tracing and waveform analysis.

Circuit Nodes

In combinational logic, the output of one gate may be connected to two or more gate inputs as shown in Figure 4–43. The interconnecting paths share a common conductive path known as a **node**. Gate 1 is driving the node, and the other gates represent loads connected to the node. A driving gate can drive a number of load gate inputs up to its specified fan-out.

FIGURE 4–43

A node in a logic circuit.

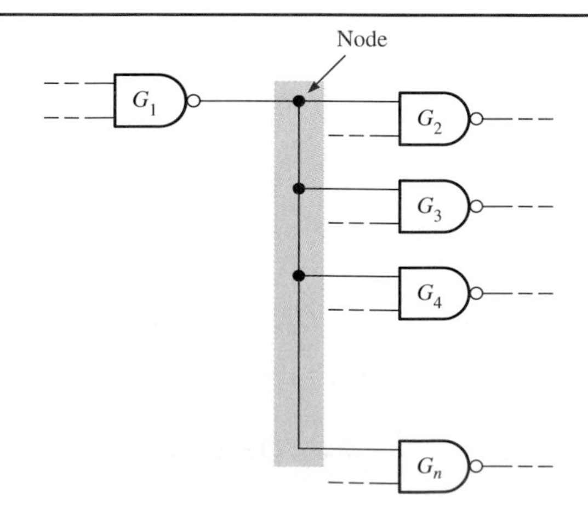

Several types of failures are possible in this type of circuit. Some of these failure modes are difficult to isolate to a single gate because all the gates connected to the node are affected. Common types of failures are the following (the term *signal* means digital waveform):

- *Open output in the driving gate* This failure will cause a loss of signal to all load gates.
- *Open input in a load gate* This failure will not affect the operation of any of the other gates connected to the node, but it will cause a loss of signal output from the faulty gate.
- *Shorted output in driving gate* This failure can cause the node to be stuck in one state or the other or produce some bizarre symptom, depending on what the output is shorted to.
- *Shorted input to a load in a load gate* Same as if the driving gate output is shorted.

Signal Tracing and Waveform Analysis

A general troubleshooting method called **signal tracing** is useful in most troubleshooting situations. Waveform measurement, as you know, is generally done with an oscilloscope or a logic analyzer.

Basically, the signal tracing method requires that you observe the waveforms and their time relationships at all accessible points in the logic circuit. You can begin at the inputs and work toward the output until a missing or incorrect waveform is found. This tells you the fault is in the last gate. You can also begin at the output and work toward the input. A general example of the signal tracing procedure is as follows:

1. Within a system identify the section of logic that is suspected of being faulty, based on your knowledge of how the circuit should operate.
2. Start at the inputs to the section of logic under examination. Assume that the input waveforms to this section of logic have been found to be correct.
3. For each gate, beginning at the input and working toward the output of the logic circuit, observe the output waveform of the gate and compare it with the input waveforms.
4. Apply your knowledge of gate operation to determine if the output waveform is correct.
5. If the output is correct, go to the next gate. Continue checking each gate until you find an incorrect waveform.
6. If the output is incorrect, the gate you are testing may be faulty. However, the fault may be outside of the IC in which the gate is located such as a shorted or open pin or circuit board trace. Carefully inspect the circuit in the area of the suspicious IC. Check the dc supply voltage (V_{CC}) and ground directly on the IC pins in case there is a bad solder joint or open on the printed circuit board that is making the IC act faulty. If you cannot find anything, remove the IC and check it independently of the rest of the circuit. If the IC is faulty, replace it with a new one because an IC cannot be repaired.

SAFETY NOTE

Never wear rings or any type of metallic jewelry while working on a circuit. These items may accidentally come in contact with the circuit, causing shock and/or damage to the circuit.

Review Questions

36. What is a node in a logic circuit?
37. What is signal tracing?
38. What is meant by the term *driving gate*?
39. What instruments are used for waveform measurement?
40. If a circuit is not working properly, what should you do first?

CHAPTER REVIEW

Key Terms

AND-OR A logic circuit that consists of any number of AND gates with outputs connected to the inputs of a single OR gate.

AND-OR-Invert A logic circuit that consists of AND gates with outputs that are connected to a NOR gate.

Combinational logic Logic gates connected together to perform some specific function in which the output level is at all times dependent on the combination of input levels.

Fan-out The maximum number of gate inputs of the same type that a single logic gate can drive.

Karnaugh map A graphic tool used to simplify a Boolean expression. It is an arrangement of cells representing all product terms for a given number of Boolean variables.

Node A common electrical point in a circuit.

Power dissipation The amount of dc power used by a circuit; the product of V_{CC} and I_{CC}.

Product-of-sums A standard form of Boolean expression that is the product (AND) of two or more sum terms (OR).

Propagation delay The time delay from a change in the input level of a gate to the corresponding change in output level.

Signal tracing A method of troubleshooting that identifies the point in a circuit where the signal is first missing or incorrect.

Sum-of-products A standard form of Boolean expression that is the sum (OR) of two or more product terms (AND).

Unit load A measure of fan-out. One gate input represents a unit load to the output of a gate within the same IC family.

XNOR A logic gate that produces a 0 (LOW) if one and only one of its inputs is 1 (HIGH).

XOR A logic gate that produces a 1 (HIGH) if one and only one of its inputs is 1 (HIGH).

Important Facts

- ❑ XOR and XNOR gates can be constructed from combinations of inverters, AND gates, and OR (NOR) gates.
- ❑ AND-OR logic produces a sum-of-products output function.
- ❑ NAND gates can be used to produce a sum-of-products.
- ❑ OR-AND logic produces a product-of-sums output function.
- ❑ NOR gates can be used to produce a product-of-sums.
- ❑ Any Boolean expression can be converted into a sum-of-products or product-of-sums form.
- ❑ A Karnaugh map is used to simplifiy a Boolean expression.
- ❑ The 7451 and 7454 are examples of IC AND-OR-Invert logic.
- ❑ Power dissipation, propagation delay, and fan-out are parameters specified for logic gates on data sheets.

Equations

XOR:

$$X = A\overline{B} + \overline{A}B$$

$$X = A \oplus B$$

XNOR:

$$X = \overline{A\overline{B} + \overline{A}B}$$

$$X = \overline{A \oplus B}$$

AND-OR:

$$X = AB + CD$$

AND-OR-Invert:

$$X = \overline{AB + CD}$$

Boolean Laws

Associative law of multiplication for three variables:

$$A(BC) = (AB)C$$

Associative law of addition for three variables:

$$A + (B + C) = (A + B) + C$$

Distributive law for three variables:

$$A(B + C) = AB + AC$$

Boolean Rules

$$A + AB = A$$

$$A + \overline{A}B = A + B$$

$$(A + B)(A + C) = A + BC$$

Chapter Checkup

Answers are at the end of the chapter.

1. The expression for an XOR gate is
   (a) $X = AB + \overline{AB}$
   (b) $X = \overline{A} + AB$
   (c) $X = \overline{A}B + A\overline{B}$
2. The equivalent logic for an XOR gate consists of
   (a) two OR gates, one AND gate, two inverters
   (b) two AND gates, one OR gate, two inverters
   (c) two AND gates and one OR gate

10. Evaluate each of the following POS expressions by determining a set of 1 and 0 combinations required to make each expression equal to 1.

(a) $(A + \overline{B} + C)(A + B + C)$

(b) $(A + B + \overline{C} + D)(\overline{A} + B + C + D)$

11. Determine the number of cells in a Karnaugh map for each of the following number of variables:

(a) 2 (b) 3

(c) 4

12. How many 1s would be placed on a Karnaugh map for each expression in Problem 9?

13. A logic gate draws an average supply current of 1 mA, and the supply voltage is 5 V. Determine the average power dissipation.

14. An inverter has a t_{PLH} of 6 ns and a t_{PHL} of 8 ns. When the input changes from LOW to HIGH, how long does it take the output to change from HIGH to LOW?

15. If the fan-out of a gate is 10, how many other gate inputs can be connected to the gate output?

16. Develop the truth table for an XOR gate.

17. Develop the truth table for an XNOR gate.

Basic-Plus Problems

18. Develop the truth table for the AND-OR circuit in Figure 4–44.

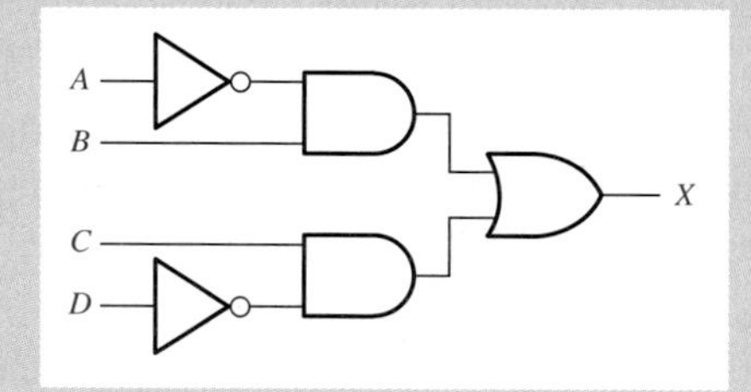

FIGURE 4–44

19. Determine the output waveform for the AND-OR logic and input waveforms shown in Figure 4–45.

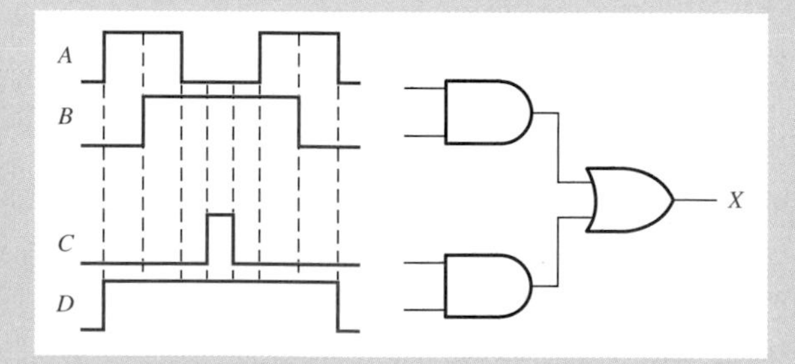

FIGURE 4–45

20. If the logic circuit in Figure 4–45 were an AND-OR-Invert, what would the output waveform look like?

21. Apply the input waveforms in Figure 4–45 to the logic circuit in Figure 4–44 and determine the output waveform.

22. Design a logic circuit for the following SOP expression:

$$X = A\overline{B}C + \overline{A}B\overline{C} + ABC + \overline{A}BC$$

23. Design a logic circuit for the following SOP expression:

$$X = \overline{A}B\overline{C}D + A\overline{B}C\overline{D} + \overline{A}\,\overline{B}\,\overline{C}\,\overline{D} + ABCD$$

24. Apply the distributive law to the following expression to convert it to an SOP expression.

$$A\overline{B}(C\overline{D} + CD) + A(B\overline{C}D + \overline{B}C\overline{D}) + ABCD$$

25. Use a 3-variable Karnaugh map to simplify, if possible, each expression:

(a) $\overline{A}\,\overline{B}\,\overline{C} + A\overline{B}C + ABC$

(b) $AB\overline{C} + \overline{A}B\overline{C} + \overline{A}BC + A\overline{B}\,\overline{C} + \overline{A}\,\overline{B}C$

26. Use a 4-variable Karnaugh map to simplify, if possible, the following expression:

$$X = \overline{A}B\overline{C}D + A\overline{B}C\overline{D} + \overline{A}\,\overline{B}\,\overline{C}\,\overline{D} + ABCD$$

27. Use a 4-variable Karnaugh map to simplify, if possible, the following expression:

$$A\overline{B}(C\overline{D} + CD) + A(B\overline{C}D + \overline{B}C\overline{D}) + ABCD$$

28. Draw the logic circuit for each simplified expression in Problem 25.

29. Draw the logic circuit for the expression in Problem 26.

30. Draw the logic circuit for the simplified expression in Problem 27.

31. A logic gate operating with a 50% duty cycle has an $I_{CCH} = 100\ \mu A$ and an $I_{CCL} = 150\ \mu A$. Determine the average power dissipation with a V_{CC} of 3.3 V.

32. If the output of the logic gate in Problem 31 stays HIGH all the time, what is the average power dissipation?

TROUBLESHOOTING PRACTICE

Use the circuits in the specified Multisim files on the website.

33. Open file TSP04-01 and troubleshoot the simulated XOR gate. Determine whether or not it is working properly. If it is not working properly, determine the fault.

34. Open file TSP04-02 and troubleshoot the simulated XNOR gate. Determine whether or not it is working properly. If it is not working properly, determine the fault.

35. Open file TSP04-03 and troubleshoot the simulated AND-OR-Invert circuit. Determine whether or not it is working properly. If it is not working properly, determine the fault.

36. Open file TSP04-04 and troubleshoot the simulated combinational logic. Determine whether or not it is working properly. If it is not working properly, determine the fault.

37. Open file TSP04-05 and troubleshoot the simulated SOP logic. Determine whether or not it is working properly. If it is not working properly, determine the fault.

The block symbol for a full-adder is shown in Figure 5–4(a). The two inputs are labeled A and B and a carry input is labeled C_{in}. There are two outputs, a sum (Σ) and a carry (C_{out}). The truth table for the full-adder is shown in part (b).

FIGURE 5–4

Symbol and truth table for a full-adder.

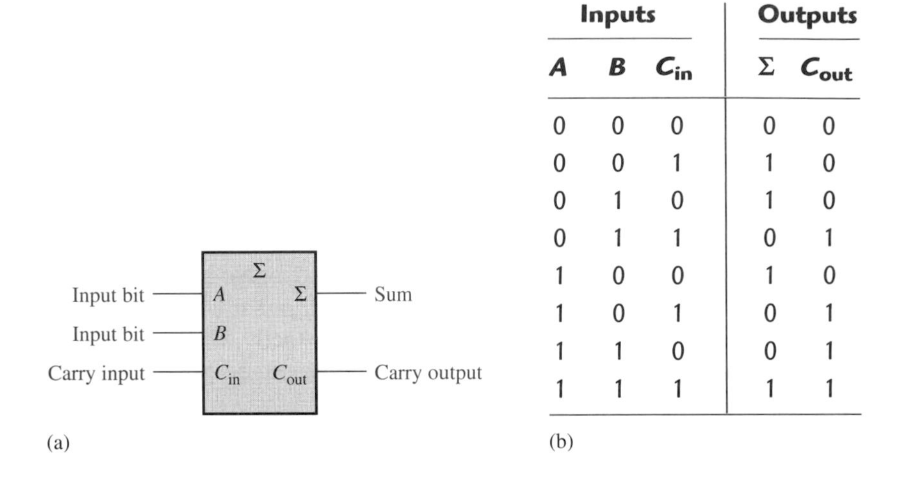

Inputs			Outputs	
A	B	C_{in}	Σ	C_{out}
0	0	0	0	0
0	0	1	1	0
0	1	0	1	0
0	1	1	0	1
1	0	0	1	0
1	0	1	0	1
1	1	0	0	1
1	1	1	1	1

Full-Adder Logic

A full-adder can be implemented with two half-adders and an OR gate, as shown in Figure 5–5. The second half-adder adds the sum output of the first half-adder and the carry input. The OR gate produces the final carry output by combining the carry outputs from each half-adder.

FIGURE 5–5

Full-adder using two half-adders and an OR gate.

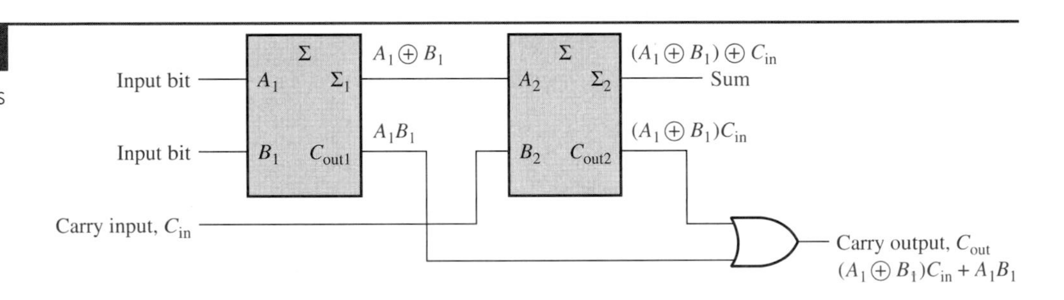

The two Boolean expressions for the sum and carry outputs of the full-adder implemented with two half-adders and an OR gate are

$$\Sigma = (A_1 \oplus B_1) \oplus C_{in}$$

$$C_{out} = (A_1 \oplus B_1)C_{in} + A_1B_1$$

where A_1 and B_1 are inputs to the first half-adder.

COMPUTER SIMULATION

Open the Multisim file F05-06DG on the website. Observe that the operation verifies the full-adder truth table in Figure 5–4(b). To apply a 1 or a 0 to an input, operate the switches with the *A* key, the *B* key, and the *C* key. This circuit is the logic for two half-adders plus a carry output OR gate.

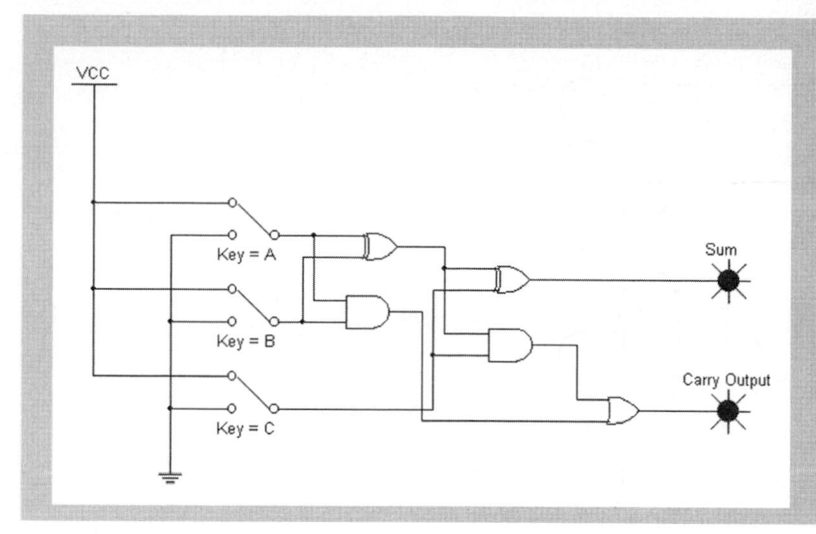

FIGURE 5–6

Another way to implement the full-adder logic is to use the truth table in Figure 5–4(b) to develop Boolean expressions for the sum and carry output. Example 5–1 illustrates full-adder logic using SOP expressions and the Karnaugh map (topics studied in Chapter 4).

EXAMPLE 5–1

Problem

Use the truth table in Figure 5–4 to implement a full-adder with AND-OR logic and inverters.

Solution

In the truth table, each combination of input variables A, B, and C_{in} that make the sum (Σ) output a 1 and the combinations of input variables that make the carry output (C_{out}) a 1 can be expressed as product terms. For example, for the combination $A = 0$, $B = 0$, $C_{in} = 1$, the product term is $\overline{A}\,\overline{B}\,C_{in}$. From the truth table, Boolean SOP expressions for the sum and carry output are

$$\Sigma = \overline{A}\,\overline{B}C_{in} + \overline{A}B\overline{C}_{in} + A\overline{B}\,\overline{C}_{in} + ABC_{in}$$

$$C_{out} = \overline{A}BC_{in} + A\overline{B}C_{in} + AB\overline{C}_{in} + ABC_{in}$$

Each Boolean expression can be implemented using inverters and AND-OR logic. To see if either expression can be simplified, use Karnaugh maps for the sum and the carry output, as shown in Figure 5–7.

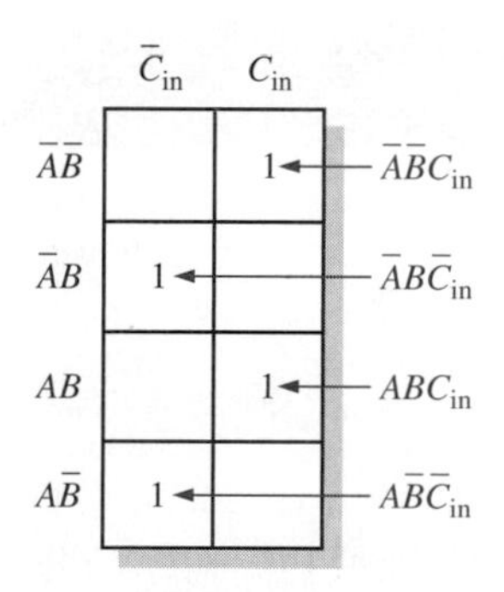

(a) Karnaugh map for Σ
$\Sigma = \overline{A}\,\overline{B}C_{in} + \overline{A}B\overline{C}_{in} + A\overline{B}\,\overline{C}_{in} + ABC_{in}$

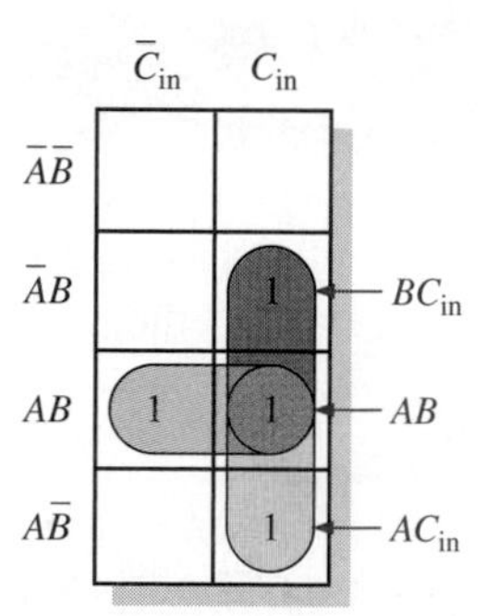

(b) Karnaugh map for C_{out}
$C_{out} = AB + BC_{in} + AC_{in}$

FIGURE 5–7

The expression for the sum cannot be simplified because none of the 1s in the Karnaugh map in Figure 5–7(a) can be grouped. The expression for the carry output can be simplified by grouping the adjacent 1s, as shown in part (b).

Figure 5–8 shows the logic for the Σ and C_{out} expressions implemented as a full-adder.

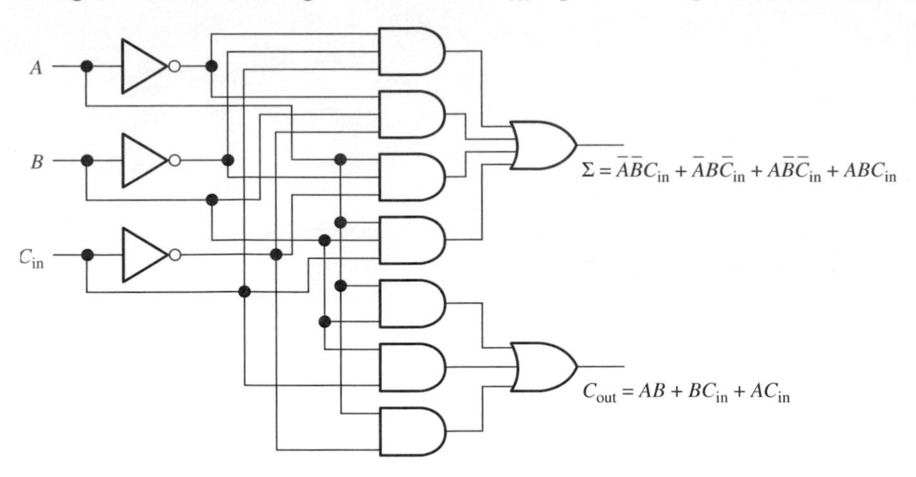

FIGURE 5–8
Logic for a full-adder derived from the truth table in SOP form.

Question*
The full-adder logic in Figure 5–6 and the logic in Figure 5–8 are not exactly the same, but they produce the same results. How can you show that both implementations of the full-adder are equivalent?

Review Questions

Answers are at the end of the chapter.

1. What does a binary half-adder do?
2. What does a binary full-adder do?
3. What is the difference between a half-adder and a full-adder in terms of their operation?
4. What are the outputs of a half-adder with a 1 on each input?
5. What are the outputs of a full-adder with a 1 on each input?

5–2 THE PARALLEL BINARY ADDER

A full-adder can add two bits at a time along with an input carry. Most applications require binary numbers with 4, 8, 16, or 32 bits to be added. To add two multibit numbers requires a full-adder for each bit in the numbers except the LSB.

In this section, you will learn how a parallel binary adder works.

A **parallel binary adder** is a logic circuit that consists of two or more full-adders that can add two binary numbers. Four full-adders are used to add two 4-bit numbers as shown in Figure 5–9.

* *Answers are at the end of the chapter.*

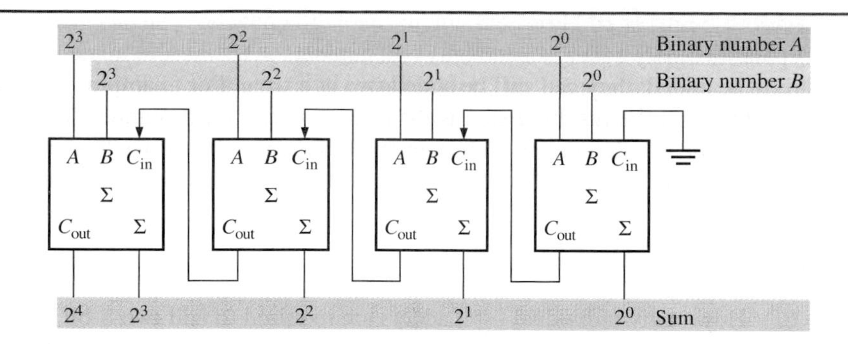

FIGURE 5–9

A 4-bit parallel binary adder.

The full-adder on the right adds the two LSBs (least significant bits). The input carry to the LSB adder is assumed to be 0, so the carry input is connected to ground. A half-adder could have been used for the LSB, but usually a multibit adder is implemented with full-adders in all positions. The adder on the left adds the two MSBs (most significant bits). The carry output from each adder goes to the carry input of the next adder because when you add two numbers, a carry from one column goes over to be added to the next column.

EXAMPLE 5–2

Problem

Add the following two 4-bit binary numbers. Then show that the parallel binary adder in Figure 5–9 adds the two numbers correctly.

$$1101 + 0111$$

Solution

Add the two numbers. The carry bits are shown in red. The final carry output becomes the most significant sum bit.

	1	1	1	0	Carry into LSB is 0
	1	1	0	1	Number A (decimal 13)
	0	1	1	1	Number B (decimal 7)
1	**0**	**1**	**0**	**0**	Sum (decimal 20)

When you apply these binary numbers to the parallel adder, the result is the same, as shown in Figure 5–10. Notice that C_{in} to the LSB adder is connected to ground to represent 0.

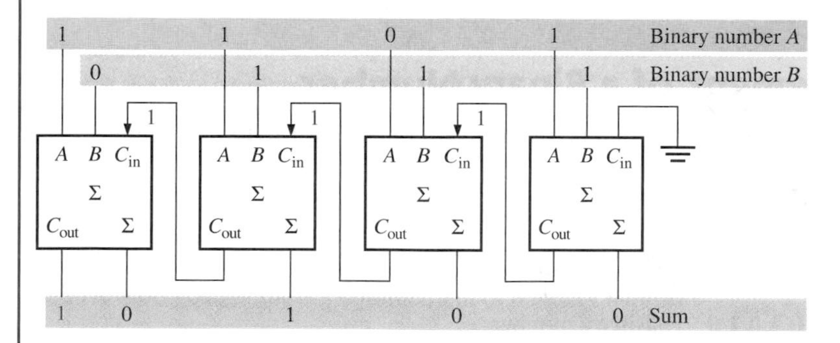

FIGURE 5–10

Question

What is the final carry out bit of the left full-adder if the numbers added are 1000 and 0101?

8 is the multiplicand, and the number 24 is the product. If you add 8 to itself 3 times, you also get 24.

$$8 + 8 + 8 = 24$$

In binary, this is

$$1000 + 1000 + 1000 = 11000$$

One algorithm for multiplication of integers is to add the multiplicand to itself a number of times equal to the multiplier. To multiply using this algorithm, a computer would first add the multiplicand to itself, then add the resulting sum to the multiplicand, then add the next sum to the multiplicand, and so on until the multiplicand is added to itself a number of times equal to the multiplier. To keep track of the number of times, the computer would subtract 1 from the multiplier each time it added the multiplicand so that when the multiplier reached zero, the process would be complete. A flow chart of this algorithm is shown in Figure 5–14.

FIGURE 5–14

Flow chart for a multiplication algorithm that adds the multiplicand to itself a number of times equal to the multiplier.

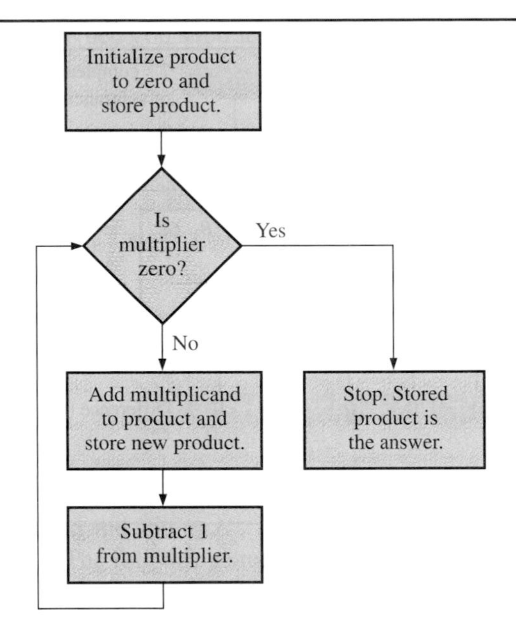

Another algorithm for multiplication uses partial products to get the final product.

1000	(8)
× 0011	(3)
1000	partial product
1000	partial product
11000	(24) final product

For each 1 in the multiplier, the partial product is the same as the multiplicand. Each succeeding partial product is shifted one place to the left and then the partial products are added to give the final product. All-zero partial products of the leading zeros in the multiplier are not shown. This algorithm can be used to multiply any two binary numbers by a series of additions and shifts. Using this algorithm, a computer would check each multiplier bit beginning with the least significant. If the bit is a 0, it would shift the partial product left, and if the bit is a 1, it would add the multiplicand to the partial product and shift left. There would be a number of additions equal to the number of 1s in the multiplier. This algorithm is much faster than repeated additions.

EXAMPLE 5–5

Problem

Multiply 0110 × 1101 using the partial products algorithm.

Solution

The multiplicand is 1101 and the multiplier is 0110.

1101	Multiplicand
0110	Multiplier
0000	1st partial product is all 0s (multiplier LSB is 0).
1101	2nd partial product is multiplicand shifted left one place.
1101	3rd partial product is multiplicand shifted left one place from the previous partial product.
0000	4th partial product is all 0s (multiplier MSB is 0).
1001110	Product is the sum of partial products.

Question

What are the decimal numbers represented in the muliplication in this example?

Binary Division

The words that describe the numbers in a division problem are the *dividend, divisor,* and *quotient.* The divisor is the number divided into the dividend, and the quotient is the final result. For example, when the dividend is 18 and the divisor is 6, you get a quotient of 3.

$$\frac{18}{6} = 3$$

One algorithm for division is basically to subtract the divisor from the dividend until you get zero or a negative number. The number of these subtractions is equal to the quotient. This process is illustrated with the same numbers as follows:

1st subtraction	$18 - 6 = 12$
2nd subtraction	$12 - 6 = 6$
3rd subtraction	$6 - 6 = 0$

Since it took three subtractions to reach a 0 result, the quotient is 3. If you reach a negative result, a remainder is indicated. The basic idea here is that division can be done by a series of subtractions, which can be basically accomplished with an adder. As in the case of multiplication, the division process can be speeded up by a shifting method.

EXAMPLE 5–6

Problem

Divide 11000 by 01000 using the subtraction algorithm. Note that the divisor and dividend have the same number of bits.

Solution

The dividend is 11000 and the divisor is 01000. Subtraction is done using 2's complement addition. The 2's complement of the divisor is 11000. Add this to the dividend. Discard the carry out.

11000
11000
10000 Result of first subtraction

Add the 2's complement of the divisor to the result of the first subtraction.

10000
11000
01000 Result of second subtraction

Add the 2's complement of the divisor to the result of the second subtraction.

01000
11000
00000 Result of third subtraction

The third subtraction results in zero. The quotient is equal to the number of subtractions.

Quotient = **00011**

Question

What are the decimal numbers represented in the division in this example?

Review Questions

21. What are the terms for the three numbers in a multiplication?
22. What are the terms for the three numbers in a division?
23. A number can be multiplied by another number by adding the first number to itself how many times?
24. What are the intermediate results of a multiplication called?
25. When dividing using subtractions, when is the division completed?

5–6 FLOATING-POINT NUMBERS

It takes lots of bits to represent very large or very small numbers. Also, in practical situations some numbers will be fractions such as 0.022157 or numbers with fractional parts, such as 23.5618. Computers can process all sizes and types of numbers because they are able to handle them in floating-point format.

In this section, you will learn to express a number in floating-point format.

Using a fixed number of bits, the **floating-point** number system can represent any positive or negative number including very large and very small numbers as well as numbers with fractional parts. A floating-point number consists of two parts plus a **sign bit**. The **mantissa** represents the digits in the magnitude of the number. The **exponent** represents the *number of places* that the decimal point is to be moved.

A decimal example will help you understand floating-point numbers. Consider the decimal number 241,506,800. The mantissa of this number is .241506800 and the exponent is 9. When a number is expressed as a floating-point number, it is normalized by moving the decimal point to the left of all the digits so that the mantissa is always a fractional number and the exponent is the power of ten. The number 241,506,800 is written in floating-point format as

$$0.241506800 \times 10^9$$

Standard binary floating-point numbers can be expressed with 32 bits (single-precision), 64 bits (double-precision), or 80 bits (extended-precision). The format for a single-precision floating-point number is shown in Figure 5–15.

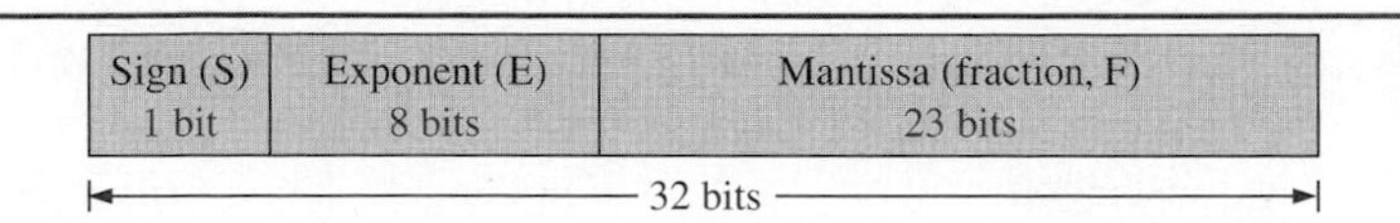

FIGURE 5–15

Binary floating-point number format.

In floating-point format, a number is written in a binary form, similar to scientific notation, where the binary point is always to the right of the most significant 1. In the mantissa or fractional part (F), the binary point (similar to decimal point in decimal numbers) is understood to be to the left of the 23 bits. Effectively, there are 24 bits in the mantissa because in any binary number the left-most (most significant) bit of the mantissa must *always* be a 1. This 1 is understood to be there although it does not occupy an actual bit position. An exception is the number 0, which is represented by zeros in all positions.

The eight bits in the exponent (E) represent a **biased exponent**, which is obtained by adding 127 to the actual exponent. The biased exponent allows positive or negative exponents without requiring a separate sign bit. Since the biased exponent allows a range of actual exponent values from -126 to $+128$, extremely large and extremely small numbers can be expressed with a fixed number of bits. For example, a 32-bit floating point binary number can replace a regular binary number with 129 bits.

To illustrate how a binary number is expressed in floating-point format, let's use the 13-bit binary number 1011010010001 (decimal 5,777) as an example. It can be expressed as 1 plus a fractional part by moving the binary point 12 places to the left and multiplying by the power of two, 2^{12}.

$$1011010010001 = 1.011010010001 \times 2^{12}$$

This is a positive number so the sign bit (S) is 0. The exponent, 12, is expressed as a biased exponent (E) by adding it to 127 and expressing the result as a binary number. You can use your calculator to convert the decimal number to binary by entering the decimal number and pressing 3rd BIN. However, the TI-36X is limited to 9-bit numbers.

$$E = 12 + 127 = 139 = 10001011$$

The mantissa is the fractional part (F) of the binary number.

$$F = .011010010001$$

Because there is always a 1 to the left of the binary point, it is understood to be there and is not included in the mantissa. The complete floating point number for this example is as follows:

S	E	F
0	10001011	01101001000100000000000

EXAMPLE 5–7

Problem

Convert the decimal number 3.248×10^4 to a single-precision floating-point binary number.

Solution

Convert the decimal number to binary. You can't convert 32,480 directly to binary on your TI-36X calculator because it is limited to 9 bits. However, you can first convert the decimal number to hexadecimal on the calculator.

The display shows 7EE0

Write the hex number in binary as follows:

7	E	E	0
0111	1110	1110	0000

$$3.248 \times 10^4 = 32{,}480 = 111111011100000 = 1.11111011100000 \times 2^{14}$$

The most significant bit (MSB) will not occupy a bit position because it is always a 1. The mantissa is the fractional 23-bit binary number 11111011100000000000000. The biased exponent is

$$14 + 127 = 141 = 10001101$$

The complete floating-point number is

S	E	F
0	10001101	11111011100000000000000

Question
What is the 32-bit floating-point format for the positive binary number 1000010110001111?

Review Questions

26. What are the three parts of a floating-point binary number?
27. What does a sign bit of 1 indicate?
28. How many bits are there in a single-precision floating-point binary number?
29. How is a biased exponent obtained?
30. What is the mantissa?

5–7 THE MAGNITUDE COMPARATOR

In addition to being added, subtracted, multiplied, and divided, numbers can be compared to determine their relative magnitudes. When the magnitudes of two numbers are compared, the result indicates if the numbers are equal or if they are not equal, which one is larger. Computers often have to perform the comparison operation when processing numbers.

In this section, you will learn to describe a magnitude comparator and explain how it works.

Magnitude Equality

The exclusive-OR gate can be used to compare two binary digits and determine whether they are equal or not equal. Recall that when the two input bits to an XOR gate are the same, the output is a 0. When the two input bits are different, the output is a 1. This is illustrated in Figure 5–16.

When two multiple-bit binary numbers are compared, the two numbers are equal only if all corresponding bits in both numbers are equal. An inequality in any bit position makes the numbers unequal.

FIGURE 5–16

The XOR gate as a 2-bit comparator.

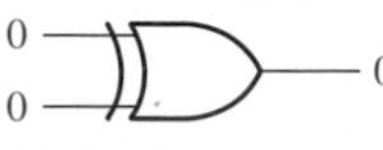

(a) Equal

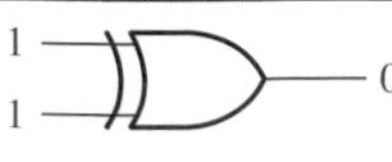

(b) Equal

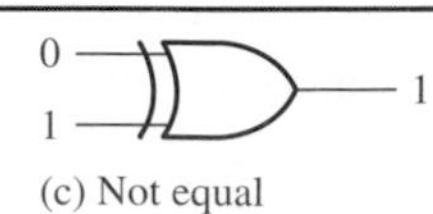

(c) Not equal

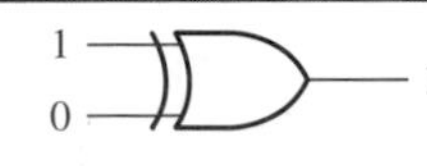

(d) Not equal

A **comparator** is a logic circuit that determines when two binary numbers are equal. A 4-bit comparator is shown in Figure 5–17. This can be expanded to accommodate any number of bits. When both numbers A and B are equal, the output of each XOR gate is 0; and the resulting output of the negative-AND (NOR) gate is 1, indicating an equality. When the numbers are unequal, the output of one or more of the XOR gates is 1; and the resulting output of the NOR (negative-AND) gate is 0, indicating an inequality.

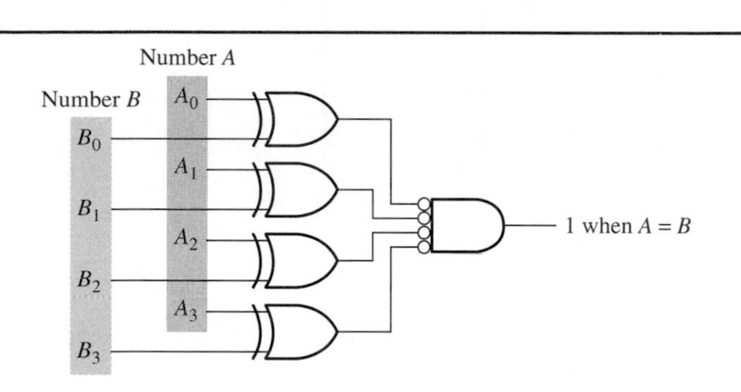

FIGURE 5–17

A 4-bit magnitude equality comparator.

EXAMPLE 5–8

Problem

Show the binary states (1s and 0s) in the 4-bit magnitude comparator for each pair of binary numbers.

(a) 1010 and 1010 (b) 1110 and 0111

Solution

Figure 5–18 shows the binary states for each of the pairs of numbers.

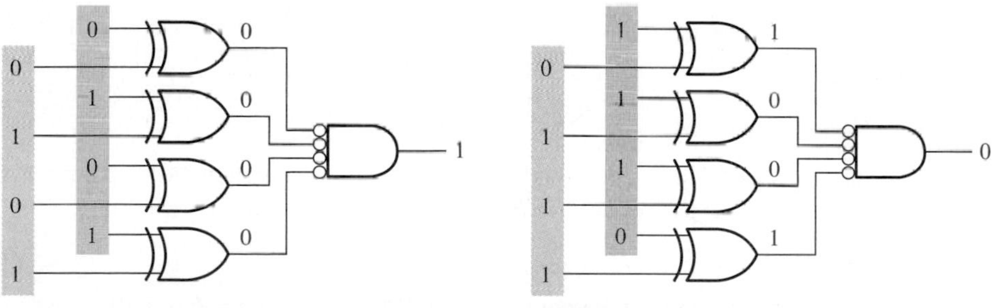

(a) The input numbers are equal.

(b) The input numbers are unequal.

FIGURE 5–18

Question

What output is produced when the inputs are 0001 and 1000?

Magnitude Inequality

The logic circuit in Figure 5–17 indicates when two numbers are not equal by a 0 on its output. In addition to an equality output ($A = B$), comparators may also have two additional outputs, $A > B$ and $A < B$. The logic required to determine which of two numbers is larger is quite complex. It must determine the highest-order position where the bits are not equal and then the number with a 1 in that position is the larger of the two numbers.

A logic symbol for a 4-bit comparator with three outputs is shown in Figure 5–19. When the $A > B$ output is a 1, number A is larger than number B; when the $A < B$ output is 1, number A is smaller than number B; and when the $A = B$ output is a 1, the numbers are equal.

FIGURE 5–19

Logic symbol for a 4-bit magnitude comparator.

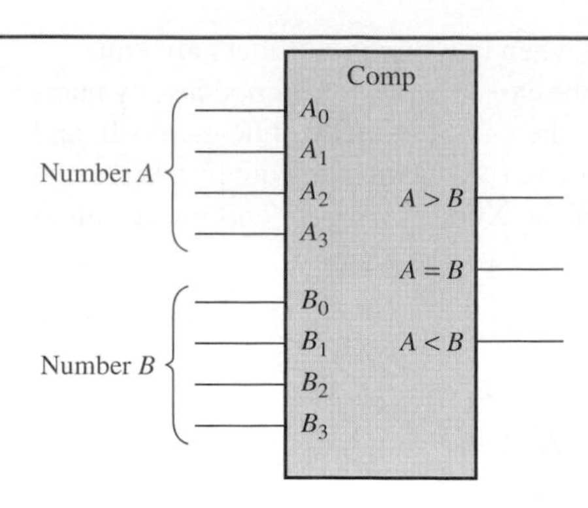

EXAMPLE 5–9

Problem

Determine the $A = B$, $A > B$, and $A < B$ output for the input numbers shown on the comparator in Figure 5–20.

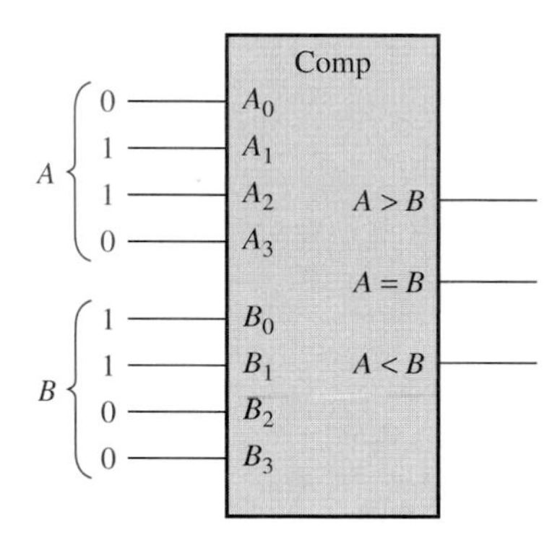

FIGURE 5–20

Solution

The number on the A inputs is 0110, and the number on the B inputs is 0011. Number A is larger than number B, so the output $A > B = 1$. The other two outputs are 0.

Question

What are the comparator outputs when number A is 1001 and number B is 1010?

Magnitude comparators would be of little use if all they could compare were 4-bit numbers. The way a magnitude comparator can compare a larger number is with cascading inputs that allow you to expand the number of bits by connecting two or more comparators together. This is described in Section 5–8.

COMPUTER SIMULATION

Open the Multisim file F05-21DG on the website. This is a 4-bit magnitude comparator like in Figure 5–17. Observe that the output is 1, as indicated by the red probe light, only when the two 4-bit numbers on the inputs are the same. When the numbers are not equal, the probe light is white, indicating a 0 output.

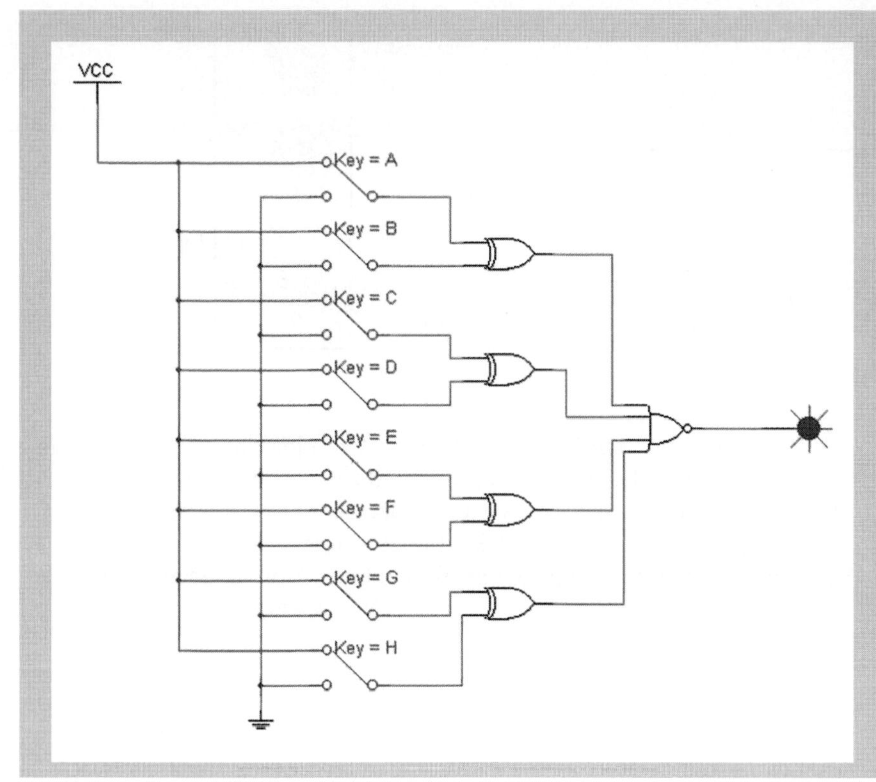

FIGURE 5–21

Review Questions

31. When are two binary numbers equal?
32. Why is the XOR gate used as the basic element in a magnitude comparator?
33. Which unsigned number is larger, 10101010 or 01010101?
34. Suppose the outputs of a comparator are $A > B = 1, A < B = 0$, and $A = B = 0$. What does this indicate?
35. Which unsigned number is smaller, 01111111 or 10000000?

INTEGRATED CIRCUITS 5–8

Addition and comparison are two important functions in digital systems. These functions are available in integrated circuit form.

In this section, you will learn about an integrated circuit parallel binary adder and a magnitude comparator.

An IC Adder

An example of an individual integrated circuit adder is the 74LS283. This device is a 4-bit full-adder that can add two 4-bit binary numbers and an input carry. It produces a 4-bit sum and an output carry. There are four full-adder circuits on the IC chip. A block diagram and the pin connection diagram are shown in Figure 5–22. The two 4-bit inputs are labeled A_1 through A_4 and B_1 through B_4, the sum outputs are labeled Σ_1 through Σ_4, the initial carry input is C_0, and the final carry output is C_4. Because the carry input C_0 represents the 0 position, the other inputs are numbered starting with 1.

FIGURE 5–22

The 74LS283 4-bit full adder.

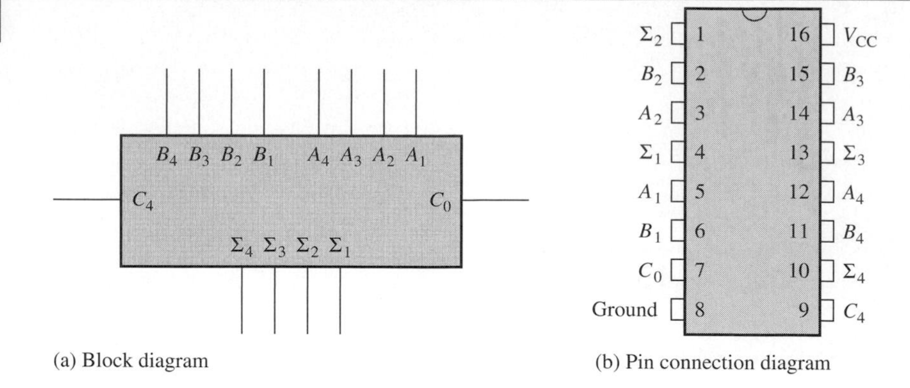

(a) Block diagram (b) Pin connection diagram

The adders can be expanded in order to add numbers with more than four bits by connecting the output carry of one adder to the input carry of the next adder, as shown in Figure 5–23, where the two 4-bit adders connected together make an 8-bit adder. One 8-bit binary number (blue) is designated B and the other 8-bit binary number (red) is designated A. There is an 8-bit sum plus the final output carry. To add two 16-bit numbers, four adders are used, and so on.

FIGURE 5–23

An 8-bit adder using two 74LS283 4-bit adders.

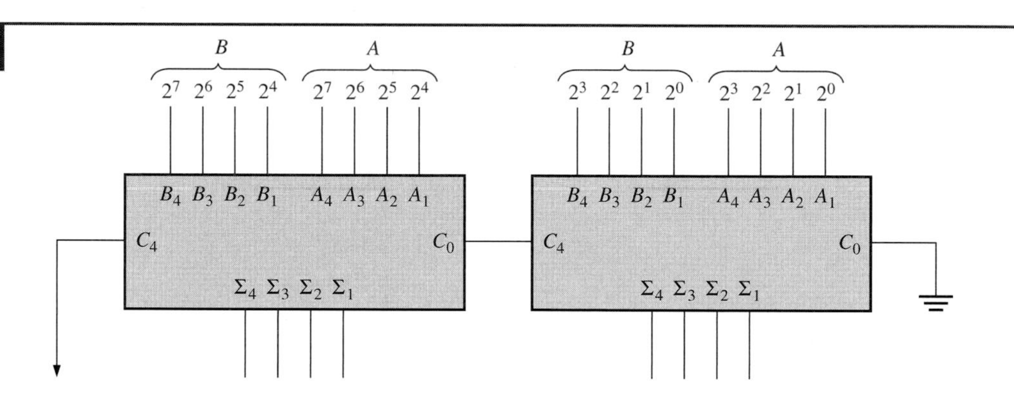

An IC Comparator

The 74LS85 is an example of an integrated circuit magnitude comparator. It can compare two 4-bit binary numbers and can be expanded to compare numbers with more bits. A block diagram and the pin connection diagram are shown in Figure 5–24.

FIGURE 5–24

The 74LS85 4-bit magnitude comparator.

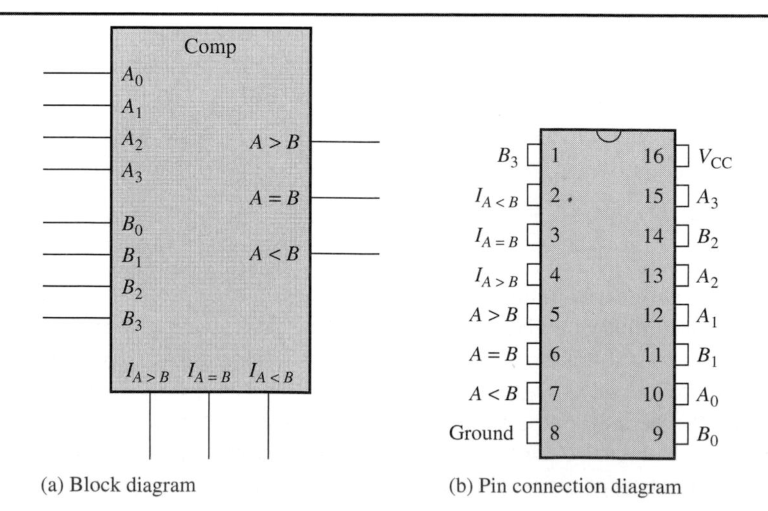

(a) Block diagram (b) Pin connection diagram

The inputs $I_{A>B}$, $I_{A<B}$, and $I_{A=B}$ are used to expand the comparator for comparison of binary numbers with any number of bits. The outputs of one comparator are connected to these expansion inputs, as shown in Figure 5–25. Notice that the unused expansion inputs of the one comparator must be connected to the specific logic levels shown. These two comparators connected together allow you to compare two 8-bit numbers. Four comparators connected in a similar way would compare two 16-bit numbers.

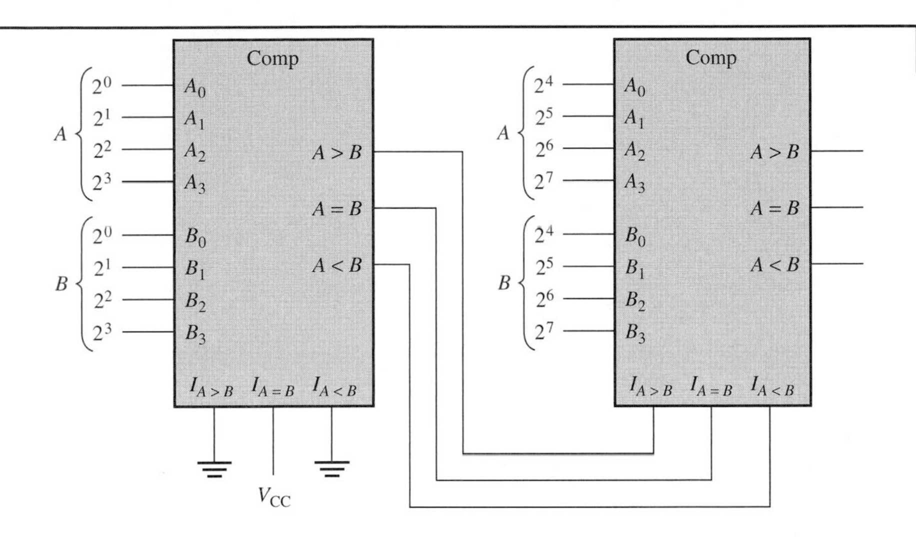

FIGURE 5–25

An 8-bit magnitude comparator using two 74LS85 4-bit comparators.

Review Questions

36. How many 4-bit parallel adder ICs are required for adding two 16-bit numbers?
37. How are the adders in Question 36 connected to each other?
38. How many full-adders are in a 74LS283?
39. Typically, what are the three outputs of a magnitude comparator and what do they indicate?
40. What is the purpose of the expansion inputs on the 74LS85 comparator?

TROUBLESHOOTING 5–9

The effects of common faults in logic gates were discussed in Chapter 2 and Chapter 3. Now we will look at specific examples of how open or shorted inputs and outputs affect the performance of an adder and a comparator.

In this section, you will learn to analyze faults in adders and comparators.

Stuck Levels

When an input or an output of a logic device never changes, it is said to be "stuck" at that level. The reason that an input or output is stuck HIGH or stuck LOW is irrelevant in most cases. These fault conditions can be caused by a short to ground, a short to the supply voltage, or a short to another input or output which can be internal or external to the faulty logic device. Also, a stuck level can be the result of an open connection that can be internal or external to the logic device. A visual inspection of the device pins and the printed circuit board interconnection may sometimes reveal an open or shorted contact.

4-Bit Adder

An open or shorted input to the adder will cause the sum output to be incorrect unless the valid level of the faulty input is the same as the stuck level. For example, assume the A_2 input of the 4-bit adder in Figure 5–26 is stuck HIGH. The adder will always see a 1 on that input, and the sum will be incorrect for those binary inputs that do not have a 1 in the A_2 position (0000, 0001, 0100, 0101, 1000, 1001, 1100, and 1101). The other numbers that have a 1 in the A_2 position will appear correct because when an input is stuck HIGH, it is the same as a 1.

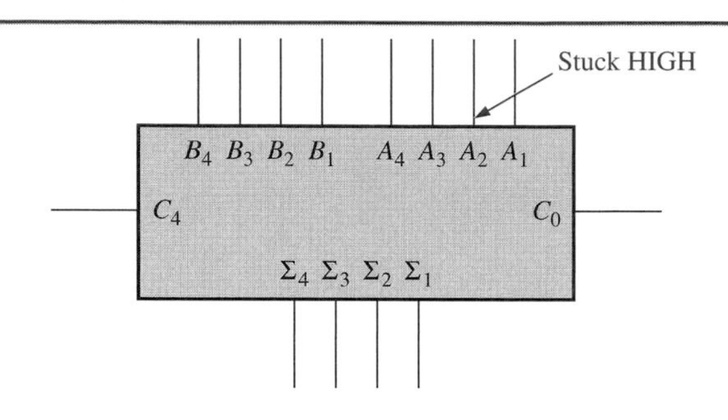

FIGURE 5–26

A 4-bit adder.

SAFETY NOTE

Do not use any electrical equipment until you have familiarized yourself with its operation, know the proper operating procedures, and are aware of any potential hazards. Use equipment with three-wire power cords (three-prong plug). Make sure that the power cord is in good condition and that the grounding pin is not bent.

Checking for a Faulty Input or Output

One way to test an adder is to make one set of binary inputs all zeros and sequence the other set of binary inputs through all possible states, then repeat for all zeros on the other set of inputs. For this test condition, the output state should always follow the input state if there are no faults because a binary number is being added to zero and the sum will equal to the number.

If there is a fault in one of the inputs or one of the outputs, the output level for that bit will be the same as the fault level (stuck HIGH or stuck LOW) and will not change no matter what the input code is. Figure 5–27 illustrates the binary sequences that you should observe for an adder with input A_2 stuck HIGH. However, any stuck HIGH or stuck LOW fault in any input or any output can be detected in a similar way.

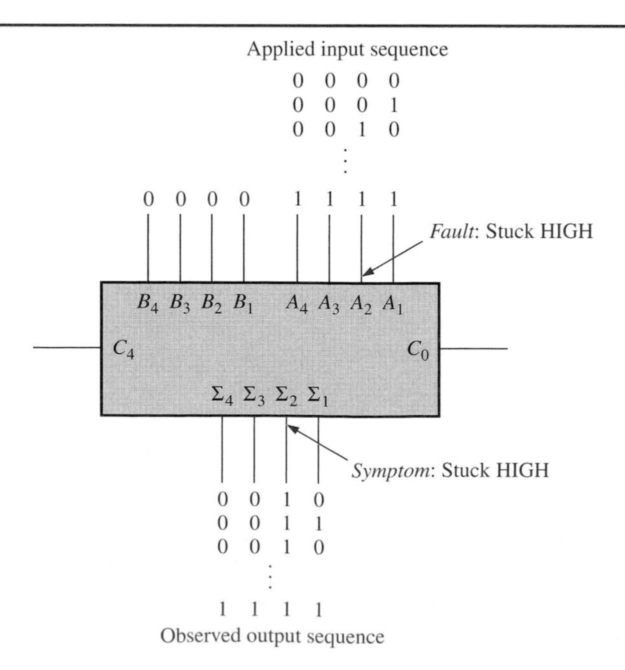

FIGURE 5–27

Illustration of troubleshooting a 4-bit adder with a specific fault.

4-Bit Comparator

An open or shorted input to the comparator will cause an incorrect output for certain binary codes. One way to test a comparator for a faulty input (assuming the $A = B$ output is not faulty) is to apply the same four bits to both the A and the B inputs and then alternate the bits so that both a 1 and a 0 are applied to each input. This can be accomplished by first applying 1010 to both A and B inputs and then 0101 to both A and B inputs, as illustrated in Figure 5–28. If there are no input faults, the $A = B$ output should be HIGH for both codes. If an input is stuck HIGH or LOW, then either the $A > B$ or the $A < B$ output will be HIGH, depending on fault.

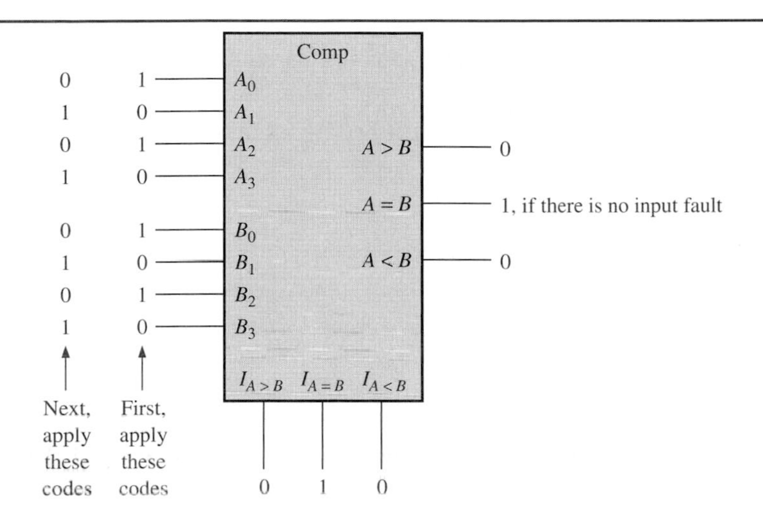

FIGURE 5–28

Illustration of testing a comparator for an input fault.

You can test for a comparator output fault as follows:

1. Apply all zeros to the A and B inputs. The $A = B$ output should be HIGH and the others LOW.
2. Apply 0001 to the A inputs with all zeros on the B inputs. The $A > B$ output should be HIGH and the others LOW.
3. Apply 0001 to the B inputs with all zeros on the A inputs. The $A < B$ output should be HIGH and the others LOW.

Of course, other codes can be used as long as the =, >, and < relationships are observed.

Review Questions

41. What are two common types of faults in logic circuits?
42. What do the terms *stuck HIGH* and *stuck LOW* mean?
43. Why were all zeros applied to one set of adder inputs in order to check for a fault?
44. Binary numbers 1010 and 0000 are applied to a 4-bit adder. If the output is 1011, can you tell what the fault is?
45. Will the comparator input fault test work if the $A = B$ output is stuck HIGH?

BINARY CODES AND DATA LOGIC

CHAPTER OUTLINE

Study aids for this chapter are available at

http://www.prenhall.com/SOE

INTRODUCTION

As you have learned, computers recognize only 1s and 0s. When we communicate with a computer in decimal digits and letters of the alphabet, they must be changed to a binary form that computers understand. Likewise, computers must communicate with us in terms we understand.

In this chapter, you will learn two important codes used in digital systems. These are the binary coded decimal (BCD) and the American Standard Code for Information Interchange (ASCII). BCD is a common interface code between a numeric input, such as found on keypads, and a digital system. ASCII is an alphanumeric (letters and numbers) code used for keyboard entry of data to a computer. You will learn how logic gates are used to implement the basic decoding function in which a code input such as BCD is converted to a decimal output. Devices that are used for this purpose are called decoders. Devices used to convert a decimal input to a code output such as BCD are called encoders.

The multiplexing operation is used in computers and digital communications for taking digital data from several lines and switching the data onto one line in a time sequence. The demultiplexing operation, which is the opposite of multiplexing, takes the data on a single line and switches it back onto several lines. Multiplexers and demultiplexers are types of data selectors.

KEY TERMS

- Binary coded decimal
- ASCII
- Parity bit
- Decoder
- 7-segment display
- Encoder
- Priority encoder
- Multiplexer
- Demultiplexer

KEY OBJECTIVES

A section number is given for each objective. After completing this chapter, you should be able to

6–1 Describe the BCD code and the ASCII code

6–2 Describe three common types of decoders

6–3 Describe a basic encoder

6–4 Explain how multiplexers and demultiplexers work

6–5 Describe decoder, encoder, and data selector ICs

6–6 Analyze and troubleshoot faults in decoders, encoders, and data selectors

COMPUTER SIMULATIONS DIRECTORY

The following figures have Multisim circuit files associated with them.

LABORATORY EXPERIMENTS DIRECTORY

The following exercises are for this chapter.

- **Experiment 8** The Seven-Segment Decoder and Display
- **Experiment 9** Combinational Logic Using a MUX

Review Questions

Answers are at the end of the chapter.

1. What does *BCD* stand for?
2. What does the BCD code represent?
3. What does *ASCII* stand for?
4. How many bits are in an ASCII code?
5. What is the purpose of a parity bit?

6–2 DECODERS

A decoder generally has inputs that are in the form of a binary code and an output that is active when a specified code is applied.

In this section, you will study three common types of decoders.

A **decoder** is a combinational logic circuit that is designed to indicate when a specified combination of bits (code) is present on its inputs. A decoder can have any number of inputs depending on the length or number of bits in the code. The number of outputs depends on the particular application. A decoder designed to detect only one combination of bits has one output to indicate when that particular combination is on the inputs. On the other hand, a decoder that is designed to detect any of several bit combinations in a given code can have up to 2^n outputs, where n is the number of bits in the code.

The Binary Decoder

A binary decoder indicates when a certain binary number is on its inputs. Suppose you need to know when the binary number 1001 is present. An AND gate and inverters can be used to accomplish this, as shown in Figure 6–3. The idea is to make all the inputs of the AND gate HIGH when the particular code is on the inputs. For the 0s (LOWs) in the code, inverters are used to make the associated AND gate inputs 1 (HIGH), as shown. For this particular decoding logic, the output is active-HIGH. That is, the output is 1 when the binary number 1001 is on its inputs, and the output is 0 when 1001 is not present. If an active-LOW output is required, a NAND gate is used instead of the AND gate.

FIGURE 6–3

Decoding logic for binary 1001.

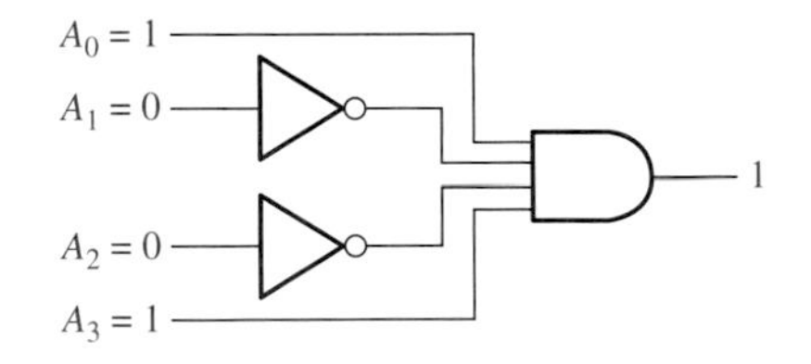

EXAMPLE 6–4

Problem

Show the decoding logic to detect the binary number 0111 on its inputs and produce an active-HIGH output.

Solution

The decoding logic is shown in Figure 6–4. Notice that one inverter is required for the 0 input.

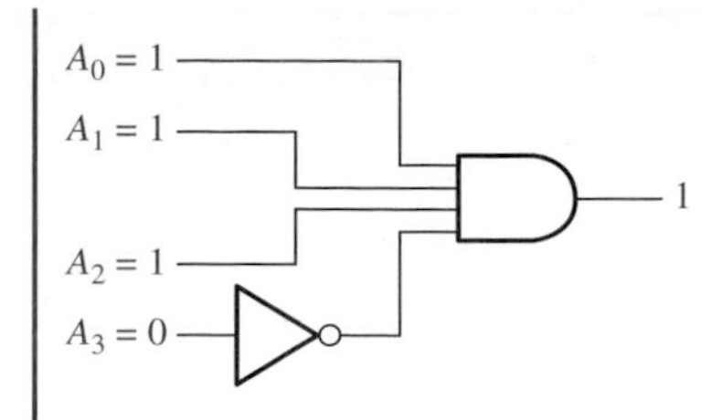

FIGURE 6–4

Question
Show the logic circuit for decoding 0101 with an active-LOW output.

The Binary-to-Decimal Decoder

A 4-bit binary-to-decimal decoder has decoding logic for each of the sixteen bit combinations that are possible with four bits ($2^4 = 16$). It has four inputs and sixteen outputs, as shown in Figure 6–5. Each of the sixteen outputs comes from an AND gate that decodes the associated binary number and produces an active-HIGH output. The logic consists of sixteen AND gates and four inverters that produce the complements of the binary inputs. Each inverter output is connected to the appropriate AND gates.

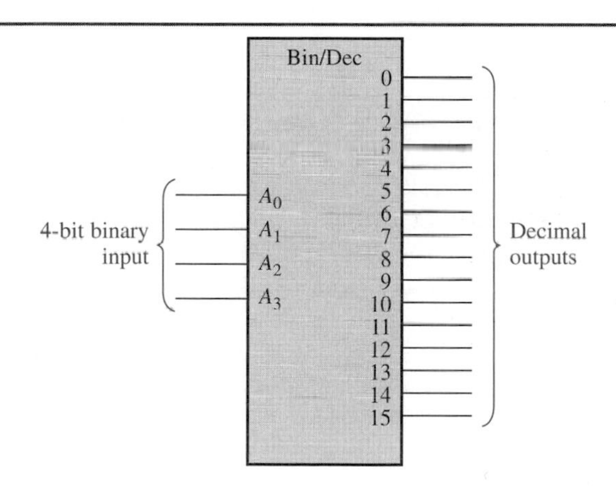

FIGURE 6–5

Basic logic symbol for a binary-to-decimal (4-line-to-16-line or 1-of-16) decoder with active-HIGH outputs.

This type of decoder is usually referred to as a binary-to-decimal decoder, a 4-line-to-16-line decoder, or a 1-of-16 decoder. The term *4-line-to-16-line* refers to the number of inputs (4) and the number of outputs (16). The term *1-of-16* refers to the fact that only one of the sixteen outputs is active for each specific input. The binary-to-decimal decoder can be expanded to decode binary numbers with more than four bits, as you will see in Section 6–5.

HISTORICAL NOTE

The IBM 1620 computer was marketed in 1959; it was a solid-state machine built around a BCD architecture. Each BCD 4-bit word included a parity bit and a flag bit that marked the beginning of a new number. The memory was 20,000 to 60,000 decimal digits, depending on how it was configured. Addition was performed by a look-up table in the memory. The 1620 was widely used in colleges and universities.

The BCD-to-Decimal Decoder

The BCD-to-decimal decoder is also called a 4-line-to-10-line decoder or a 1-of-10 decoder for obvious reasons. This decoder is basically the same as the binary-to-decimal decoder except that it has ten outputs instead of sixteen. A basic logic symbol is shown in Figure 6–6.

FIGURE 6–6

Basic logic symbol for a BCD-to-decimal decoder with active-HIGH outputs.

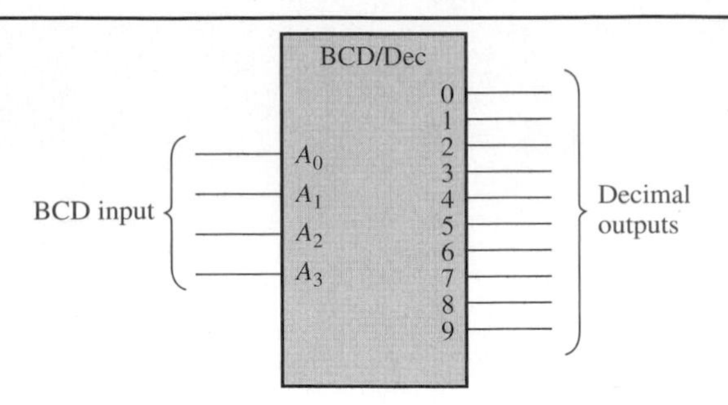

EXAMPLE 6–5

Problem

What are the output waveforms if the input waveforms in Figure 6–7(a) are applied to the BCD-to-decimal decoder in Figure 6–6?

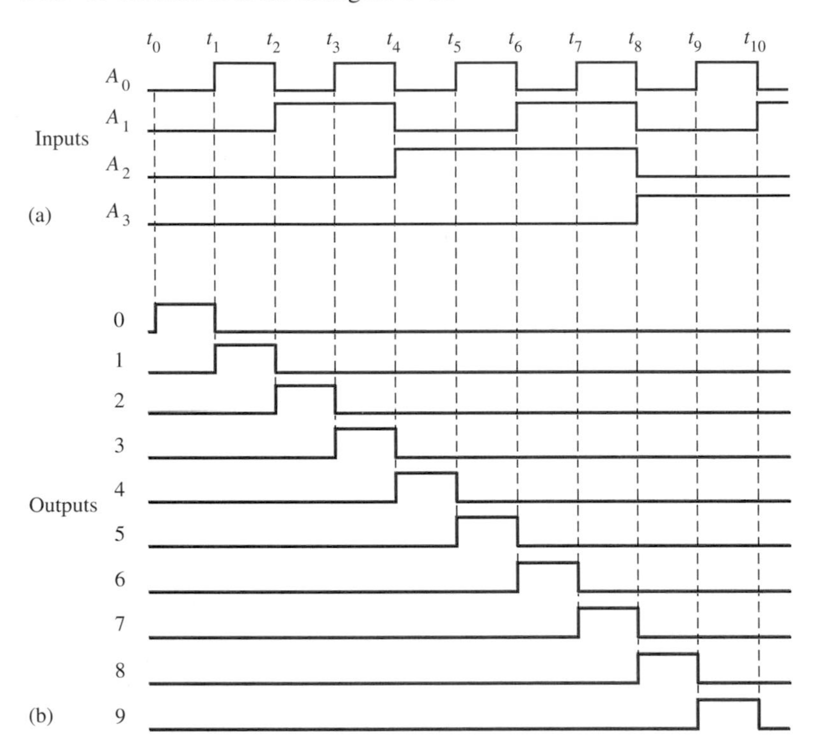

FIGURE 6–7

Solution

The output waveforms are shown in Figure 6–7(b). The input waveforms sequence through the BCD digits 0 through 9. The output waveforms indicate the BCD sequence. A 1 (HIGH) output indicates that the corresponding BCD code is on the inputs.

Question

When the "7" output is HIGH, what code is on the inputs?

COMPUTER SIMULATION

Open the Multisim file F06-08DG on the website. This is a 2-bit binary-to-decimal decoder. To apply a 1 or a 0 to an input, operate the switches with the *L* key and the *M* key. *L* is the least significant bit, and *M* is the most significant bit. The decoded number is indicated by a red probe light.

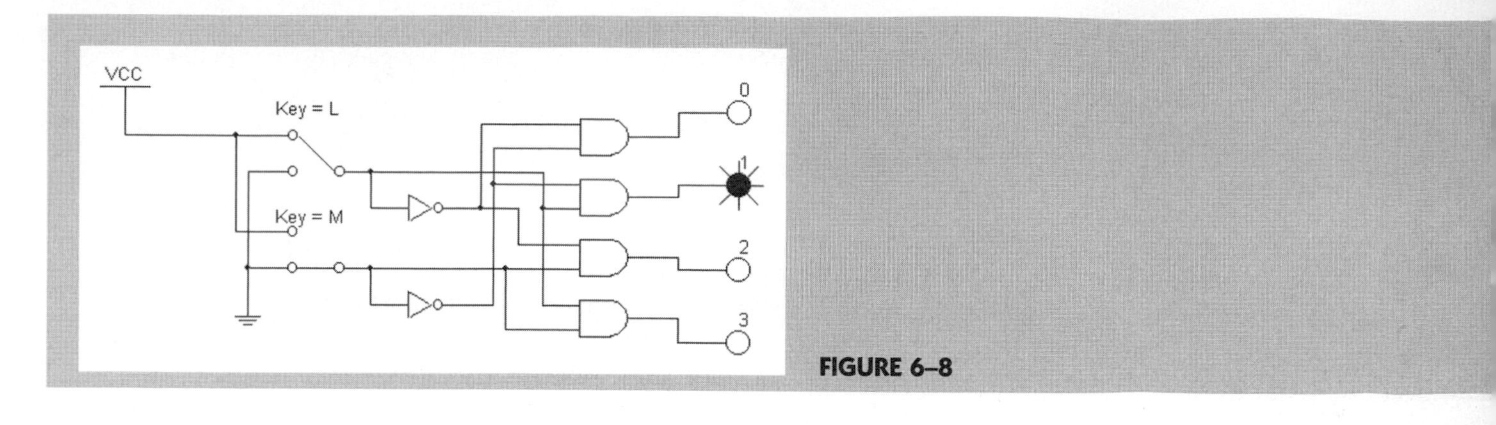

FIGURE 6–8

The BCD-to-7-Segment Decoder

The BCD-to-7-segment decoder drives a 7-segment display to produce a decimal readout corresponding to the BCD code on the inputs. Seven-segment displays are common in all types of instruments and appliances that you see every day. A **7-segment display** can be either an LED (light-emitting diode) or LCD (liquid crystal display) that consists of seven separate elements arranged to form the decimal digits 0 through 9, as shown in Figure 6–9(a). Each segment in the display is designated by a letter, as indicated in Figure 6–9(b).

(a) (b)

FIGURE 6–9 The 7-segment display showing the activated segments for each decimal digit.

A BCD-to-7-segment decoder connected to a 7-segment display is shown in Figure 6–10. There are seven decoder outputs that are connected to each of the seven segments in the display element. For each 4-bit BCD code, the decoder activates the appropriate segments in the display. For example, segments *b* and *c* are activated when a 0001 (decimal 1) is on the decoder inputs. Segments *a, b, c, d,* and *g* are activated when a 0011 (decimal 3) is on the decoder inputs.

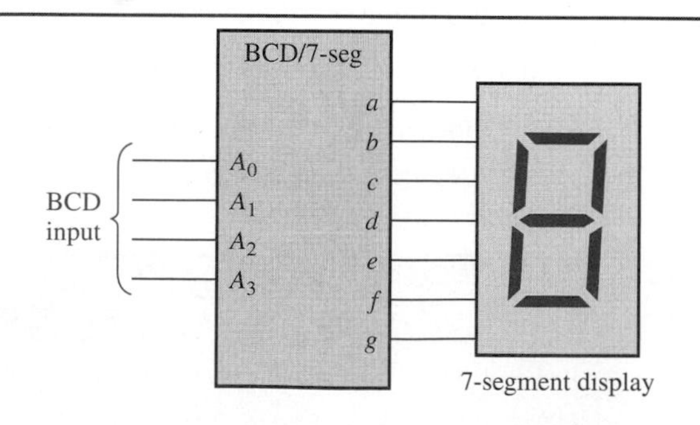

FIGURE 6–10 Block diagram of a BCD-to-7-segment decoder and display.

There are seven decoding circuits, one for each segment, in the BCD-to-7-segment decoder. To understand the decoding logic, let's go through the logic for one segment. Notice in Figure 6–9, that segment *a* is used in eight of the digits (0, 2, 3, 5, 6, 7, 8, and 9). A Boolean expression for the segment-*a* logic in terms of the BCD inputs is written as follows:

$$a = 0000 + 0010 + 0011 + 0101 + 0110 + 0111 + 1000 + 1001$$
$$= \overline{A}_3\overline{A}_2\overline{A}_1\overline{A}_0 + \overline{A}_3\overline{A}_2A_1\overline{A}_0 + \overline{A}_3\overline{A}_2A_1A_0 + \overline{A}_3A_2\overline{A}_1A_0 + \overline{A}_3A_2A_1\overline{A}_0 + \overline{A}_3A_2A_1A_0 + A_3\overline{A}_2\overline{A}_1\overline{A}_0 + A_3\overline{A}_2\overline{A}_1A_0$$

This expression is simplified on a Karnaugh map, as shown in Figure 6–11(a), and the resulting logic is shown in Figure 6–11(b).

The simplified expression for the segment-*a* logic from Figure 6–11 is

$$a = \overline{A}_3A_1 + \overline{A}_3\overline{A}_2A_1 + \overline{A}_3A_2A_0 + A_3\overline{A}_2\overline{A}_1$$

The logic in Figure 6–11(b) is just for segment *a* of the 7-segment decoder. Logic for each of the other six segments can be developed in a similar way.

FIGURE 6–11

Segment-*a* logic for the BCD-to-7-segment decoder.

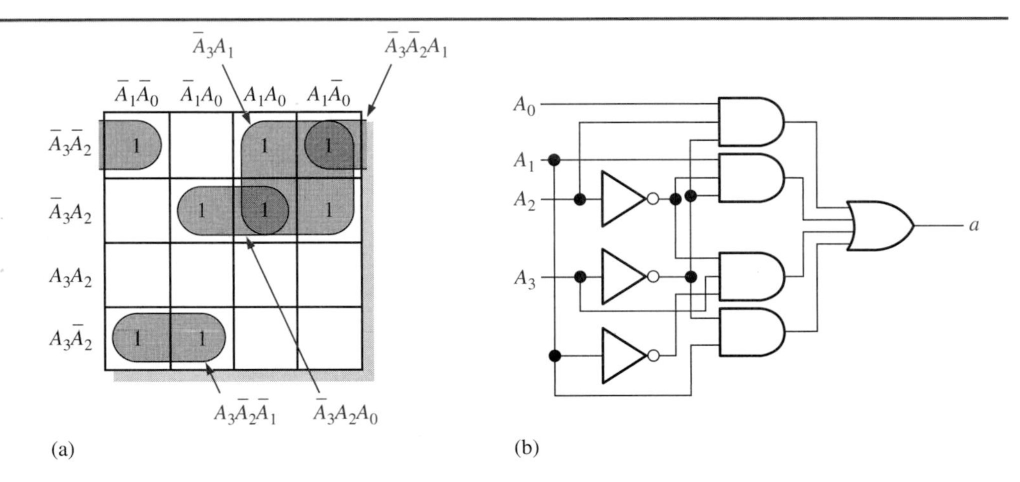

In many types of applications, a binary code must be converted to BCD for display with a 7-segment element. Code converters of this type are fairly complex and are generally implemented with programmable logic devices (PLDs). A PLD that is preprogrammed by the manufacturer to perform a specified function, such as binary-to-BCD conversion, is called an ASIC (Application Specific Integrated Circuit).

Review Questions

6. What does a binary-to-decimal decoder do?
7. What is a 4-line-to-16 line decoder?
8. What is a 1-of-10 decoder?
9. What does a BCD-to-7-segment decoder do?
10. Which segments of a 7-segment display are activated to form a 3?

6–3 ENCODERS

Encoders produce coded representations of input numbers or characters. A common type of encoder is one that converts decimal inputs to BCD outputs.

In this section, you will learn how an encoder works.

An **encoder** is a combinational logic circuit that produces a code representing the particular input that is active. It is designed to perform essentially a "reverse" decoder operation. An encoder generally has inputs that represent a number system such as decimal or octal and outputs that produce a code representing the numerical input.

The Decimal-to-BCD Encoder

This type of encoder has nine inputs, one for each decimal digit except 0, and a 4-bit BCD output, as shown in Figure 6–12. This encoder is sometimes referred to as a 10-line-to-4-line encoder because it can encode each of the ten decimal digits.

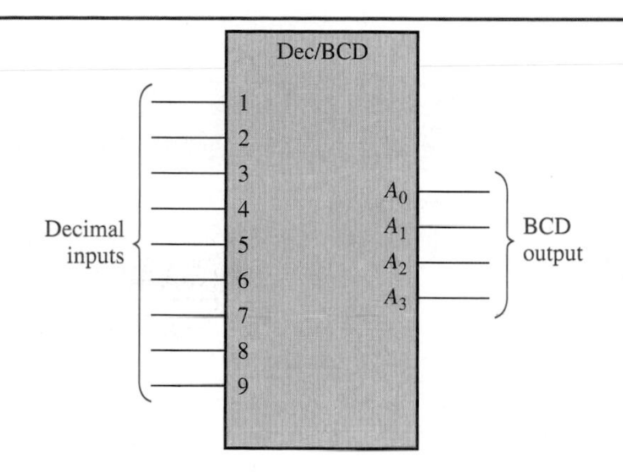

FIGURE 6–12

Logic symbol for a decimal-to-BCD encoder. The decimal input for 0 is not shown. It is not necessary because if none of the inputs is active, the BCD output is automatically zero.

The basic encoder logic is fairly simple. By determining which decimal inputs cause each bit of the BCD output to be a 1, you can understand the logic circuit that is required. This process is as follows, based on Table 6–4:

TABLE 6–4

Decimal digit	A_3	A_2	A_1	A_0
0	0	0	0	0
1	0	0	0	1
2	0	0	1	0
3	0	0	1	1
4	0	1	0	0
5	0	1	0	1
6	0	1	1	0
7	0	1	1	1
8	1	0	0	0
9	1	0	0	1

- The least significant BCD bit, A_0, is a 1 for decimal digits 1, 3, 5, 7, or 9. The OR expression for this output is

$$A_0 = 1 + 3 + 5 + 7 + 9$$

- The next BCD bit A_1 is a 1 for decimal digits 2, 3, 6, or 7. The OR expression for this output is

$$A_1 = 2 + 3 + 6 + 7$$

- The BCD bit A_2 is a 1 for decimal digits 4, 5, 6, or 7. The OR expression for this output is

$$A_2 = 4 + 5 + 6 + 7$$

- The most significant BCD bit, A_3, is a 1 for decimal digits 8 or 9. The OR expression for this output is

$$A_3 = 8 + 9$$

The encoder can be implemented from these expressions using four OR gates, as shown in Figure 6–13.

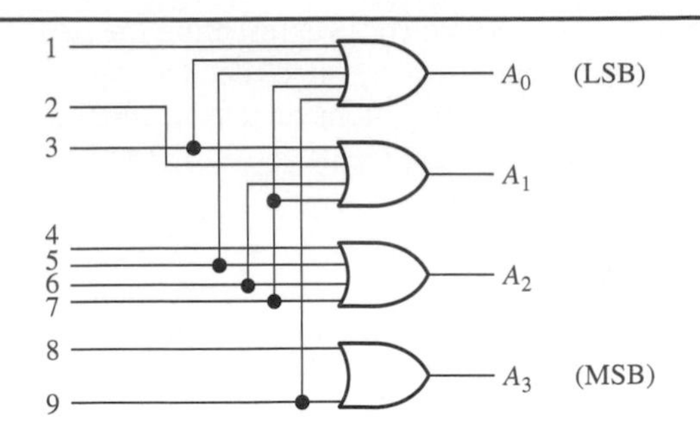

FIGURE 6–13

Logic diagram for a decimal-to-BCD encoder. The decimal digit 0 is not included because the BCD output bits are all 0 when all inputs 1 through 9 are 0.

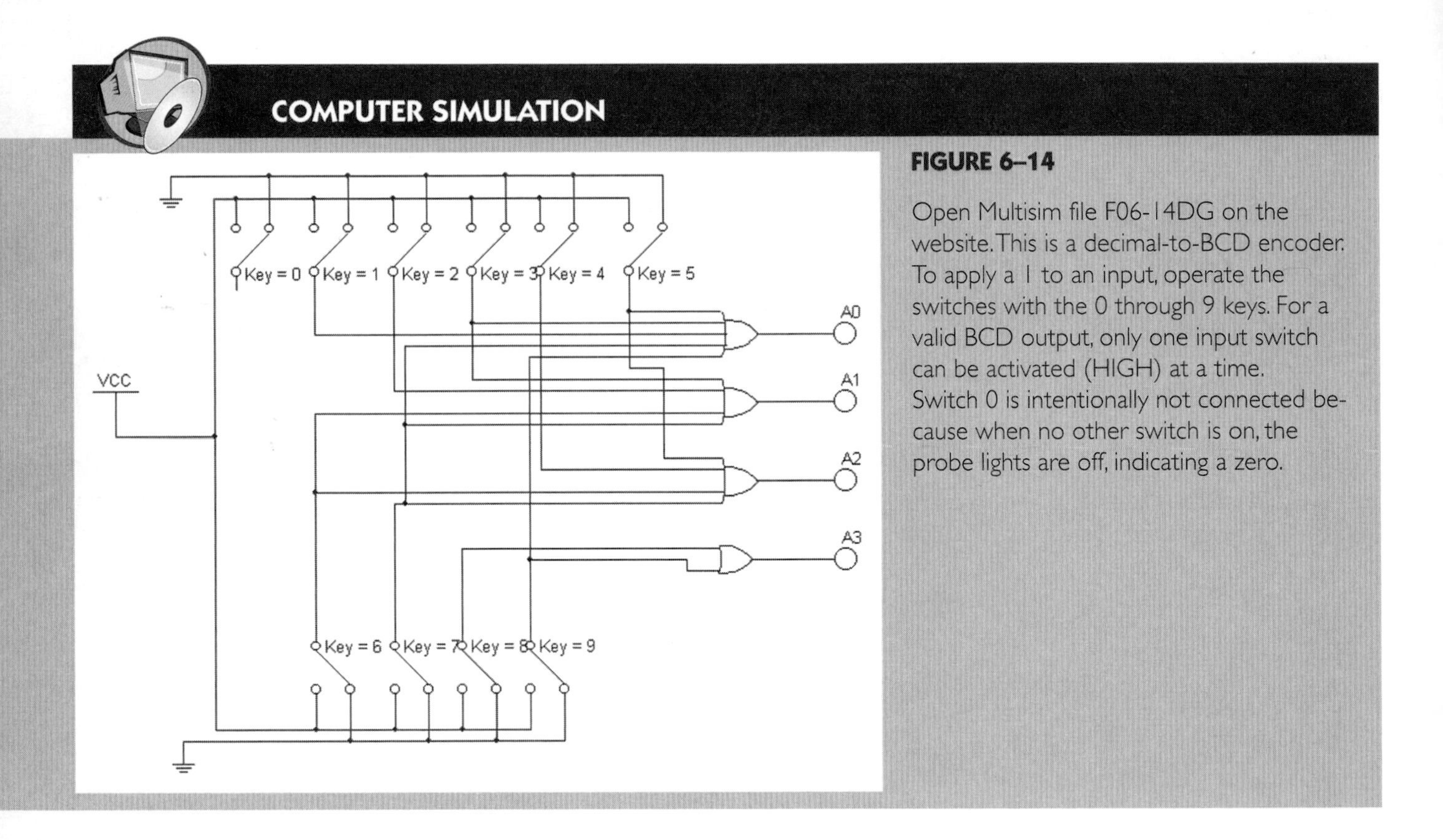

FIGURE 6–14

Open Multisim file F06-14DG on the website. This is a decimal-to-BCD encoder. To apply a 1 to an input, operate the switches with the 0 through 9 keys. For a valid BCD output, only one input switch can be activated (HIGH) at a time. Switch 0 is intentionally not connected because when no other switch is on, the probe lights are off, indicating a zero.

The Priority Encoder

As you have seen, the decimal-to-BCD encoder in Figure 6–13 produces a valid BCD output if one and only one of the decimal inputs is active (1). If more than one input is active at the same time, the output is invalid. Integrated circuit encoders usually include an additional logic circuit, called priority logic, that allows the BCD output to show only the highest-value decimal input. Any other lower-value decimal input that may be activated at the same time will be ignored.

A **priority encoder** is an encoder that ignores all active inputs except the one with the highest value. For example, if decimal inputs 8 and 5 are active at the same time, the BCD output will indicate 1000, which represents the 8. The logic circuits required for this make the priority encoder much more complex than the one in Figure 6–13. A logic symbol for a decimal-to-BCD priority encoder is shown in Figure 6–15. The HPRI (high priority) label means that the highest-value decimal input has priority over any others that may be active.

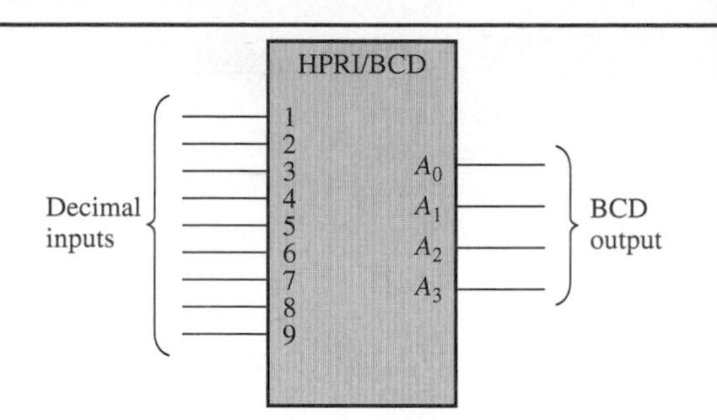

FIGURE 6–15

Logic symbol for a decimal-to-BCD priority encoder.

EXAMPLE 6–6

Problem

(a) Determine the output of a decimal-to-BCD encoder like the one in Figure 6–13 if the 7 and the 9 inputs are active.

(b) Determine the output of a decimal-to-BCD priority encoder if the 7 and the 9 inputs are active.

Solution

(a) When more than one input of a decimal-to-BCD encoder are active, the output is *invalid.* In this case, the output would be $A_3A_2A_1A_0 = 1111$, reflecting both inputs.

(b) For the priority encoder, the 9 input is the highest priority. The output is $A_3A_2A_1A_0 = 1001$, which is the valid BCD code for 9. The 7 input is ignored in this case.

Question

The 1 and 2 inputs of a priority encoder are active. What is the BCD output?

Review Questions

11. What is the purpose of an encoder?
12. What is a 10-line-to-4-line encoder?
13. What is a priority encoder?
14. If the 1 and 5 inputs of a priority encoder are active, what is the BCD output?
15. A priority encoder has 0111 on its outputs. Which input is active?

DATA SELECTORS 6–4

In general, a data selector is a combinational logic circuit that switches digital data from two or more sources to one line or from one line to two or more destinations.

In this section, you will learn about two types of data selectors, the multiplexer and the demultiplexer.

Two basic types of data selectors are the multiplexer (MUX) and the demultiplexer (DEMUX). Multiplexing/demultiplexing allows data to be switched from many parallel lines by the multiplexer and transmitted over a single line to a destination point where the data are then switched by the demultiplexer back onto parallel lines. Switching of data is controlled on a time basis by a process called time-division multiplexing, where the data from each source are switched at different times.

The Multiplexer

A **multiplexer** is a type of data selector that accepts data on two or more data input lines and routes that data, in a time sequence, to a single output line. The time sequence is determined by a set of data select inputs. A multiplexer can be thought of as a parallel-to-serial data converter because it takes data in parallel on the input lines and produces a serial stream of data on the output line.

For illustration, the 4-input multiplexer shown in Figure 6–16 has two data select lines because it takes two bits to select any of the four inputs ($2^2 = 4$). The D_0 input is selected when the data select inputs are $S_1S_0 = 00$, and the data on that input appears on the output. Similarly, the D_1 input is selected when $S_1S_0 = 01$, the D_2 input is selected when $S_1S_0 = 10$, and the D_3 input is selected when $S_1S_0 = 11$. The Boolean expression for this particular logic circuit is

$$\text{Output} = D_0\overline{S}_1\overline{S}_0 + D_1\overline{S}_1S_0 + D_2S_1\overline{S}_0 + D_3S_1S_0$$

FIGURE 6–16

A 4-input multiplexer.

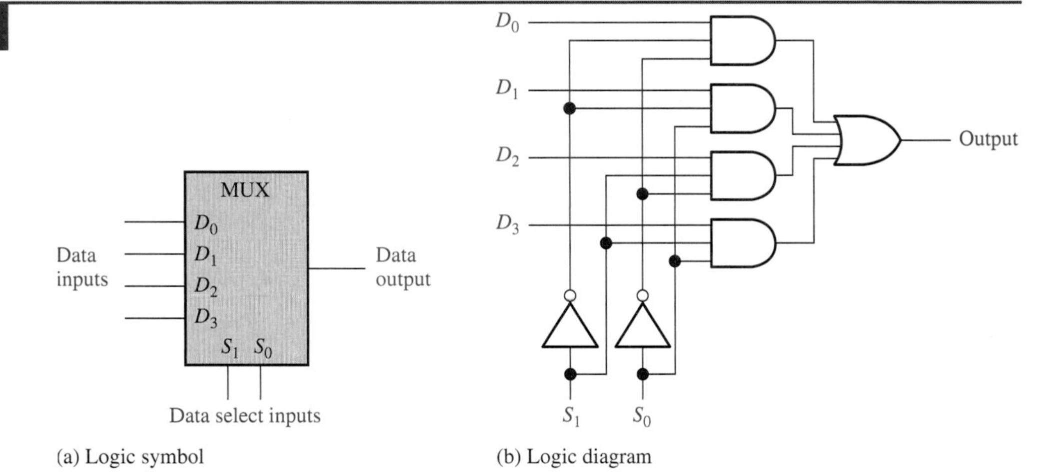

EXAMPLE 6–7

Problem

The data inputs and data select inputs shown in Figure 6–17(a) are applied to the multiplexer in Figure 6–16. Determine the output waveform showing the proper time relationship to the inputs.

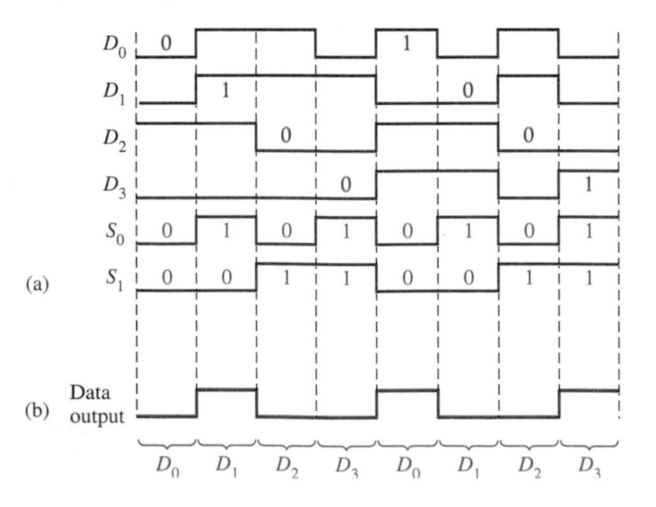

FIGURE 6–17

Solution

The binary states of the data select inputs during each time interval (marked by dashed lines) determine which data input is selected. In this case, the data select inputs go through a binary sequence 00, 01, 10, 11 and then repeat. The resulting output waveform is shown in Figure 6–17(b).

Question

If the multiplexer data inputs are $D_0 = 1$, $D_1 = 0$, $D_2 = 1$, and $D_3 = 0$ and the data select inputs are $S_0 = 1$ and $S_1 = 0$, what is the output?

Another application for a MUX is to implement a combinational logic circuit directly from a truth table. Each output row on a truth table has a corresponding input line on the MUX. If the output shown on any line on the truth table is logic 1, the input is connected to a HIGH; otherwise, it is connected to ground (logic 0). Figure 6–18 illustrates the idea. The first line of the truth table is highlighted. This line is selected when the inputs (*CBA*), which are connected to the Select lines on the MUX, are all LOW (000). The truth table indicates the output should be a 1 for this case, so the D_0 line is connected to a HIGH. The input variables choose D_0, and it is routed to the output as required. The enable input (*EN*) is commonly used on ICs to disable or enable the device. Sometimes *E* is used to designate an enable input.

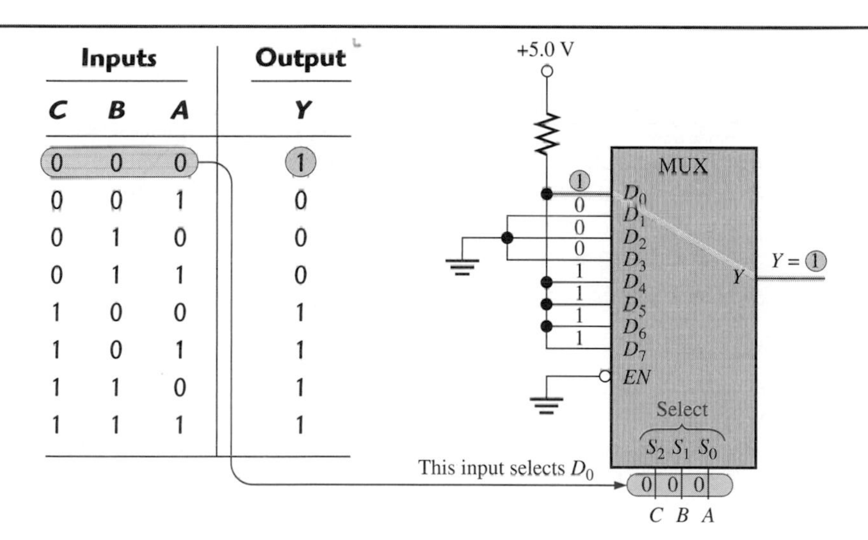

Inputs			Output
C	B	A	Y
0	0	0	1
0	0	1	0
0	1	0	0
0	1	1	0
1	0	0	1
1	0	1	1
1	1	0	1
1	1	1	1

FIGURE 6–18
Implementing combinational logic with a MUX.

When the input logic variables (*CBA*) change, a different line is selected but still corresponds to the truth table. Thus, the truth table is implemented in a single integrated circuit. The complete SOP expression for *Y* is

$$\overline{C}\,\overline{B}\,\overline{A} + C\overline{B}\,\overline{A} + C\overline{B}A + CB\overline{A} + CBA$$

You will have an opportunity to investigate this in Experiment 9 of the lab exercises.

COMPUTER SIMULATION

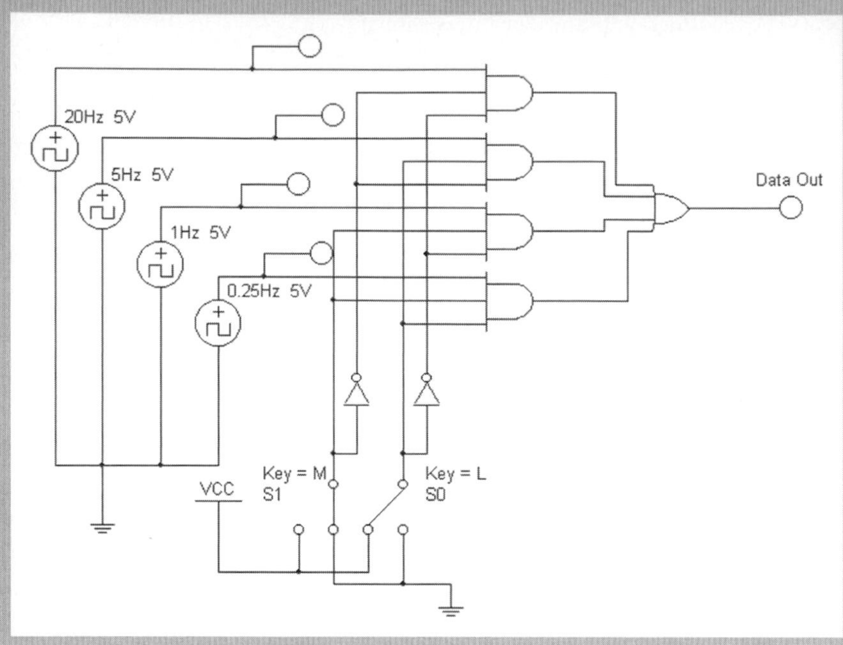

FIGURE 6–19

Open the Multisim file F06-19DG on the website. This is a 4-input multiplexer. The data inputs are pulses at different frequencies which are low enough to be visible via the blinking probe lights. Select the data select inputs by the *M* and *L* keys. You can select any one of the four inputs and observe that the output probe light blinks at the same rate as the selected inputs. This visually shows how a multiplexer works.

The Demultiplexer

A **demultiplexer** accepts data from one data input line and routes that data, in a time sequence, to two or more output lines. Like in the multiplexer, the time sequence is determined by a set of data select inputs. A demultiplexer can be considered a serial-to-parallel data converter because it takes a serial stream of data on the input line and switches the data to parallel output lines. For a MUX/DEMUX system, the data select inputs for the DEMUX must be synchronized with those of the MUX.

For illustration, the 4-output demultiplexer shown in Figure 6–20 has two data select lines because it takes two bits to select any of the four outputs ($2^2 = 4$). When the data select inputs are $S_1S_0 = 00$, the input data are switched to the D_0 output. Similarly, when $S_1S_0 = 01$, the input data are switched to the D_1 output. When $S_1S_0 = 10$, D_2 is selected; and when $S_1S_0 = 11$, D_3 is selected.

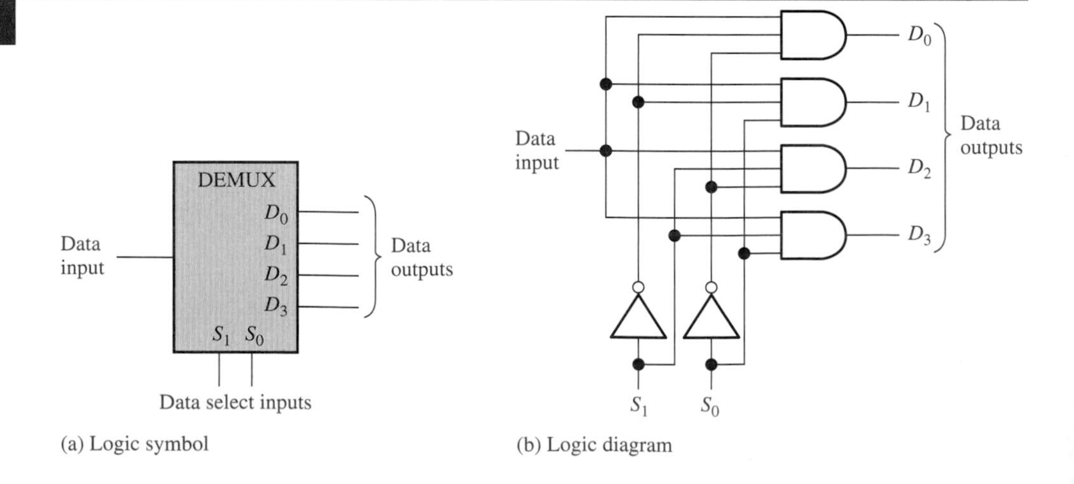

FIGURE 6–20

A 4-output demultiplexer.

EXAMPLE 6–8

Problem

The data input and data select inputs shown in Figure 6–21(a) are applied to the demultiplexer in Figure 6–20. Determine the data output waveforms showing the proper time relationship to the data input and data select inputs.

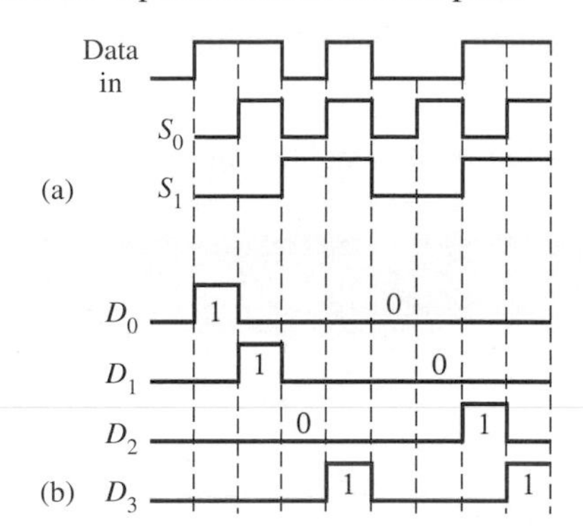

FIGURE 6–21

Solution

The binary states of the data select inputs during each time interval (marked by dashed lines) determine which data output is selected. In this case, the data select inputs go through a binary sequence 00, 01, 10, 11 and then repeat. The resulting output waveforms are shown in Figure 6–21(b).

Question

If the demultiplexer data input is constantly 1 and the data select inputs are $S_0 = 0$ and $S_1 = 1$, what are the outputs?

COMPUTER SIMULATION

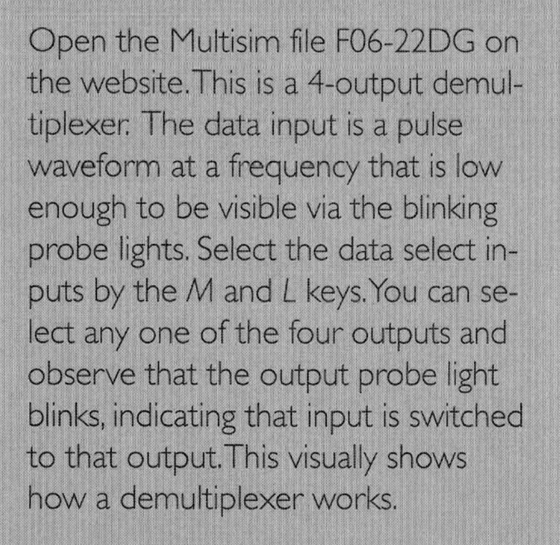

FIGURE 6–22

Open the Multisim file F06-22DG on the website. This is a 4-output demultiplexer. The data input is a pulse waveform at a frequency that is low enough to be visible via the blinking probe lights. Select the data select inputs by the *M* and *L* keys. You can select any one of the four outputs and observe that the output probe light blinks, indicating that input is switched to that output. This visually shows how a demultiplexer works.

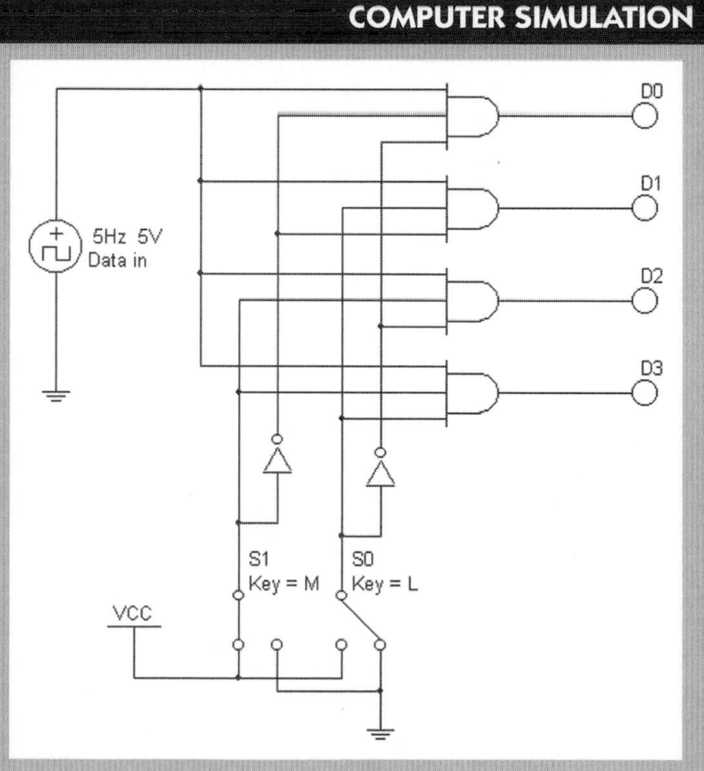

Review Questions

16. What does a multiplexer do?
17. What is the purpose of the data select inputs?
18. How many select inputs are required for an 8-bit multiplexer?
19. What does a demultiplexer do?
20. What are the abbreviations for multiplexer and demultiplexer?

6–5 INTEGRATED CIRCUITS

Several types of decoders, encoders, and data selectors are available as ICs. Generally, they come in several logic families, including LS, HC, and AHC. Certain devices may be available in only one family.

In this section, you will learn about decoder, encoder, and data selector ICs and observe computer simulations of decoder and MUX/DEMUX operations.

A BCD-to-Decimal Decoder

The logic symbol and the pin diagram for a 74LS42 BCD-to-decimal (1-of-10) decoder are shown in Figure 6–23. Notice that the outputs are all active-LOW, as indicated by the bubbles. When a given BCD code is on the inputs, the associated output is LOW and the other outputs are HIGH.

FIGURE 6–23

The 74LS42 BCD-to-decimal (1-of-10) decoder.

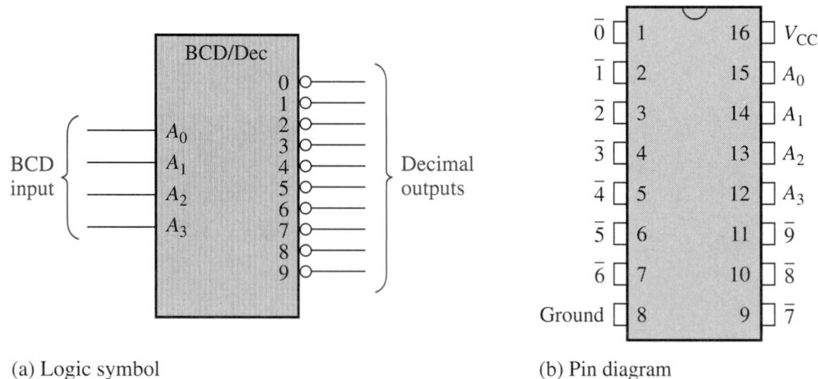

A BCD-to-7-Segment Decoder/Driver

The 74LS47 is one example of a BCD-to-7-segment decoder. The logic symbol and pin diagram are shown in Figure 6–24. The outputs are active-LOW and are designed to drive a

FIGURE 6–24

The 74LS47 BCD-to-7-segment decoder/driver.

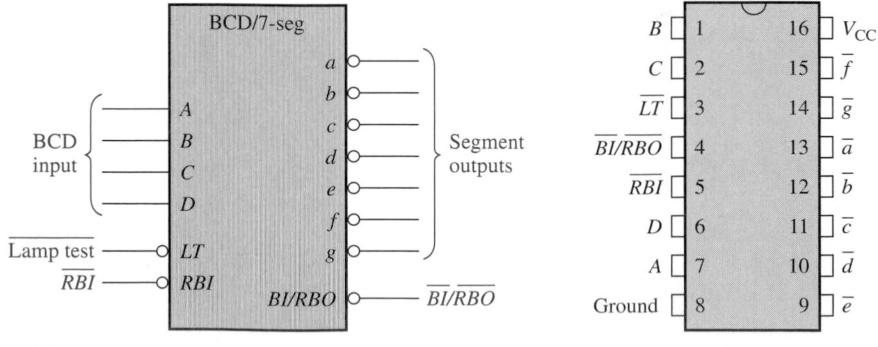

7-segment display. In addition to the BCD input lines and the segment outputs, there are three other lines: the *LT, RBI,* and *BI*/*RBO*. The *LT* is the lamp test that is used to light all display segments. The *RBI* input and *BI*/*RBO* output are for blanking out leading or trailing zeros in a multiple display application.

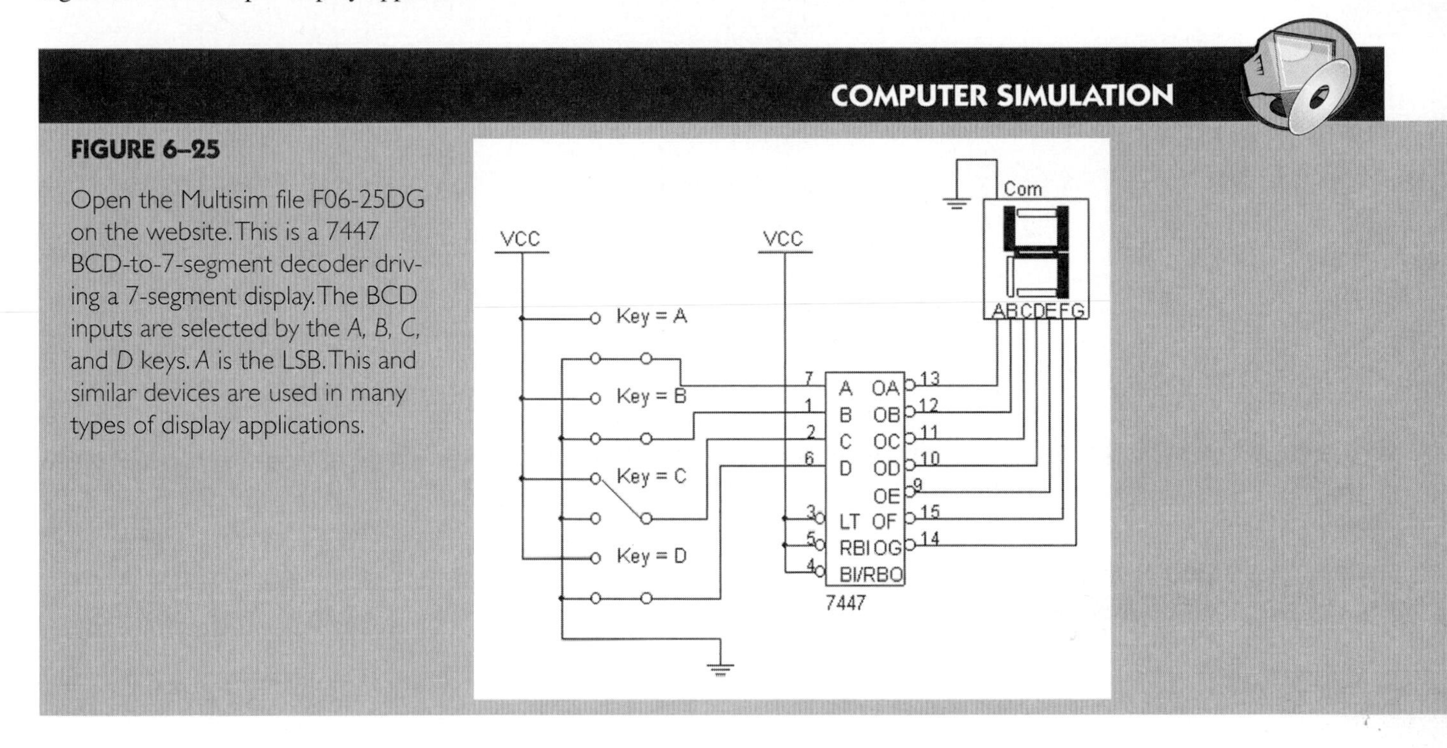

FIGURE 6–25

Open the Multisim file F06-25DG on the website. This is a 7447 BCD-to-7-segment decoder driving a 7-segment display. The BCD inputs are selected by the *A, B, C,* and *D* keys. *A* is the LSB. This and similar devices are used in many types of display applications.

Priority Encoders

The 74HC147 is an example of a priority encoder, which as you recall, produces the BCD code for the highest active decimal input. For this particular device, shown in Figure 6–26, the inputs and outputs are active-LOW, as indicated by the bubbles. There are nine actual inputs because a decimal zero input is implied when all the nine inputs are HIGH (not active).

Another type of priority encoder available in IC form is the 74LS148, which is an 8-line-to-3-line encoder for octal to 3-bit binary encoding.

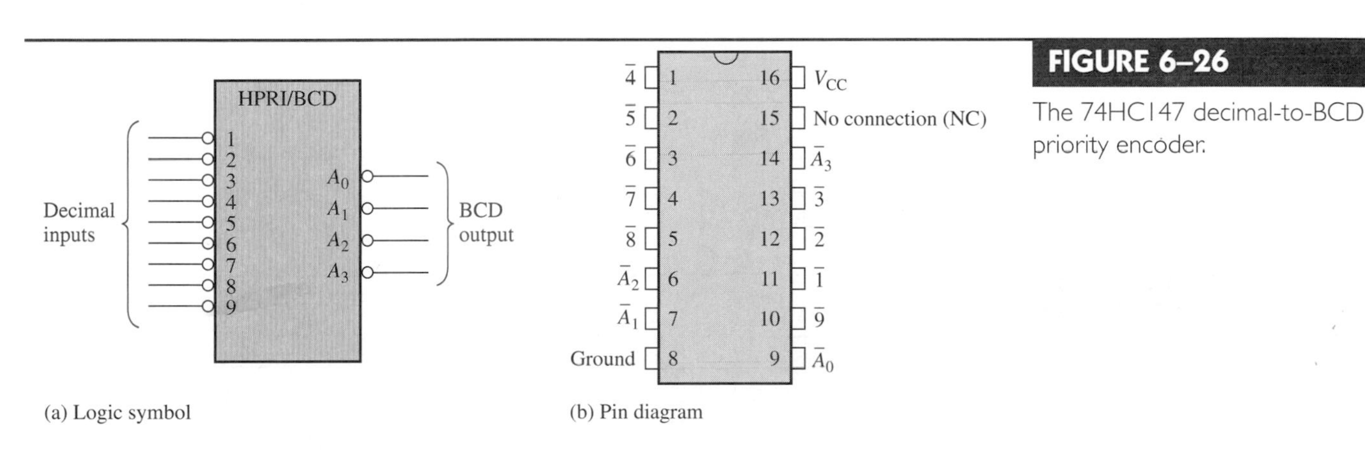

FIGURE 6–26

The 74HC147 decimal-to-BCD priority encoder.

An 8-Input Multiplexer

As shown in Figure 6–27, the 74LS151 multiplexer has eight input lines. It takes three data select lines to select any one of the eight inputs ($2^3 = 8$). Also, this device has an enable input (E) for completely enabling or disabling the chip. A LOW enables the multiplexer. In addition to the single output line, there is a complemented output.

FIGURE 6–27

The 74LS151 8-input multiplexer.

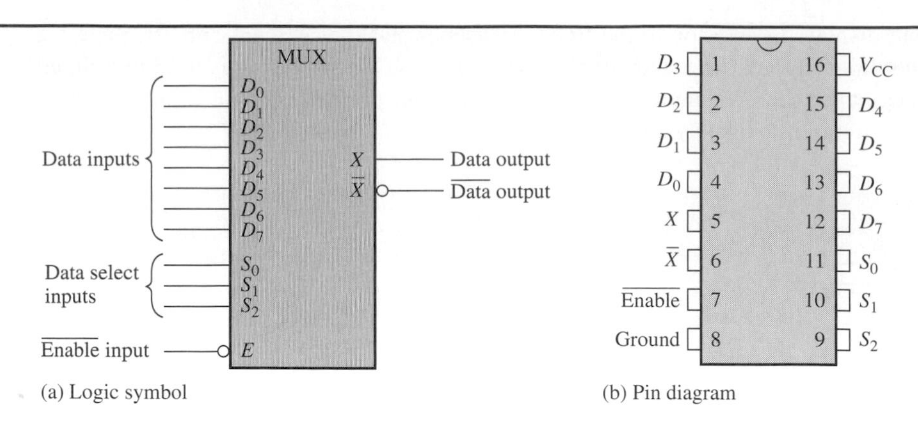

(a) Logic symbol

(b) Pin diagram

Multiplexer Expansion

Two 74LS151 8-input multiplexers can be connected to form a 16-input multiplexer, as shown in Figure 6–28. To select any one of the sixteen inputs, four data select lines are required ($2^4 = 16$). The enable input (E) is used as the fourth select input. When E is 0, MUX 1 is enabled and MUX 2 is disabled; so the data select inputs S_2, S_1, and S_0 select the inputs to MUX 1. When E is 1, MUX 1 is disabled and MUX 2 is enabled; so S_2, S_1, and S_0 select the inputs to MUX 2. The outputs from each multiplexer are ORed to form the single output.

FIGURE 6–28

Two 74LS151 multiplexers connected to form a 16-input multiplexer.

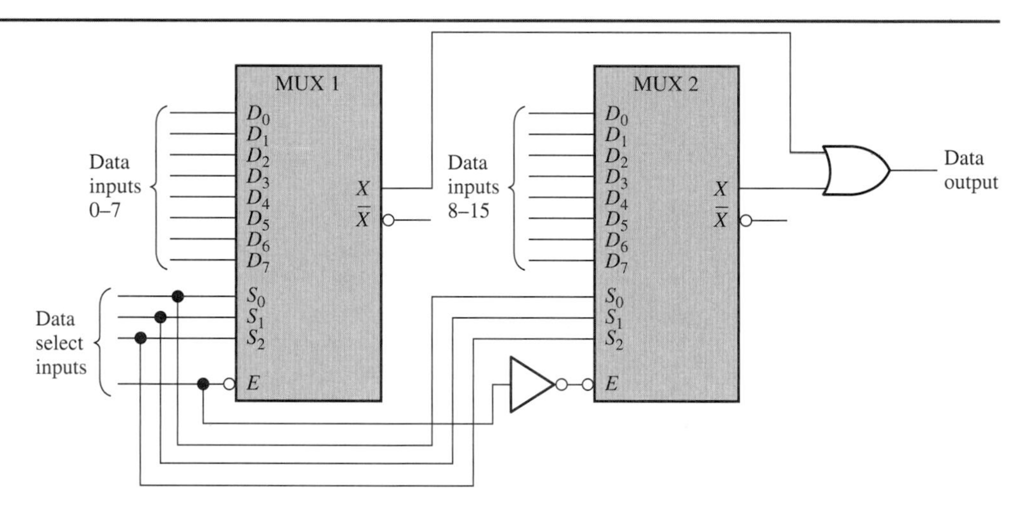

An 8-Output Demultiplexer

The 74LS138 demultiplexer has two active-LOW enable inputs in addition to the three data select inputs, the basic data input (E_3) and the eight active-LOW outputs (D_0–D_7), as shown in Figure 6–29. Both enable inputs, $\overline{E}_1$ and $\overline{E}_2$, must be LOW to enable the outputs. When they are HIGH, the outputs are held HIGH regardless of what any of the other inputs are.

FIGURE 6–29

The 74LS138 8-output demultiplexer.

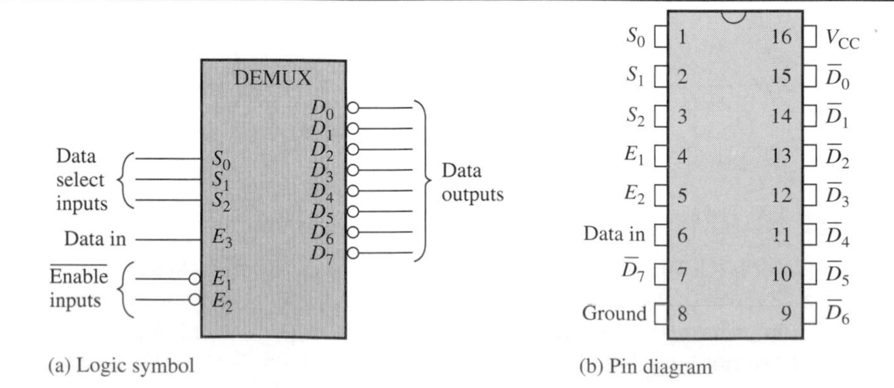

(a) Logic symbol

(b) Pin diagram

The 74LS138 also functions as a 1-of-8 decoder when E_1 and E_2 are LOW and E_3 is HIGH. When used as a decoder, E_3 functions as an active-HIGH enable input; but when used as a demultiplexer, E_3 functions as the data input.

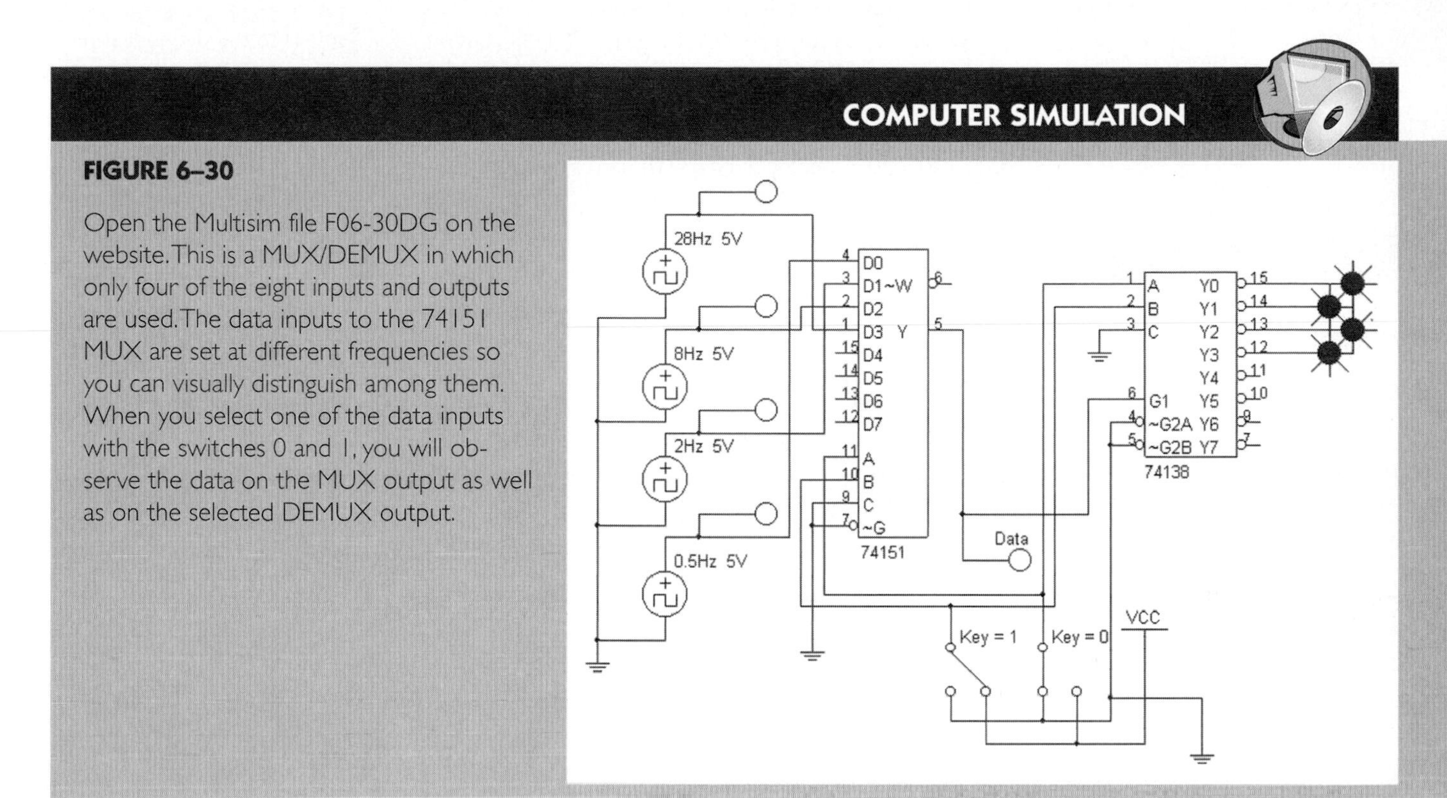

FIGURE 6–30

Open the Multisim file F06-30DG on the website. This is a MUX/DEMUX in which only four of the eight inputs and outputs are used. The data inputs to the 74151 MUX are set at different frequencies so you can visually distinguish among them. When you select one of the data inputs with the switches 0 and 1, you will observe the data on the MUX output as well as on the selected DEMUX output.

Review Questions

21. To which pins do V_{CC} and ground connect on most 16-pin ICs?
22. How many data select inputs would be required for a 16-input multiplexer?
23. What is the purpose of an enable input?
24. What does the lamp test input on a BCD-to-7-segment decoder do?
25. If the data select inputs on the 74LS138 demultiplexer are $S_0 = 0$, $S_1 = 1$, $S_2 = 1$, on which output is the data?

TROUBLESHOOTING 6–6

As you know, opens and shorts in logic circuits affect the circuit operation. Generally, when an input or output is stuck HIGH or LOW, its effect can always be observed on the output.

In this section, you will learn how to recognize faults in decoders, encoders, and data selectors so that you can apply these observations to logic circuits in general.

Decoder Faults

When a decoder input is stuck HIGH or LOW, certain outputs will never be active because the faulty input will limit the number of code combinations that can be decoded. One way to test a decoder is to apply all possible code combinations to the inputs and observe the

SAFETY NOTE

When working with electrical circuits, always wear shoes and keep them dry. Do not stand on metal or wet floors. Never handle instruments when your hands are wet.

outputs. For example, if you apply all ten code combinations sequentially to a BCD-to-decimal decoder, you should see each output go to its active state when the corresponding code is on the inputs. A fault in any one of the inputs will result in one or more incorrect outputs at some point in the code sequence.

Figure 6–31 shows BCD code (0 through 9) sequentially applied to the decoder inputs as waveforms. Part (a) indicates the correct output waveforms, and part (b) shows the output waveforms if the least significant bit A_0 is stuck LOW. Notice that outputs 1, 3, 5, 7, and 9 never go to the active-HIGH state because the A_0 input is never HIGH. The other outputs, 0, 2, 4, 6, and 8 are each active-HIGH for two input codes.

FIGURE 6–31

The effect of a faulty input on decoder operation.

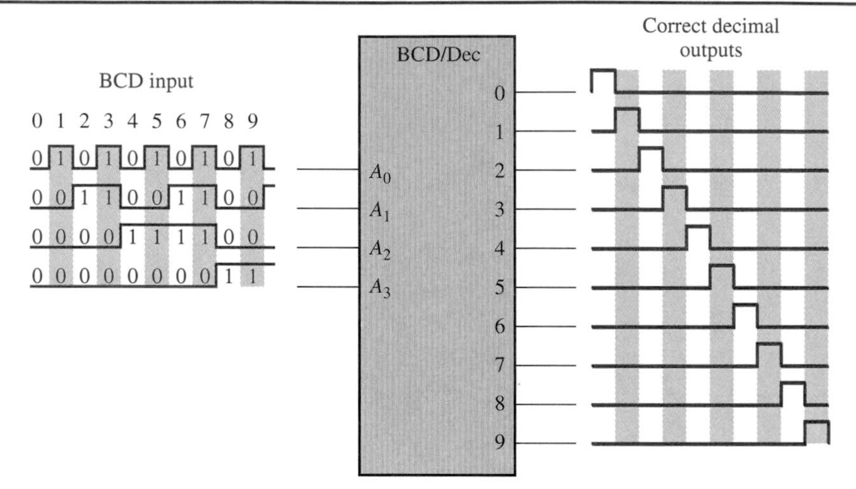

(a) Decoder operating properly

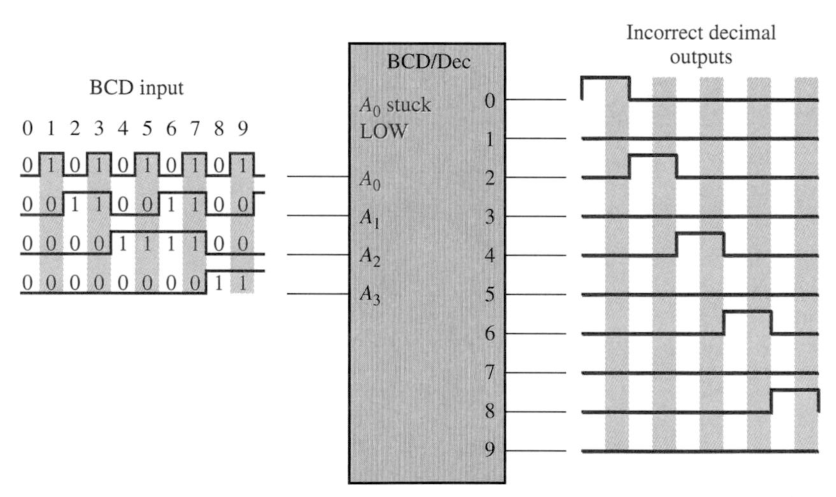

(b) Decoder with input A_0 stuck LOW

If a decoder output is faulty, only that particular output is affected. The other outputs will be correct. For example, if output 6 is stuck HIGH you will never see a pulse on it, but you will observe pulses on the other inputs when the corresponding codes are applied to the inputs.

Encoder Faults

Now, let's look at how a faulty input can affect the operation of a decimal-to-BCD priority encoder. The output code will be incorrect if the input fault results in an active level and it has a higher priority than the actual input. For example, suppose an encoder has active

HIGH inputs and the input for decimal 5 is stuck HIGH. The output code will be incorrect if the actual input is less than 5, but it will be correct if the actual input is 5 or greater. Figure 6–32 illustrates the encoder operation for this particular case.

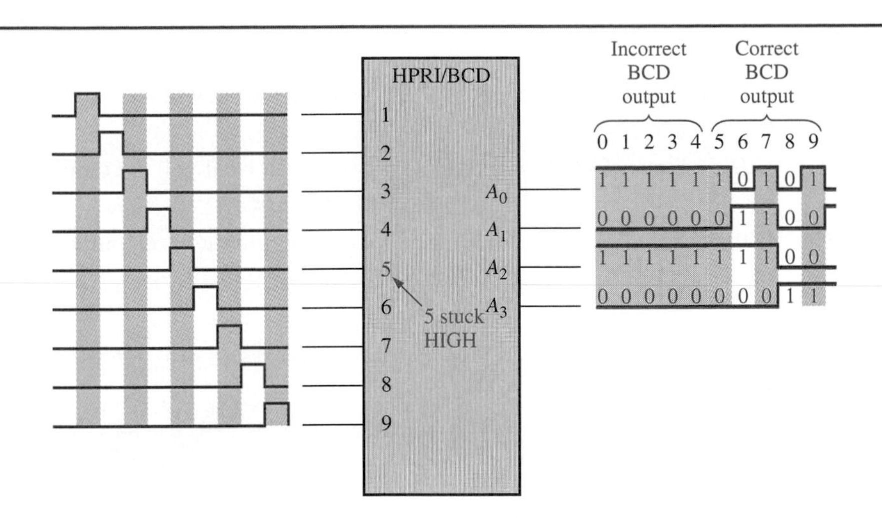

FIGURE 6–32

The effect of a faulty input on encoder operation.

As with other types of logic circuits, if a fault occurs in an encoder output, the faulty line is constantly HIGH or LOW and will not change no matter what the inputs are.

Data Selector Faults

As you know, multiplexers and demultiplexers are the basic types of data selectors. The multiplexer switches data on its inputs to the output in a time sequence based on the state of the select inputs. If a data input is stuck HIGH or stuck LOW, a constant level will be observed on the output when that data input is selected. If a select input is faulty, incorrect data will be switched to the output at certain times, depending on which select input is faulty.

To illustrate, let's examine a case where the S_1 select line in Figure 6–33 is stuck LOW. Assume that pulse waveforms of different frequencies are applied to the data inputs as shown. Although the four binary states, 00, 01, 10, and 11 are applied to the select inputs, the stuck LOW fault at S_1 results in only the 00 and the 01 being effective. By observing the data output, you can see that the multiplexer is switching back and forth between D_0 and D_1 without ever switching to D_2 and D_3.

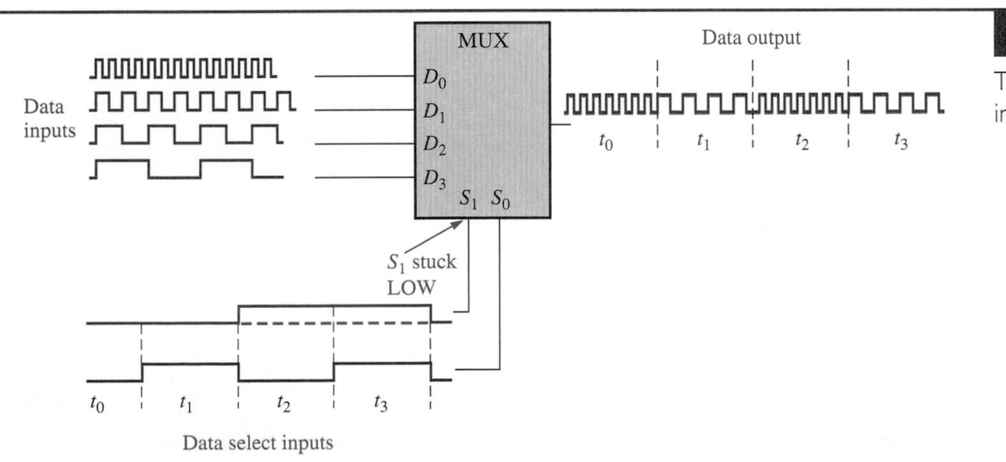

FIGURE 6–33

The effect of a faulty select input on multiplexer operation.

Review Questions

26. If you apply 0000 to the inputs of the decoder in Figure 6–31(a) and observe that all outputs are LOW, what are possible faults?
27. If you apply a HIGH to the 0 input of a priority encoder like the one shown in Figure 6–32 and observe that the outputs are binary 4 (0100), what are the possible faults?
28. If D_0 in Figure 6–33 is stuck LOW (S_1 normal), what will the output look like for the input and select waveforms shown?
29. If D_0 in Figure 6–33 is stuck HIGH (S_1 normal), what will the output look like for the input and select waveforms shown?
30. If the S_1 input in Figure 6–33 is stuck HIGH, what will the output be for the inputs shown?

CHAPTER REVIEW

Key Terms

ASCII The American Standard Code for International Interchange is a 7-bit binary code that represents both numeric and alphabetic characters as well as symbols and commands. It is commonly used as a keyboard entry code for computers.

Binary coded decimal Abbreviated BCD, this is a binary code that uses four bits to represent each of the ten decimal digits.

Decoder A logic circuit that produces an output corresponding to the particular code on its inputs.

Demultiplexer A logic circuit (abbreviated DEMUX) that moves serial data from a single input line onto several parallel output lines.

Encoder A logic circuit that produces a code representing the particular input that is active.

Multiplexer A logic circuit (abbreviated MUX) that moves data from several parallel input lines to a single output for serial transmission.

Parity bit A bit added to a given code to make the total number of bits either odd or even and used for error detection.

Priority encoder An encoder that ignores all active inputs except the one with the highest value.

7-segment display A display device that consists of seven segments that can be activated to form each of the ten decimal digits.

Important Facts

- ❑ Each digit in a decimal number can be converted directly into a 4-bit binary code known as BCD, which stands for binary coded decimal.
- ❑ ASCII (pronounced "askee") is an alphanumeric code used in computers.
- ❑ ASCII consists of 128 characters and symbols represented by a 7-bit code.
- ❑ A BCD-to-decimal decoder is also known as a 4-line-to-10 line or 1-of-10 decoder.

- A BCD-to-7-segment decoder has seven outputs that drive a 7-segment display.
- A 7-segment display can be either LED (light-emitting diode) or LCD (liquid crystal display).
- A priority encoder allows more than one input to be active at a time; however, the input with the highest value has priority, which means that it is the only one encoded.
- A multiplexer essentially converts parallel data from two or more input lines to serial data on one output line.
- A demultiplexer essentially converts serial data from one input line to parallel data on two or more output lines.

Chapter Checkup

Answers are at the end of the chapter.

1. BCD is used to represent
   (a) binary numbers (b) decimal numbers
   (c) hexadecimal numbers
2. The abbreviation BCD means
   (a) binary coded digit
   (b) binary complemented digit
   (c) binary coded decimal
3. To represent a decimal digit using BCD, it takes
   (a) four bits
   (b) two bits
   (c) a number of bits that depends on the digit
4. 12-bit BCD numbers can be used to represent decimal numbers up to
   (a) 4,095 (b) 1,024
   (c) 999
5. The abbreviation ASCII means
   (a) all-purpose serial code for integrated input
   (b) American standard code for information interchange
   (c) American secret code for international interchange
6. ASCII has
   (a) seven bits (b) twelve bits
   (c) four bits
7. ASCII is a type of
   (a) numeric code
   (b) alphanumeric code
   (c) alphabetic code
8. A BCD-to-decimal decoder has
   (a) 10 inputs and 4 outputs
   (b) 4 inputs and 10 outputs
   (c) 4 inputs and 16 outputs

9. Basic decoder logic can consist of
   (a) only inverters and OR gates
   (b) only inverters and AND gates
   (c) inverters, OR gates, and AND gates
10. A BCD-to-decimal decoder is also known as a
    (a) 1-of-10 decoder
    (b) 1-of-4 decoder
    (c) 10-line-to-4-line decoder
11. A BCD-to-7-segment decoder has
    (a) 4 inputs and 10 outputs
    (b) 4 inputs and 7 outputs
    (c) 7 inputs and 4 outputs
12. When more than one input of a priority encoder is active, the output code is
    (a) a combination of all the active inputs
    (b) equivalent to the lowest-value input
    (c) equivalent to the highest-value input
13. In a multiplexer, the process for switching the input data onto the single output is called
    (a) time-division multiplexing
    (b) time-sequence switching
    (c) time conversion
14. A multiplexer can be considered as a
    (a) decoder
    (b) parallel-to-serial converter
    (c) serial-to-parallel converter
15. A demultiplexer can be considered as a
    (a) a reverse multiplexer
    (b) serial-to-parallel converter
    (c) both (a) and (b)

Questions

Answers to odd-numbered questions are at the end of the book.

1. Which of the following are valid BCD codes?
   (a) 1010 (b) 0010
   (c) 1100 (d) 0111
2. What is the BCD for the decimal digit 6?
3. What is the BCD for decimal number 39?
4. Referring to Table 6–2, what is the ASCII code in binary for the letter B?
5. Referring to Table 6–2, what is the ASCII code in binary for a right parenthesis?
6. What parity bit would you add to the code 1011011 to make it even parity?
7. Is there an error in the odd parity code 01000100?
8. How can you tell if the input or output of a logic device is active-HIGH or active-LOW?

9. What does a decoder do?
10. If a BCD-to-decimal decoder has active-HIGH outputs, what are the states (HIGH or LOW) of all the outputs if a 0101 code is on the input?
11. What does an encoder do?
12. If a 10-line-to-4-line encoder has active-LOW inputs and outputs, what are the output states if there is a LOW only on the input for decimal 5?
13. If the 2, 4, and 7 inputs of a decimal-to-BCD priority encoder are all active at the same time, what code is represented on the output? Assume active-HIGH inputs and outputs.
14. What does a multiplexer do?
15. What is the function of the S_2, S_1, and S_0 inputs of an 8-input multiplexer?
16. What does a demultiplexer do?

Basic Problems

PROBLEMS

Answers to odd-numbered problems are at the end of the book.

1. Write each of the ten 4-bit BCD codes?
2. Write the decimal digit represented by each of the BCD codes.
3. Write the BCD code for each of the following decimal numbers:
   (a) 24 (b) 39
   (c) 57
4. Write the decimal number for each BCD code.
   (a) 00010111 (b) 10001001
   (c) 10010100
5. The ASCII code has 7 bits. What is the maximum number of digits, letters, and symbols that it can represent?
6. Refer to Table 6–2. Look up the ASCII code for each of the following and add the proper even parity bit to each.
   (a) 2 (b) F
   (c) ?
7. Refer to Table 6–2. Determine what number is represented by each of the following ASCII codes and show the BCD value for the number.
   (a) 0110000 (b) 0110100
   (c) 0111000
8. Include an even parity bit in each of the 8-bit codes in Problem 4.
9. Include an odd parity bit in each ASCII code in Problem 7.
10. Determine the active output of a 1-of-16 binary-to-decimal decoder for each of the following binary numbers applied to the inputs:
    (a) $A_3A_2A_1A_0 = 0001$ (b) $A_3A_2A_1A_0 = 0100$
    (c) $A_3A_2A_1A_0 = 1110$
11. Determine the active output of a BCD-to-decimal decoder for each of the following input codes:
    (a) $A_3A_2A_1A_0 = 0111$ (b) $A_3A_2A_1A_0 = 0101$
    (c) $A_3A_2A_1A_0 = 0110$

CHAPTER 7

LATCHES, FLIP-FLOPS, AND TIMERS

CHAPTER OUTLINE

Study aids for this chapter are available at

http://www.prenhall.com/SOE

INTRODUCTION

As you have learned in your study of combinational logic, the output of any combinational logic circuit depends directly on the inputs. Beginning in this chapter, you will learn about sequential logic in which devices are used that can store a logic state (1 or 0). Generally, in a sequential logic circuit, the output is dependent not only on the input states but also on the stored state. The latch is used for the temporary storage of a data bit. The flip-flop forms the basis for most types of sequential logic, such as registers and counters. Also, the one-shot and 555 timer are two types of timing circuits that you will study.

KEY TERMS

- Latch
- Bistable
- SET
- RESET
- Storage
- Enable
- Flip-flop
- Edge-triggering
- Toggle
- Synchronous
- Asynchronous
- One-Shot
- Monostable
- Astable
- Timer
- Duty cycle

KEY OBJECTIVES

A section number is given for each objective. After completing this chapter, you should be able to

7–1 Describe the operation of the S-R latch

7–2 Describe the operation of two types of gated latches

7–3 Explain edge-triggering and describe the operation of the D flip-flop

7–4 Describe the operation of the J-K flip-flop

7–5 Discuss the one-shot and its operation

7–6 Use the 555 timer to produce one-shot and astable operation

7–7 Describe some latch and flip-flop ICs

7–8 Analyze latches, flip-flops, one-shots, and timers for faulty operation

COMPUTER SIMULATIONS DIRECTORY

The following figures have Multisim circuit files associated with them.

- Figure 7–4 Page 196
- Figure 7–14 Page 201
- Figure 7–20 Page 206
- Figure 7–25 Page 209

LABORATORY EXPERIMENTS DIRECTORY

The following exercises are for this chapter.

- **Experiment 10** The D Latch and the D Flip-Flop
- **Experiment 11** The J-K Flip-Flop

EXAMPLE 7–2

Problem

The *S*, *R*, and *EN* input waveforms shown in Figure 7–7(a) are applied to a gated S-R latch like in Figure 7–6 that is initially in the RESET state. Determine the Q and $\overline{Q}$ output waveforms.

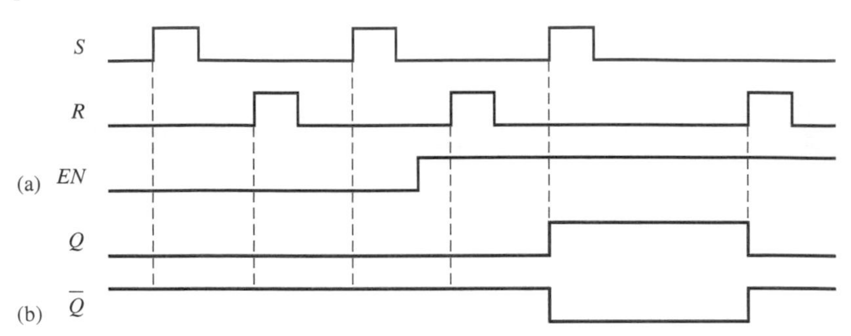

FIGURE 7–7

Solution

The output waveforms are shown in Figure 7–7(b).

Question

Why doesn't the output change on the first two pulses in the *S* waveform and the first pulse in the *R* waveform?

The D Latch

The **D latch** is a type of gated latch that is a variation of the S-R latch, as shown in Figure 7–8(a). It has a *D* input and an enable input. When the *EN* input is HIGH, a HIGH on the *D* input will set the latch and a LOW on the *D* input will reset the latch. When the *EN* input is LOW, the *D* input does not affect the state of the latch. This latch is usually indicated by a single logic symbol, as shown in Figure 7–8(b).

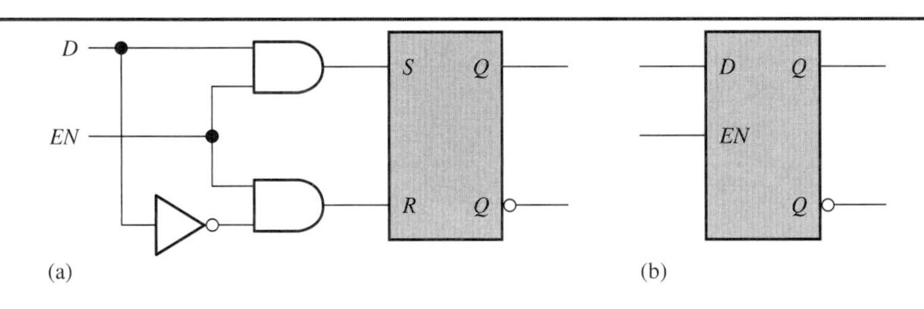

FIGURE 7–8

The D latch.

EXAMPLE 7–3

Problem

The *D* and *EN* input waveforms shown in Figure 7–9(a) are applied to a D latch that is initially in the RESET state. Determine the Q and $\overline{Q}$ output waveforms.

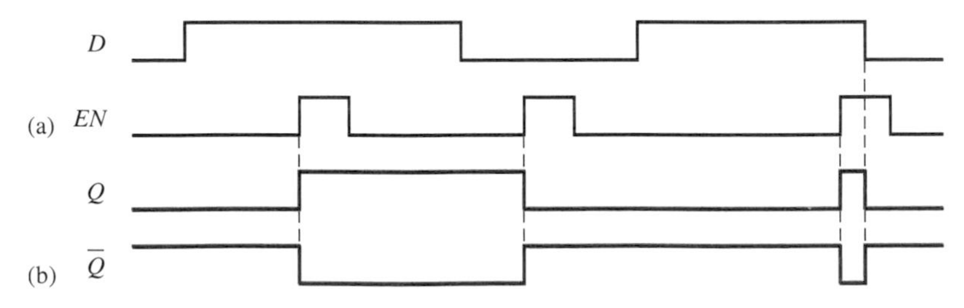

FIGURE 7–9

Solution
The output waveforms are shown in Figure 7–9(b).

Question
What would be the Q waveform if EN stayed LOW all the time?

EXAMPLE 7–4

Problem
The D and EN input waveforms shown in Figure 7–10(a) are applied to a D latch that is initially in the RESET state. Determine the Q output waveform.

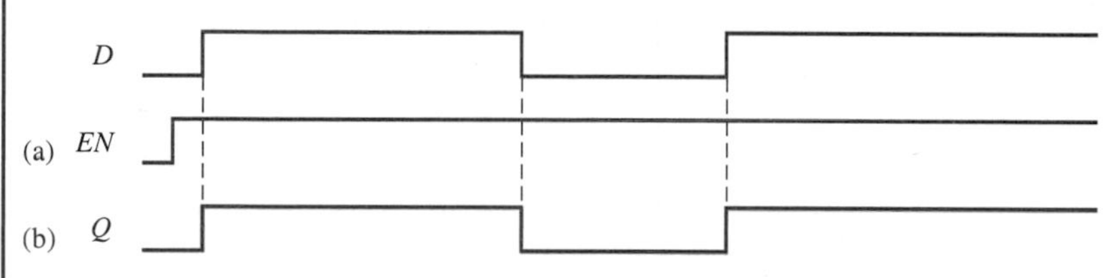

FIGURE 7–10

Solution
The output waveform is the same as the D input when the EN is held HIGH as shown in Figure 7–10(b).

Question
What is the Q output after EN goes LOW while D is HIGH?

Review Questions

6. What are the inputs of a D latch?
7. What is the purpose of the EN input for a D latch?
8. What is the difference between a D latch and a gated S-R latch?
9. If the D input of a latch is HIGH and the latch is enabled, what is the Q output?
10. What happens to the Q output if the latch is not enabled and the D input changes?

THE D FLIP-FLOP 7–3

Flip-flops have numerous applications including shift registers, counters, and some static random access memories (SRAMs). One common type of flip-flop has a D input, just like the D latch which determines the state to which it will go when a clock pulse is applied.

In this section, you will learn what edge-triggering means and how a D flip-flop works.

A **flip-flop** is a bistable logic circuit similar to a latch except for the way in which it changes its state from SET to RESET or from RESET to SET. Flip-flops have an input called a clock, which is used to change the state of the flip-flop. When a pulse is applied to the clock input, the flip-flop can change state only on a transition of the pulse. A method of changing the state of a flip-flop on one of the edges (transitions) of a pulse is known as **edge-triggering**. Nearly all flop-flops use this method.

The **D flip-flop** stores data based on the state of the *D* input at the triggering edge of a clock pulse. The logic symbol for a D flip-flop is shown in Figure 7–11. As you can see, it has a *D* input, a clock (*C*) input, a *Q* output, and a $\overline{Q}$ output. The small triangle at the *C* input indicates that the flip-flop is edge-triggered. When there is no bubble with the triangle, the flip-flop can change state only on the positive-going edge of a clock pulse. A bubble with the triangle means that the flip-flop can change state only on the negative-going edge of the clock pulse.

FIGURE 7–11

Logic symbols for edge-triggered D flip-flops.

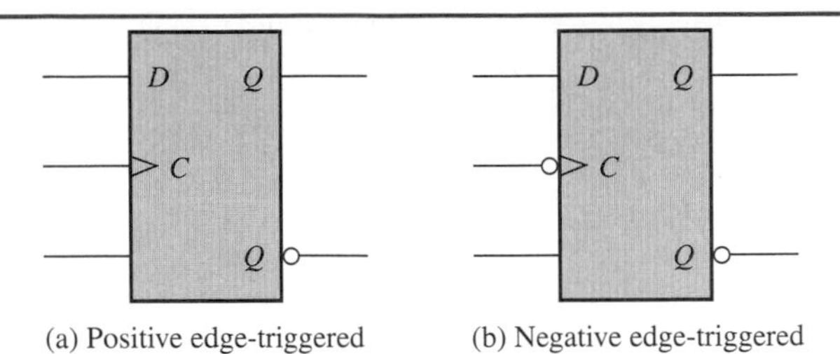

D Flip-Flop Operation

When a clock pulse is applied, the output of a flip-flop depends on the state of the *D* input. When the *D* input is HIGH, the flip-flop will go to the SET state (*Q* HIGH) on the triggering edge of the clock pulse. If it is already SET, it will remain SET. When the *D* input is LOW, the flip-flop will go to the RESET state (*Q* LOW) on the triggering edge of the clock pulse. If the flip-flop is already RESET, it will remain RESET. This operation is illustrated in Figure 7–12 for both positive edge-triggering and negative edge-triggering. Notice that *Q* follows *D* only on the active edge of the clock.

FIGURE 7–12

Operation of positive and negative edge-triggered flip-flops.

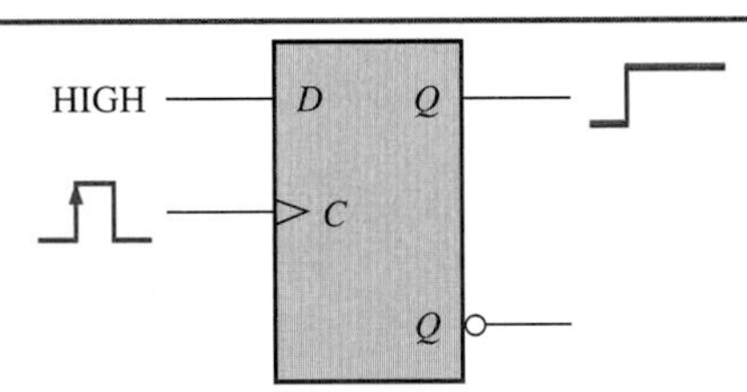

(a) The arrow on the clock pulse indicates the flip-flop triggers on the positive-going edge and *Q* changes to the HIGH state because *D* is HIGH.

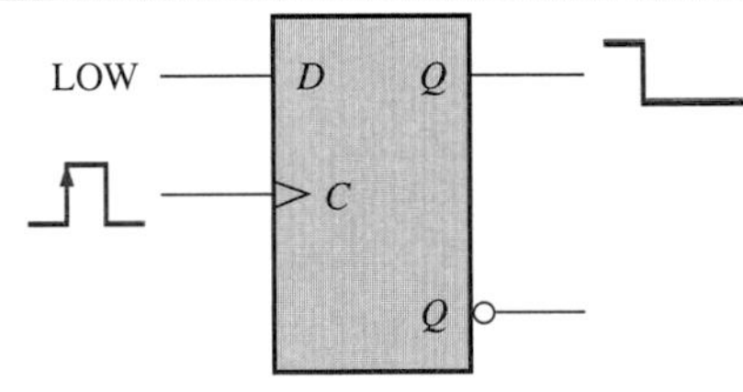

(b) The arrow on the clock pulse indicates the flip-flop triggers on the positive-going edge and *Q* changes to the LOW state because *D* is LOW.

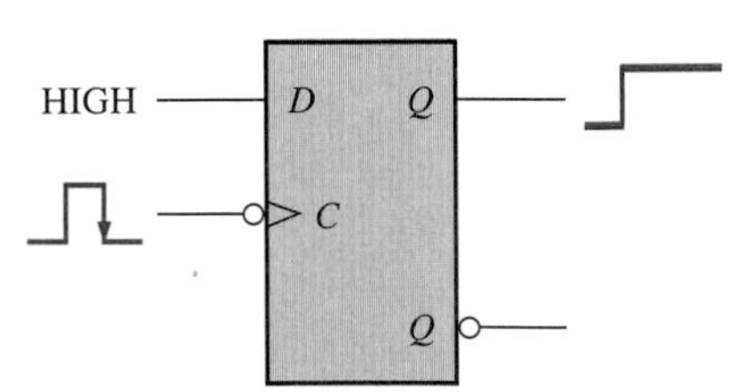

(a) The arrow on the clock pulse indicates the flip-flop triggers on the negative-going edge and *Q* changes to the HIGH state because *D* is HIGH.

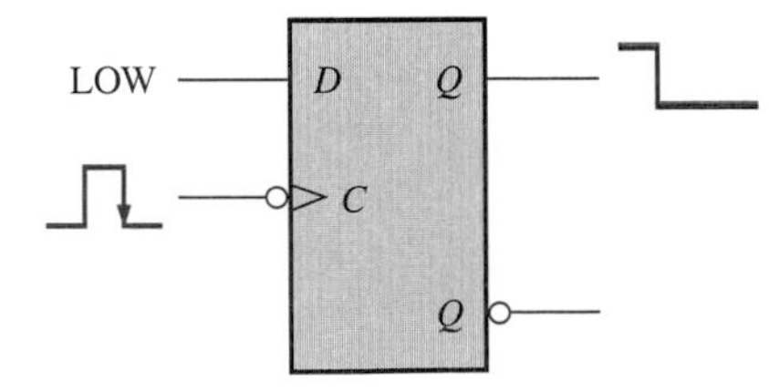

(b) The arrow on the clock pulse indicates the flip-flop triggers on the negative-going edge and *Q* changes to the LOW state because *D* is LOW.

A truth table that summarizes the operation of a positive edge-triggered D flip-flop is shown in Table 7–3. When *Q* is LOW, a 0 is stored; and when *Q* is HIGH, a 1 is stored.

Inputs		Outputs		
D	*C*	*Q*	$\overline{Q}$	Comments
LOW	↑	LOW	HIGH	RESET (stores a 0)
HIGH	↑	HIGH	LOW	SET (stores a 1)

TABLE 7–3

Truth table for a positive edge-triggered D flip-flop.

EXAMPLE 7–5

Problem

The D and clock input waveforms shown in Figure 7–13(a) are applied to a positive edge-triggered D flip-flop that is initially in the RESET state. Determine the Q and $\overline{Q}$ output waveforms.

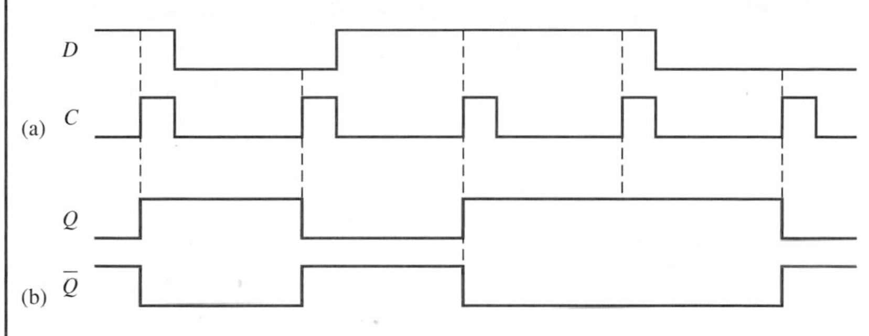

FIGURE 7–13

Solution

The output waveforms are shown in Figure 7–13(b).

Question

What would be the Q waveform if D stayed HIGH all the time?

COMPUTER SIMULATION

FIGURE 7–14

Open the Multisim file F07-14DG on the website. The pulse generator produces one clock pulse every two seconds. The Q output of the D flip-flop will go to the state of the D input on the next clock pulse. As you change the D switch, you can see the Q output changing on the clock pulse.

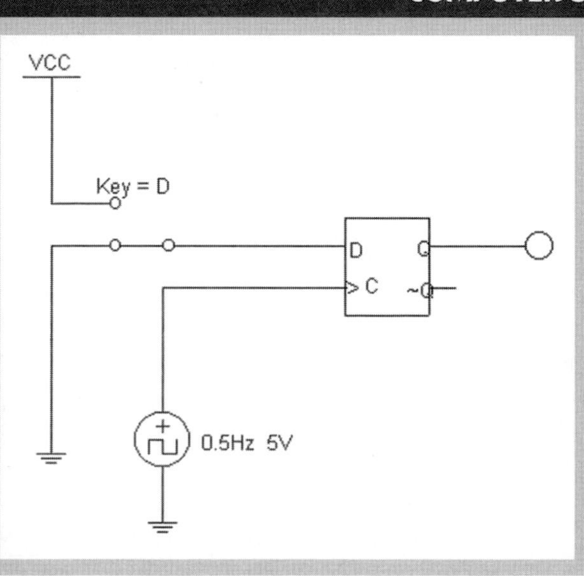

Review Questions

11. What are the inputs of an edge-triggered D flip-flop?
12. What indicates on a flip-flop symbol that it is edge-triggered?
13. When can the output of a positive edge-triggered D flip-flop change?
14. How can you tell that a flip-flop is negative edge-triggered?
15. If the *D* input is HIGH and continuous clock pulses are applied, what does the *Q* output do if the flip-flop is initially RESET?

7–4 THE J-K FLIP-FLOP

The J-K flip-flop is another popular type that has two control inputs in addition to the clock input. These control inputs are called *J* and *K* (in honor of Jack Kilby, who invented the integrated circuit). Like the D flip-flop, the J-K is also edge-triggered.

In this section, you will learn how a J-K flip-flop operates.

J-K Flip-Flop Operation

The **J-K flip-flop** stores data based on the states of the *J* and *K* inputs at the triggering edge of a clock pulse. When a clock pulse is applied, the output of a flip-flop depends on the state of the *J* and *K* inputs. When the *J* input is HIGH and the *K* input is LOW, the flip-flop will go to the SET state (*Q* HIGH) on the triggering edge of the clock pulse. If it is already SET, it will remain SET. When the *J* input is LOW and the *K* input is HIGH, the flip-flop will go to the RESET state (*Q* LOW) on the triggering edge of the clock pulse. If it is already RESET, it will remain RESET. If the the *J* and *K* inputs are both HIGH, the flip-flop will go to the opposite state (this is called **toggle** operation). If the *J* and *K* inputs are both LOW, the flip-flop will not change from the state that it happens to be in (this is called a no-change or latched condition). This operation is illustrated in Figure 7–15 for a positive edge-triggered J-K flip-flop.

FIGURE 7–15 The four conditions of J-K flip-flop operation.

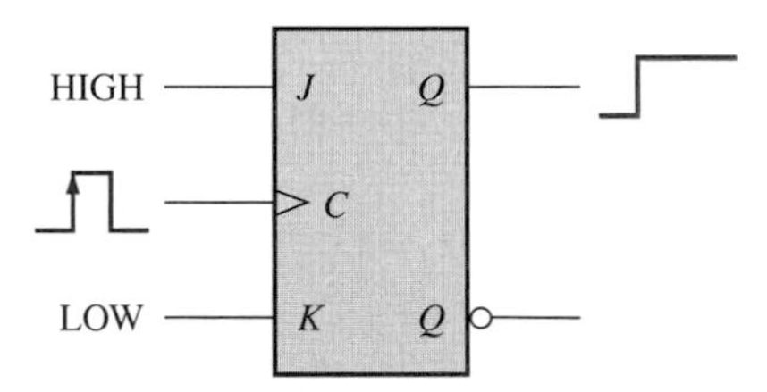

(a) SET condition: *Q* output goes HIGH.

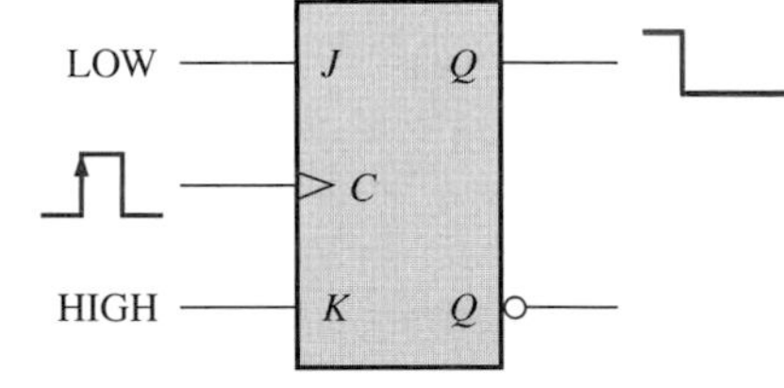

(b) RESET condition: *Q* output goes LOW.

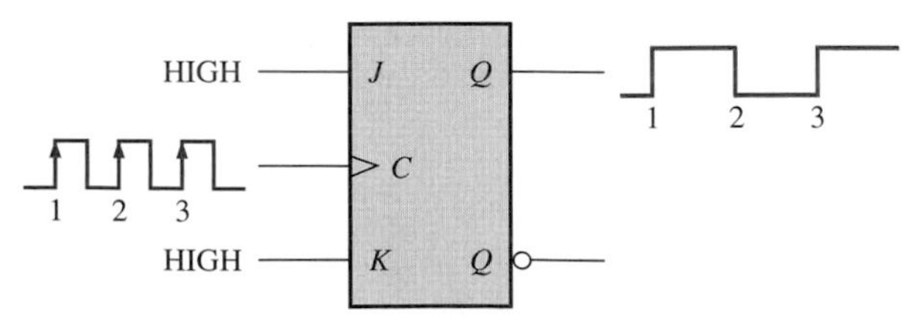

(c) Toggle condition: *Q* output changes on each clock pulse.

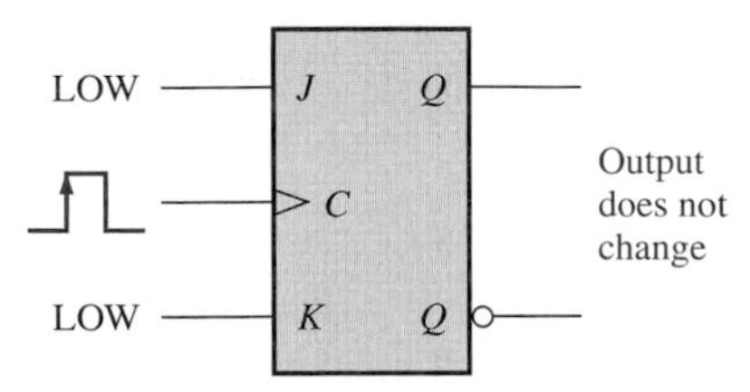

(d) No-change condition

A truth table that summarizes the operation of a positive edge-triggered J-K flip-flop is shown in Table 7–4.

Inputs			Outputs		
J	*K*	*C*	*Q*	$\overline{Q}$	Comments
LOW	LOW	↑	Remains the same		No change
LOW	HIGH	↑	LOW	HIGH	RESET
HIGH	LOW	↑	HIGH	LOW	SET
HIGH	HIGH	↑	Changes state		Toggle

TABLE 7–4

Truth table for a positive edge-triggered J-K flip-flop.

EXAMPLE 7–6

Problem

The *J, K,* and clock input waveforms shown in Figure 7–16(a) are applied to a positive edge-triggered J-K flip-flop that is initially in the RESET state. Determine the *Q* output waveform.

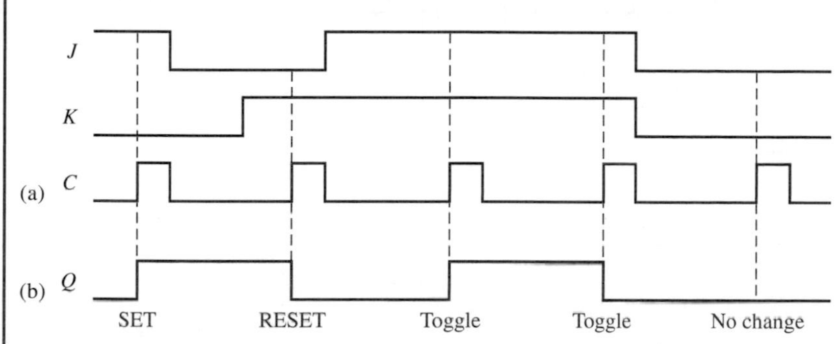

FIGURE 7–16

Solution

The output waveform is shown in Figure 7–16(b) with the condition at each clock pulse.

Question

What would be the *Q* waveform if *J* and *K* stayed HIGH all the time?

Preset and Clear Inputs

For the D flip-flop and the J-K flip-flop, the *D, J,* and *K* inputs are called **synchronous** inputs because the state of these inputs control the output only on the triggering edge of a clock pulse; that is, they are synchronized with the clock. Most integrated circuit flip-flops also have **asynchronous** inputs that can change the output without a clock pulse; that is, they work independently of the clock.

The two asynchronous inputs are called preset (*PRE*) and clear (*CLR*) or, in some cases, direct set (S_D) and direct reset (R_D). When the preset input is active, the flip-flop is SET, regardless of the other inputs. When the clear input is active, the flip-flop is RESET, regardless of the other inputs. Usually, the asynchronous inputs are active-LOW inputs, indicated with an overbar on the variable and a bubble on the flip-flop symbol.

EXAMPLE 7–7

Problem

The C, $\overline{PRE}$, and $\overline{CLR}$ input waveforms shown in Figure 7–17(a) are applied to a positive edge-triggered J-K flip-flop. The flip-flop is initially in the RESET state and is in the toggle condition with J and K HIGH. Determine the Q output waveform.

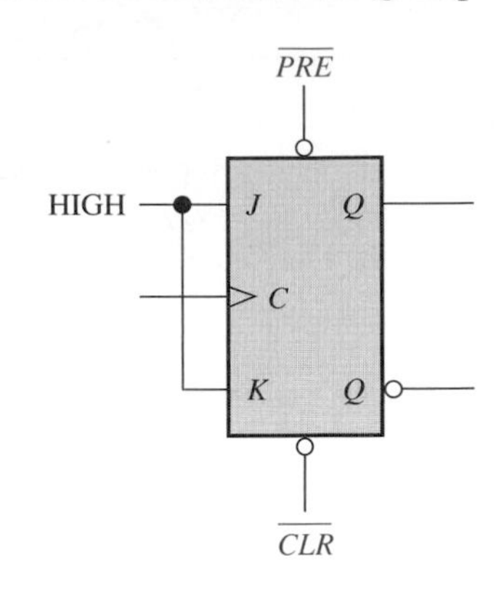

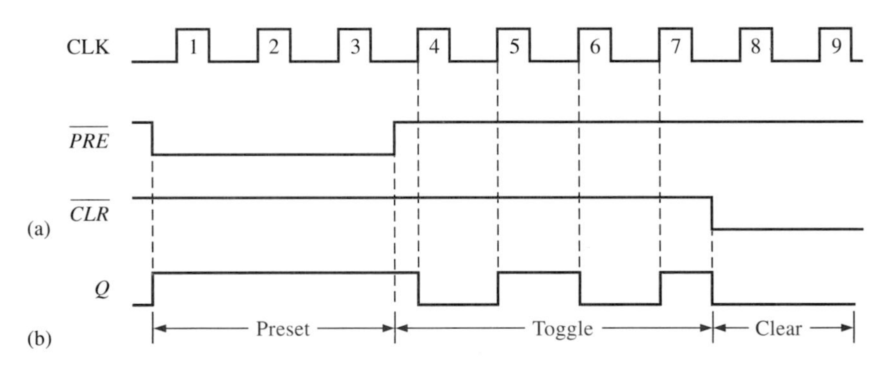

FIGURE 7–17

Solution

During clock pulses 1, 2, and 3 the asynchronous preset ($\overline{PRE}$) is LOW, keeping the flip-flop SET, regardless of the J and K inputs.

For clock pulses 4, 5, 6, and 7 toggle operation occurs because both J and K are HIGH and both $\overline{PRE}$ and $\overline{CLR}$ are HIGH (not active).

For clock pulses 8 and 9 the asynchronous clear ($\overline{CLR}$) input is LOW, keeping the flip-flop RESET, regardless of the J and K inputs. The resulting Q output waveform is shown in Figure 7–17(b).

Question

What would be the Q waveform if $\overline{PRE}$ and $\overline{CLR}$ stayed HIGH all the time?

Review Questions

16. What are the inputs of an edge-triggered J-K flip-flop?
17. What must the J and K inputs be for the flip-flop to toggle?
18. When do the J and K inputs control the output of a negative edge-triggered J-K flip-flop?
19. What is the purpose of the preset input to a flip-flop?
20. What is the purpose of the clear input to a flip-flop?

THE ONE-SHOT 7–5

The one-shot is a type of multivibrator circuit, just as the latch and flip-flop. However, the one-shot has only one stable state whereas the latch and flip-flop have two stable states. The primary use of one-shots is to produce single pulses for time delay and other applications.

In this section, you will learn what a one-shot is and how it operates.

A **one-shot** is a type of triggered logic circuit that has one stable state. Because it has only one stable state, the one-shot is classified as a **monostable** multivibrator. A basic one-shot has a trigger input and Q and $\overline{Q}$ outputs, as shown in Figure 7–18. When a pulse is applied to the trigger input, a single pulse is produced on the Q output. After the pulse ends, the one-shot must be triggered again to produce another output pulse. The small triangle indicates that triggering occurs on the pulse edge. The width or duration of the output pulse depends on an *RC* circuit (either internal or external) associated with the one-shot.

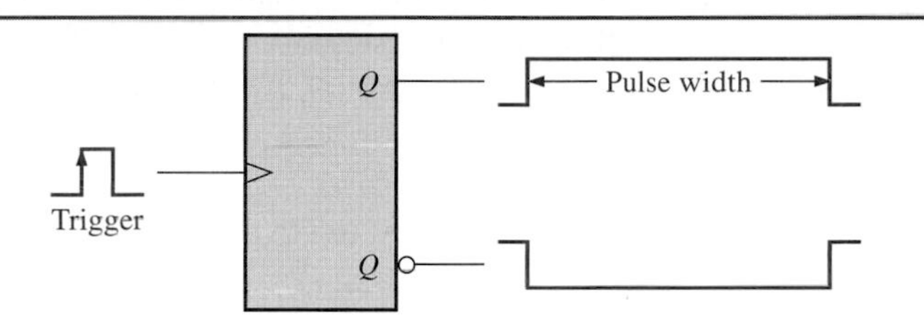

FIGURE 7–18

Logic symbol for a basic one-shot showing its operation.

Most integrated circuit one-shots provide connections for an external resistor and capacitor for setting the output pulse width. Figure 7–19 shows the basic one-shot symbol with an external resistor, R_{EXT}, and external capacitor, C_{EXT}, connected. By choosing the values of R_{EXT} and C_{EXT}, you can set the time constant of the circuit to get the desired pulse width. The two basic categories of integrated circuit one-shots are nonretriggerable and retriggerable.

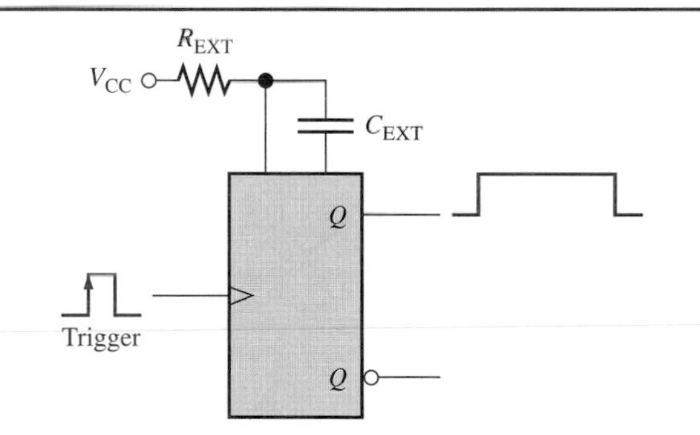

FIGURE 7–19

A one-shot with an external resistor and capacitor to set the output pulse width.

The Nonretriggerable One-Shot

Once a nonretriggerable one-shot is triggered, it cannot be triggered again until after it times out, which is at the end of the output pulse. If another trigger input occurs during the output pulse, the one-shot will not recognize the trigger. The 74121 is an example of a nonretriggerable IC one-shot. The output pulse width, t_W, of this device is determined by the following formula when the external components are used.

$$t_W = 0.7R_{EXT}C_{EXT}$$

where t_W is in seconds, R_{EXT} is in ohms, and C_{EXT} is in farads.

The Retriggerable One-Shot

A retriggerable one-shot can be retriggered at any time. If it is retriggered while timing out, the output pulse is extended from the time it is retriggered by an additional time equal to the pulse width. The 74LS122 is an example of this type of IC one-shot. The output pulse width of this device is calculated a little differently than the pulse width of the 74121, due to differences in internal circuitry.

$$t_W = 0.32R_{\text{EXT}}C_{\text{EXT}}\left(1 + \frac{700}{R_{\text{EXT}}}\right)$$

where t_W is in seconds, R_{EXT} is in ohms, and C_{EXT} is in farads. The formula may be shown differently, depending on the units used.

EXAMPLE 7–8

Problem

(a) Determine the output pulse width for a 74121 one-shot for external component values of $R_{\text{EXT}} = 100\text{ k}\Omega$ and $C_{\text{EXT}} = 0.1\ \mu\text{F}$.

(b) Calculate the value of C_{EXT} required for a 74LS122 one-shot to produce an output pulse width of 1.0 ms, using an R_{EXT} of 22 kΩ.

Solution

(a) $t_W = 0.7R_{\text{EXT}}C_{\text{EXT}} = 0.7(100\text{ k}\Omega)(0.1\ \mu\text{F}) = \mathbf{7\ ms}$

(b) $t_W = 0.32R_{\text{EXT}}C_{\text{EXT}}(1 + 700/R_{\text{EXT}})$

$$C_{\text{EXT}} = \frac{t_W}{0.32R_{\text{EXT}}\left(1 + \frac{700}{R_{\text{EXT}}}\right)} = \frac{1.0\text{ ms}}{0.32(22\text{ k}\Omega)\left(1 + \frac{700}{22\text{ k}\Omega}\right)} = \mathbf{0.138\ \mu F}$$

Question

What would be the C_{EXT} for a 74122 if $R_{\text{EXT}} = 100\text{ k}\Omega$ and the output pulse width is 5 ms?

COMPUTER SIMULATION

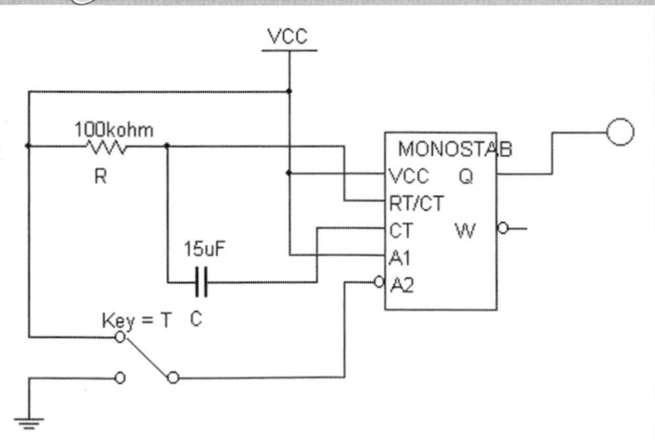

FIGURE 7–20

Open the Multisim file F07-20DG on the website. The one-shot is triggered when you move the switch to ground using the *T* key. The probe light connected to the *Q* output will turn on for a short time each time the one-shot is triggered. The duration of the output can be changed by changing the values of *R* or *C*.

Review Questions

21. How many stable states does a one-shot have?
22. How many pulses does a one-shot produce when it is triggered?
23. What determines the duration of the output pulse of a one-shot?
24. What is a nonretriggerable one-shot?
25. What is a retriggerable one-shot?

THE 555 TIMER 7–6

The 555 timer is a popular device because it can be set up to operate as either a monostable multivibrator (one-shot) or as an astable multivibrator.

In this section, you will learn how to connect external components to make a 555 timer work in either mode.

An **astable** multivibrator has no stable states and therefore changes back and forth between its two unstable states, without external triggering. The astable multivibrator is a type of oscillator that is used to produce pulse waveforms.

A **timer** is a circuit that can be used as a one-shot or as an oscillator. A block diagram of a 555 timer is shown in Figure 7–21. The threshold (THRESH) and the discharge (DISCH) inputs are connected to an external *RC* circuit for setting the output pulse width when used as a one-shot or for setting the frequency when used as an astable multivibrator. The numbers in parentheses are the pin numbers of the IC package.

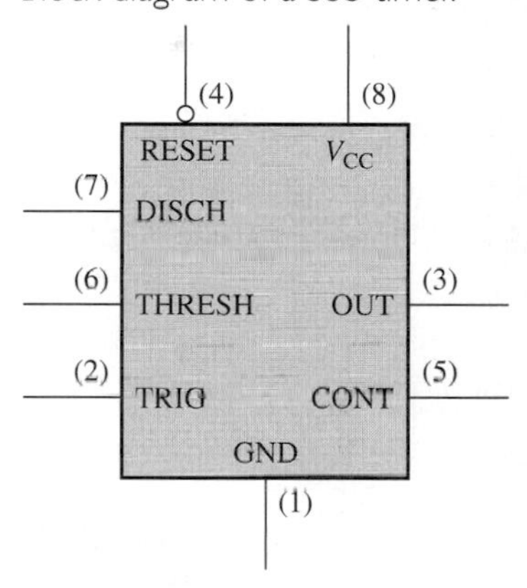

FIGURE 7–21

Block diagram of a 555 timer.

One-Shot Operation

By connecting the 555 timer as shown in Figure 7–22, the timer becomes a one-shot with the output pulse width determined by the values of the external resistor and capacitor, as expressed by the following formula:

$$t_W = 1.1R_{EXT}C_{EXT}$$

where t_W is in seconds, R_{EXT} is in ohms, and C_{EXT} is in farads.

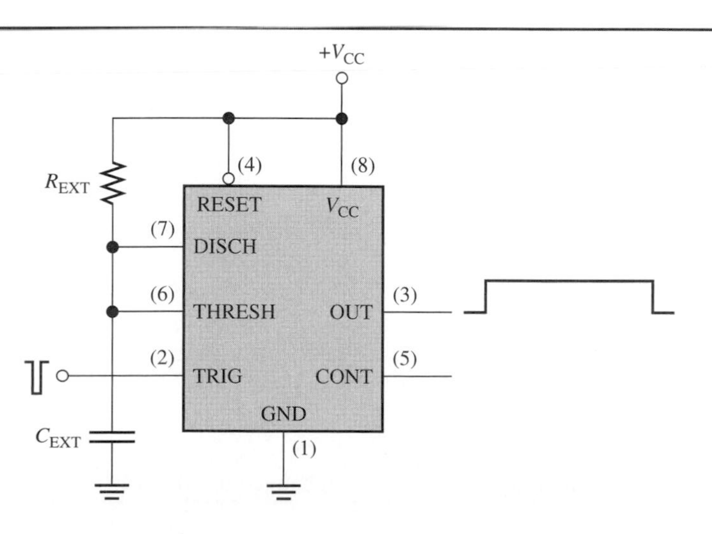

FIGURE 7–22

The 555 timer set up to operate as a one-shot. It is triggered on the negative-going edge of the trigger pulse.

Before a trigger pulse is applied, the output is LOW and C_{EXT} is discharged. When the trigger pulse occurs, the output goes HIGH; at this time C_{EXT} begins charging through R_{EXT}, and the output pulse begins. When the capacitor charges to a certain level, the output goes back LOW and the output pulse ends. The charging rate of C_{EXT} determines how long the output is HIGH.

Astable Operation

By connecting the 555 timer as shown in Figure 7–23, the timer becomes an oscillator with the frequency of the output pulses determined by the values of two external resistors and a capacitor, as expressed by the following formula:

$$f = \frac{1.44}{(R_{EXT1} + 2R_{EXT2})C_{EXT}}$$

FIGURE 7–23

A 555 timer set up to operate as an astable multivibrator (oscillator).

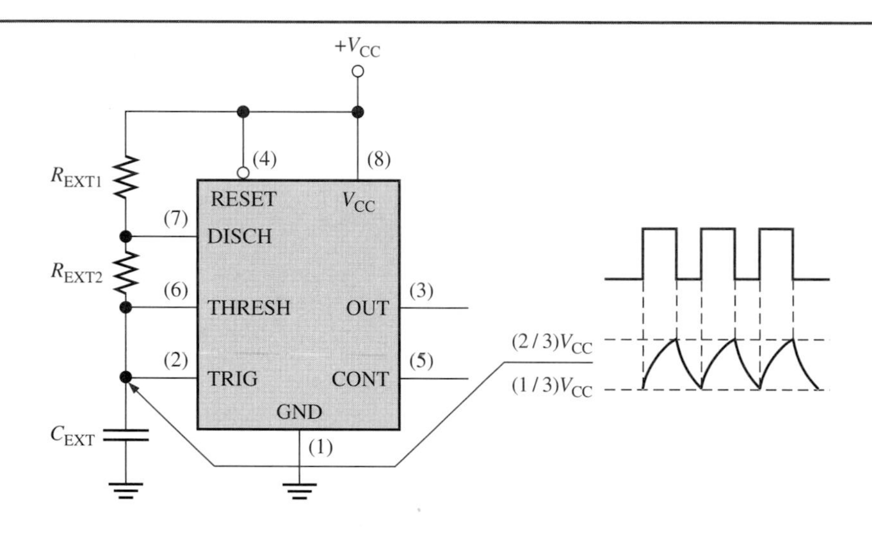

Initially, when the power is turned on, C_{EXT} is uncharged and the trigger voltage is 0 V. C_{EXT} begins charging through R_{EXT1} and R_{EXT2}; and when a voltage equal to $(2/3)V_{CC}$ is reached, the capacitor discharges back to a voltage equal to $(1/3)V_{CC}$. The internal circuitry of the timer produces an output pulse as a result of the charging and discharging of the capacitor. The charging and discharging cycle continues to repeat, resulting in a continuous pulse waveform on the output, as indicated in Figure 7–23.

FIGURE 7–24

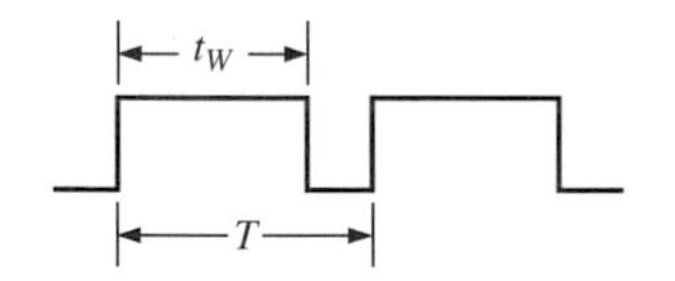

The **duty cycle** of the output waveform is the ratio of the time that the pulse is HIGH, t_W, to the period, T, of the waveform, as shown in Figure 7–24. The value of the duty cycle is set by R_{EXT1} and R_{EXT2} and is greater than 50% but can be made close to 50% if R_{EXT2} is much greater than R_{EXT1}.

EXAMPLE 7–9

Problem

What is the output frequency of a 555 timer set up as an oscillator if $R_{EXT1} = 5600\ \Omega$, $R_{EXT2} = 560\ \Omega$, and $C_{EXT} = 0.1\ \mu F$?

Solution

$$f = \frac{1.44}{(R_{EXT1} + 2\,R_{EXT2})C_{EXT}} = \frac{1.44}{(5600\ \Omega + 1120\ \Omega)0.1\ \mu F} = \mathbf{2.14\ kHz}$$

Question

What is the frequency if C_{EXT} is changed to 1000 pF?

COMPUTER SIMULATION

FIGURE 7–25

Open the Multisim file F07-25DG on the website. The 555 timer configured for astable operation will oscillate at a low frequency which is the rate at which the probe light flashes. To view the pulse output waveform or the capacitor waveform, click on the oscilloscope icon. The frequency of the output waveform can be changed by changing the external R and C values.

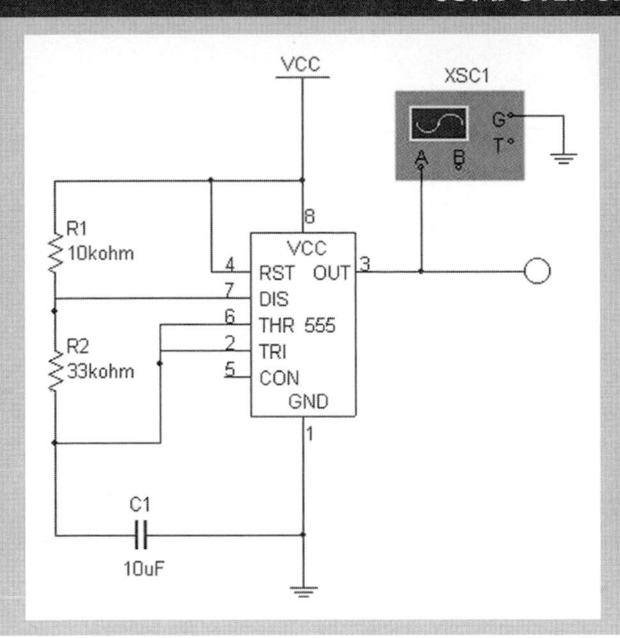

Review Questions

26. What are the two modes of operation for the 555 timer?
27. What determines the pulse width of the 555 timer when it is set up as a one-shot?
28. What determines the frequency of the output waveform when the timer is set up to operate as an oscillator?
29. How would you decrease the frequency of a 555 timer circuit?
30. How is the duty cycle of a 555 timer set?

INTEGRATED CIRCUITS 7–7

Several types of latches, flip-flops, one-shots, and the 555 timer are available as integrated circuits. Remember that most devices are available in several IC families including both CMOS and TTL. Specific IC one-shots and the 555 timer have already been covered.

In this section, you will see some examples of other specific ICs.

Latches

Several types of latches are available in IC form. In all configurations, there are typically two or more latches in an IC package. An example is the 74LS279 quad $\overline{S}$-$\overline{R}$ latch, which has four SET-RESET latches with active-LOW inputs in a package, as shown in Figure 7–26. The standard logic diagram is in part (a), and the pin diagram is in part (b). The number that precedes the S, R, and Q labels is the latch number (1 through 4). Notice that latches 1 and 3 have two SET inputs, S_1 and S_2, for more flexibility.

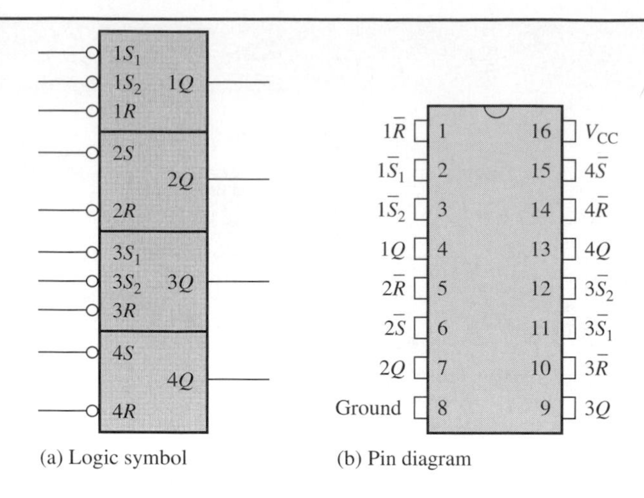

FIGURE 7–26

The 74LS279 quad $\overline{S}$-$\overline{R}$ latch.

Another example of an IC latch is the 74LS75, which is a quad gated D latch, as shown in Figure 7–27.

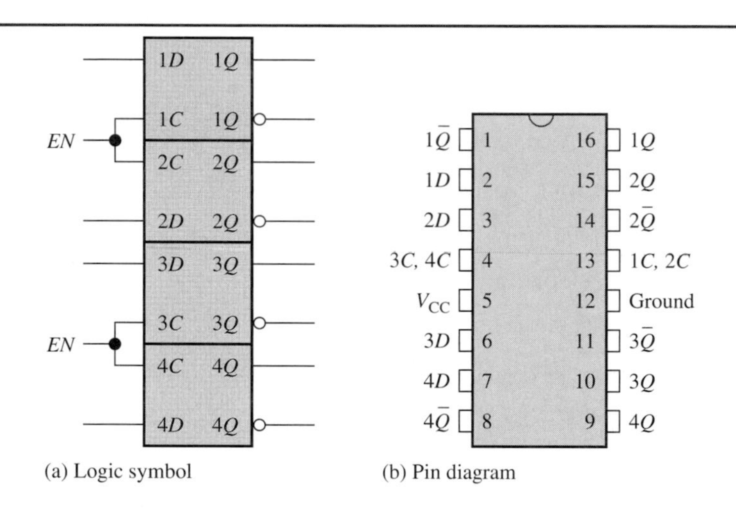

FIGURE 7–27

The 74LS75 quad gated D latch.

Flip-Flops

IC flip-flops, like latches, normally have two or more devices in a single package. An example of a D flip-flop is the 74LS74A, which contains two devices, as shown in Figure 7–28. This flip-flop has active-LOW preset and clear inputs, as indicated by the bubbles. It is a positive edge-triggered device, as indicated by the absence of a bubble at the clock (C) input.

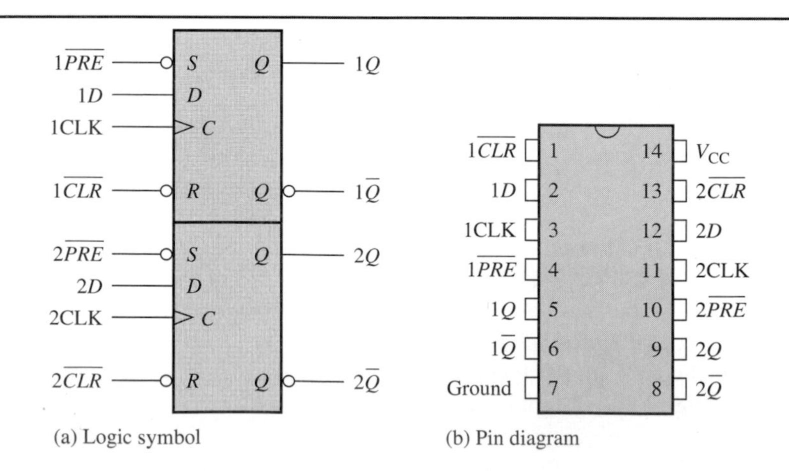

FIGURE 7–28

The 74LS74A dual positive-edge triggered D flip-flop.

The 74LS112 is an example of an integrated circuit *J-K* flip-flop. As shown in Figure 7–29, there are two negative edge-triggered flip-flops with active-LOW preset and clear.

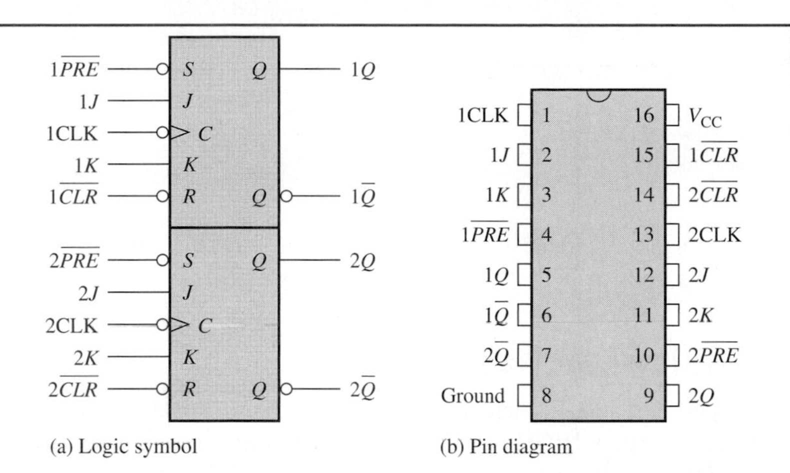

FIGURE 7–29

The 74LS112 dual negative edge-triggered J-K flip-flop.

EXAMPLE 7–10

Problem

The 1*J*, 1*K*, 1CLK, $1\overline{PRE}$, and $1\overline{CLR}$ waveforms in Figure 7–30(a) are applied to one of the negative edge-triggered flip-flops in a 74LS112. Determine the 1*Q* output waveform.

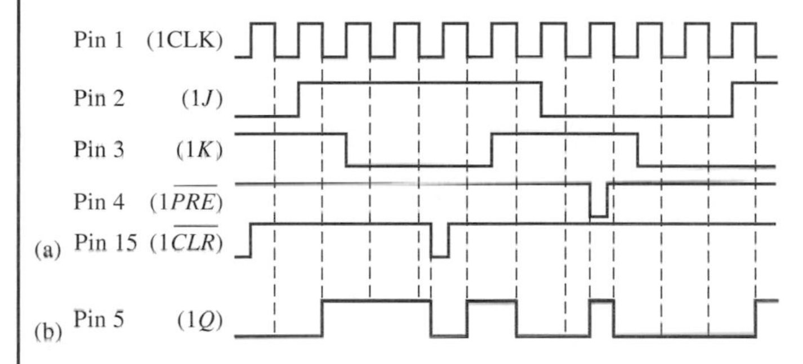

FIGURE 7–30

Solution

The resulting 1*Q* waveform is shown in Figure 7–30(b). Each time a LOW is applied to the preset and clear inputs, the flip-flop is SET or RESET regardless of the states of the other inputs.

Question

What is the 1*Q* waveform if the waveforms for the preset and clear are interchanged?

Review Questions

31. What is a quad latch package?
32. What is the difference between a latch and a flip-flop?
33. In a 74LS112, what inputs go to both flip-flops commonly?
34. What is the logic function of a 74LS279?
35. What is the logic function of a 74LS74A?

7–8 TROUBLESHOOTING

Basic troubleshooting of latches and flip-flops is much the same as for gate and other logic circuits. A latch or flip-flop can be stuck in one logic state so that its output never changes. When one-shots and timers fail, they normally will stop producing pulse outputs or the characteristics of the pulse outputs will be altered.

In this section, you will learn how faults affect operation of latches, flip-flops, one-shots, and timers.

SAFETY NOTE

If another person cannot let go of an energized conductor, switch the power off immediately. If that is not possible, use any available nonconductive material to try to separate the body from the contact point. Seek medical help right away for any electrical burns and be ready to perform CPR.

When any circuit does not operate properly, always first check the power and ground connections to the IC and look for any signs of shorts or open contacts.

Faulty Latches

Recall that an S-R latch is controlled by the set *(S)* and reset *(R)* inputs. If either of these inputs is stuck HIGH or LOW, the latch will not store a bit of data. For example, Figure 7–31(a) illustrates how the latch behaves if the *R* input is stuck HIGH (in TTL, this could be due to an open input). The faulty HIGH on the *R* input causes the *Q* output to stay LOW. If pulses are applied to the *S* input, the $\overline{Q}$ output will also have pulses as shown because the top input of the lower NOR gate is LOW. If the latch were working properly, it would be in the SET state after the first pulse on *S*, so *Q* would remain HIGH and $\overline{Q}$ LOW.

FIGURE 7–31 Examples of faults in a latch.

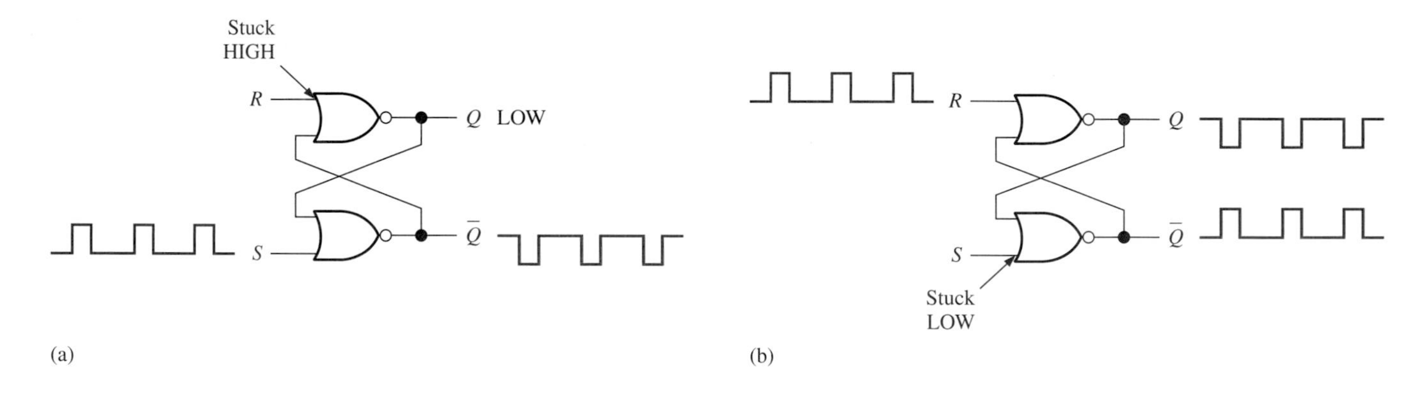

In Figure 7–31(b), the *S* input is stuck LOW. The latch goes to the RESET state after the first pulse on *R*. The $\overline{Q}$ remains LOW and *Q* remains HIGH on successive pulses on *R*, as they should. If pulses are applied to the *S* input, the latch will not SET because of the stuck LOW condition.

You have seen examples of how certain faults can cause a latch to behave strangely. Of course, there are other possible faults and each may cause unique symptoms.

FIGURE 7–32 Basic J-K flip-flop.

Faulty Flip-Flops

A J-K flip-flop is shown in Figure 7–32. Faults in a flip-flop, like the latch, can cause it to behave erratically or to hang up in one state. As you know, flip-flops have a clock input that latches do not have; and if the clock input is stuck either HIGH or LOW, the flip-flop will not change its state. If the *D* input of a flip-flop is stuck in either state, the *Q* output will go to the failed state on the first clock pulse and remain there. Several possibilities exist for input failures in a basic J-K flip-flop, as summarized in Table 7–5.

Fault	Symptom
J stuck HIGH	If *K* is LOW, *Q* stays HIGH. If *K* is HIGH, *Q* toggles and operation appears correct.
J stuck LOW	If *K* is HIGH, *Q* stays LOW. If *K* is LOW, *Q* stays in either state.
K stuck HIGH	If *J* is LOW, *Q* stays LOW. If *J* is HIGH, *Q* toggles and operation appears correct.
K stuck LOW	If *J* is LOW, *Q* stays in either state. If *J* is HIGH, *Q* stays HIGH.
Clock stuck LOW	*Q* stays in one state.
Clock stuck HIGH	*Q* stays in one state.

TABLE 7–5

J-K flip-flop input faults. The faults shown are based on truth table operation.

Faulty One-Shots and Timers

A faulty one-shot or timer will usually have no output or the pulse width or the frequency of the output will be incorrect. As you have learned, a one-shot or timer usually requires an external resistor and capacitor, either of which can fail or there can be an open or shorted path. Also, the one-shot or timer circuit itself can fail.

For example, if the resistor or the capacitor opens, the output pulse width of a one-shot will be less than its nominal value or there will be no output, depending on the type of one-shot. If the trigger input is stuck either HIGH or LOW, there will be no pulse output.

When a 555 timer is configured for astable operation as shown in Figure 7–23, an open in any of the external components will result in no output pulses. If the value of either resistor or the capacitor changes value, the frequency of the output pulse waveform will change. If the DISCH, THRESH, or TRIG are stuck in one state, there will be no output pulses.

Review Questions

36. If the *S* input of the latch in Figure 7–31(b) is stuck HIGH, what will you observe on the Q and $\overline{Q}$ outputs?
37. If the *J* input of a flip-flop is stuck HIGH, what is the *Q* output if *K* is HIGH and clock pulses are applied?
38. If the *K* input of a flip-flop is stuck LOW, what is the *Q* output if *J* is HIGH and clock pulses are applied?
39. What happens to the flip-flop if the clock input is open?
40. What could cause the output pulse width of a 555 one-shot to be greater than it should be?

CHAPTER REVIEW

Key Terms

Astable Having no stable state.

Asychronous Having no fixed time relationship.

Bistable Having two stable states.

Duty cycle The ratio of pulse width to the period, usually expressed as a percentage.

Edge-triggering The method used to cause a flip-flop to store a data bit on the occurrence of a clock edge, based on the state of its inputs. If the flip-flop triggers on the rising edge of a clock pulse, it is called a positive edge-triggered flip-flop. If it triggers on the falling edge, it is called a negative edge-triggered flip-flop.

Enable An input on a logic circuit that prevents or allows response to another input, depending on the level of the enable input.

Flip-flop A bistable logic circuit similar to a latch except that it can change state only on the occurrence of one edge of a clock pulse.

Latch A bistable logic circuit that can store a binary 1 or 0.

Monostable Having only one stable state.

One-shot A type of triggered logic circuit that has only one stable state and is also referred to as a monostable multivibrator.

RESET One state of a latch or flip-flop in which the Q output is LOW and the device is effectively storing a 0.

SET One state of a latch or flip-flop in which the Q output is HIGH and the device is effectively storing a 1.

Storage The property that allows a device to retain a 1 or 0 indefinitely.

Synchronous Having a fixed time relationship.

Timer A term that refers to either a monostable or an astable multivibrator. A monostable multivibrator is a circuit that produces a single pulse when triggered. An astable multivibrator is a type of oscillator that produces a continuous pulse waveform.

Toggle The action of a flip-flop when it changes state on each clock pulse.

Important Facts

- ❑ A basic latch consists of two gates with the output of each gate connected back to the input of the other gate, which is a form of feedback.
- ❑ The S-R latch has two inputs (S and R) and two outputs (Q and $\overline{Q}$). A gated S-R latch has an additional enable (EN) input.
- ❑ The D latch has two inputs (D and EN) and two outputs (Q and $\overline{Q}$).
- ❑ Latches and flip-flops are storage devices for binary data.
- ❑ A flip-flop cannot store a data bit until the triggering edge of a clock pulse occurs.
- ❑ There are two main types of flip-flops, the D and the J-K.
- ❑ A "bubble" at the clock input of a flip-flop symbol indicates negative edge-triggering. The absence of a "bubble" at the clock input indicates positive edge-triggering.
- ❑ The D or J-K inputs of a flip-flop are synchronous inputs because they control the state of the flip-flop only when a triggering clock edge occurs.
- ❑ The preset (PRE) input of a flip-flop is used to set the flip-flop independent of the clock pulse. The clear (CLR) input of a flip-flop is used to reset the flip-flop independent of the clock pulse. These inputs are asynchronous because they control the state of the flip-flop by overriding the clock.
- ❑ A one-shot produces a single output pulse when triggered. The width of the output pulse depends on the values of internal or external resistors and a capacitor.
- ❑ The 555 timer can be set up to operate as either a one-shot (monostable) or a pulse oscillator (astable).

Equations

Pulse width of 74121:

$$t_W = 0.7R_{\text{EXT}}C_{\text{EXT}}$$

Pulse width of 74LS122:

$$t_W = 0.32R_{\text{EXT}}C_{\text{EXT}}\left(1 + \frac{700}{R_{\text{EXT}}}\right)$$

Pulse width and frequency for the 555 timer:

$$t_W = 1.1R_{\text{EXT}}C_{\text{EXT}}$$

$$f = \frac{1.44}{(R_{\text{EXT1}} + 2R_{\text{EXT2}})C_{\text{EXT}}}$$

Chapter Checkup

Answers are at the end of the chapter.

1. A latch is a type of
   (a) monostable multivibrator (b) bistable multivibrator
   (c) astable multivibrator
2. A bistable device has
   (a) one stable state (b) no stable states
   (c) two stable states
3. When a latch is in the SET state, it is storing
   (a) a binary 1 (b) a binary 0
   (c) neither a 1 nor a 0
4. When a latch is RESET, it is
   (a) storing a binary 1 (b) storing a binary 0
   (c) cleared of all data
5. A gated D latch stores the data appearing on its D input
   (a) as soon as the data appears on the D input
   (b) when the data appears on the D input and the enable input is active
   (c) when the data appears on the D input and the enable input is not active
6. The basic difference between a latch and a flip-flop is
   (a) a flip-flop has no clock input
   (b) a flip-flop has more states
   (c) a flip has a clock input
7. A positive edge-triggered flip-flop can change its state only at
   (a) the rising edge of a clock pulse
   (b) the falling edge of a clock pulse
   (c) either the rising or falling edge of a clock pulse
8. When the D input of a D flip-flop is HIGH, the Q output after the next clock pulse will be
   (a) LOW
   (b) HIGH
   (c) determined by state of the Q output before the clock pulse

9. When both the J and K inputs of a J-K flip flop are HIGH and the Q output is LOW, the Q output after the next clock pulse will be

   (a) LOW

   (b) HIGH

   (c) determined by the Q output before the clock pulse

   (d) both (b) and (c)

10. When a flip-flop is set up for toggle operation, the Q output will

    (a) remain LOW (b) change state on each clock pulse

    (c) remain HIGH

11. When a flip-flop is cleared, it is in the

    (a) RESET state (b) SET state

    (c) no-change condition

12. An astable multivibrator has

    (a) one stable state (b) two stable states

    (c) no stable states

Questions

Answers to odd-numbered questions are at the end of the book.

1. Under what input conditions does an S-R latch assume the SET state?
2. When a latch is in the RESET state, is the Q output HIGH or LOW?
3. What is the difference between a gated and a nongated latch?
4. What are two basic types of latches as defined by their inputs?
5. How does a flip-flip differ from a latch?
6. What are the two basic types of flip-flops as defined by their inputs?
7. When can a positive edge-triggered flip-flop change state?
8. If $D = 1$, what is Q after a clock pulse occurs?
9. If $D = 0$, what is Q after a clock pulse occurs?
10. If $J = 1$ and $K = 0$, what is Q after a clock pulse occurs?
11. If $J = 0$ and $K = 1$, what is Q after a clock pulse occurs?
12. If $J = 0$ and $K = 0$, what is Q after a clock pulse occurs?
13. If $J = 1$ and $K = 1$, what is Q after a clock pulse occurs?
14. When a one-shot is triggered, what is its Q output?
15. What determines the output frequency of a 555 timer set up to operate in the astable mode?

PROBLEMS

Basic Problems

Answers to odd-numbered problems are at the end of the book.

1. For an S-R latch, determine the output for the following inputs if Q is initially HIGH.

   (a) S = LOW, R = LOW

   (b) S = LOW, R = HIGH

   (c) S = HIGH, R = LOW

2. Determine the Q output of a gated D latch for the following inputs if Q is initially LOW.
   (a) S = LOW, R = LOW, EN = HIGH
   (b) S = LOW, R = HIGH, EN = HIGH
   (c) S = HIGH, R = LOW, EN = HIGH
   (d) S = HIGH, R = LOW, EN = LOW
3. Identify each of the symbols in Figure 7–33.

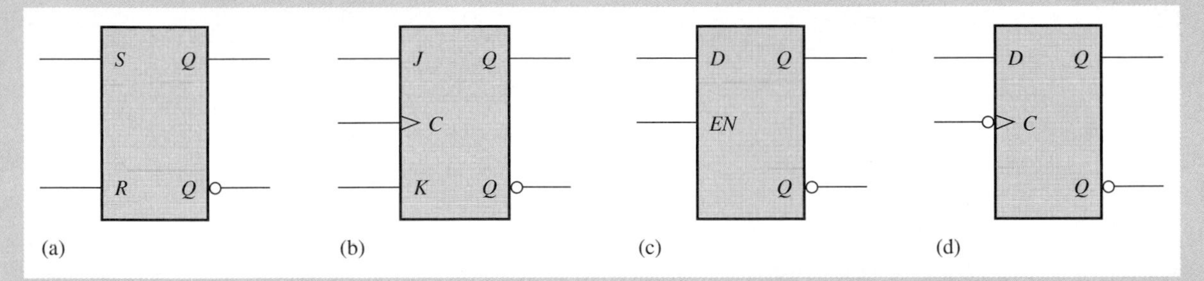

FIGURE 7–33

4. Determine the $\overline{Q}$ output of a D flip-flop after one clock pulse for the following inputs:
   (a) $D = 1$ (b) $D = 0$
5. Determine the $\overline{Q}$ output of a J-K flip-flop after one clock pulse for the following inputs. Assume that $\overline{Q}$ is initially LOW.
   (a) $J = 0, K = 0$ (b) $J = 0, K = 1$
   (c) $J = 1, K = 0$ (d) $J = 1, K = 1$
6. What is the Q output of a J-K flip-flop with $J = 1$ and $K = 1$ after three clock pulses? Assume Q starts out LOW (0).
7. If a D flip-flop has active-LOW preset ($\overline{PRE}$) and clear ($\overline{CLR}$) inputs, determine the Q output after one clock pulse for the following inputs.
   (a) $D = 0, \overline{CLR} = 1, \overline{PRE} = 0$
   (b) $D = 1, \overline{CLR} = 0, \overline{PRE} = 1$
   (c) $D = 1, \overline{CLR} = 1, \overline{PRE} = 1$
8. Sketch the Q output waveform for each flip-flop in Figure 7–34 in relation to the clock waveform shown. Assume Q is LOW to start.

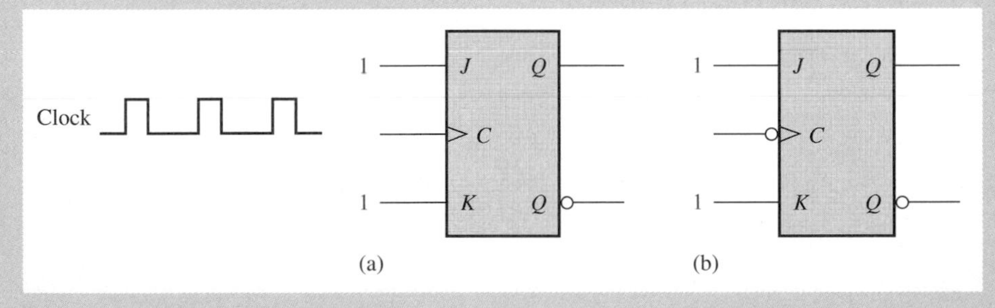

FIGURE 7–34

9. A 74121 one-shot has external component values of $R_{EXT} = 10\ k\Omega$ and $C_{EXT} = 0.22\ \mu F$. Determine the width of the output pulse.
10. Calculate the pulse width if the one-shot in Problem 9 is a 74LS122.
11. Calculate the pulse width for a 555 timer configured as a one-shot as shown in Figure 7–22 with $R_{EXT} = 2.2\ k\Omega$ and $C_{EXT} = 1.5\ \mu F$.

12. Determine the frequency of a 555 timer connected for astable operation as shown in Figure 7–23 for $R_{EXT1} = 1.0\ k\Omega$, $R_{EXT2} = 5.6\ k\Omega$, and $C_{EXT} = 1\ \mu F$.

Basic-Plus Problems

13. Show the Q output waveform in relation to the S and R inputs shown in Figure 7–35. Q starts out LOW.

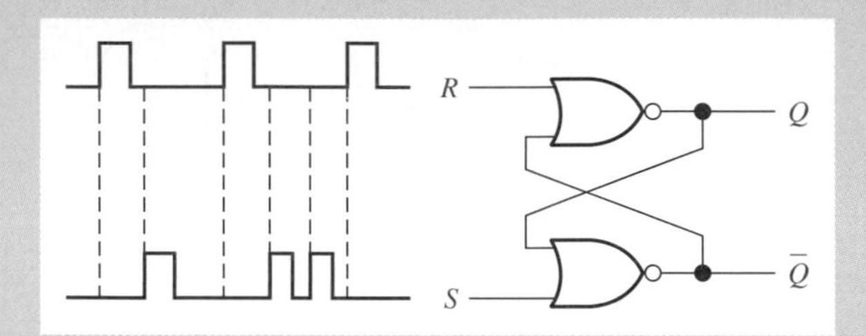

FIGURE 7–35

14. The waveforms in Figure 7–36 are applied to a gated S-R latch like the one in Figure 7–6. Show the Q and $\overline{Q}$ output waveforms. The latch is initially RESET.

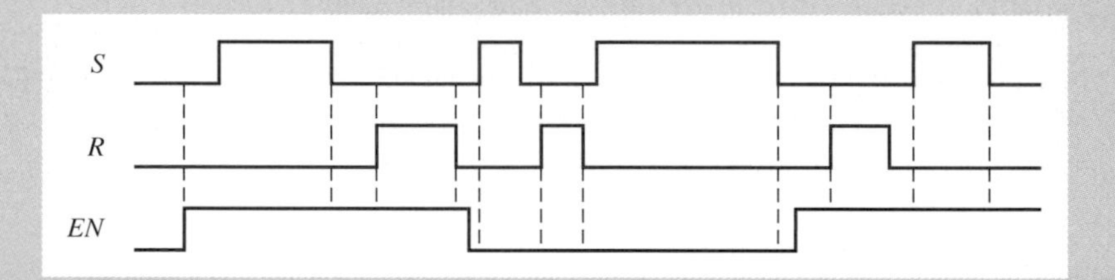

FIGURE 7–36

15. Show the Q and $\overline{Q}$ output waveforms for the latch in Figure 7–37 in relation to the input waveforms shown. The latch starts out RESET.

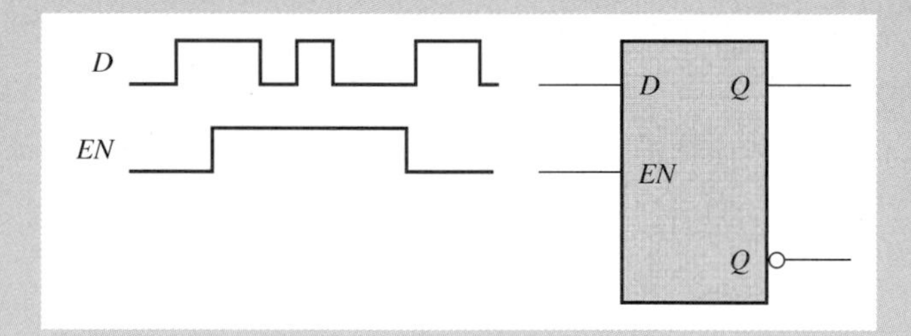

FIGURE 7–37

16. Determine the Q and $\overline{Q}$ outputs of the flip-flop in Figure 7–38 for the input waveforms shown. Q starts out LOW.

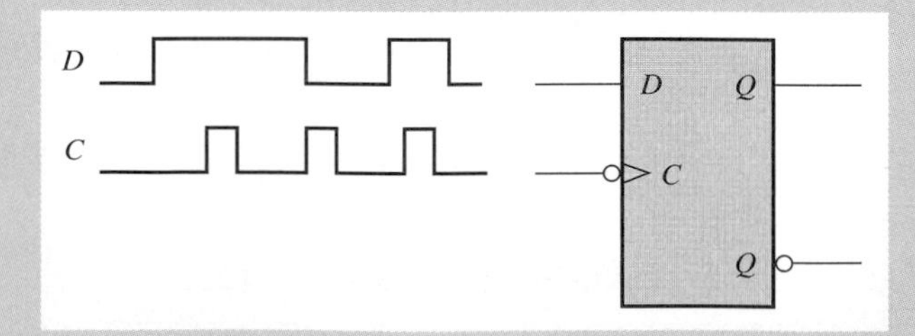

FIGURE 7–38

17. Complete the timing diagram in Figure 7–39 by drawing the Q output waveform of the flip-flop for the input waveforms shown. The flip-flop is initially RESET.

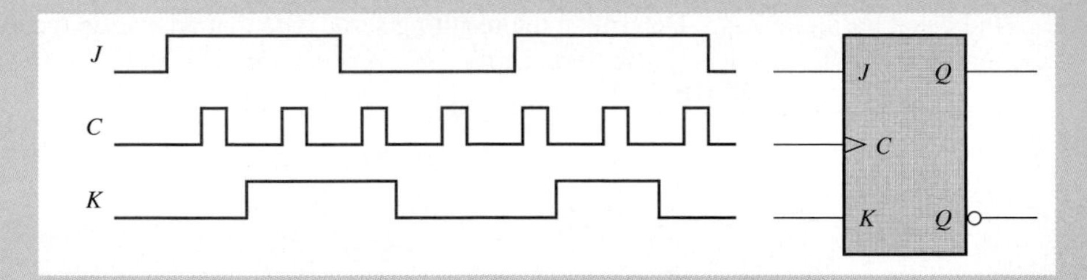

FIGURE 7–39

18. For the input waveforms in Figure 7–40, show the flip-flop's Q waveform. Q is initially 0.

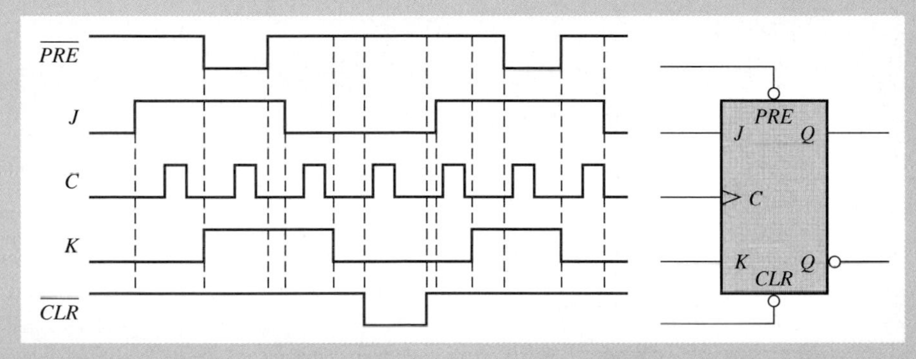

FIGURE 7–40

19. Draw the Q output waveform relative to the clock for a positive edge-triggered D flip-flop with the input waveforms shown in Figure 7–41. The flip-flop is initially in the RESET state.

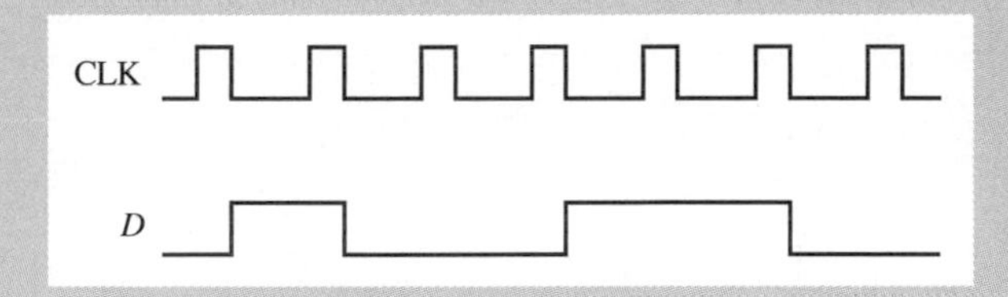

FIGURE 7–41

20. Show the Q waveform relative to the clock for a positive edge-triggered J-K flip-flop with the inputs in Figure 7–42 and starting in the RESET state.

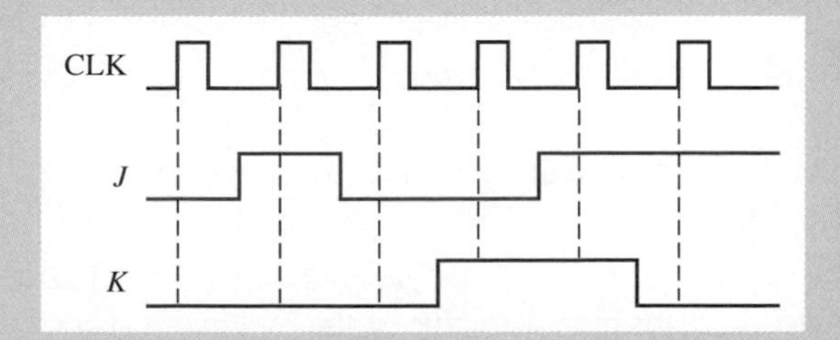

FIGURE 7–42

21. Design a circuit using a J-K flip-flop that divides the input clock frequency by 2 and produces a square wave.

22. The following serial data are applied to the flip-flop through the AND gates as indicated in Figure 7–43. Determine the resulting serial data that appear on the Q output. There is one clock pulse for each bit time. Assume that Q is initially 0 and $\overline{PRE}$ and $\overline{CLR}$ are both HIGH. The leftmost bit is applied first.

J_1: 1 0 1 0 0 1 1

J_2: 0 1 1 1 0 1 0

J_3: 1 1 1 1 0 0 0

K_1: 0 0 0 1 1 1 0

K_2: 1 1 0 1 1 0 0

K_3: 1 0 1 0 1 0 1

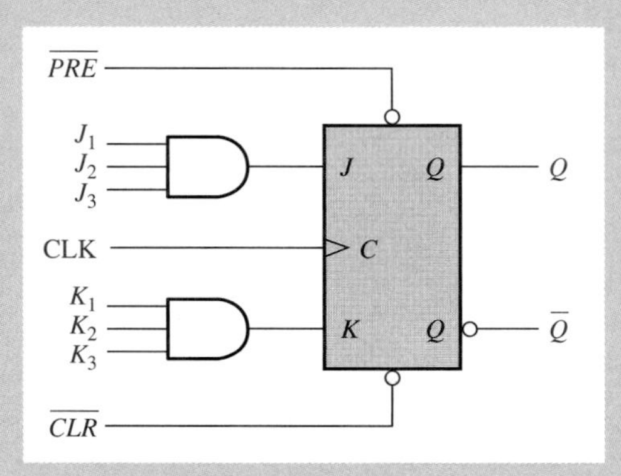

FIGURE 7–43

23. The D flip-flop in Figure 7–44 is connected as shown. Determine the Q output in relation to the clock if Q is initially 0.

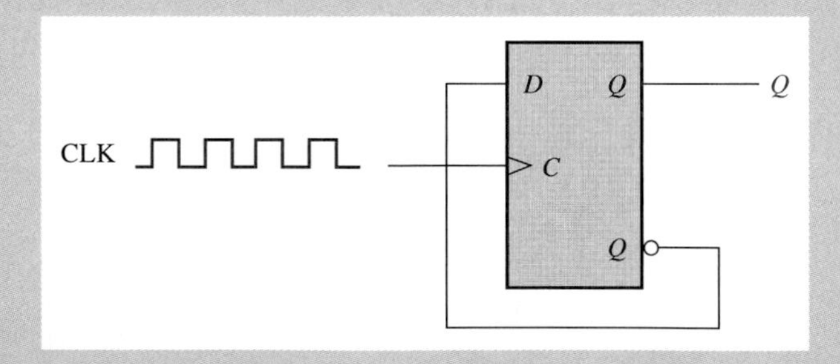

FIGURE 7–44

24. Determine the capacitor value for a 74121 one-shot if the external resistor is 5.6 kΩ and the pulse width is to be 5.1 μs.

25. An output pulse with a pulse width of 0.5 ms is to be generated by a 74LS122 one-shot. Using a 10,000 pF capacitor, determine the value of the external resistor.

26. Design a 555 timer to operate as a one-shot that will produce a 0.25 s output pulse.

27. A 555 timer is connected to operate in the astable mode as shown in Figure 7–45. Determine the output frequency.

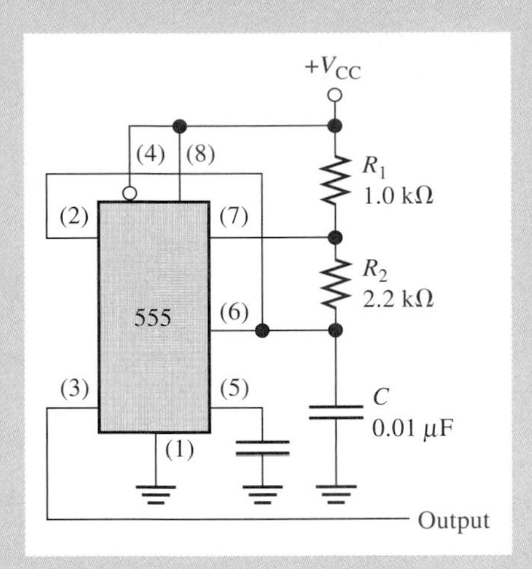

FIGURE 7–45

TROUBLESHOOTING PRACTICE

Use the circuits in the specified Multisim files on the website.

28. Open file TSP07-01 and troubleshoot the simulated S-R latch. Determine whether or not it is working properly. If it is not working properly, determine the most likely fault.

29. Open file TSP07-02 and troubleshoot the simulated gated S-R latch. Determine whether or not it is working properly. If it is not working properly, determine the most likely fault.

30. Open file TSP07-03 and troubleshoot the simulated gated D latch. Determine whether or not it is working properly. If it is not working properly, determine the most likely fault.

31. Open file TSP07-04 and troubleshoot the simulated D flip-flop. Determine whether or not it is working properly. If it is not working properly, determine the most likely fault.

32. Open file TSP07-05 and troubleshoot the simulated J-K flip-flop. Determine whether or not it is working properly. If it is not working properly, determine the most likely fault.

33. Open file TSP07-06 and troubleshoot the simulated 555 timer operating as an astable multivibrator. Determine whether or not it is working properly. If it is not working properly, determine the most likely fault.

ANSWERS

Example Questions

7–1: Because the latch is already in the SET state.

7–2: Because the latch is not enabled (*EN* is LOW).

7–3: The Q output would not change.

7–4: The Q output would be HIGH (equal to D).

7–5: Q would go HIGH on the first clock pulse and stay HIGH.

7–6: Q would change state at each clock pulse (toggle).

7–7: Q would change state at each clock pulse (toggle).

7–8: $C_{EXT} = 0.155\ \mu F$

7–9: $f = 214$ kHz

7–10: See Figure 7–46.

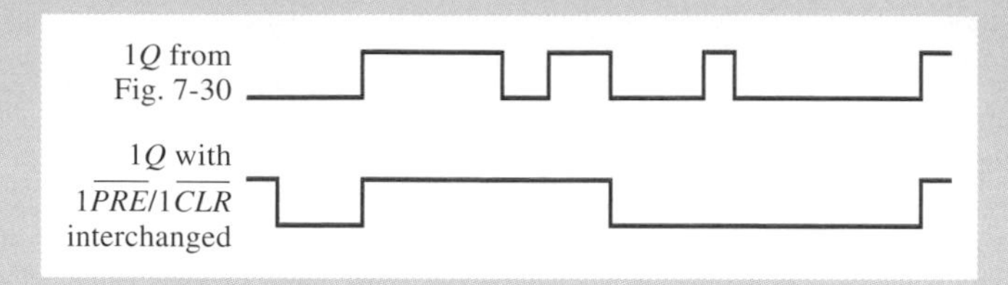

FIGURE 7–46

Review Questions

1. A latch stores a binary 1 or 0.
2. S is SET and R is RESET.
3. Q is HIGH (1) when SET.
4. Q is LOW (0) when RESET.
5. S = HIGH and R = HIGH is invalid.
6. The inputs of a D latch are D and EN (enable).
7. The EN input enables the latch so that Q follows the D input.
8. The D latch is a variation of the gated S-R latch with the D input connected directly to S and to R through an inverter.
9. Q is HIGH when D is HIGH and the latch is enabled.
10. The Q output does not change.
11. The inputs are D and C (clock).
12. A triangular symbol at the clock input indicates edge-triggering.
13. On the rising edge of a clock pulse
14. A bubble at the clock input indicates negative edge-triggering.
15. The Q goes HIGH on the first clock pulse and remains HIGH.
16. The inputs are J, K, and C (clock).
17. Both J and K must be HIGH (1) for toggle operation.
18. On the falling edge of a clock pulse
19. The preset input is used to set the flip-flop independent of the clock.
20. The clear input is used to reset the flip-flop independent of the clock.
21. A one-shot has one stable state.
22. A one-shot produces one pulse when triggered.
23. The external resistor and capacitor values determine the duration (width) of the output pulse.

24. A nonretriggerable one-shot, once triggered cannot be triggered again until it times out.
25. A retriggerable one-shot can be triggered before the output pulse ends in order to extend the pulse width.
26. Monostable and astable
27. The external resistor and capacitor values determine the width of the output pulse.
28. The external resistor and capacitor values determine the frequency.
29. Increase the external resistor or capacitor values.
30. By the external resistors
31. A quad latch package contains four latches.
32. A latch does not operate with a synchronous clock input as a flip-flop does.
33. V_{CC} and ground
34. Quad $\overline{S}$-$\overline{R}$ latch.
35. Dual positive edge-triggered D flip-flop
36. Q HIGH, $\overline{Q}$ LOW
37. The flip-flop will toggle.
38. Q will go HIGH and stay there.
39. Q will not change.
40. Incorrect values of R_{EXT} or C_{EXT}

Chapter Checkup

1. (b)	2. (c)	3. (a)	4. (b)	5. (b)
6. (c)	7. (a)	8. (b)	9. (d)	10. (b)
11. (a)	12. (c)			

COUNTERS

CHAPTER OUTLINE

Study aids for this chapter are available at

http://www.prenhall.com/SOE

INTRODUCTION

This chapter introduces counters, one of the most important types of circuits in digital electronics. Counters are used not only for counting events but also to set the order of operations in a sequential logic circuit. Counters are also applied to diverse circuits such as analog-to-digital converters, arithmetic circuits, waveform generators, keyboard encoders, and frequency dividers.

Digital counters are made of interconnected flip-flops. The basic function of a counter is to advance through a prescribed series of binary states in response to a series of input pulses called the clock. The number of flip-flops and the way they are connected determine the number of states in a counter's sequence. Integrated circuit counters incorporate the flip-flops into a single integrated circuit and frequently have added features such as parallel load inputs, clear, and enable functions. You will learn about these features in this chapter.

KEY OBJECTIVES

A section number is given for each objective. After completing this chapter, you should be able to

- **8–1** Count with binary numbers
- **8–2** Describe the operation of a binary counter and explain frequency division
- **8–3** Describe the operation of decade counters and explain cascaded counters
- **8–4** Explain how a counter is decoded
- **8–5** Understand the basic operation of a digital clock
- **8–6** Describe two specific IC counters
- **8–7** Troubleshoot basic faults in counters

COMPUTER SIMULATIONS DIRECTORY

The following figures have Multisim circuit files associated with them.

- Figure 8–3 Page 229
- Figure 8–8 Page 232
- Figure 8–12 Page 235
- Figure 8–23 Page 244

LABORATORY EXPERIMENTS DIRECTORY

The following exercise is for this chapter.

- **Experiment 12** Synchronous Counters

KEY TERMS

- Binary counter
- Synchronous counter
- Asynchronous counter
- Modulus
- Frequency divider
- Terminal count
- Decade counter
- Cascade
- Counter decoding

ON THE JOB. . .

(Getty Images)

All valuable employees are reliable. It is important to arrive at your workplace on time and be ready to work. If there is a valid reason you can't be at work or will be late, call your supervisor and explain the situation. Most supervisors will appreciate knowing about a problem before you are absent or late rather than afterward.

The traditional method of analyzing biomolecules such as proteins or DNA uses a test tube approach, which is somewhat limited. Scientists are now working to create "labs on chips" and other nanoscale devices for the purpose of sorting, measuring, and counting the various molecules of life. This research may make it possible to perform far more accurate genetic and diagnostic tests and greatly expand the methods for analyzing very small amounts of biochemicals.

8–1 COUNTING WITH BINARY NUMBERS

Binary numbers were introduced in Chapter 1. Counting with binary numbers forms the basis for understanding the operation of counters and how they are used.

In this section, you will learn how to count using binary numbers.

Recall that a binary number is made up of bits (1s and 0s) that are weighted depending on the position of the bit in the number. For example, an n-bit whole number has a weight structure where the value of the weights increase from right to left. The left-most bit is the **MSB** (most-significant bit) and has a weight of 2^n. The right-most bit is the **LSB** (least-significant bit) and has a weight of 2^0.

$$2^n \cdots 2^4\ 2^3\ 2^2\ 2^1\ 2^0$$

MSB LSB

The number of bits in a binary number determines the range and highest number that can be represented. With n bits, you can count from 0 to $2^n - 1$. For example, with three bits you can count from 0 to 7, with four bits you can count from 0 to 15, and so on. Table 8–1 shows the range of numbers and the highest number for various numbers of bits.

TABLE 8–1

Number of bits	Range of numbers	Highest number ($2^n - 1$)
2	0 to 3	3
4	0 to 15	15
8	0 to 255	255
16	0 to 65,535	65,535
32	0 to 4,294,967,295	4,294,967,295

When you count through a sequence of binary numbers, each bit changes according to a specific pattern. The LSB changes on every number, the next bit to the left (2^1) changes every two numbers, the next bit to the left (2^2) changes every four numbers, and so on up to the MSB (2^n) which changes only once. For example, the following 3-bit binary sequence shows the pattern.

	0	0	0	LSB changes on every other number
	0	0	1	
	0	1	0	
	0	1	1	
	1	0	0	Next bit changes every two numbers
MSB changes once in the sequence	1	0	1	
	1	1	0	
	1	1	1	

Review Questions

Answers are at the end of the chapter.

1. What is the range of numbers and the highest number using four bits?
2. What is the range of numbers and the highest number using five bits?
3. How many times does the LSB change when counting with a complete 4-bit sequence?
4. How many times does the MSB change when counting with a complete 4-bit sequence?
5. In the binary number 100, what is the weight of the 1?

THE BINARY COUNTER 8–2

Counters are used in digital applications for producing number sequences, counting events such as the occurrence of pulses, and dividing frequencies down to lower values.

In this section, you will learn how a binary counter operates and be able to explain frequency division.

A **binary counter** is a digital circuit that uses flip-flops connected together to produce a sequence of states that represents binary numbers. Recall that a flip-flop is in the SET state when its Q output is HIGH, representing a 1. A flip-flop is in the RESET state when its Q output is LOW, representing a 0. From a binary point of view, we will refer to the SET state as the 1 state and the RESET state as the 0 state.

All counters have a clock input that causes the counter to sequence through its states. Counters can be clocked in two basic ways — synchronously and asynchronously. In a **synchronous counter**, the clock input is connected to each flip-flop so that the flip-flops are triggered at the same time. In an **asynchronous counter**, the clock input is connected only to the least significant flip-flop. The remaining flip-flops in the counter are clocked by the output of the preceding flip-flop instead of by a common clock pulse, so they are not triggered at exactly the same time due to propagation delays. We will restrict the coverage to synchronous counters because most IC counters are of that type. However, an asynchronous counter using J-K flip-flops was included in Experiment 11 in the lab manual to illustrate this method.

A 2-Bit Counter

When two J-K flip-flops, designated FF0 and FF1 are connected together as shown in Figure 8–1, you have a 2-bit synchronous binary counter. FF0 is the LSB and FF1 is the MSB. The clock input (CLK) is connected to both flip-flops, and the J and K inputs of FF0 are connected to a HIGH. The Q output of FF0 is connected to the J and K inputs of FF1.

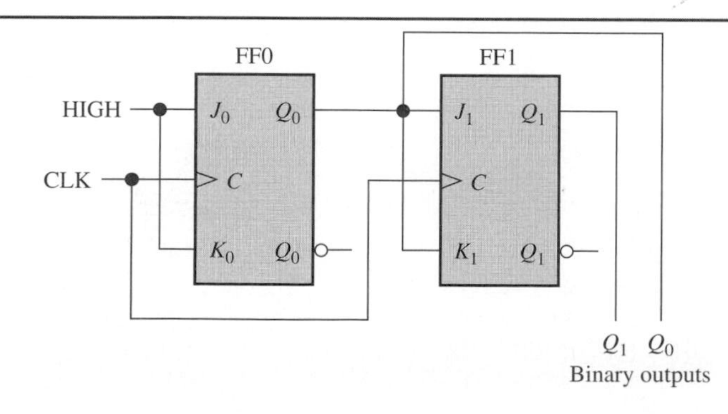

FIGURE 8–1
A 2-bit synchronous binary counter.

TABLE 8–2

State table for a 2-bit binary counter.

Clock pulse (CLK)	Q_1	Q_0
Initial state	0	0
After CLK 1	0	1
After CLK 2	1	0
After CLK 3	1	1

Table 8–2 shows the state table for the 2-bit binary counter. The states of Q_1 and Q_0 represent the binary number sequence. The operation is described as follows:

First clock pulse Let's assume that both flip-flops are initially RESET to the 0 state, so the counter is in the 00 state ($Q_1 = 0$ and $Q_0 = 0$). When the triggering edge of first clock pulse occurs, FF0 will change to the 1 state because its J_0 and K_0 inputs are both HIGH, creating a toggle condition. FF1 will not change because, at the time of the clock edge, its J_1 and K_1 inputs are both 0, creating a no-change condition. The counter is now in the 01 state ($Q_1 = 0$ and $Q_0 = 1$).

Second clock pulse When the second clock pulse occurs, FF0 toggles back to the 0 state. Because $J_1 = 1$ and $K_1 = 1$ at the time of the triggering clock edge, FF1 will toggle to the 1 state. The counter is now in the 10 state ($Q_1 = 1$ and $Q_0 = 0$).

Third clock pulse When the third clock pulse occurs, FF0 toggles back to the 1 state and, because $J_1 = 0$ and $K_1 = 0$ at the time of the triggering clock edge, FF1 will remain in the 1 state. The counter is now in the 11 state ($Q_1 = 1$ and $Q_0 = 1$). This is the maximum binary count and is equivalent to 3.

Recycle When the fourth clock pulse occurs, FF0 toggles back to the 0 state and, because $J_1 = 1$ and $K_1 = 1$ at the time of the clock edge, FF1 will also toggle back to the 0 state. The counter is now back in the 00 state ($Q_1 = 0$ and $Q_0 = 0$). The counter has *recycled* from its maximum state, 11, back to its minimum state, 00.

The counter goes through a sequence of states representing four binary numbers from 00 (0) to 11 (3). The number of unique states through which a counter sequences is called the **modulus**. The 2-bit binary counter has a modulus of 4.

Figure 8–2 shows the timing diagram of the 2-bit counter in Figure 8–1. It ideally shows that the Q outputs change at the same instant that the triggering edge of the clock pulse occurs. However, you should realize that there is a propagation delay time from the triggering edge of the clock pulse to the actual transition of a Q output. This delay time is due to the propagation through the flip-flops and is usually only a few nanoseconds. At the positive edge of the second clock pulse, Q_0 changes from 1 to 0. In reality it takes a few nanoseconds for this to happen, so the 1 on Q_0 remains on the J_1 and K_1 inputs of FF1 long enough for Q_1 to toggle from 0 to 1 on the same clock pulse. Similarly, at the triggering edge of the third clock pulse, the 0 on Q_0 remains on the J_1 and K_1 inputs of FF1 just long enough to prevent Q_1 from changing. The propagation delay time through a flip-flop is crucial for a counter such as the one in Figure 8–1 to operate properly.

FIGURE 8–2

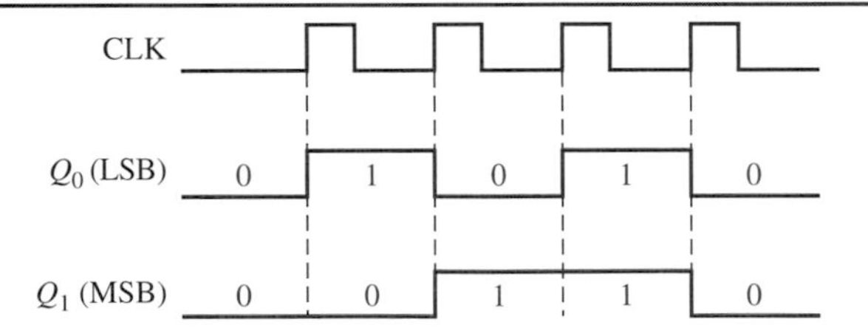

Timing diagram for the 2-bit binary counter in Figure 8–1. The timing diagram agrees with the state table in Table 8–2.

COMPUTER SIMULATION

Open the Multisim file F08-03DG on the website. The pulse generator produces one pulse per second and is the clock for the counter. The counter will go through its binary sequence 00, 01, 10, 11. You can observe this operation by following the states of the probe lights. Notice that in this counter, the flip-flops are positive edge triggered.

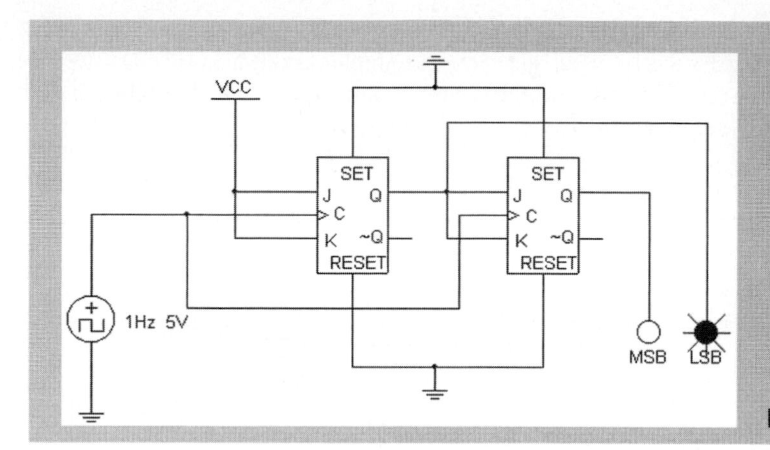

FIGURE 8–3

A 4-Bit Binary Counter

Table 8–3 shows the state table for a 4-bit binary counter. The counter has $2^4 = 16$ states and it counts through a binary sequence from 0000 (0) to 1111 (15). Because the 4-bit counter has sixteen states, its modulus is 16.

TABLE 8–3

State table for a 4-bit binary counter. Toggle conditions are indicated by the blue areas.

Clock pulse (CLK)	Q_3	Q_2	Q_1	Q_0
Initial state	0	0	0	0
After CLK 1	0	0	0	1
After CLK 2	0	0	1	0
After CLK 3	0	0	1	1
After CLK 4	0	1	0	0
After CLK 5	0	1	0	1
After CLK 6	0	1	1	0
After CLK 7	0	1	1	1
After CLK 8	1	0	0	0
After CLK 9	1	0	0	1
After CLK 10	1	0	1	0
After CLK 11	1	0	1	1
After CLK 12	1	1	0	0
After CLK 13	1	1	0	1
After CLK 14	1	1	1	0
After CLK 15	1	1	1	1

As you can see from Table 8–3, Q_0 changes state (toggles) on each clock pulse edge. Q_1 always changes state (toggles) on the next clock pulse after Q_0 is 1. Q_2 always changes state on the next clock pulse after both Q_0 and Q_1 are 1. Q_3 always changes state on the next clock pulse after Q_0, Q_1, and Q_2 are all 1. These conditions, indicated by the blue areas in the table, are described by the following Boolean equations:

$$J_0 = K_0 = 1$$

$$J_1 = K_1 = Q_0$$

$$J_2 = K_2 = Q_1Q_0$$

$$J_3 = K_3 = Q_2Q_1Q_0$$

The counter in Figure 8–4 goes through the binary sequence shown in Table 8–3. The conditions just described by the Boolean equations are implemented using AND gates in addition to the flip-flops. AND gate G_1 detects when Q_0 and Q_1 are both 1. AND gate G_2 detects when Q_0 and Q_1 and Q_2 are all 1.

FIGURE 8–4

A 4-bit binary counter.

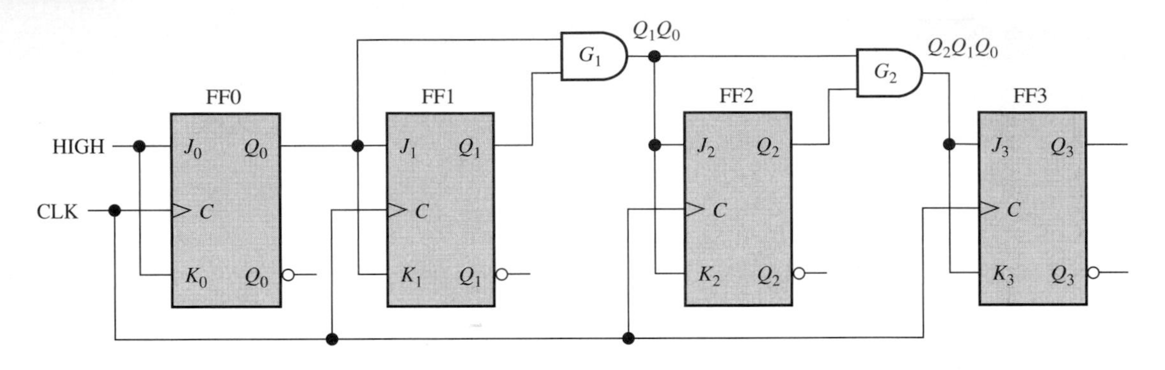

Figure 8–5 shows how the Q outputs of the positive edge-triggered flip-flops change with respect to the clock pulse. Q_0 changes every clock pulse, Q_1 changes every other clock pulse, Q_2 changes on every fourth clock pulse, and Q_3 changes every eighth clock pulse.

FIGURE 8–5

Timing diagram for the 4-bit binary counter in Figure 8–4. The red labels indicate when the AND gate outputs are HIGH.

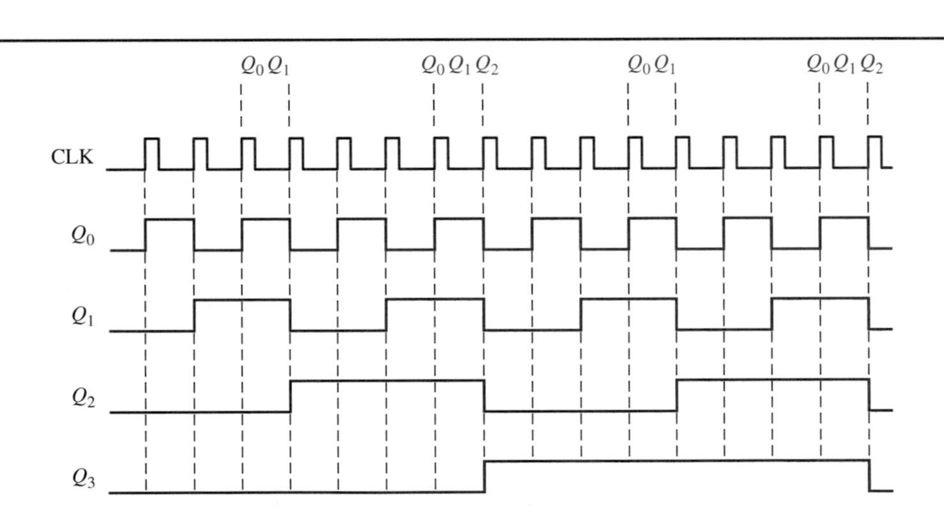

Frequency Division

A counter used specifically to divide the input clock frequency by a specified amount is a **frequency divider**. In the timing diagram of Figure 8–5, you can see that Q_0 is half the frequency of the clock, Q_1 is one-fourth the clock frequency, Q_2 is one-eighth the clock frequency, and Q_3 is one-sixteenth the clock frequency. The maximum division factor defines the counter, so a 4-bit binary counter is a divide-by-16 counter.

EXAMPLE 8–1

Problem

Show the timing diagram for a 3-bit binary counter with negative edge-triggering.

Solution

All the Q waveforms change on the negative edge of a clock pulse. Q_0 changes every clock pulse, Q_1 changes every other clock pulse, and Q_2 changes every fourth clock pulse. The Q waveforms are shown in Figure 8–6.

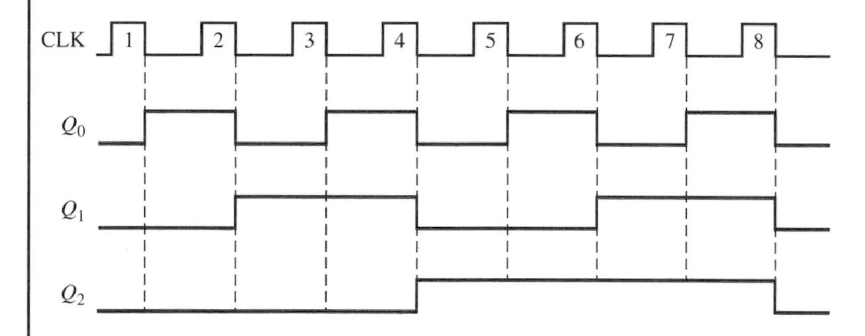

FIGURE 8–6

Question*

Under what condition does Q_2 change?

Instead of showing all the flip-flops, a counter is usually shown as a block symbol. For example, the block symbol for a basic 4-bit counter is shown in Figure 8–7. The DIV 16 label identifies the counter as a 4-bit binary counter by indicating its "divide-by" factor.

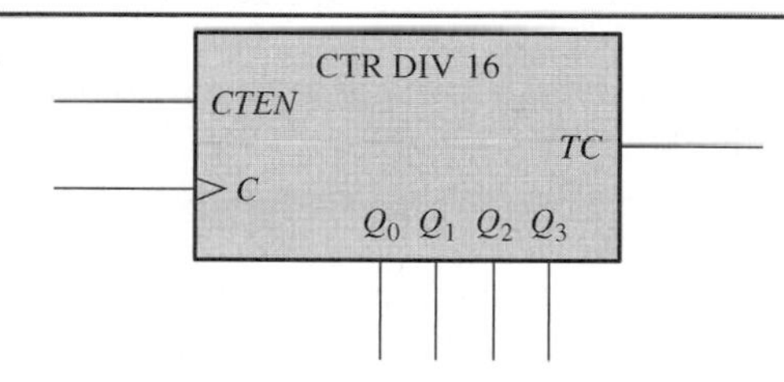

FIGURE 8–7

General block symbol for a 4-bit counter.

Most counters provide the Qs of all the flip-flop as outputs. The *CTEN* (count enable) input actually goes to the J and K inputs of the first flip-flop and must be 1 (HIGH) in this case, for the counter to operate. The **terminal count** (*TC*) is the final state in a counter's sequence. The *TC* output is sometimes called ripple counter output, *RCO,* and is common on most counters. A 1 (HIGH) on this output indicates that the counter has reached its final state or terminal count. For a 4-bit binary counter the terminal count is 15 and is indicated by decoding the 1111 state. The terminal count decoder can be implemented with an AND gate with Q_0, Q_1, Q_2, and Q_3 on its inputs, so its output is 1 when all the inputs are 1s.

* *Answers are at the end of the chapter.*

COMPUTER SIMULATION

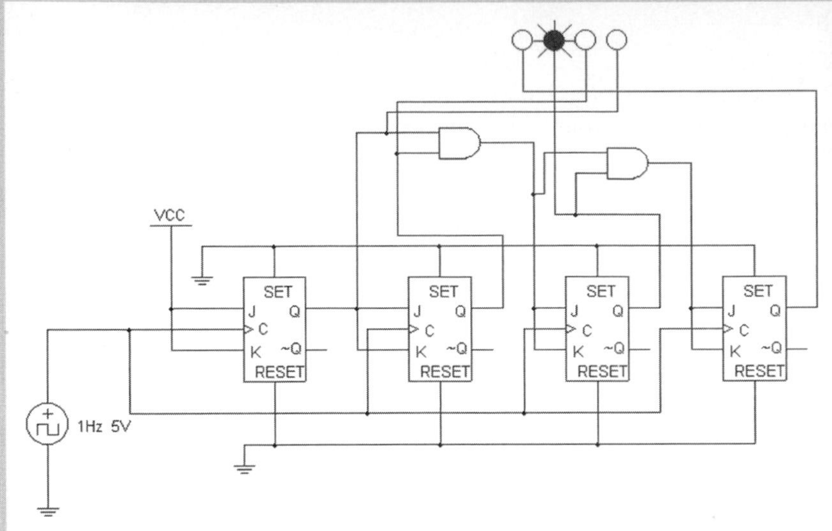

FIGURE 8–8

Open the Multisim file F08-08DG on the website. The pulse generator produces one pulse per second and is the clock for the counter. The counter will go through its binary sequence 0000 to 1111. You can observe this operation by following the states of the probe lights. Notice that in this counter, the flip-flops are positive edge triggered.

Higher-Modulus Counters

As you have learned, the modulus of a counter is defined to be the number of states in its sequence. By using a sufficient number of flip-flops, a counter with any modulus 2^n can be made, where n is the number of flip-flops. The number of flip-flops in a counter must equal the power-of-two required to produce a number equal to or greater than the modulus of the counter. For example, a binary counter with five flip-flops has a modulus of $2^5 = 32$ and can count from 0 to 31 and a counter with six flip-flops has a modulus of $2^6 = 64$ and can count from 0 to 63.

EXAMPLE 8–2

Problem
A certain binary counter is required to count up to 255. What is its modulus and how many flip-flops are required?

Solution
The counter counts from 0 to 255 so it has **256** states in its sequence, which is its modulus. It takes **eight** flip-flops to implement a binary counter with a modulus of 256 because $2^8 = 256$.

Question
What is the maximum count possible for a binary counter with 10 flip-flops?

Review Questions

6. How many states does a 2-bit binary counter have?
7. What is the modulus of a 4-bit binary counter?
8. What is the highest binary number for an 8-bit binary counter?
9. What is the terminal count of a counter?
10. What is the difference between a synchronous and an asynchronous counter?

THE DECADE COUNTER 8–3

A decade counter has ten states. Because ten is not a power of two, the modulus of a 4-bit binary counter must be modified to achieve a counter with ten states. Keep in mind that any modulus can be obtained in a counter with the proper logic.

In this section, you will learn how a binary counter can be modified to make a decade counter and also how to cascade counters.

The BCD Decade Counter

Any counter having ten states in its sequence is a **decade counter**. The most common decade counter has a binary coded decimal (BCD) sequence. The BCD decade counter is a 4-bit counter with a sequence that goes from 0 to 9 (0000 to 1001). It is the same sequence as the 4-bit binary counter up to 1001, but then it recycles back to 0000, as shown in Table 8–4.

TABLE 8–4

State table for a BCD decade counter.

Clock pulse (CLK)	Q_3	Q_2	Q_1	Q_0
Initial state	0	0	0	0
After CLK 1	0	0	0	1
After CLK 2	0	0	1	0
After CLK 3	0	0	1	1
After CLK 4	0	1	0	0
After CLK 5	0	1	0	1
After CLK 6	0	1	1	0
After CLK 7	0	1	1	1
After CLK 8	1	0	0	0
After CLK 9	1	0	0	1

A decade counter must have four flip-flops to have ten states because three flip-flops can produce a maximum of only eight states ($2^3 = 8$). As you have seen with the 4-bit binary counter, four flops-flops can produce sixteen states. Therefore, the decade counter must be forced to recycle to 0000 instead of going to 1010 when it is at the 1001 state. This can be done with logic gates.

Table 8–4 shows that Q_0 changes on each clock pulse, even when it recycles back to 0000. Therefore, FF0 must toggle on each clock pulse, so the Boolean expression for its J and K inputs is

$$J_0 = K_0 = 1$$

This is implemented by connecting J_0 and K_0 to a HIGH.

Next, notice in Table 8–4 that Q_1 changes on the next clock pulse each time that $Q_0 = 1$ except when the counter is in the 1001 state. It can be stated that Q_1 changes each time that $Q_0 = 1$ and $Q_3 = 0$ but not when $Q_3 = 1$. To set up the proper toggle condition for FF1, the Boolean expression for its J and K inputs is

$$J_1 = K_1 = \overline{Q}_3Q_0$$

This is implemented by connecting Q_0 and $\overline{Q_3}$ to the inputs of an AND gate and connecting the gate output to the J_1 and K_1 inputs of FF1.

Table 8–4 shows that Q_2 changes on the next clock pulse each time $Q_0 = 1$ and $Q_1 = 1$. Therefore, the Boolean expression for the J and K inputs of FF2 is

$$J_2 = K_2 = Q_1Q_0$$

This is implemented by ANDing Q_0 and Q_1 and connecting the gate output to the J_2 and K_2 inputs of FF2.

Finally, as you can see in Table 8–4, Q_3 changes on the next clock pulse when $Q_0 = 1$, $Q_1 = 1$, and $Q_2 = 1$ (state 7) or it also changes when $Q_0 = 1$ and $Q_3 = 1$ (state 9). This condition is expressed as

$$J_3 = K_3 = Q_2Q_1Q_0 + Q_3Q_0$$

This condition is implemented with AND/OR logic connected to the J_3 and K_3 inputs of FF3. The complete logic diagram of a BCD decade counter is shown in Figure 8–9.

FIGURE 8–9 The BCD decade counter.

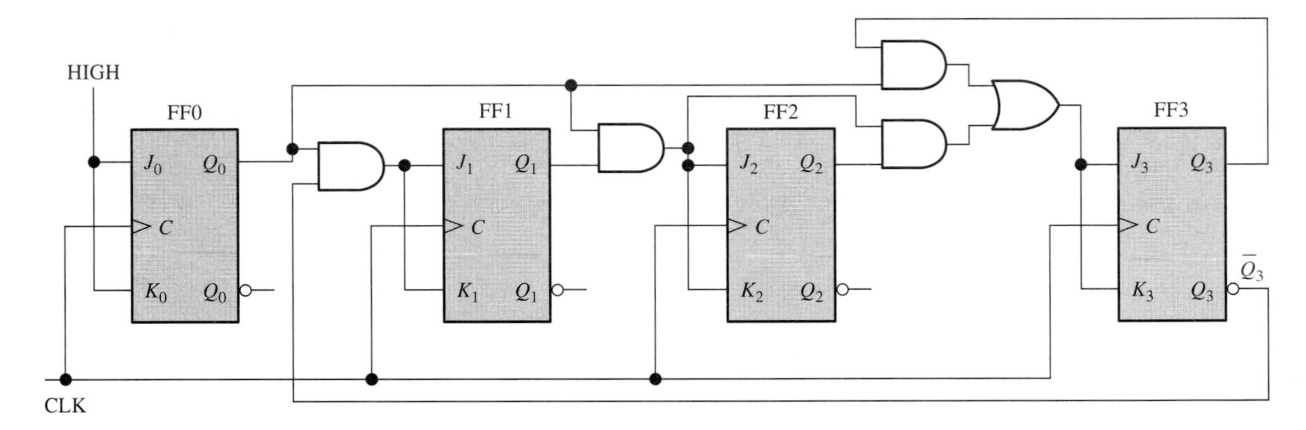

Figure 8–10 shows how the Q outputs of the positive edge-triggered flip-flops change with respect to the clock pulse. Notice that the BCD decade counter goes through the same binary sequence as the 4-bit binary counter up to 1001. However, instead of going to count 1010, the BCD decade counter recycles back to 0000 because of its shortened (truncated) sequence.

FIGURE 8–10 Timing diagram for the BCD decade counter.

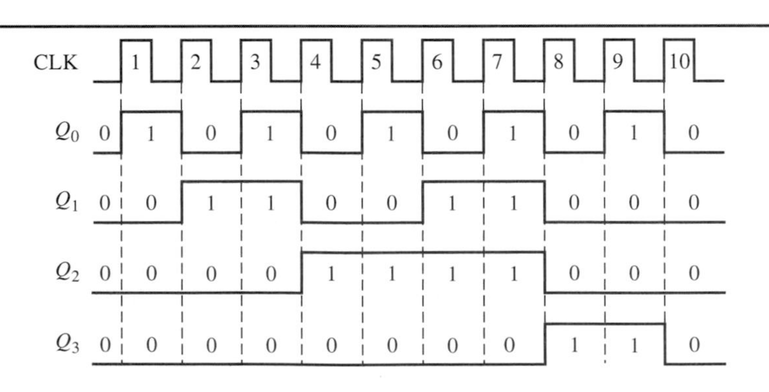

The block symbol for a decade counter is shown in Figure 8–11. The DIV 10 label identifies the counter as a decade counter by indicating its "divide-by" factor. The terminal count (TC) 1001 can be decoded with an AND gate with Q_0 and Q_3 on its inputs because this is the only state in the sequence in which both of these bits are 1.

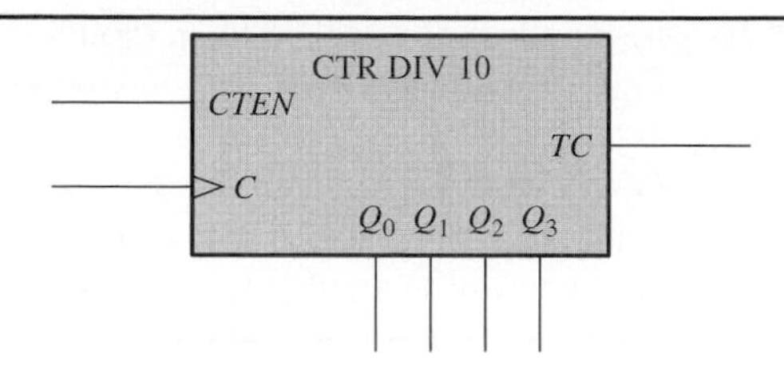

FIGURE 8–11

Basic block symbol for a decade counter.

COMPUTER SIMULATION

FIGURE 8–12

Open the Multisim file F08-12DG on the website. The pulse generator produces one pulse per second and is the clock for the counter. The decade counter will go through its BCD sequence 0000 through 1001. You can observe this operation by following the states of the probe lights. Notice that in this counter, the flip-flops are positive edge triggered.

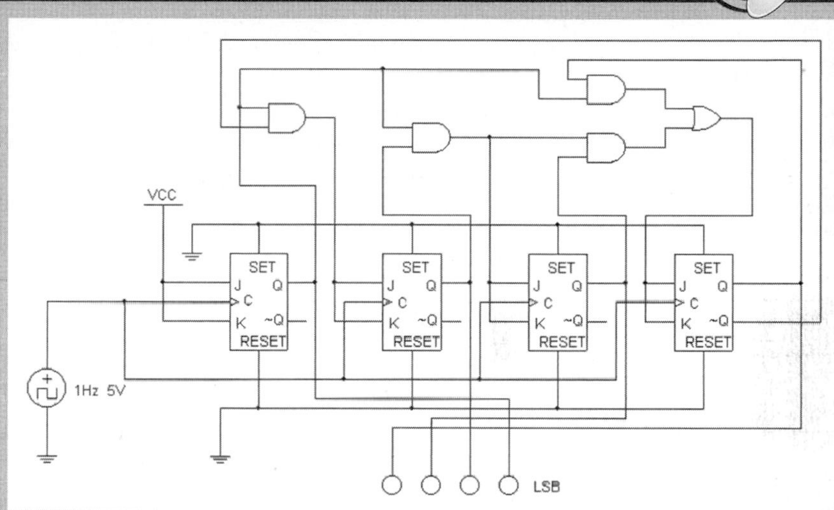

Cascading Counters

Counters can be connected in cascade to achieve higher-modulus operation. **Cascade** means that the *TC* (terminal count) output of one counter is connected to the *CTEN* (count enable) input of the next counter, as shown in Figure 8–13 for two decade counters. The *TC* of counter 1 is connected to the *CTEN* of counter 2. Counter 2 is inhibited by the 0 on its *CTEN* until counter 1 reaches its terminal count and its *TC* output goes to a 1. On the next clock pulse, counter 2 will advance from 0000 to 0001 and will remain there until counter 1 goes through another complete cycle. Counter 1 must go through ten complete cycles before Counter 2 completes one cycle.

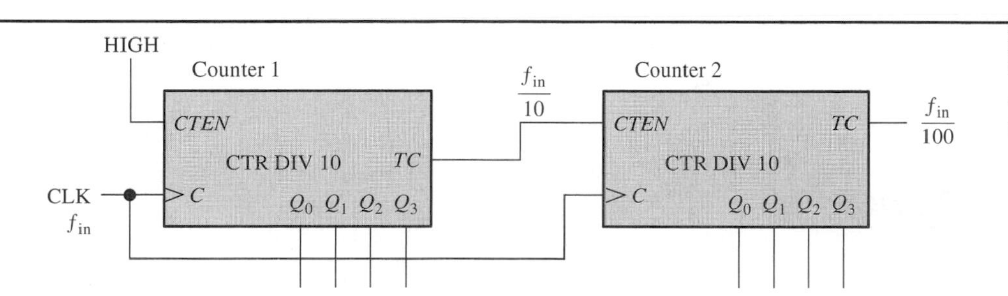

FIGURE 8–13

A modulus-100 counter using two cascaded decade counters.

When counters are cascaded, the overall modulus is the product of each counter's individual modulus. For example, when two decade counters are cascaded, the result is an overall modulus of $10 \times 10 = 100$. Since the "divide-by" factor is the same as the modulus, you also have a divide-by-100 counter. Notice in Figure 8–13 that the waveform at the *TC* of counter 1 has a frequency of $f_{in}/10$ and the waveform at the *TC* of counter 2 has a frequency of $f_{in}/100$.

When a 4-bit binary counter and a decade counter are cascaded, the result is an overall modulus of $16 \times 10 = 160$ and the *TC* output has a frequency of $f_{in}/160$. This idea can be extended to any number of cascaded counters to generate a lower frequency from a higher one. The higher and lower frequencies are exact multiples of each other.

EXAMPLE 8–3

Problem

Show how to obtain a divide-by-1000 counter. If the clock frequency is 1 MHz, what is the frequency of the *TC* output from the last counter?

Solution

Cascade three decade counters by connecting the *TC* of one counter to the *CTEN* of the next. The overall modulus is

$$10 \times 10 \times 10 = 1000$$

The frequency of the third counter is

$$f = \frac{f_{in}}{1000} = \frac{1\text{ MHz}}{1000} = \mathbf{1\text{ kHz}}$$

Figure 8–14 shows the basic diagram.

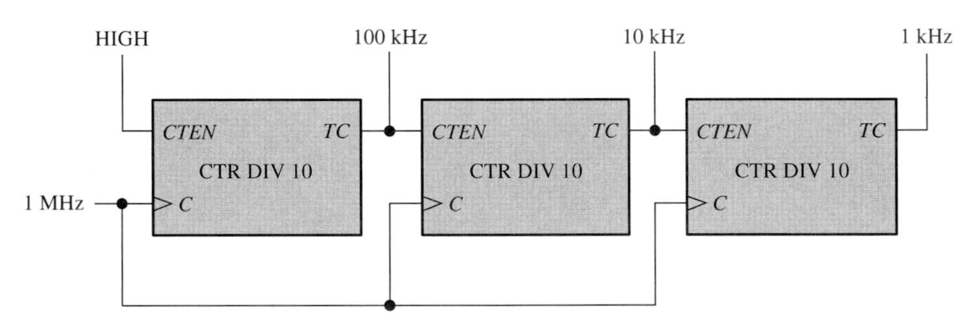

FIGURE 8–14

Question

How many 4-bit binary counters would it take to make a modulus-256 counter?

Review Questions

11. How many states does a decade counter have?
12. What is the modulus of a decade counter?
13. How many flip-flops does it take to make a decade counter?
14. What is the overall modulus of three decade counters in cascade?
15. What does *CTEN* stand for?

COUNTER DECODING 8–4

In many applications, it is necessary that some or all of the counter states be decoded. Counter states are decoded using logic gates or decoders.

In this section, you will learn how a counter is decoded.

Counter decoding is a process used to determine when a counter is in a certain binary state in its sequence. The terminal count function is a single decoded state (the last state) in the counter sequence, but any state in a counter's sequence can be decoded. Suppose that you want to decode binary state 6 (110) of a 3-bit binary counter. When $Q_2 = 1$, $Q_1 = 1$, and $Q_0 = 0$, a HIGH appears on the output of the decoding gate, indicating that the counter is in state 6. This can be done as shown in Figure 8–15. This is called active-HIGH decoding. Replacing the AND gate with a NAND gate provides active-LOW decoding.

One purpose for using a single-gate decoder, such as shown, is to change the modulus by using the decoded state to reset the counter (all 0s). Although not shown on the diagram, most flip-flops have an asynchronous clear (*CLR*) input. In this case, if the *CLR* inputs are active-HIGH, the output of the decoding AND gate can be connected to the *CLR* input of each flip-flop. When the counter reaches state 6 (110), it is immediately reset to 000 by the output of the decoding AND gate and does not stay in state 6. Thus, the counter recycles after 5 (101) and therefore, has a modulus of 6. If the *CLR* inputs are active-LOW, a NAND gate should be used for decoding.

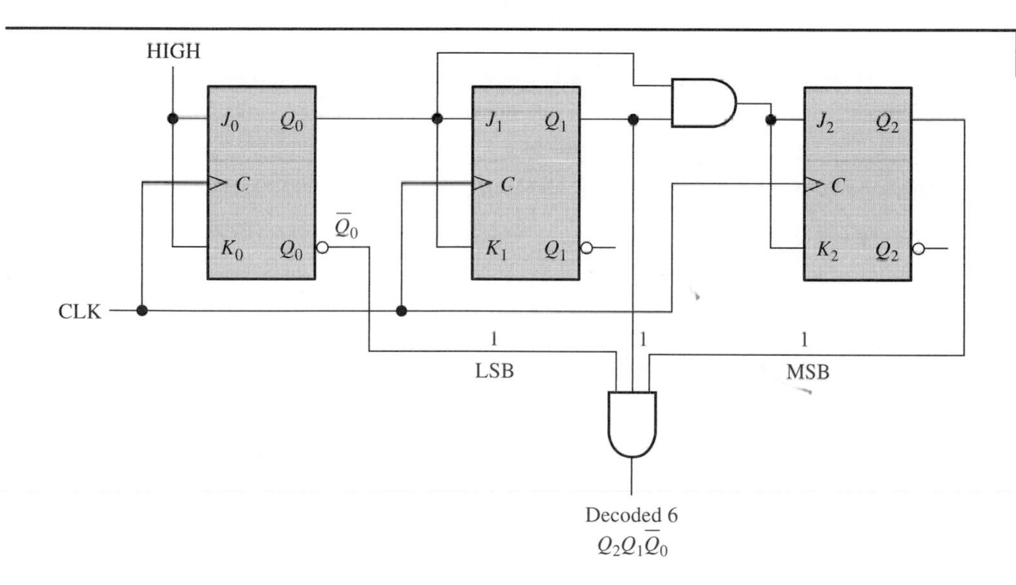

FIGURE 8–15

Decoding state 6 (110).

EXAMPLE 8–4

Problem

Implement the decoding of binary state 2 and binary state 7 of a 3-bit counter. Show the complete counter timing diagram and the output waveforms of the decoding gates.

Solution

See Figure 8–16. When $Q_2Q_1Q_0 = 010$, the output is binary 2. In this case, $\overline{Q}_2 = 1$, $Q_1 = 1$, and $\overline{Q}_0 = 1$. ANDing these terms produces the decoded output for 2. Binary 7 is decoded by ANDing $Q_2Q_1Q_0$ as shown.

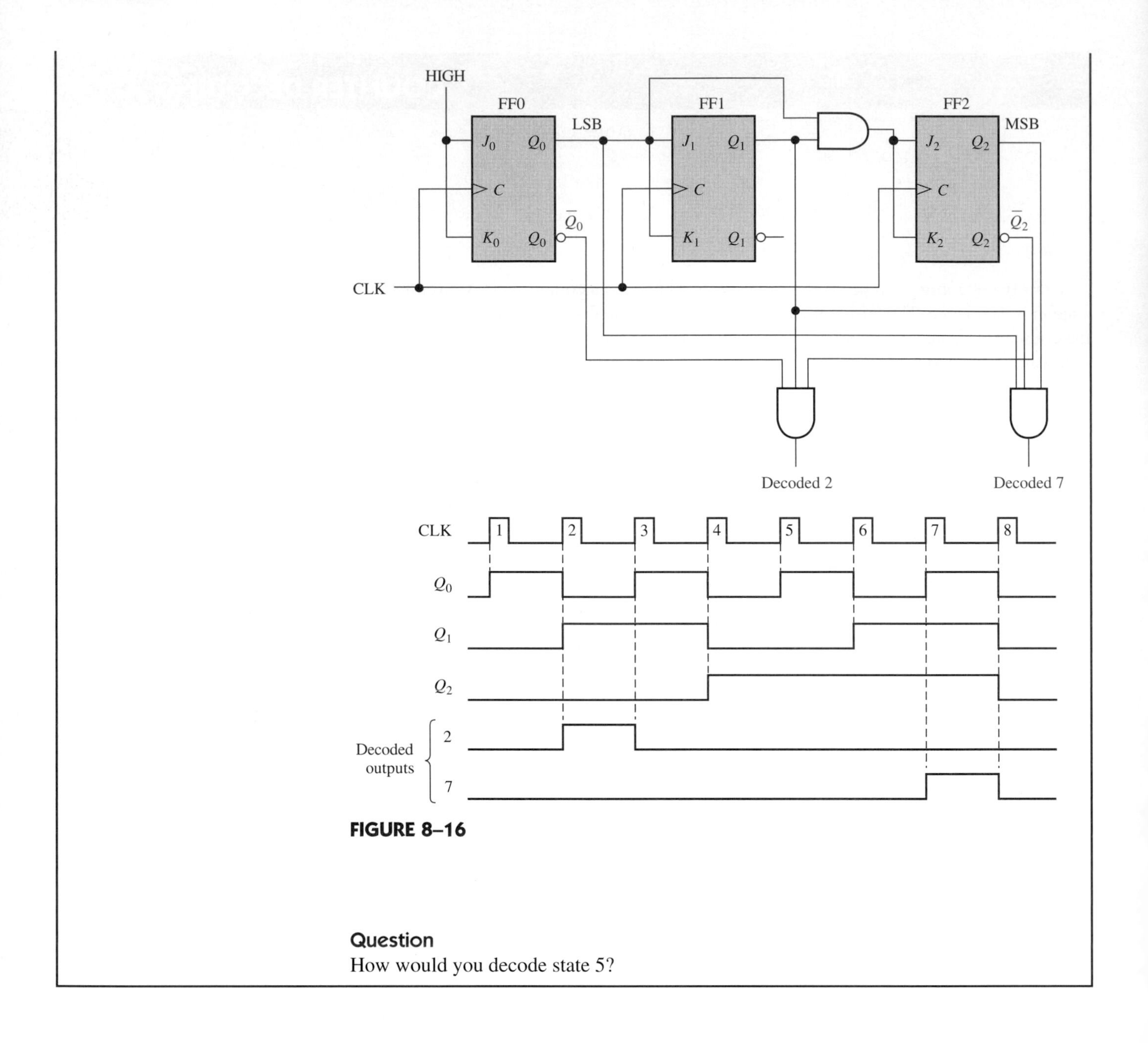

FIGURE 8–16

Question

How would you decode state 5?

Review Questions

16. How many gates are required to decode one counter state?
17. To decode all of the states of a BCD decade counter, how many gates are required?
18. What 3-bit binary counter outputs are connected to an AND gate to decode state 4 with an active-HIGH output?
19. How do you decode the terminal count of a 3-bit binary counter?
20. What 4-bit binary counter outputs are connected to a gate to decode state 15?

8–5 A COUNTER APPLICATION

Counters have many applications in digital systems. One good example of the application of counters is in a digital clock.

In this section, you will learn how counters can be used in a digital clock.

Complete digital clocks are available in a single IC. The particular digital clock that we will discuss requires both decade and divide-by-6 counters. The divide-by-six (modulus 6) counter is basically a 3-bit binary counter with logic for recycling after state 5 (101) instead of state 7. A block diagram for a digital clock is shown in Figure 8–17 with many details such as time setting and alarm logic omitted for simplicity.

FIGURE 8–17 Simplified block diagram for a digital clock.

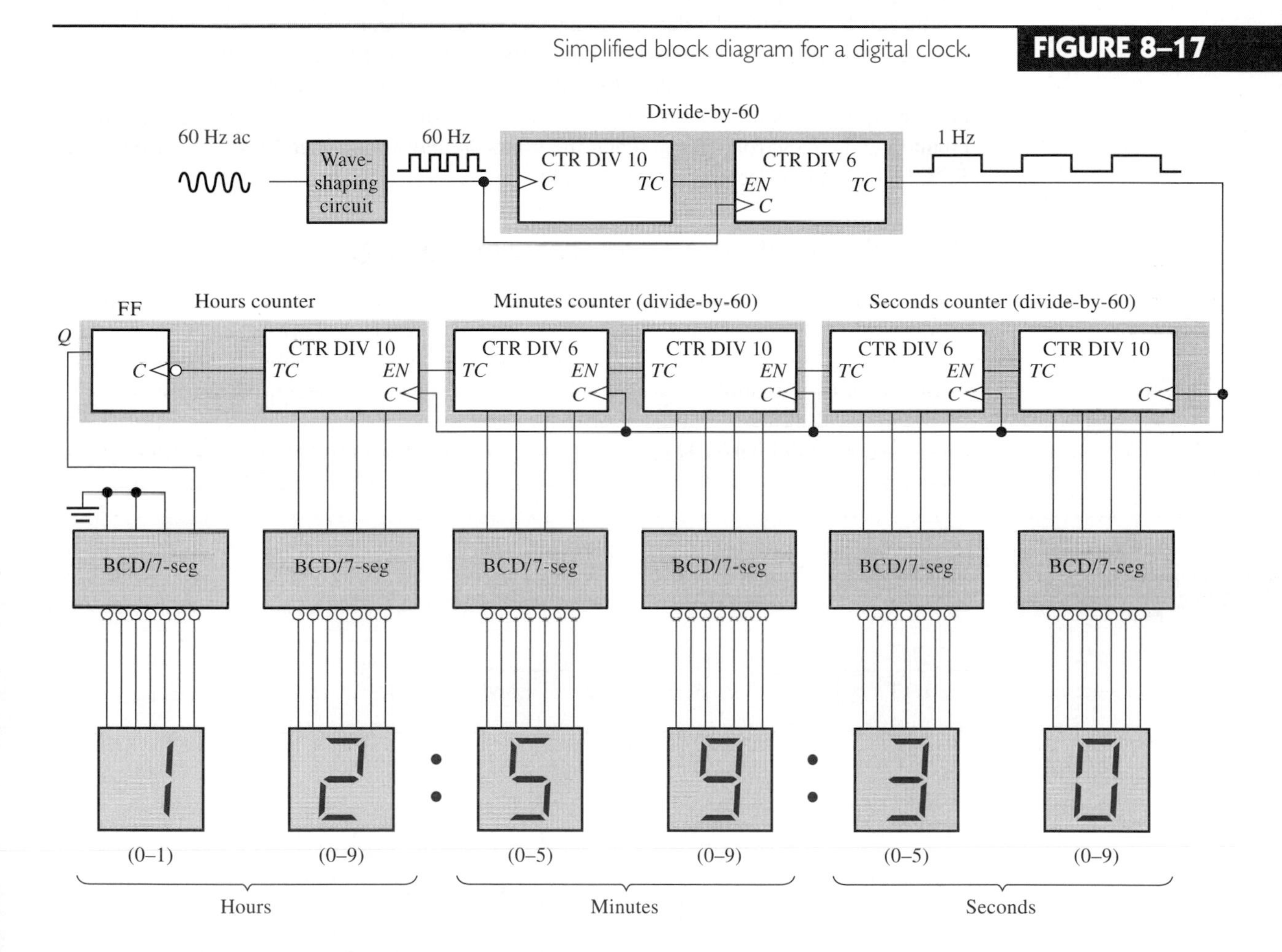

A 60 Hz square wave is derived from the 60 Hz ac power line by a wave-shaping circuit and divided down by 60 to obtain a 1 Hz waveform (1 pulse every second). The divide-by-60 counter is formed by cascading a decade (divide-by-10) counter and a divide-by-6 counter.

The seconds counter and the minutes counter are also divide-by-60 counters that count from 0 to 59 and then recycle back to 0. The states of these counters are decoded by the BCD-to-7 segment decoders and displayed on the 7-segment displays.

When the clock is running, you will see the seconds display change as it goes through the sequence 00, 01, 02, ··· 58, 59, and then recycle back to 00, continually repeating the sequence. It will take one minute for it to go through its sixty states. The decade counter produces the digits 0 through 9 required for the units digit of the seconds display. The divide-by-6 counter produces the digits 0 through 5 required for the tens digit of the seconds display.

The terminal count of the seconds counter enables the minutes counter each time the seconds counter reaches a count of 59. The minutes counter will advance once for each complete cycle of the seconds counter; that is, it will change state once every minute, and it will take one hour for it to go through its sixty states from 00 to 59. The decade counter produces the digits 0 through 9 required for the units digit of the minutes display. The divide-by-6 counter produces the digits 0 through 5 required for the tens digit of the minutes display.

When the minutes counter reaches a count of 59, its terminal count enables the hours counter, which changes state once every hour. The hours counter is composed of a decade counter, a flip-flop, and some logic gates. The decade counter produces the digits 0 through 9 required for the units digit of the hours display. Since the hours display requires numbers 1 through 12, the tens digit can only be a 0 or a 1. The flip-flop is clocked to its SET state causing the tens digit to be a 1 when the decade counter goes from count 9 to count 0. After the flip-flop is SET, the decade counter is 0 and the hours display shows 10. When the minutes counter goes from 59 to 00, the hours decade counter advances to count 1 and the hours display shows 11. The next time the minutes counter goes from 59 to 00, the hours decade counter advances to count 2 and the hours display shows 12. When the minutes counter again goes from 59 to 00 and the hours counter is at count 12, the flip-flop is RESET, and the hours decade counter goes to count 1.

Details of the Divide-by-60 Counters

As you have seen, three divide-by-60 counters are used in the digital clock. One counter produces the 1 Hz clock pulses, one produces the seconds display by counting the 1 Hz clock pulses, and one produces the minutes display by counting the number of seconds as indicated by the number of terminal count pulses from the seconds counter.

Figure 8–18 shows a detailed logic diagram of the divide-by-60 counter. It is formed by a decade counter with a BCD sequence (DIV 10) cascaded with a 4-bit counter that is modified to have six states (DIV 6). These two counters have an active-LOW $\overline{CLR}$ input.

FIGURE 8–18 Logic diagram of the divide-by-60 counter.

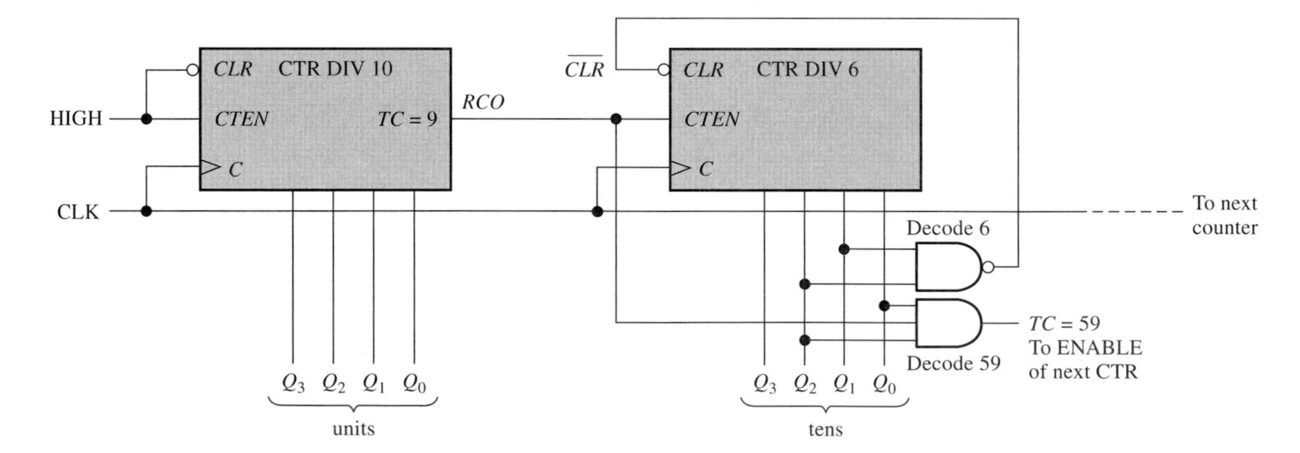

The $\overline{CLR}$ input forces the divide-by-6 counter to 0 as soon as it goes to count 6. This is accomplished by the NAND gate that decodes state 0110 (6). Notice that only two of the counter outputs go to the NAND gate because the first time that these two outputs are both 1 the counter is in state 0110. The output of the NAND gate is connected to the $\overline{CLR}$ input of the counter so that immediately after the counter goes to 0110, the output of the NAND goes LOW and clears the counter to state 0000. The counter is in state 0110 for a very brief time (probably nanoseconds), so it is not seen on the display.

The AND gate is used to decode count 59 and produce a terminal count (*TC*) output pulse. When the *TC* output of the divide-by-10 counter is HIGH at count 9 and the divide-by-6 counter is in state 0101 (5), the AND gate output goes HIGH to indicate terminal count.

Details of the Hours Counter

The hours counter is more complicated because the modulus of the decade counter must be altered, depending on the state of the flip-flop. A logic diagram is shown in Figure 8–19.

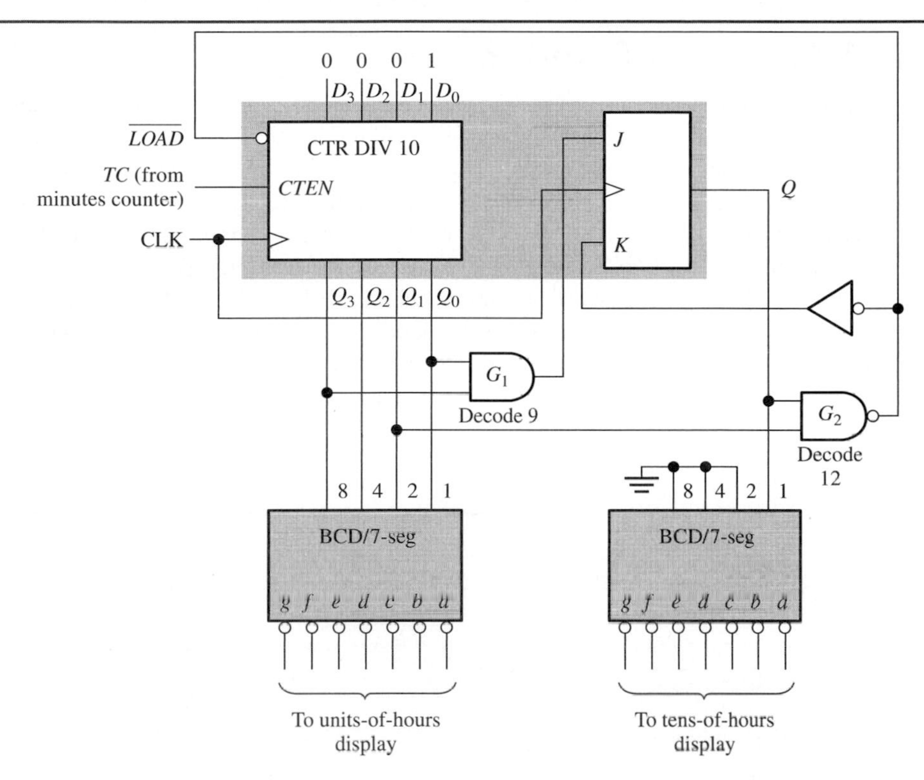

FIGURE 8–19

Logic diagram for the hours counter.

When the divide-by-10 counter reaches state 1001 (count 9), the output of AND gate, G_1, goes HIGH. On the next clock pulse, the counter recycles back to state zero and the flip-flop goes to the SET state, illuminating a 1 on the tens-of-hours display. At this point, the total count is ten (the decade counter is in the zero state and the flip-flop is SET). Next, the total count advances to eleven and then to twelve. In state 12 the Q_1 output of the decade counter is HIGH and the flip-flop is still SET; therefore, the decode-12 gate, G_2, output is LOW. This activates the $\overline{LOAD}$ input of the decade counter. On the next clock pulse, the decade counter is preset to state 0001 by the parallel data inputs and the flip-flop is RESET. The logic always causes the counter to recycle from state 12 back to state 1 instead of back to zero.

Review Questions

21. What are the two recycle conditions for the hours counter in Figure 8–17?
22. What is the purpose of the "Decode 6" gate in Figure 8–18?
23. What is the purpose of the "Decode 59" gate in Figure 8–18?
24. What is the purpose of the "Decode 9" gate in Figure 8–19?
25. In the hours counter of Figure 8–19, when does the flip-flop go from the SET state to the RESET state?

8–6 INTEGRATED CIRCUITS

Several types of counters are available in integrated circuit form. A typical IC 4-bit binary counter and a BCD decade counter are introduced here. Most IC counters have additional functions that were not included on the basic counters that were covered in the previous sections. Keep in mind that although the LS family of logic has been used for illustration, identical logic functions in other families such as F, ALS, and HC can be used as direct replacements.

In this section, you will learn about two specific IC counters.

A 4-Bit Binary Counter

The 74LS163A is an example of an integrated circuit counter. A logic symbol is shown in Figure 8–20(a), and a package pin diagram is shown in part (b). The 74LS163A counter has four data inputs (D_3, D_2, D_1, D_0) in addition to the Q outputs. Using these four data inputs, the counter can be loaded with any 4-bit number. Recall that a "bubble" indicates an active-LOW input. When $\overline{LOAD}$ is LOW, the counter will assume the state of the data inputs on the next clock pulse because it is a synchronous input. This allows the counter to begin a sequence with any 4-bit binary number. When $\overline{CLR}$ is LOW, the counter is cleared or reset to 0000 when the active clock edge occurs because $\overline{CLR}$ is a synchronous input. *ENP* and *ENT* are enable inputs, and they must both be HIGH for the counter to count. The *RCO* (ripple carry output) is the same as terminal count (*TC*) and is HIGH when the counter is in the 1111 state.

FIGURE 8–20

The 74LS163A 4-bit binary counter.

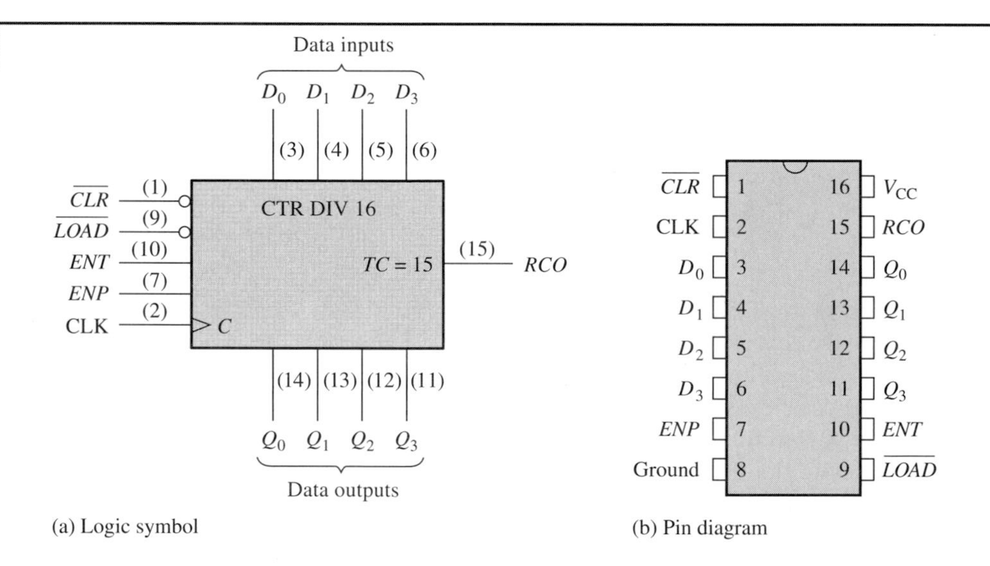

Figure 8–21 shows an example timing diagram for the 74LS163A. This timing diagram illustrates the counter being loaded or preset to twelve (1100) and then counting up to its terminal count of fifteen (1111).

First, the LOW pulse on the $\overline{CLR}$ input causes all the Q outputs to be LOW (0) when the active clock edge occurs. Next, the LOW pulse on the $\overline{LOAD}$ input enters the data that is on the data inputs (D_3, D_2, D_1, D_0) into the counter. These data appear on the Q outputs on the first positive-going clock pulse edge after the $\overline{LOAD}$ input goes LOW. Q_0 is LOW, Q_1 is LOW, Q_2 is HIGH, and Q_3 is HIGH which represents a binary 12. Remember that D_0 is the least significant bit.

The counter now advances through states 1101, 1110 and 1111 on the next three clock pulses and then produces an *RCO* pulse. Then, it recycles to 0000 and continues to 0001

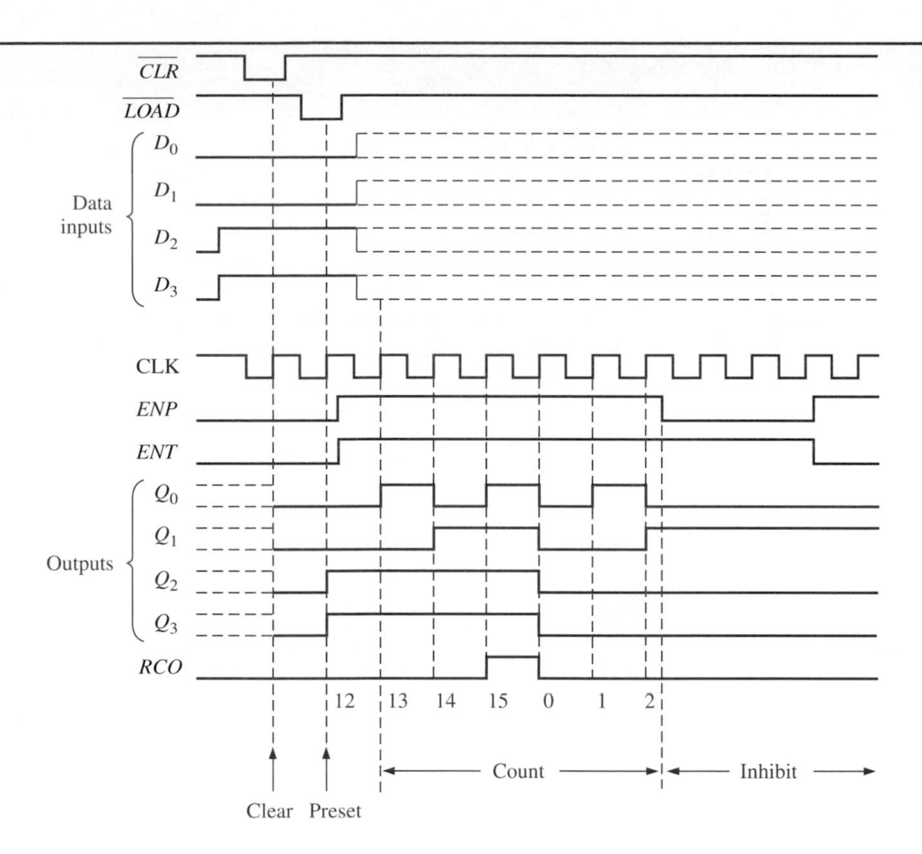

FIGURE 8–21

Timing example for a 74LS163A.

and 0010. It would continue to count but, in the example shown, *ENP* goes LOW. This causes the counter to be inhibited and remain in the binary 2 (0010) state.

A BCD Decade Counter

The 74LS160A is another example of an integrated circuit counter. A logic symbol is shown in Figure 8–22(a), and a package pin diagram is shown in part (b).

The 74LS160A counter also has four data inputs (D_3, D_2, D_1, D_0) in addition to the Q outputs. It also has all of the inputs and outputs that were described for the 74LS163A 4-bit binary counter except the $\overline{CLR}$ input is asynchronous in the 74LS160A. The states ten through fifteen (1010 through 1111) are invalid states for a BCD decade counter. Therefore, the counter cannot be loaded or preset to one of these invalid states. *RCO* is *HIGH* when the terminal count of 1001 is reached.

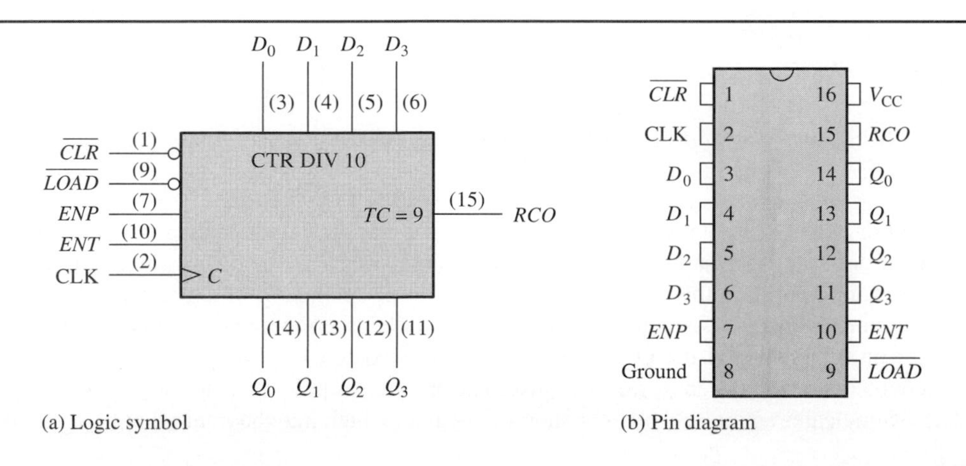

FIGURE 8–22

The 74LS160A BCD decade counter.

COMPUTER SIMULATION

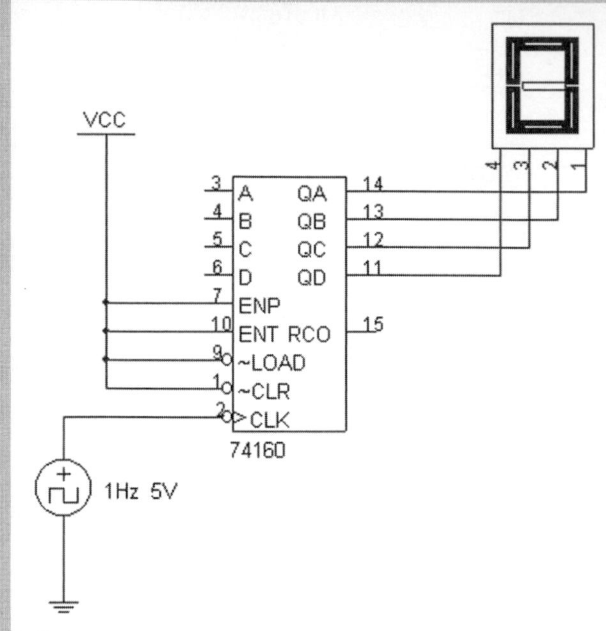

FIGURE 8–23

Open the Multisim file F08-23DG on the website. The pulse generator produces one pulse per second and is the clock for the counter. The decade counter will go through its BCD sequence 0000 through 1001. You can observe the count on the 7-segment decoded display. Normally, a decoder is required but the Multisim program combines the decoder in the display.

Review Questions

26. What is the 74LS163A?
27. What is the 74LS160A?
28. What is the purpose of the $\overline{LOAD}$ input?
29. What is the purpose of the *ENT* and *ENP* inputs?
30. What is *RCO*?

8–7 TROUBLESHOOTING

When any circuit does not operate properly, always first check the power and ground connections to the IC and look for any signs of shorts or open contacts. Troubleshooting counters can be simple or quite involved, depending on the type of counter and the type of fault.

In this section, you will learn to troubleshoot counters.

For a faulty counter with a straightforward sequence, such as a binary counter that is not controlled by external logic, about the only thing to check (other than power, ground, and bad connections) is the possibility of open or shorted inputs or outputs. An IC counter almost never alters its sequence of states because of an internal fault, so you need only check for pulse activity on the *Q* outputs to detect the existence of an open or short when the counter is being driven by a clock pulse. The absence of pulse activity on one of the *Q* outputs indicates an internal open or short on the line, which may be internal or external to the

IC. Absence of pulse activity on all the Q outputs indicates that the clock input is faulty, the clear or load inputs are stuck in the active state, or the enable input is not active.

To check the clear input, apply a constant active level. For a synchronous clear input, the counter must be clocked. You should observe a LOW on each of the Q outputs if it is functioning properly. A parallel load feature on a counter can be checked by activating the $\overline{LOAD}$ input and exercising each state as follows. Apply LOWs to the parallel data inputs, pulse the clock input once, and check for LOWs on all the Q outputs. Next, apply HIGHs to all the parallel data inputs, pulse the clock input once, and check for HIGHs on all the Q outputs.

SAFETY NOTE

Liquids and electricity don't mix! You should never place a drink or other liquid where it could be accidentally spilled onto an electrical circuit.

Cascaded Counters

A failure in one of the counters in a chain of cascaded counters can affect all the counters that follow it. For example, if a count enable input opens, it effectively acts as a HIGH (for TTL logic), and the counter is always enabled. This type of failure in one of the counters will cause the counter to run at the full clock rate and will also cause all of the succeeding counters to run at higher than normal rates. This condition is illustrated in Figure 8–24 for a divide-by-1000 cascaded-counter arrangement where an open count enable input (*CTEN*) acts as a HIGH and continuously enables the second counter. Other faults that can affect "down stream" counters are open or shorted clock input or terminal count outputs. In some of these situations, pulse activity can be observed, but it may be at the wrong frequency. Exact frequency measurements must be made to isolate the fault.

FIGURE 8–24

Example of a failure that affects following counters in a cascaded arrangement.

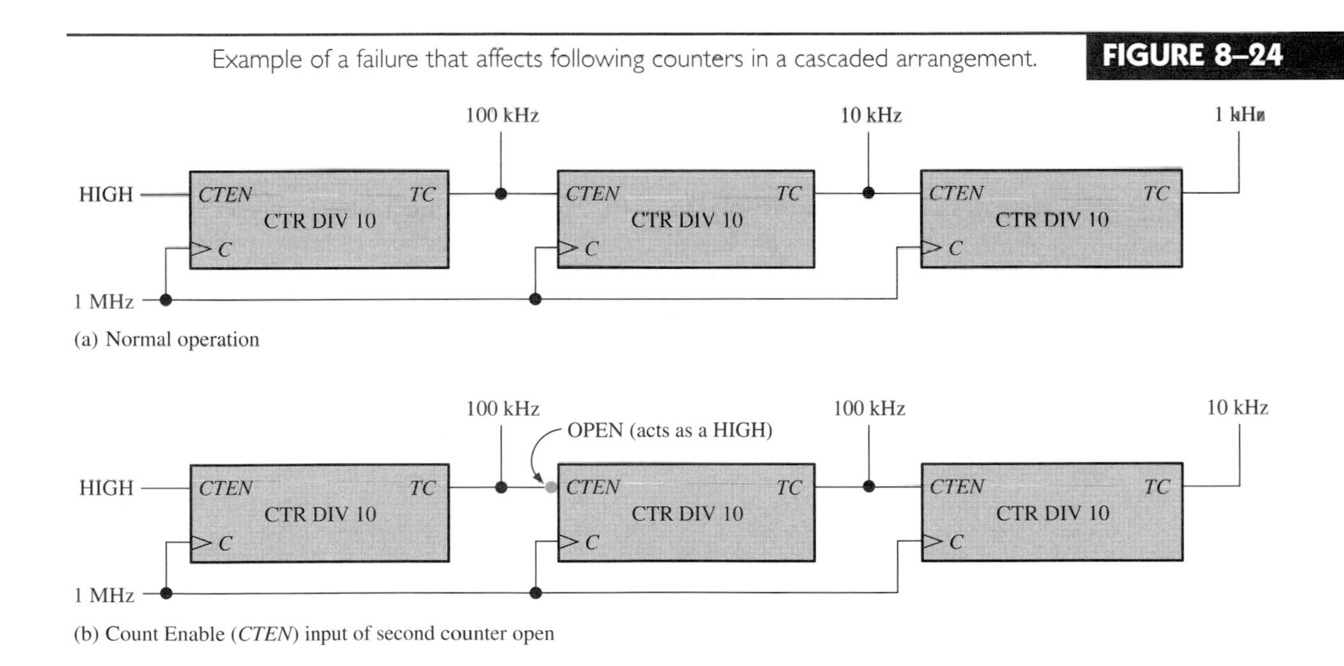

(a) Normal operation

(b) Count Enable (*CTEN*) input of second counter open

Counters Implemented with Individual Flip-Flops

Counters implemented with individual flip-flops and gates are sometimes more difficult to troubleshoot than IC counters. There are many more inputs and outputs with external connections than there are in an IC counter. The sequence of a counter can be altered by a single open or short on an input or output, as Example 8–5 shows.

CHAPTER

SHIFT REGISTERS

CHAPTER OUTLINE

Study aids for this chapter are available at

http://www.prenhall.com/SOE

INTRODUCTION

Shift registers are a type of sequential logic circuit closely related to digital counters. Registers are used primarily for the storage of digital data and typically do not possess a characteristic internal sequence of states as counters do. An exception is the shift register counter.

Shift registers are used in a variety of applications, but two of the most important are converting numbers from parallel to serial form or vice-versa and as shift register counters. As data converters, they have application in digital communication systems; for example, it is necessary to convert serial data sent over the telephone line to parallel data for a computer to use. A shift register counter is useful for generating a repeating sequence of binary numbers, which in effect is a digital waveform generator.

In this chapter, the basic types of shift registers are studied and several applications are presented. Also, an important troubleshooting method is introduced.

KEY TERMS

- Shift register
- Stage
- Shift
- Load
- Bidirectional
- Shift register counter

KEY OBJECTIVES

A section number is given for each objective. After completing this chapter, you should be able to

9–1 Identify the basic types of data movement and storage capacity in shift registers

9–2 Explain how serial in/serial out and serial in/parallel out shift registers operate

9–3 Explain how parallel in/serial out and parallel in/parallel out shift registers operate

9–4 Describe the operation of a bidirectional shift register

9–5 Describe the operation of two types of shift register counters

9–6 Discuss several IC shift registers

9–7 Discuss the applications of shift registers

9–8 Use test patterns to troubleshoot a digital system

COMPUTER SIMULATIONS DIRECTORY

The following figures have Multisim circuit files associated with them.

- Figure 9–8 Page 263
- Figure 9–11 Page 265
- Figure 9–14 Page 267
- Figure 9–30 Page 278

LABORATORY EXPERIMENTS DIRECTORY

The following exercise is for this chapter.

- **Experiment 13** Shift Register Counters

ON THE JOB. . .

(Getty Images)

In many types of technical jobs, you will be required to write reports on projects. Reporting may be anything from simple logbook entries to formal technical reports. If you are involved in writing reports, find out the policy of the company. Some may require entries in ink and signed; others may have less formal requirements. If you are writing an explanation, keep in mind that an illustration of a problem may be a valuable addition to your report. If you are not a good speller, write out your report on a word processor and don't forget to use the spell checker!

A single-molecule logic circuit has been experimentally developed. Researchers have essentially turned a nanotube, which is a sheet of carbon atoms rolled into a super tiny straw, into an inverter or NOT circuit. Carbon nanotubes are the prime candidate to replace silicon when current chip features cannot be made any smaller. This is expected to occur in about 10 to 15 years and will inevitably lead to unimagined miniaturization of computers and other electronic circuits. Researchers also observed that this carbon nanotube inverter exhibited current gain, which is essential to most circuit designs.

9–1 BASIC SHIFT REGISTER FUNCTIONS

Shift registers consist of arrangements of flip-flops and are important in applications involving the storage and transfer of data in a digital system. A register, unlike a counter, has no specified sequence of states, except in certain specialized applications. A register, in general, is used for storing and shifting data (1s and 0s) entered into it from an external source and typically possesses no characteristic internal sequence of states.

In this section, you will learn the basic types of data movement and storage capacity in shift registers.

A **shift register** is a digital circuit with two basic functions: data storage and data movement. The storage capability of a register makes it an important type of memory device. Figure 9–1 illustrates the concept of storing a 1 or a 0 in a D flip-flop. A 1 is applied to the data input as shown, and a clock pulse is applied that stores the 1 by setting the flip-flop. When the 1 on the input is removed, the flip-flop remains in the SET state, thereby storing the 1. A similar procedure applies to the storage of a 0 by resetting the flip-flop, as also illustrated in Figure 9–1.

FIGURE 9–1 The flip-flop as a storage element.

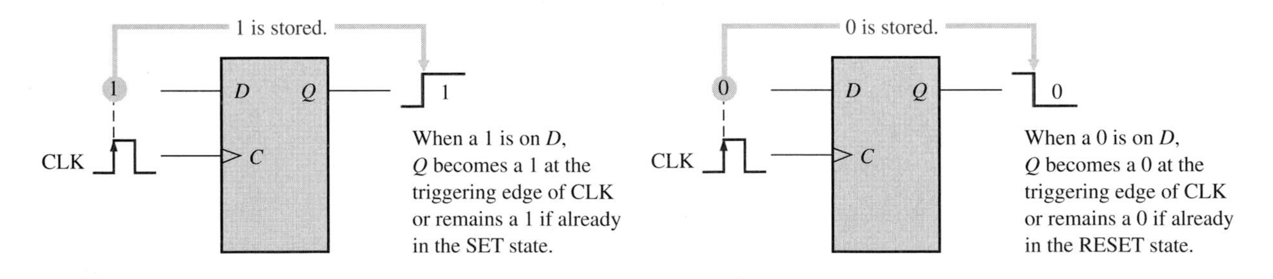

The storage capacity of a shift register is the total number of bits (1s and 0s) of digital data it can retain. Each **stage** (flip-flop) in a shift register represents one bit of storage capacity; therefore, the number of stages in a register determines its storage capacity.

The **shift** capability of a register permits the movement of data from stage to stage within the register or into or out of the register upon application of clock pulses. Figure 9–2 illustrates the types of data movement in shift registers. The block represents any arbitrary 4-bit register, and the arrows indicate the direction of data movement.

FIGURE 9–2

Basic data movement in shift registers. (Four bits are used for illustration. The bits move in the direction of the arrows.)

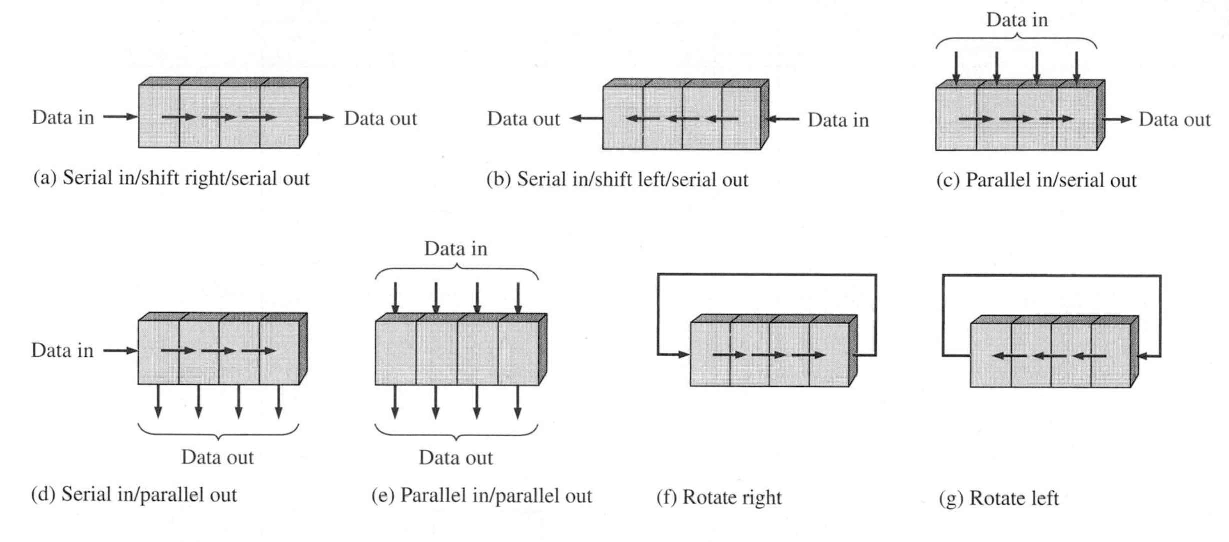

Review Questions

Answers are at the end of the chapter.

1. Generally, what is the difference between a counter and a shift register?
2. What two principal functions are performed by a shift register?
3. What is *storage capacity*?
4. What is a stage in a shift register?
5. How does a flip-flop store a 1? How does it store a 0?

SERIAL INPUT SHIFT REGISTERS 9–2

The serial input shift register accepts data serially—that is, one bit at a time on a single line. Depending on the configuration, it can output the stored data either in serial form or in parallel form.

In this section, you will learn how serial input shift registers work.

Serial In/Serial Out Shift Registers

Figure 9–3 shows a 4-bit serial in/serial out shift register implemented with D flip-flops. With four stages, this register can store up to four bits of data.

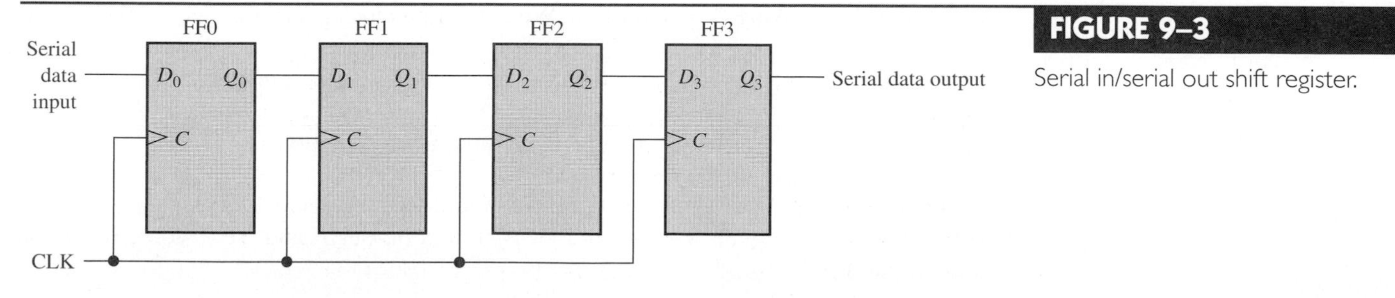

FIGURE 9–3

Serial in/serial out shift register.

FIGURE 9–4 Four bits (1010) being entered serially into the register.

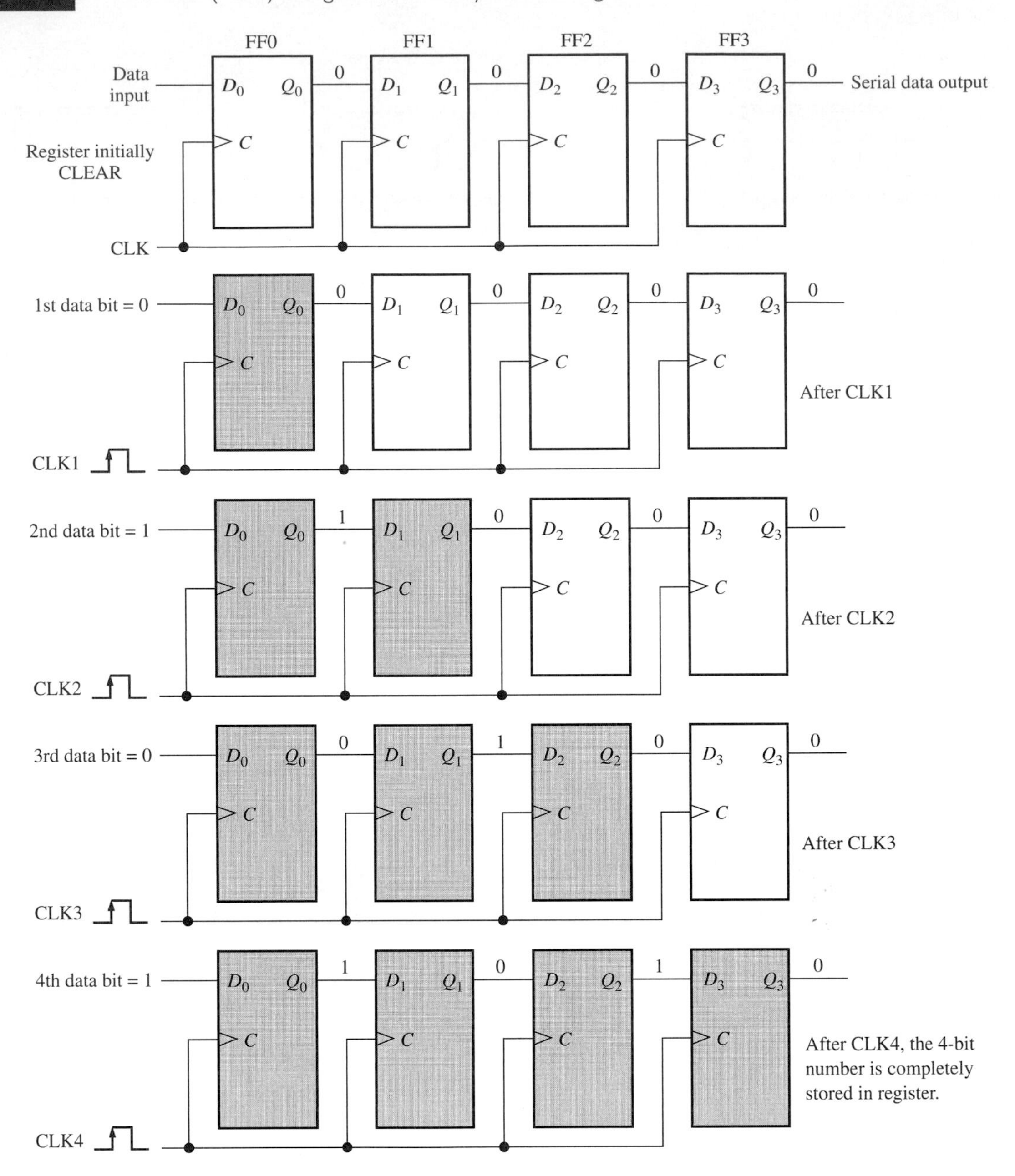

Figure 9–4 illustrates entry of the four bits 1010 into the register, beginning with the right-most bit. The register is initially clear and, thus, stores no data. The 0 is put onto the data input line, making $D_0 = 0$ for FF0. When the first clock pulse is applied, FF0 is reset, thus storing the 0.

Next the second bit, which is a 1, is applied to the data input, making $D_0 = 1$ for FF0 and $D_1 = 0$ for FF1 because the D_1 input of FF1 is connected to the Q_0 output. When the second clock pulse occurs, the 1 on the data input is shifted into FF0, causing FF0 to set; and the 0 that was in FF0 is shifted into FF1.

The third bit, a 0, is now put onto the data-input line, and a clock pulse is applied. The 0 is entered into FF0; the 1 stored in FF0 is shifted into FF1; and the 0 stored in FF1 is shifted into FF2.

The last bit, a 1, is now applied to the data input, and a clock pulse is applied. This time the 1 is entered into FF0; the 0 stored in FF0 is shifted into FF1; the 1 stored in FF1 is shifted into FF2; and the 0 stored in FF2 is shifted into FF3. This completes the serial entry of the four bits into the shift register, where they can be stored for any length of time as long as the flip-flops have dc power.

If you want to get the data out of the register, the bits must be shifted out serially and taken off the Q_3 output, as Figure 9–5 illustrates. After CLK4 in the data-entry operation just described, the right-most bit, 0, appears on the Q_3 output. When clock pulse CLK5 is applied, the second bit appears on the Q_3 output. Clock pulse CLK6 shifts the third bit to the output, and CLK7 shifts the fourth bit to the output. While the original four bits are being shifted out, more bits can be shifted in. All zeros are shown being shifted in.

FIGURE 9–5

Four bits (1010) being serially shifted out of the register and replaced by all zeros.

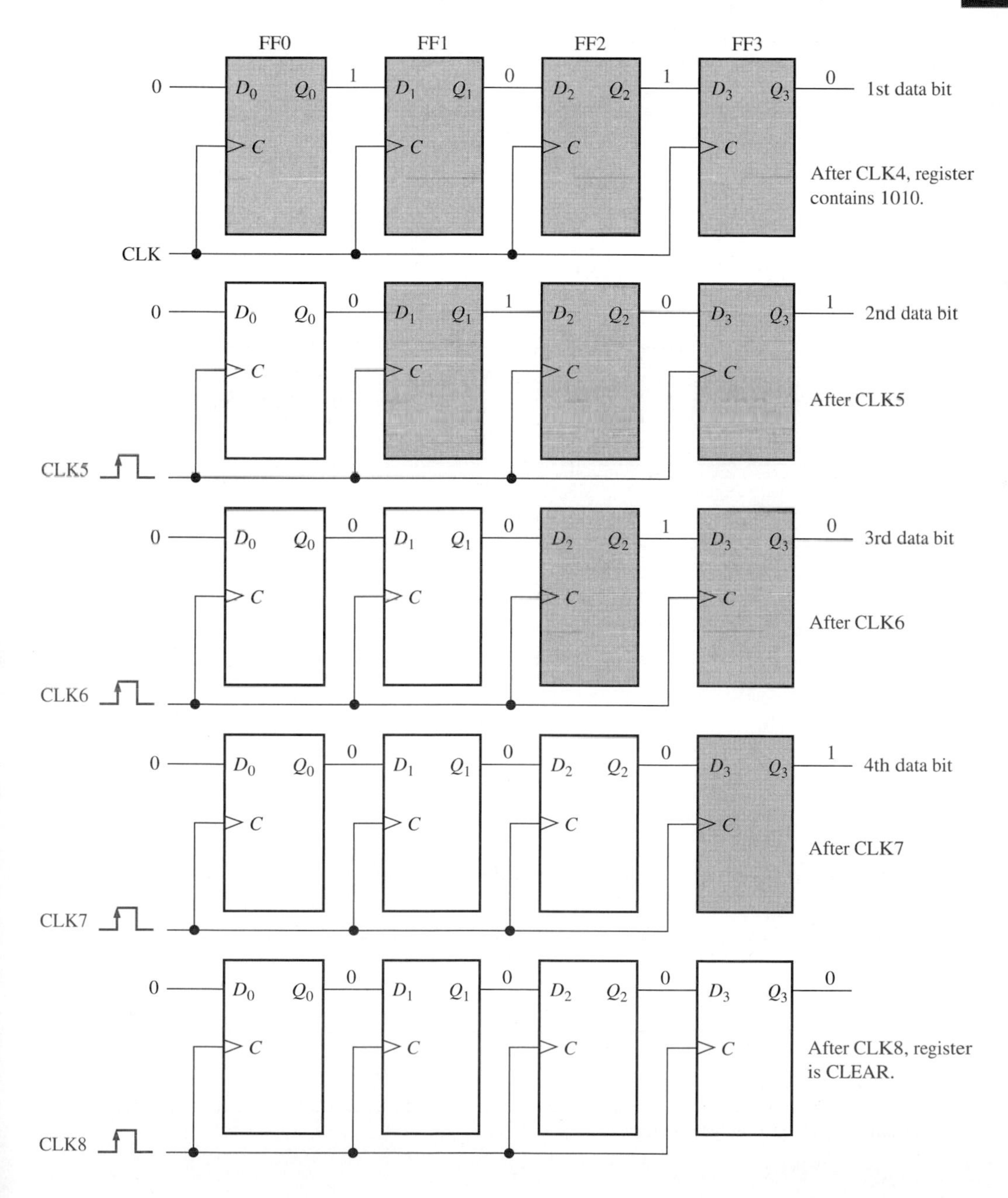

EXAMPLE 9–1

Problem

Show the states of the 5-bit register in Figure 9–6(a) for the specified data input and clock waveforms. Assume that the register is initially cleared (all 0s).

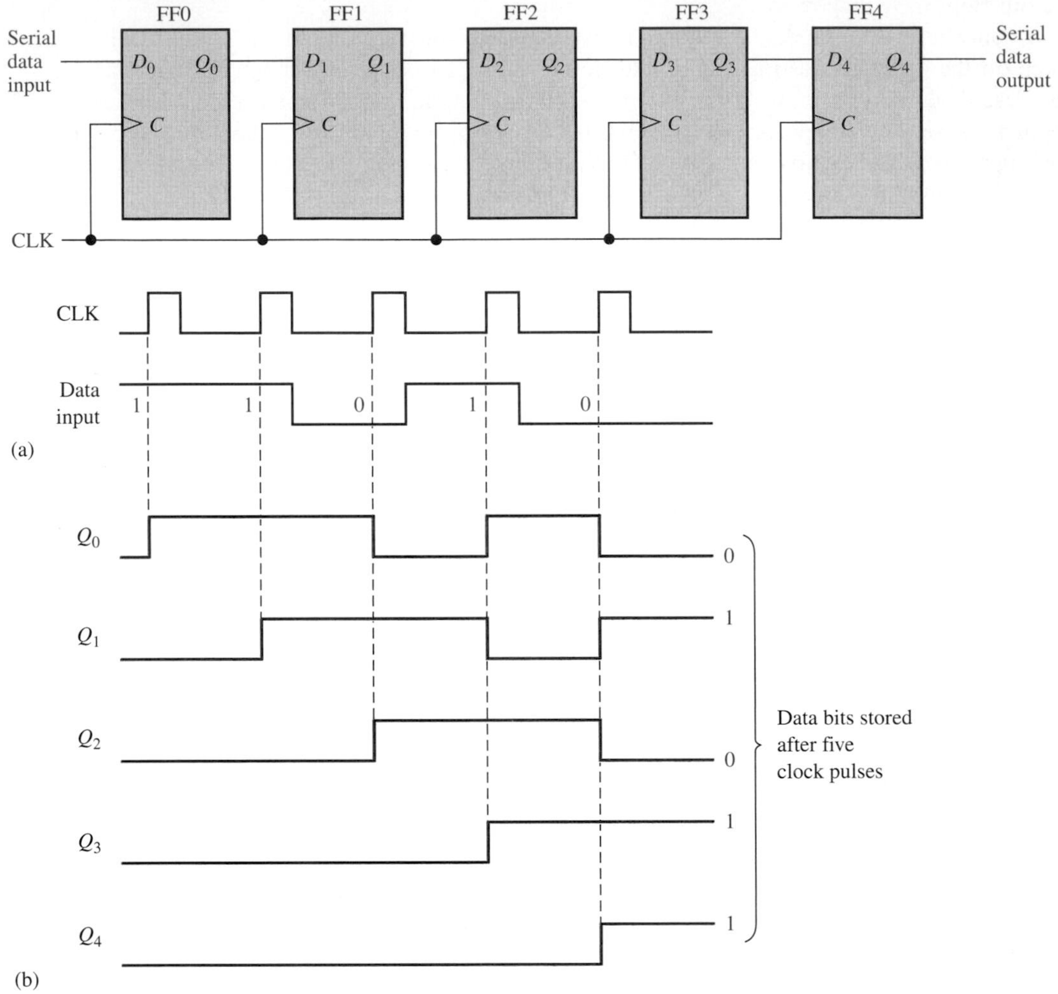

FIGURE 9–6

Solution

The first data bit (1) is entered into the register on the first clock pulse and then shifted from left to right as the remaining bits are entered and shifted. The register contains $Q_4Q_3Q_2Q_1Q_0 = 11010$ after five clock pulses. See Figure 9–6(b).

Question*

What are the Q outputs of the register if the data input is inverted? The register is initially cleared.

* *Answers are at the end of the chapter.*

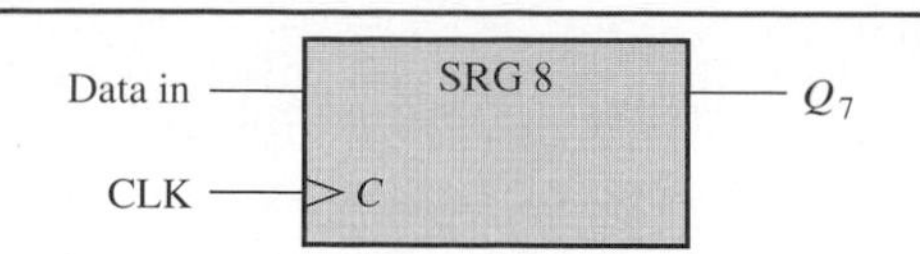

FIGURE 9–7

Logic symbol for an 8-bit serial in/serial out shift register.

A traditional logic block symbol for an 8-bit serial in/serial out shift register is shown in Figure 9–7. The "SRG 8" designation indicates a shift register (SRG) with an 8-bit capacity.

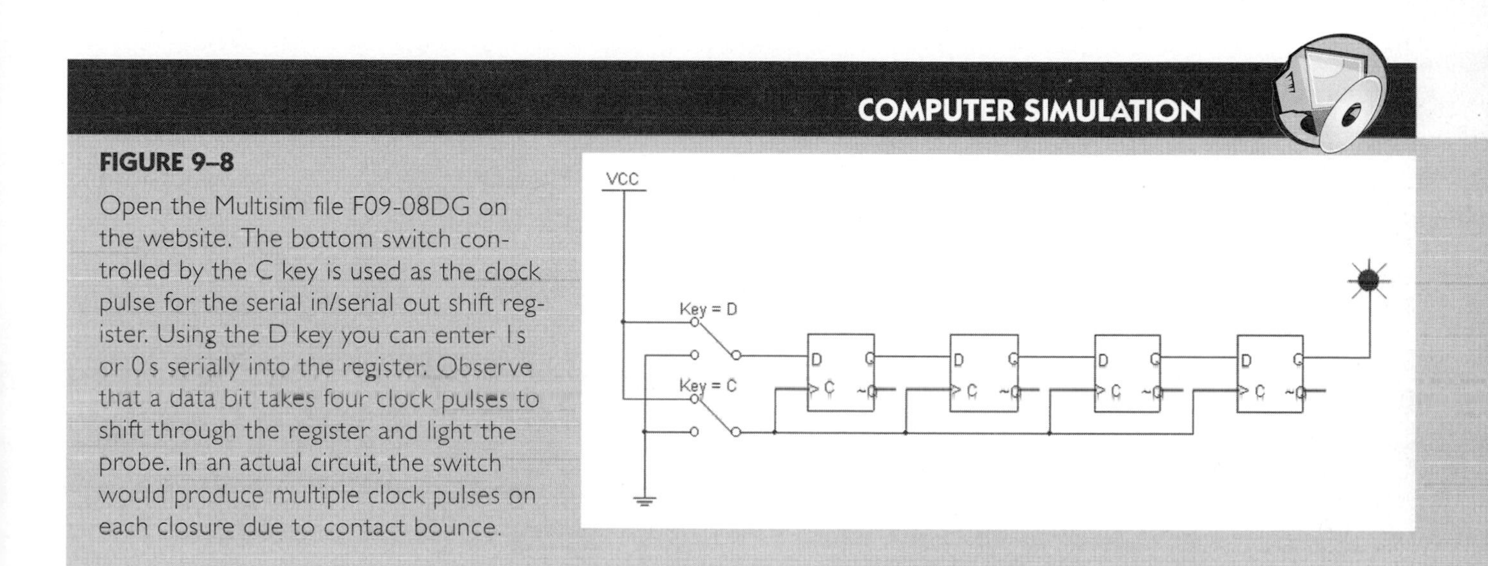

FIGURE 9–8

Open the Multisim file F09-08DG on the website. The bottom switch controlled by the C key is used as the clock pulse for the serial in/serial out shift register. Using the D key you can enter 1s or 0s serially into the register. Observe that a data bit takes four clock pulses to shift through the register and light the probe. In an actual circuit, the switch would produce multiple clock pulses on each closure due to contact bounce.

Serial In/Parallel Out Shift Registers

Data bits are entered into a serial in/parallel out shift register in the same manner as in the serial in/serial out shift register. The difference is that the serial in/parallel out shift register has all outputs of the flip-flops available. Once the data are stored, each bit appears on its respective output line, and all bits are available simultaneously, rather than on a bit-by-bit basis as with the serial output.

A serial in/parallel out shift register implemented with D flip-flops is shown in Figure 9–9(a). Its logic symbol is shown in Figure 9–9(b). Compare the logic drawing with the one for a serial in/serial out shift register in Figure 9–3 and the logic symbol with Figure 9–7. The inputs are identical for the two shift registers; only the available outputs differ. Some integrated circuit shift registers will also include another input labeled strobe. *Strobe* is essentially an enable function that either enables or disables the clock. By disabling the clock, the data can be stored indefinitely (as long as power remains on the circuit) because the data will not move through the register without a clock.

FIGURE 9–9 A serial in/parallel out shift register.

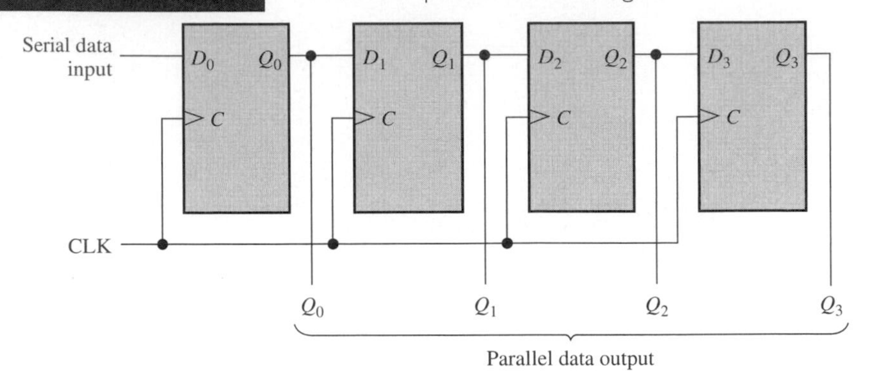

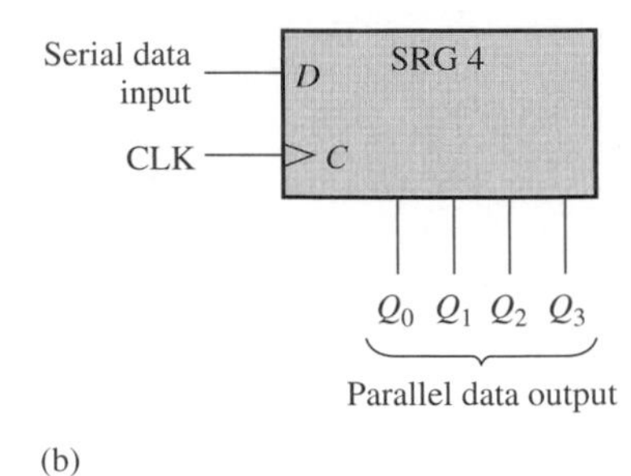

EXAMPLE 9–2

Problem

Show the states of the 4-bit register (SRG 4) for the data input and clock waveforms in Figure 9–10(a). The register initially contains all 1s.

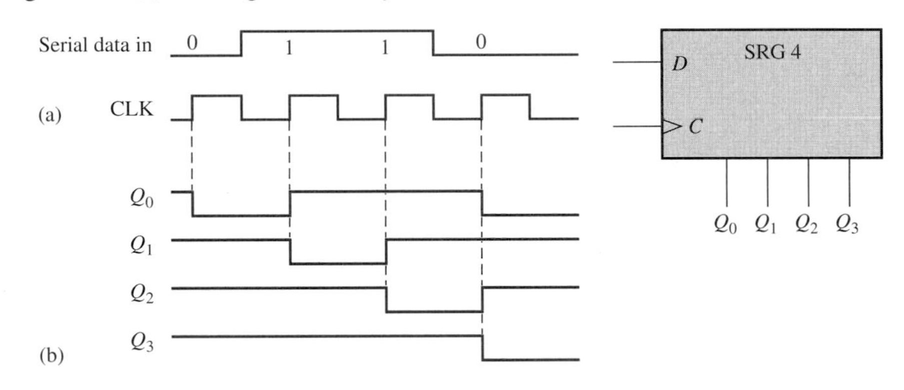

FIGURE 9–10

Solution

The register contains 0110 after four clock pulses. See Figure 9–10(b).

Question

If the data input remains 0 after the fourth clock pulse, what is the state of the register after three additional clock pulses?

COMPUTER SIMULATION

Open the Multisim file F09-11DG on the website. The pulse generator produces one pulse per second so that you can observe data on the parallel outputs as they shift through the register. Use the *D* key to enter either a 1 or a 0 on the serial data input.

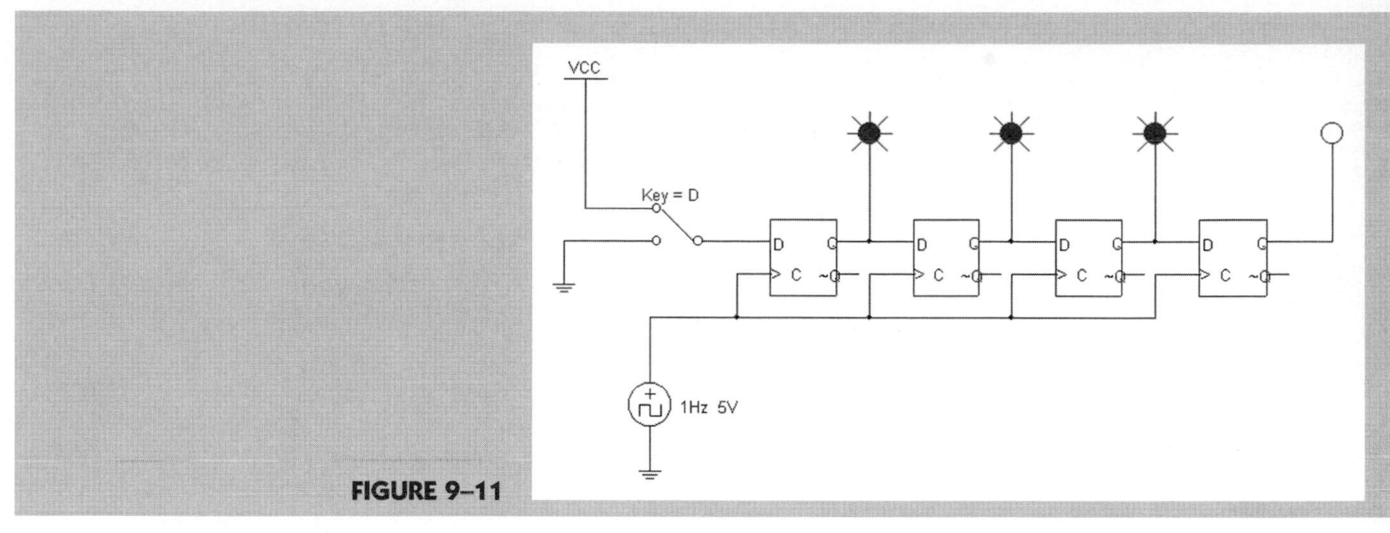

FIGURE 9–11

Review Questions

6. How many clock pulses are required to enter a byte of data serially into an 8-bit shift register?
7. What is meant by the term *cleared* in relation to a shift register?
8. What does *SRG* stand for?
9. The bit sequence 1101 is serially entered (right-most bit first) into a 4-bit parallel out shift register that is initially clear. What are the Q outputs after one clock pulse?
10. Referring to Question 9, what are the Q outputs after two clock pulses?

PARALLEL INPUT SHIFT REGISTERS 9–3

For a register with parallel data inputs, the bits are entered simultaneously into their respective stages on parallel lines rather than on a bit-by-bit basis as with serial data inputs. The data can be output in either serial or parallel form once the data are completely stored in the register.

In this section, you will learn how parallel input shift registers work.

Parallel In/Serial Out Shift Registers

Figure 9–12 illustrates a 4-bit parallel in/serial out shift register and a typical logic symbol. There are four data-input lines, (D_0, D_1, D_2, and D_3) and a $SHIFT/\overline{LOAD}$ input, which allows four bits of data to **load** in parallel into the register.

When $SHIFT/\overline{LOAD}$ is LOW, gates G_1 through G_3 are enabled, allowing each data bit to be applied to the D input of its respective flip-flop. When a clock pulse is applied, the flip-flops with $D = 1$ will set and those with $D = 0$ will reset, thereby storing all four bits simultaneously.

When $SHIFT/\overline{LOAD}$ is HIGH, gates G_1 through G_3 are disabled and gates G_4 through G_6 are enabled, allowing the data bits to shift right from one stage to the next. The OR gates allow either the normal shifting operation or the parallel data-entry operation, depending on which AND gates are enabled by the level on the $SHIFT/\overline{LOAD}$ input. Also, when the $SHIFT/\overline{LOAD}$ input is HIGH, a new data bit is entered into FF0 each time a shift occurs. Therefore, D_0 serves also as a serial input for this register.

FIGURE 9–12 A 4-bit parallel in/serial out shift register.

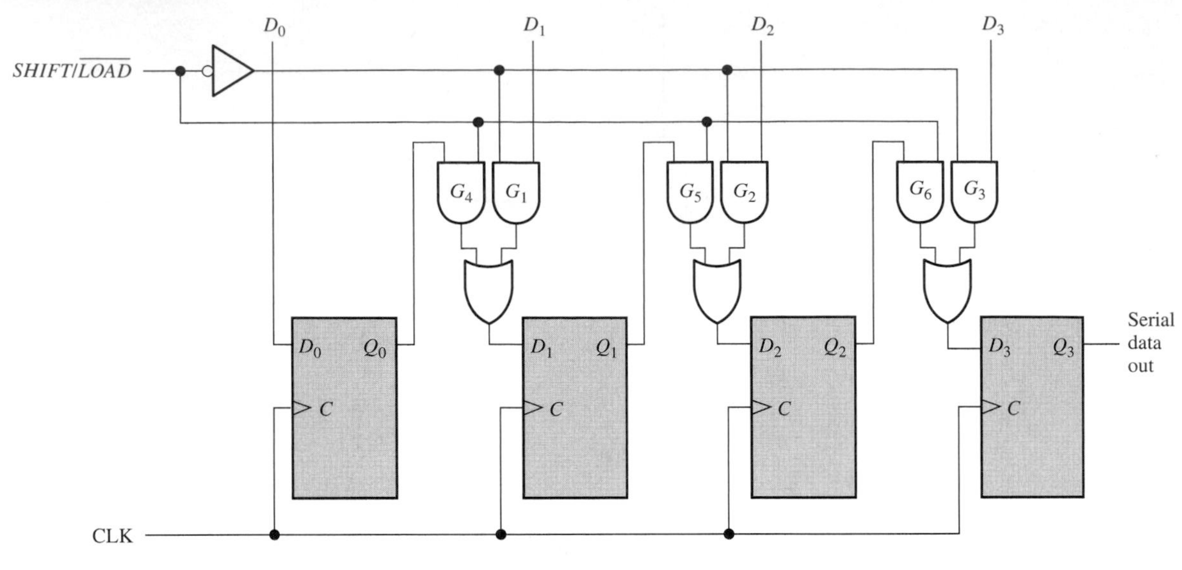

(a) Logic diagram

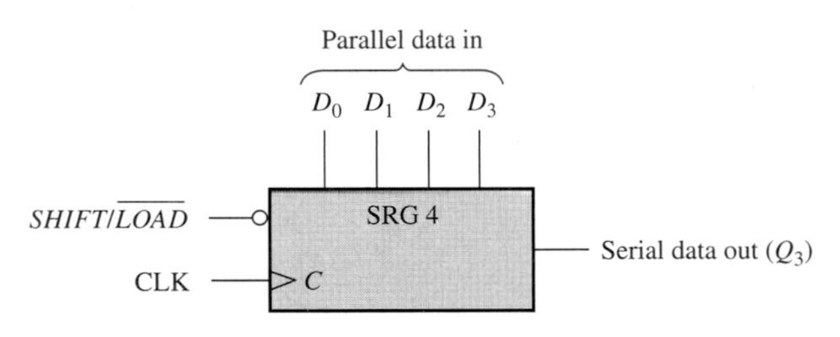

(b) Logic symbol

EXAMPLE 9–3

Problem

Show the data-output waveform for the 4-bit shift register with the parallel input data and the clock and *SHIFT*/$\overline{LOAD}$ waveforms given in Figure 9–13(a). Refer to Figure 9–12(a) for the logic diagram.

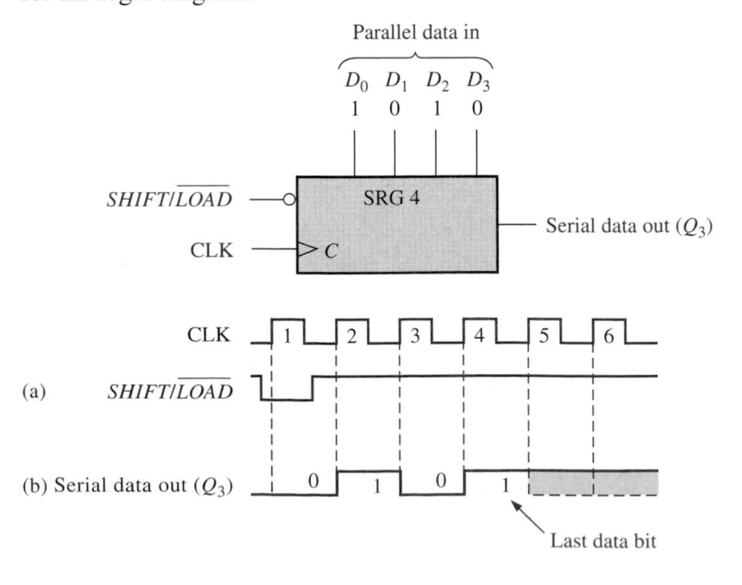

FIGURE 9–13

Solution

On clock pulse 1, the parallel data ($D_0D_1D_2D_3 = 1010$) are loaded into the register, making Q_3 a 0. On clock pulse 2, the 1 from Q_2 is shifted onto Q_3; on clock pulse 3, the 0 is shifted onto Q_3; on clock pulse 4, the last data bit (1) is shifted onto Q_3; and on clock pulse 5, all data bits have been shifted out, and only 1s remain in the register (assuming the D_0 input remains a 1). See Figure 9–13(b).

Question

What is the data-output waveform for the clock and $SHIFT/\overline{LOAD}$ inputs shown in Figure 9–13(a) if the parallel data are $D_0D_1D_2D_3 = 0101$?

COMPUTER SIMULATION

FIGURE 9–14

Open Multisim file F09-14DG on the website. The pulse generator produces one pulse per second so that you can observe data on the parallel inputs and as they shift through the register. The switch is used to either parallel load or to serially shift the data. The parallel data inputs are 1100.

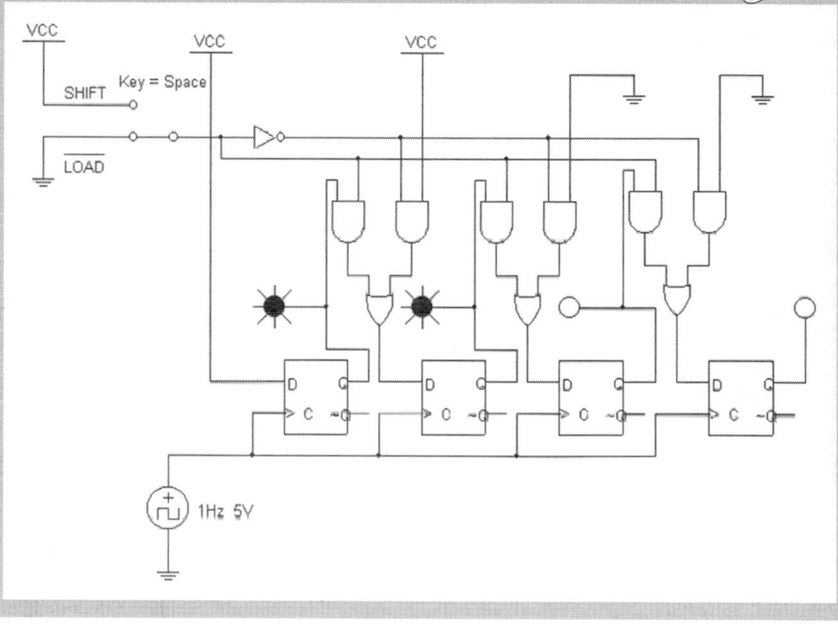

Parallel In/Parallel Out Shift Registers

Parallel in and parallel out operation may seem to be unnecessary. After all, if the data is in parallel form, why use the shift register? The answer depends on the circuit requirements. Sometimes it is necessary to delay data for a clock cycle to allow other operations to occur within the system. The main purpose of a parallel in/parallel out shift register is for temporary storage of data until the next clock pulse.

Figure 9–15 shows a parallel in/parallel out register. In this type of shift register, the data on the parallel inputs are clocked into the register and stored until the next clock pulse.

FIGURE 9–15

A parallel in/parallel out register.

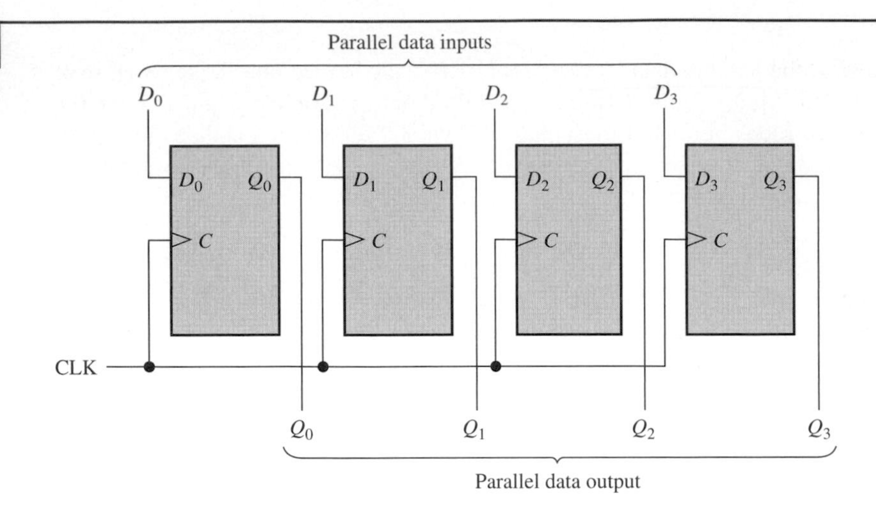

EXAMPLE 9–4

Problem

Show the Q outputs for the register in Figure 9–15, given the data inputs in Figure 9–16(a).

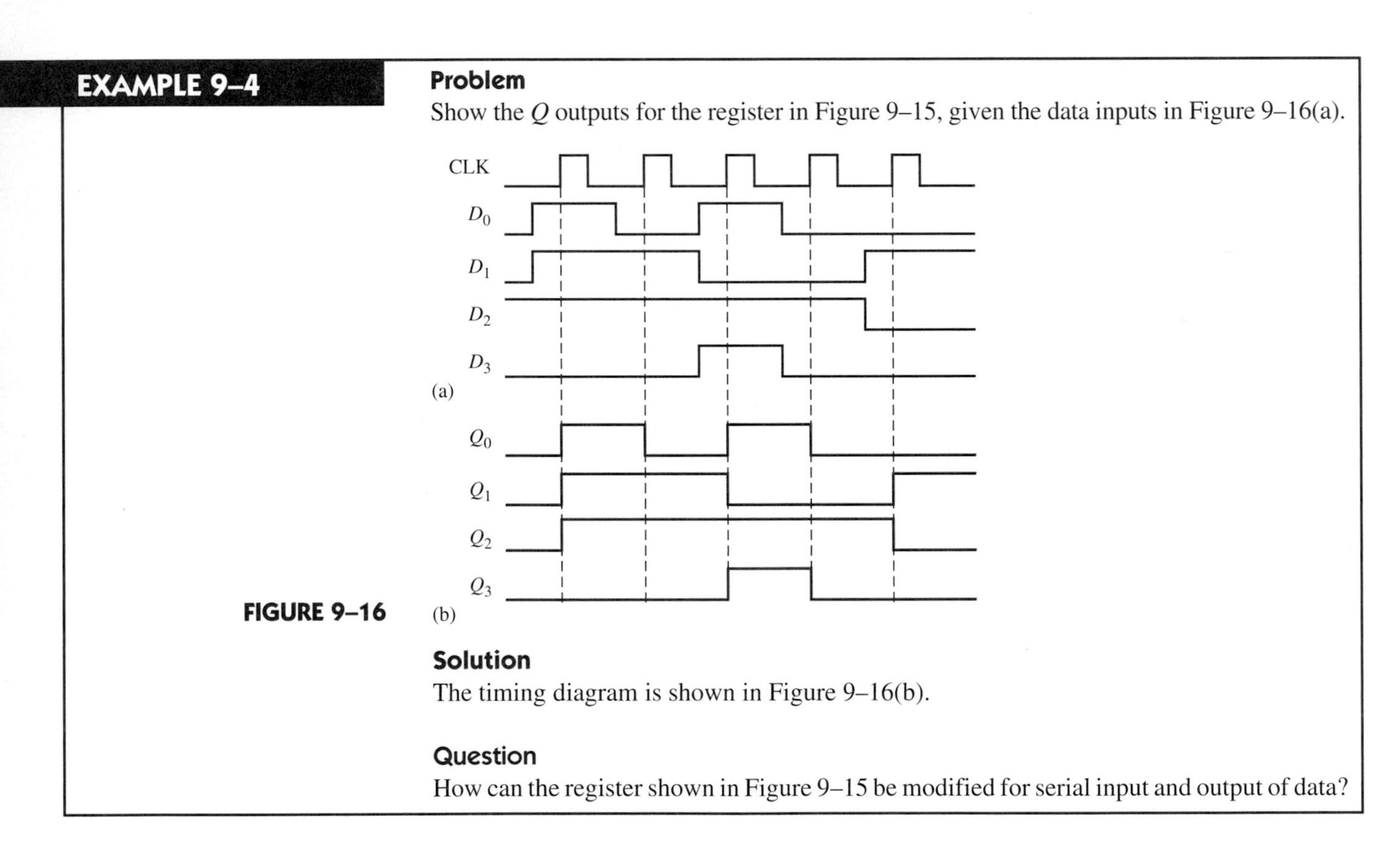

FIGURE 9–16

Solution

The timing diagram is shown in Figure 9–16(b).

Question

How can the register shown in Figure 9–15 be modified for serial input and output of data?

Review Questions

11. What is the function of the $SHIFT/\overline{LOAD}$ input?
12. When the parallel load operation in a shift register is synchronous, what does this mean?
13. What is the purpose of the AND-OR logic in Figure 9–12?
14. When the $SHIFT/\overline{LOAD}$ input is HIGH and a clock pulse occurs, what happens to the data in the shift register?
15. What does the label *SRG 8* mean?

BIDIRECTIONAL SHIFT REGISTERS 9–4

A bidirectional shift register is one in which the data can be shifted either left or right. It can be implemented by using gating logic that enables the transfer of a data bit from one stage to the next stage to the right or to the left, depending on the level of a control line.

In this section, you will learn how a bidirectional shift register works.

A 4-bit **bidirectional** shift register is shown in Figure 9–17. A HIGH on the $RIGHT/\overline{LEFT}$ control input allows data bits inside the register to be shifted to the right, and a LOW enables data bits inside the register to be shifted to the left. An examination of the gating logic will make the operation apparent.

FIGURE 9–17 Four-bit bidirectional shift register.

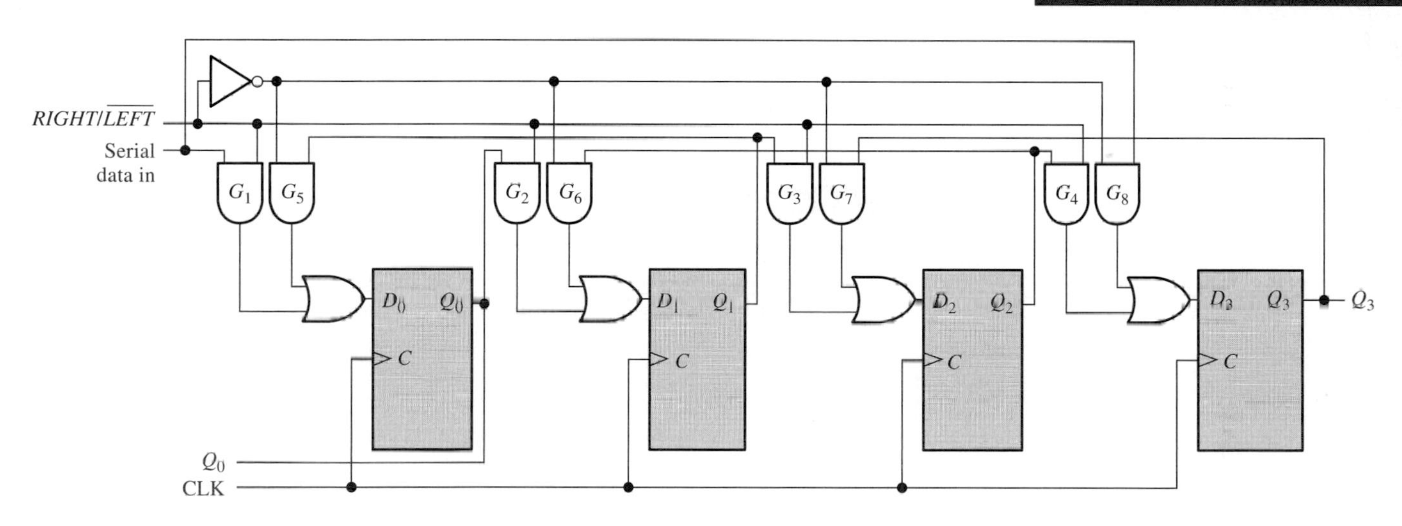

When the $RIGHT/\overline{LEFT}$ control input is HIGH, gates G_1 through G_4 are enabled, and the state of the Q output of each flip-flop is passed through to the D input of the following flip-flop. When a clock pulse occurs, the data bits are shifted one place to the *right*.

When the $RIGHT/\overline{LEFT}$ control input is LOW, gates G_5 through G_8 are enabled, and the Q output of each flip-flop is passed through to the D input of the preceding flip-flop. When a clock pulse occurs, the data bits are then shifted one place to the *left*.

EXAMPLE 9–5

Problem

Determine the state of the shift register of Figure 9–17 after each clock pulse for the given $RIGHT/\overline{LEFT}$ control input waveform in Figure 9–18(a). Assume that $Q_0 = 1$, $Q_1 = 1$, $Q_2 = 0$, and $Q_3 = 1$ and that the serial data-input line is LOW.

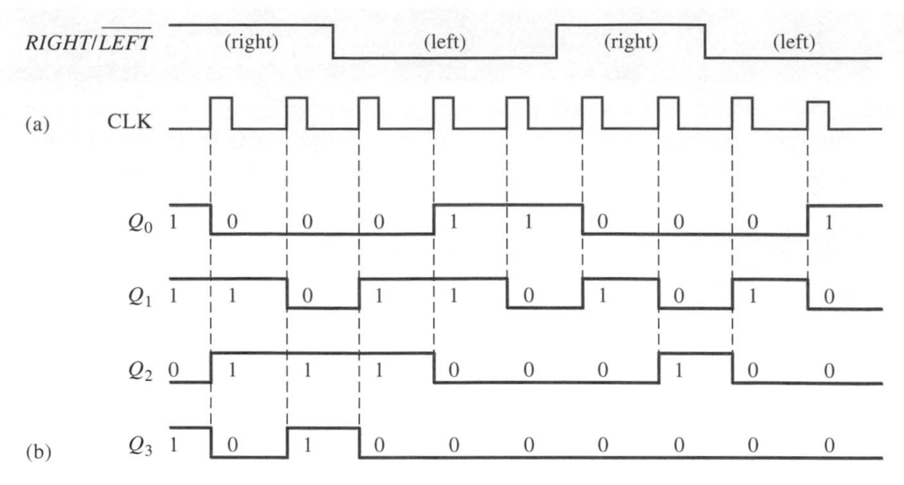

FIGURE 9–18

Solution
See Figure 9–18(b).

Question
If the $RIGHT/\overline{LEFT}$ waveform is inverted, what are the Q outputs of the shift register in Figure 9–17 after each clock pulse?

Review Questions

16. What is a bidirectional shift register?
17. Which input in Figure 9–17 controls the direction in which data are shifted?
18. Assume that the 4-bit bidirectional shift register in Figure 9–17 has the following contents: $Q_0 = 1$, $Q_1 = 1$, $Q_2 = 0$, and $Q_3 = 0$. There is a 1 on the serial data-input line. If $RIGHT/\overline{LEFT}$ is HIGH for three clock pulses and LOW for two more clock pulses, what are the contents after the fifth clock pulse?
19. Repeat Question 18 for $Q_0 = 0$, $Q_1 = 0$, $Q_2 = 1$, and $Q_3 = 1$.
20. Repeat Question 18 for $Q_0 = 1$, $Q_1 = 0$, $Q_2 = 1$, and $Q_3 = 0$.

9–5 SHIFT REGISTER COUNTERS

Although they are shift registers, shift register counters are often classified as counters because they exhibit a specified sequence of states. Two of the most common types of shift register counters are the Johnson counter and the ring counter.

In this section, you will learn how these two types of shift register counters work.

A **shift register counter** is basically a shift register with the serial output connected back to the serial input to produce special sequences.

The Johnson Counter

In a **Johnson counter** the complement of the output of the last flip-flop is connected back to the D input of the first flip-flop (it can be implemented with other types of flip-flops as well). This feedback arrangement produces a characteristic sequence of states, as shown in Table 9–1 for a 4-bit device and in Table 9–2 for a 5-bit device. Notice that the 4-bit sequence has a total of eight states, or bit patterns, and that the 5-bit sequence has a total of ten states. In general, a Johnson counter will produce a number of states equal to $2n$, where n is the number of stages in the counter.

TABLE 9–1

Four-bit Johnson sequence.

Clock pulse	Q_0	Q_1	Q_2	Q_3
0	0	0	0	0
1	1	0	0	0
2	1	1	0	0
3	1	1	1	0
4	1	1	1	1
5	0	1	1	1
6	0	0	1	1
7	0	0	0	1

TABLE 9–2

Five-bit Johnson sequence.

Clock pulse	Q_0	Q_1	Q_2	Q_3	Q_4
0	0	0	0	0	0
1	1	0	0	0	0
2	1	1	0	0	0
3	1	1	1	0	0
4	1	1	1	1	0
5	1	1	1	1	1
6	0	1	1	1	1
7	0	0	1	1	1
8	0	0	0	1	1
9	0	0	0	0	1

The implementations of the 4-stage and 5-stage Johnson counters are shown in Figure 9–19. The implementation of a Johnson counter is very straightforward and is the same regardless of the number of stages. The Q output of each stage is connected to the D input of the next stage (assuming that D flip-flops are used). The single exception is that the $\overline{Q}$ output of the last stage is connected back to the D_0 input of the first stage. As the sequences in Table 9–1 and 9–2 show, the counter will "fill up" with 1s from left to right, and then it will "fill up" with 0s again.

Diagrams of the timing operations of the 4-bit and 5-bit counters are shown in Figures 9–20 and 9–21, respectively.

The Ring Counter

The **ring counter** utilizes one flip-flop for each state in its sequence. It has the advantage that decoding gates are not required because there is a unique output for each state. In the case of a 10-bit ring counter, there is a unique output for each decimal digit.

FIGURE 9–19

Four-bit and 5-bit Johnson counters.

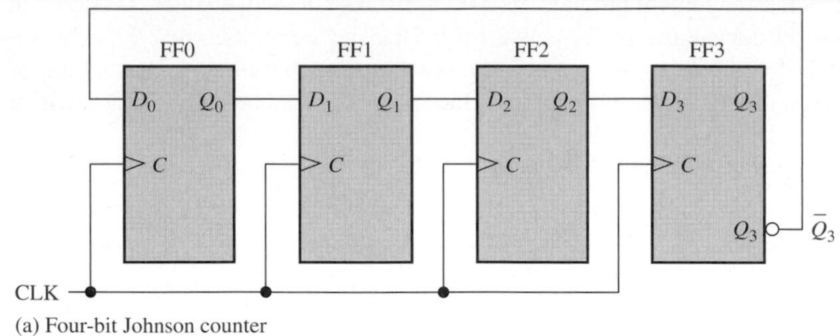

(a) Four-bit Johnson counter

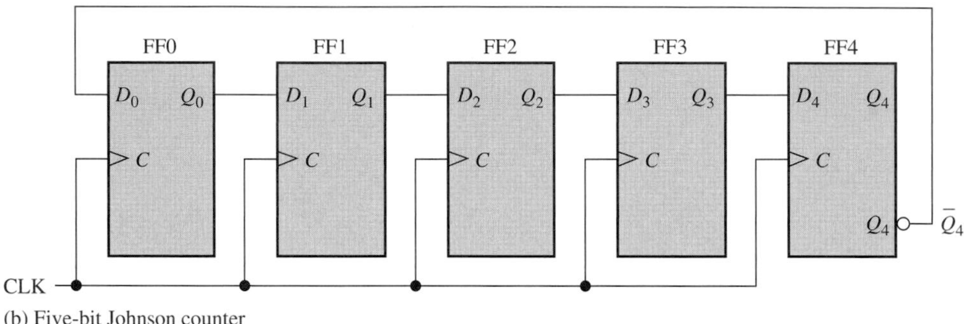

(b) Five-bit Johnson counter

FIGURE 9–20

Timing sequence for a 4-bit Johnson counter.

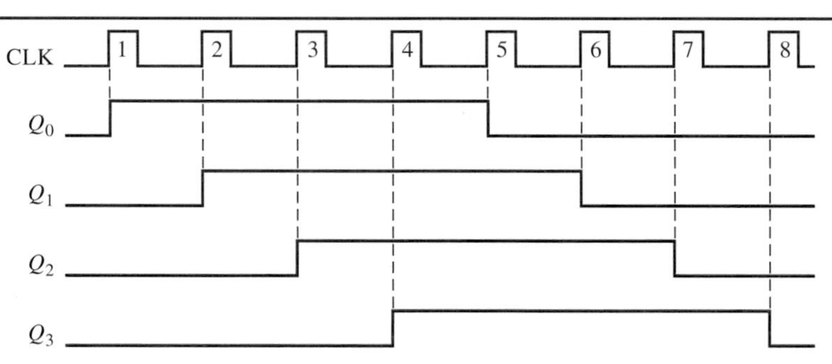

FIGURE 9–21

Timing sequence for a 5-bit Johnson counter.

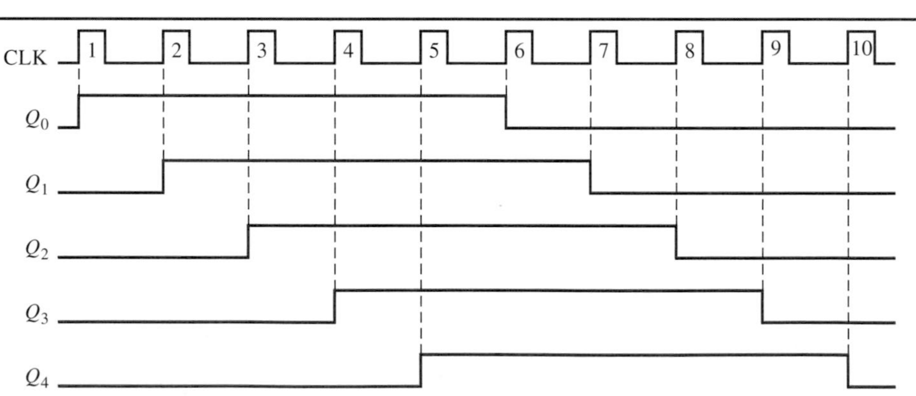

A logic diagram for a 10-bit ring counter is shown in Figure 9–22. The sequence for this ring counter is given in Table 9–3. Initially, a 1 is preset into the first flip-flop, and the rest of the flip-flops are cleared. The interstage connections are the same as those for a Johnson counter, except that Q rather than $\overline{Q}$ is fed back from the last stage. The ten outputs of the counter indicate directly the decimal count of the clock pulse. For instance, a 1 on Q_0 represents a zero, a 1 on Q_1 represents a one, a 1 on Q_2 represents a two, a 1 on Q_3 represents a three, and so on. Notice that the 1 is always retained in the counter and simply shifted "around the ring," advancing one stage for each clock pulse.

Modified sequences can be achieved by having more than a single 1 in the counter, as illustrated in Example 9–6.

FIGURE 9–22 A 10-bit ring counter.

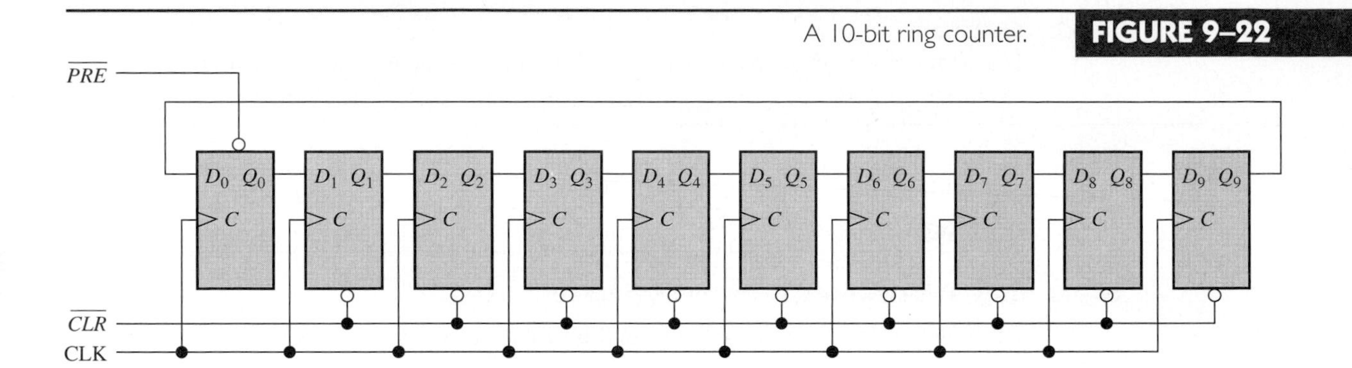

TABLE 9–3 Ten-bit ring counter sequence.

Clock pulse	Q_0	Q_1	Q_2	Q_3	Q_4	Q_5	Q_6	Q_7	Q_8	Q_9
0	1	0	0	0	0	0	0	0	0	0
1	0	1	0	0	0	0	0	0	0	0
2	0	0	1	0	0	0	0	0	0	0
3	0	0	0	1	0	0	0	0	0	0
4	0	0	0	0	1	0	0	0	0	0
5	0	0	0	0	0	1	0	0	0	0
6	0	0	0	0	0	0	1	0	0	0
7	0	0	0	0	0	0	0	1	0	0
8	0	0	0	0	0	0	0	0	1	0
9	0	0	0	0	0	0	0	0	0	1

EXAMPLE 9–6

Problem

If a 10-bit ring counter similar to Figure 9–22 has the initial state 1010000000, determine the waveform for each of the Q outputs.

Solution

See Figure 9–23.

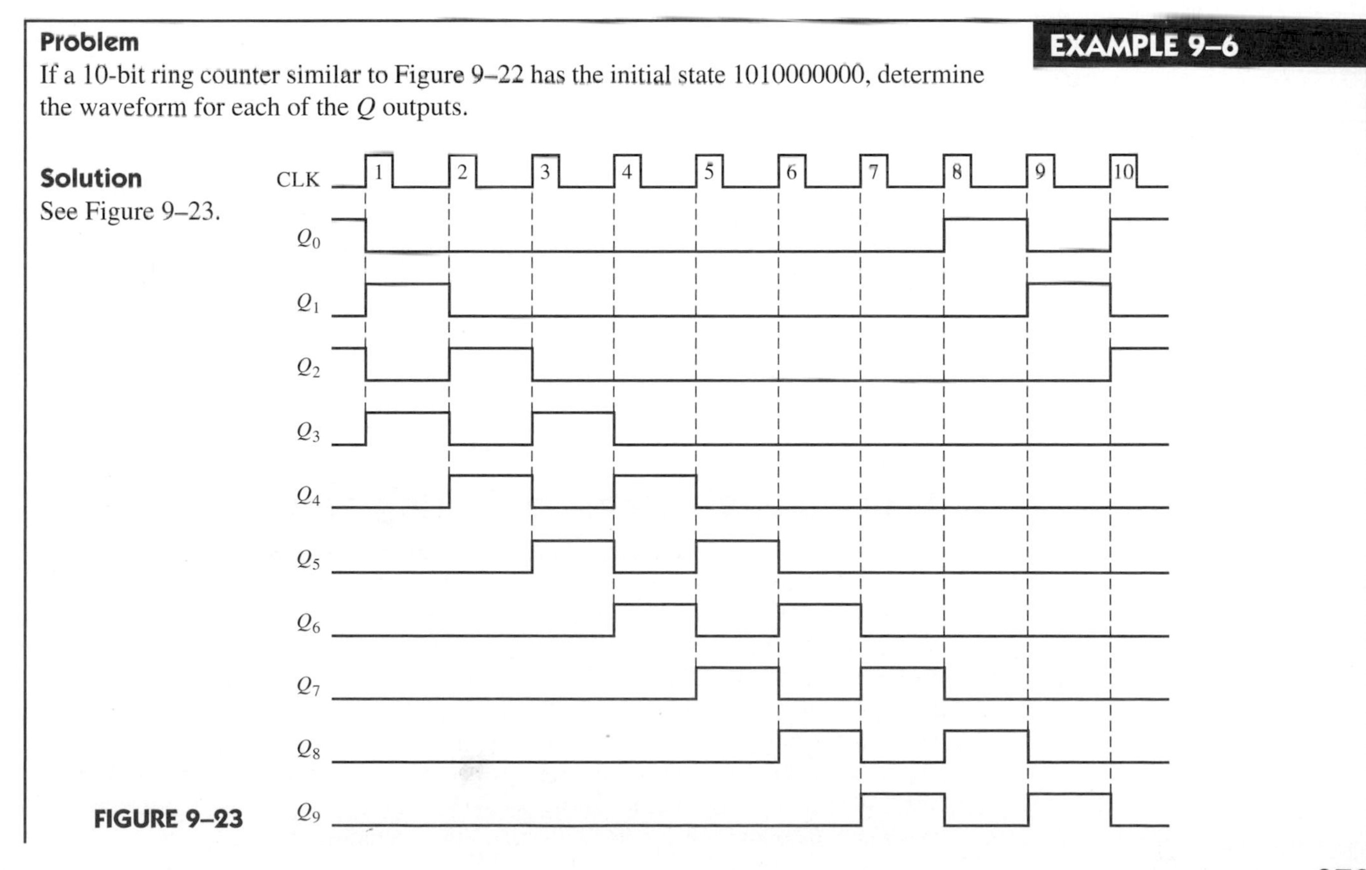

FIGURE 9–23

Question
If a 10-bit ring counter has an initial state 0101001111, determine the waveform for each Q output.

Review Questions

21. What are two types of shift register counters?
22. How many states are there in an 8-bit Johnson counter sequence?
23. What is the sequence of states for a 3-bit Johnson counter starting with 000?
24. How many states does a 10-bit ring counter have?
25. What is the sequence of binary states in the timing diagram of Figure 9–20?

9–6 INTEGRATED CIRCUITS

Several types of shift registers are available in integrated circuit form. As with other IC devices, most IC shift registers have additional functions that were not included on the basic registers that were covered in the previous sections. Occasionally, there may be limited availability of some IC logic devices in certain families. However, any particular device is normally compatible with the same device in another family and can be directly substituted.

In this section, you will learn about several specific IC shift registers.

An 8-Bit Serial In/ Parallel Out Shift Register

The 74LS164 is an example of an IC shift register having serial in/parallel out operation. The logic diagram is shown in Figure 9–24(a), and a typical logic block symbol is shown in part (b). Notice that this device has two serial inputs, A and B, and a clear $(\overline{CLR})$ input that is active-LOW. One of the serial inputs can be used as an enable. The parallel outputs are Q_0 through Q_7.

A sample timing diagram for the 74LS164 is shown in Figure 9–25. Notice that the serial input data on input A are shifted into and through the register after input B goes HIGH.

A 4-Bit Parallel-Access Shift Register

The 74LS195A can be used for parallel in/parallel out operation. Because it also has a serial input, it can be used for serial in/serial out and serial in/parallel out operations. It can be used for parallel in/serial out operation by using Q_3 as the output. A typical logic block symbol is shown in Figure 9–26.

When the $SHIFT/\overline{LOAD}$ input $(SH/\overline{LD})$ is LOW, the data on the parallel inputs are entered synchronously on the positive transition of the clock. When $SH/\overline{LD}$ is HIGH, stored data will shift right (Q_0 to Q_3) synchronously with the clock. Inputs J and $\overline{K}$ are the serial data inputs to the first stage of the register (Q_0); Q_3 can be used for serial output data. The active-LOW clear input is asynchronous.

The timing diagram in Figure 9–27 illustrates the operation of this register.

FIGURE 9–24

The 74LS164 8-bit serial in/parallel out shift register.

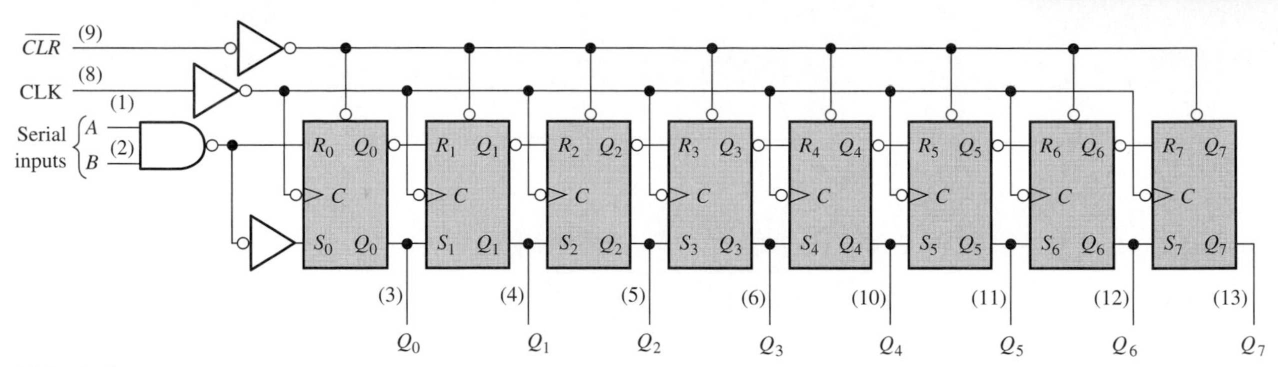

(a) Logic diagram

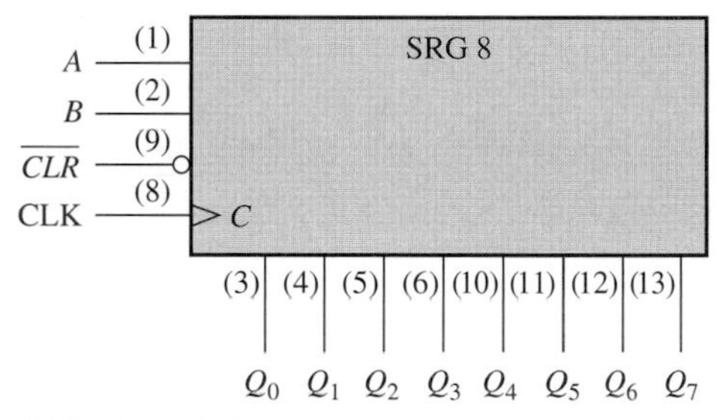

(b) Logic symbol

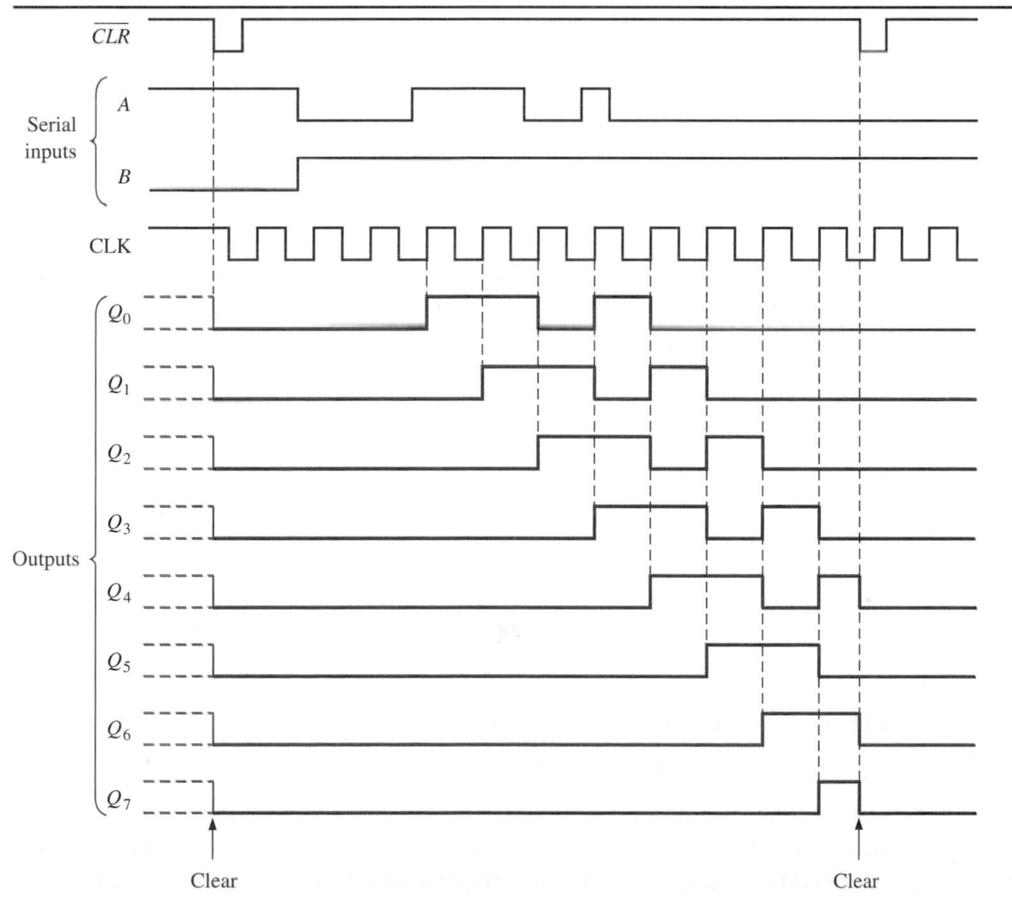

FIGURE 9–25

Sample timing diagram for a 74LS164 shift register.

FIGURE 9–26

The 74LS195A 4-bit parallel access shift register.

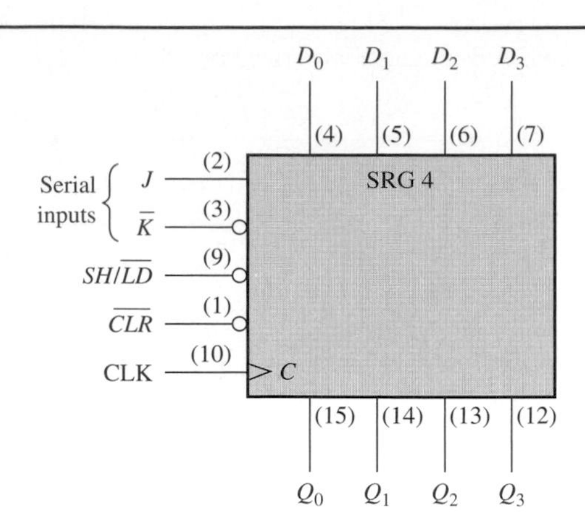

FIGURE 9–27

Sample timing diagram for a 74LS195A shift register.

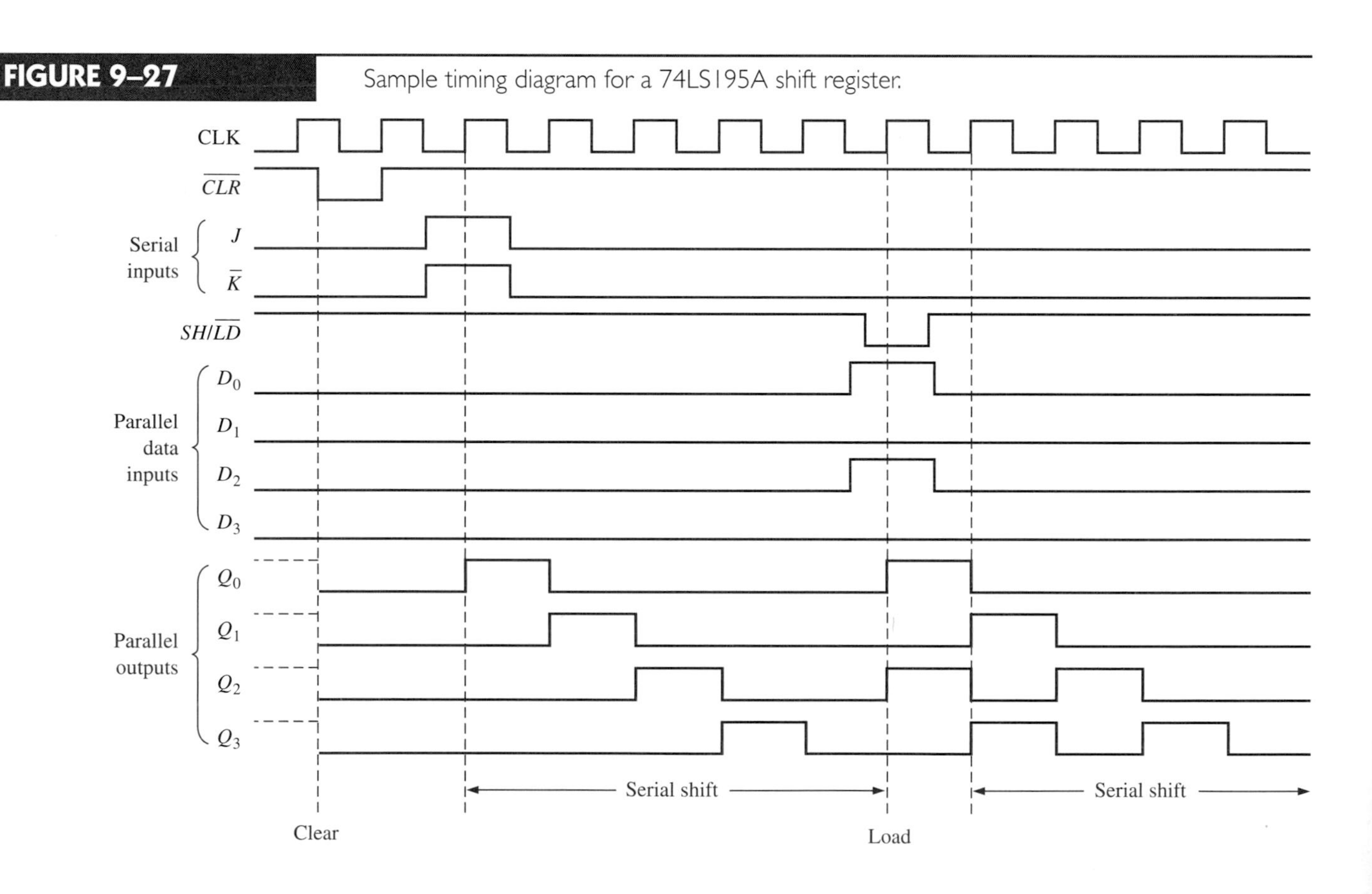

A 4-Bit Bidirectional Universal Shift Register

The 74LS194A is an example of a universal bidirectional shift register in integrated circuit form. A **universal shift register** has both serial and parallel input and output capability. A logic block symbol is shown in Figure 9–28, and a sample timing diagram is shown in Figure 9–29.

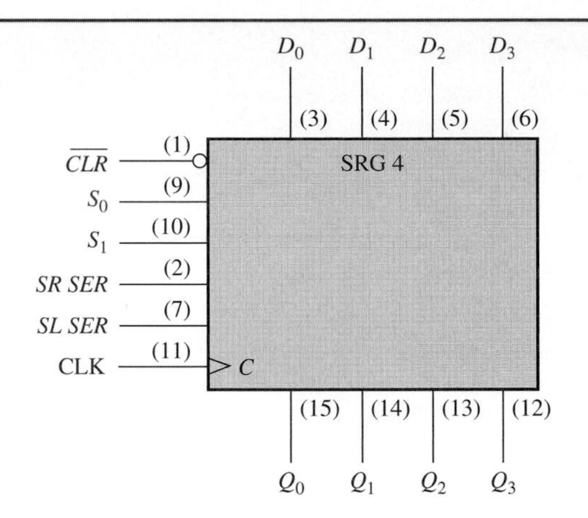

FIGURE 9–28

The 74LS194A 4-bit bidirectional universal shift register.

FIGURE 9–29

Sample timing diagram for a 74LS194A shift register.

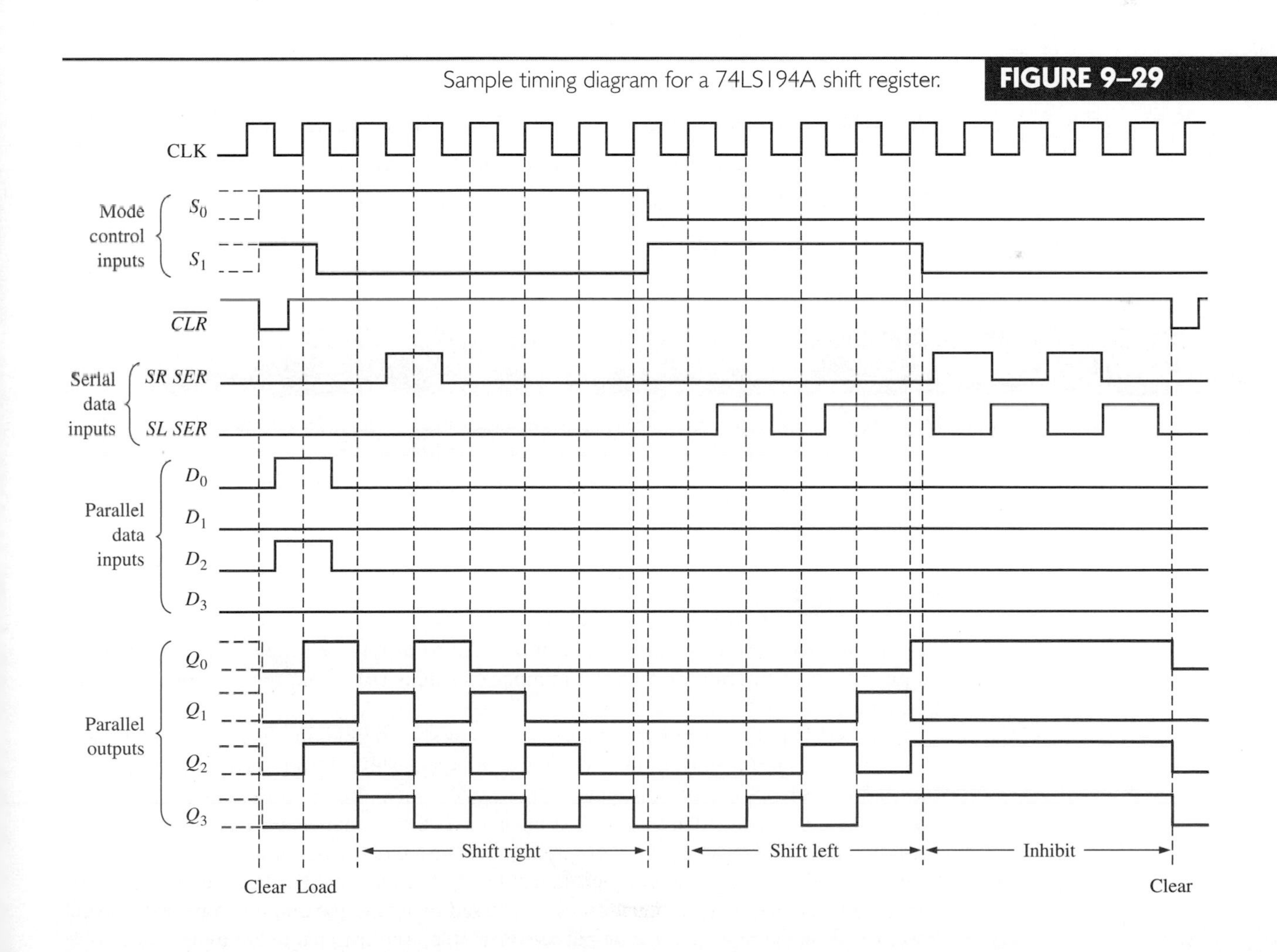

COMPUTER SIMULATION

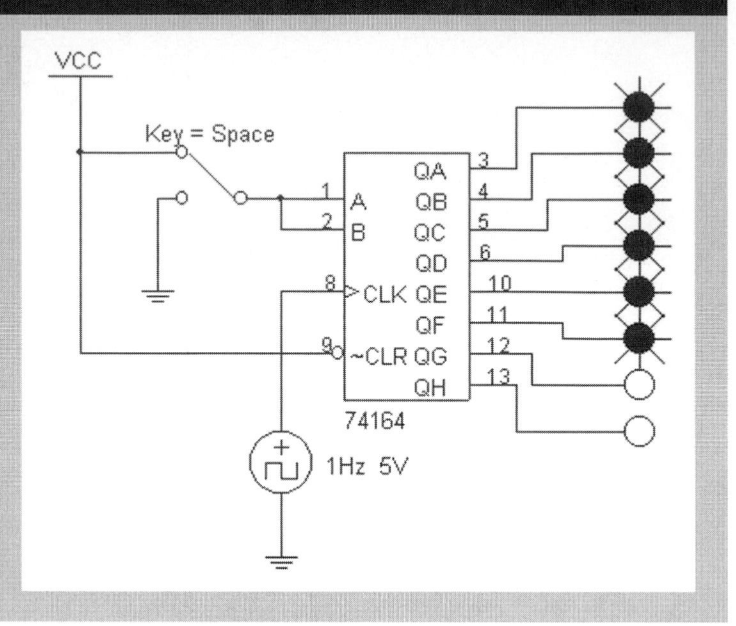

FIGURE 9–30 Open the Multisim file F09-30DG on the website. Use the space bar to enter 1s and watch the shift register fill up with the 1s as indicated by the probe lights. Then switch to 0 inputs and watch the register fill up with 0s.

Review Questions

26. For the 74LS164, to which pin (number) is the clock applied?
27. Which clock pulse edge will trigger the 74LS164 shift register?
28. How can a single serial input be accommodated on a 74LS164?
29. How can the 74LS195A be used for serial output operation?
30. For a 74LS195A, $SH/\overline{LD} = 1$, $J = 1$, and $\overline{K} = 1$. What is Q_0 after one clock pulse?

9–7 APPLICATIONS

Shift registers are found in many types of applications. Three applications discussed in this section are time delay, data conversion, and keyboard encoding.

In this section, you will learn how shift registers can be used.

Time Delay

The serial in/serial out shift register can be used to provide a time delay from input to output that is a function of both the number of stages (n) in the register and the clock frequency.

When a data pulse is applied to the serial input as shown in Figure 9–31, it enters the first stage on the triggering edge of the clock pulse. It is then shifted from stage to stage on each successive clock pulse until it appears on the serial output n clock periods later. This time-delay operation is illustrated in Figure 9–31 in which an 8-bit serial in/serial out shift register is used with a clock frequency of 1 MHz to achieve a time delay (t_d) of 8 μs (8 × 1 μs). This time can be adjusted up or down by changing the clock frequency. The time delay can also be increased by cascading shift registers and decreased by taking the outputs from successively lower stages in the register if the outputs are available, as illustrated in Example 9–7.

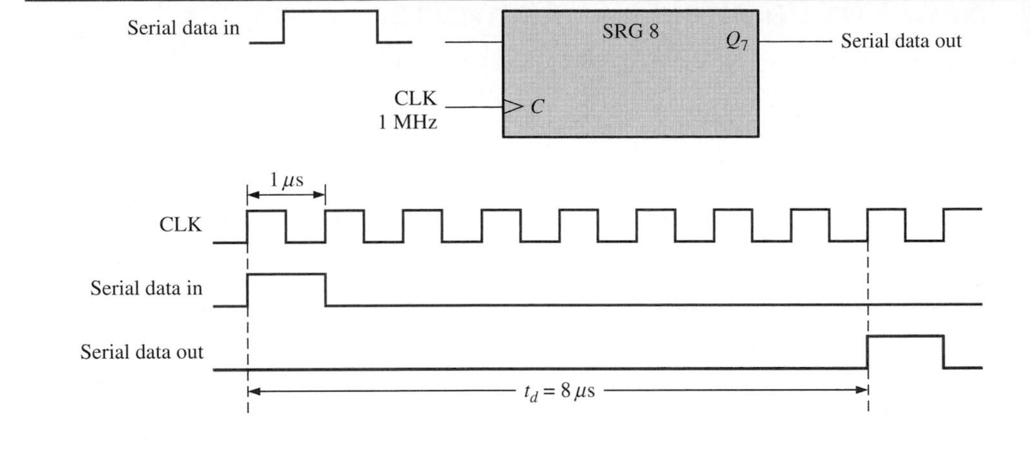

FIGURE 9–31
The shift register as a time-delay device.

EXAMPLE 9–7

Problem

Determine the amount of time delay between the serial input and each output in Figure 9–32. Show a timing diagram to illustrate.

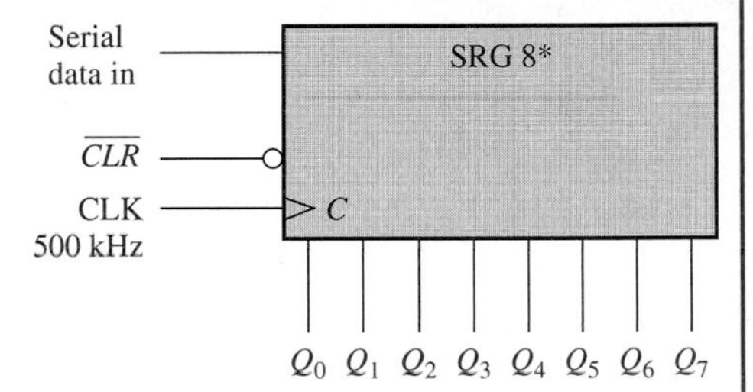

* Data shifts from Q_0 toward Q_7.

FIGURE 9–32

Solution

The clock period is 2 μs. Thus, the time delay can be increased or decreased in 2 μs increments from a minimum of 2 μs to a maximum of 16 μs, as illustrated in Figure 9–33.

Question

What clock frequency is required to obtain a time delay of 24 μs to the Q_7 output in Figure 9–32?

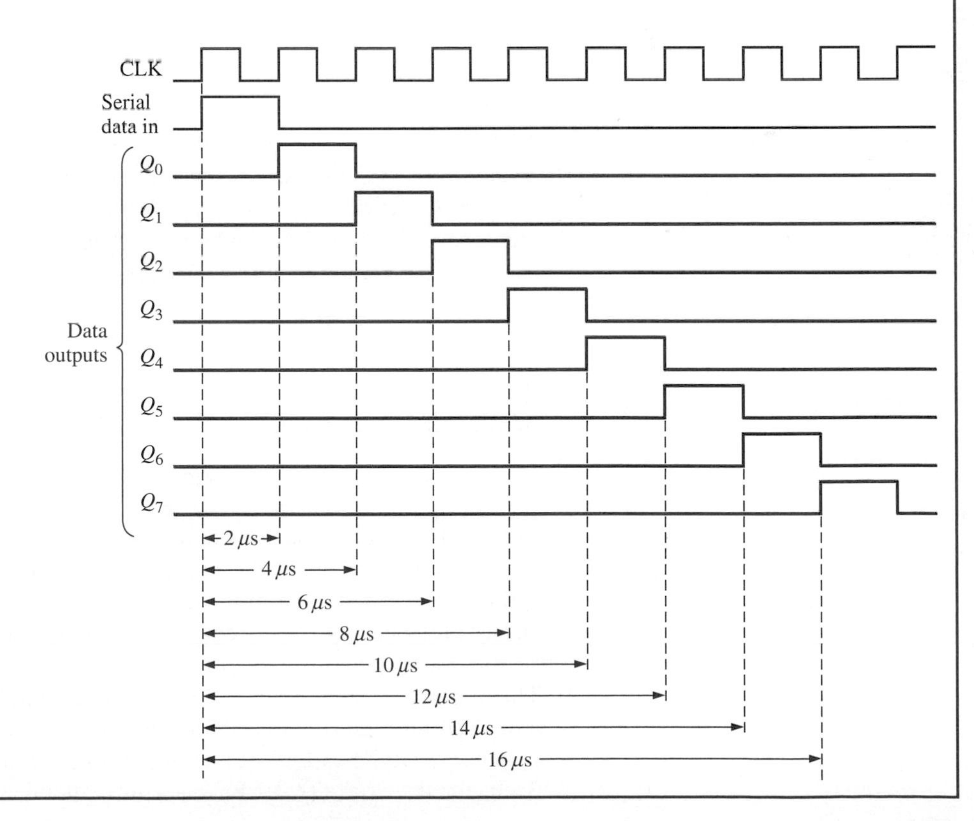

FIGURE 9–33
Timing diagram showing time delays for the register in Figure 9–32.

Serial-to-Parallel Data Converter

Serial data transmission from one digital system to another is commonly used to reduce the number of wires in the transmission line. For example, eight bits can be sent serially over one wire, but it takes eight wires to send the same data in parallel.

A computer or microprocessor-based system commonly requires incoming data to be in parallel format, thus the requirement for serial-to-parallel conversion. A simplified serial-to-parallel data converter, in which two types of shift registers are used, is shown in Figure 9–34.

FIGURE 9–34 Simplified logic diagram of a serial-to-parallel converter.

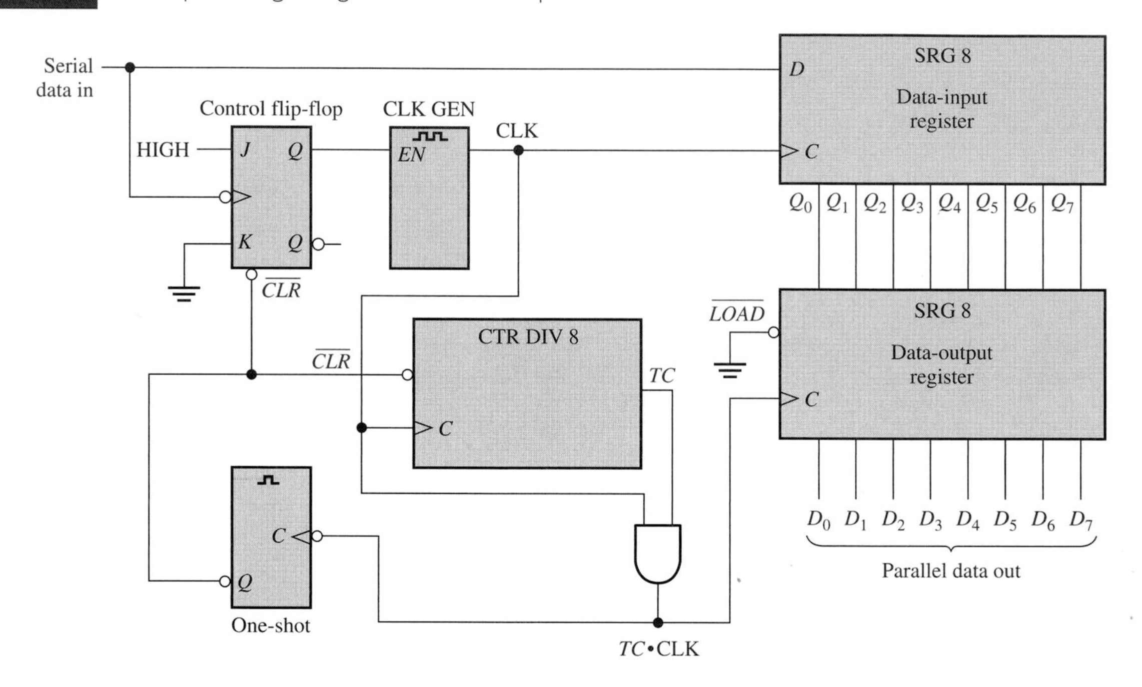

To illustrate the operation of this serial-to-parallel converter, the serial data format shown in Figure 9–35 is used. It consists of eleven bits. The first bit (start bit) is always 0 and always begins with a HIGH-to-LOW transition. The next eight bits (D_7 through D_0) are the data bits (one of the bits can be parity), and the last two bits (stop bits) are always 1s. When no data are being sent, there is a continuous 1 on the serial data line.

FIGURE 9–35 Serial data format.

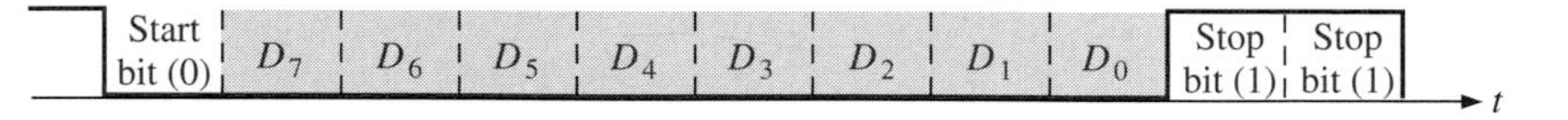

The HIGH-to-LOW transition of the start bit sets the control flip-flop, which enables the clock generator. After a fixed delay time, the clock generator begins producing a pulse waveform, which is applied to the data-input register and to the divide-by-8 counter. The clock has a frequency precisely equal to that of the incoming serial data, and the first clock pulse after the start bit occurs in the center of the first data bit.

The timing diagram in Figure 9–36 illustrates the following basic operation: The eight data bits (D_7 through D_0) are serially shifted into the data-input register. After the eighth clock pulse, a HIGH-to-LOW transition of the terminal count (*TC*) output of the counter ANDed with the clock (*TC*·CLK) loads the eight bits that are in the data-input register into

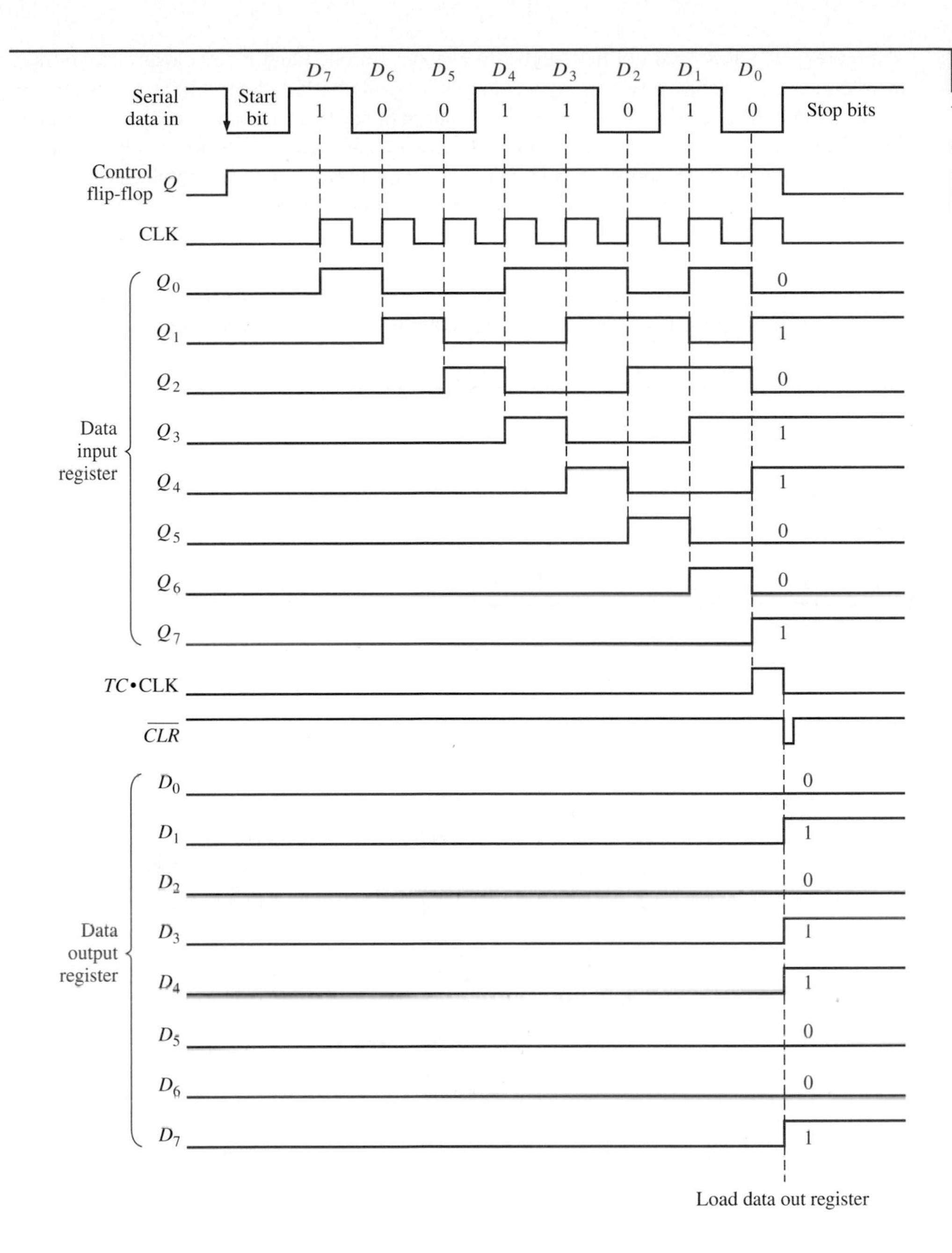

FIGURE 9–36 Timing diagram illustrating the operation of the serial-to-parallel data converter in Figure 9–34.

the data-output register. This same transition also triggers the one-shot, which produces a short-duration pulse to clear the counter and reset the control flip-flop and thus disable the clock generator. The system is now ready for the next group of eleven bits, and it waits for the next HIGH-to-LOW transition at the beginning of the start bit.

By reversing the process just stated, parallel-to-serial data conversion can be accomplished. However, since the serial data format must be produced, additional requirements must be taken into consideration.

Keyboard Encoder

The keyboard encoder is a good example of the application of a shift register used as a ring counter in conjunction with other devices to rotate data from the last stage back to the first.

Figure 9–37 shows a simplified keyboard encoder for encoding a key closure in a 64-key matrix organized in eight rows and eight columns. Two 74LS195A 4-bit shift registers are connected as an 8-bit ring counter with a fixed bit pattern of seven 1s and one 0 preset into it when the power is turned on. Two 74LS147 priority encoders are used as eight-line-to-three-line encoders (9 input HIGH, 8 output unused) to encode the ROW and COLUMN lines of the keyboard matrix. An IC containing six independent flip-flops (74LS174A) is used as a parallel in/parallel out register in which the ROW/COLUMN code from the priority encoders is stored.

The basic operation of the keyboard encoder in Figure 9–37 is as follows: The ring counter "scans" the rows for a key closure as the clock signal shifts the 0 around the counter at a 5 kHz rate. The 0 (LOW) is sequentially applied to each ROW line, while all other

FIGURE 9–37 Simplified keyboard encoding circuit.

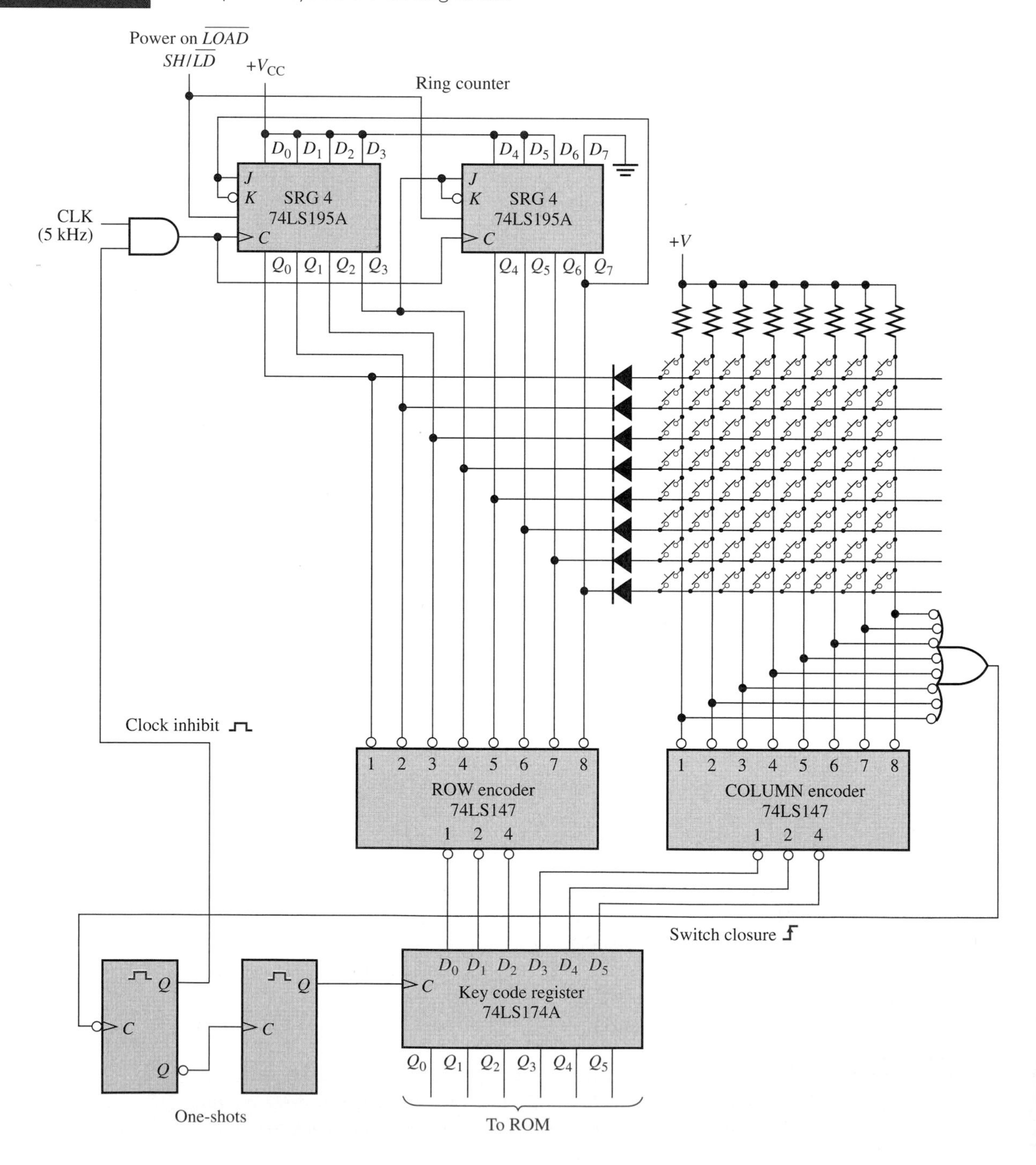

ROW lines are HIGH. All the ROW lines are connected to the ROW encoder inputs, so the 3-bit output of the ROW encoder at any time is the binary representation of the ROW line that is LOW. When there is a key closure, one COLUMN line is connected to one ROW line. When the ROW line is taken LOW by the ring counter, that particular COLUMN line is also pulled LOW. The COLUMN encoder produces a binary output corresponding to the COLUMN in which the key is closed. The 3-bit ROW code plus the 3-bit COLUMN code uniquely identifies the key that is closed. This 6-bit code is applied to the inputs of the key code register. When a key is closed, the two one-shots produce a delayed clock pulse to parallel-load the 6-bit code into the key code register. This delay allows the contact bounce to die out. Also, the first one-shot output inhibits the ring counter to prevent it from scanning while the data are being loaded into the key code register.

The 6-bit code in the key code register is now applied to a ROM (read-only memory) to be converted to an appropriate alphanumeric code that identifies the keyboard character. ROMs are studied in Chapter 11.

Review Questions

31. How much time delay will a 16-bit shift register produce using a 1 MHz clock?
32. In the keyboard encoder, at what rate does the ring counter scan for key closures?
33. What is the 6-bit ROW/COLUMN code (key code) for the top row and the left-most column in the keyboard encoder?
34. What is the purpose of the two one-shots in the keyboard encoder?
35. Why is serial transmission frequently used instead of parallel data transmission to transfer data between two systems?

TROUBLESHOOTING 9–8

One basic method of troubleshooting sequential logic and other more complex digital systems uses a procedure of "exercising" a circuit under test. A fault can be detected by applying a known input waveform (stimulus) and then observing the output for the proper bit pattern.

In this section, you will learn how test patterns are used to troubleshoot a digital system.

The serial-to-parallel data converter in Figure 9–34 is used to illustrate the "exercising" procedure. The main objective in exercising the circuit is to force all elements (flip-flops and gates) into all of their states to be certain that nothing is stuck in a given state as a result of a fault. The input test pattern, in this case, must be designed to force each flip-flop in the registers into both states, to clock the counter through all of its eight states, and to take the control flip-flop, the clock generator, the one-shot, and the AND gate through their paces.

The input test pattern that accomplishes this objective for the serial-to-parallel data converter is based on the serial data format in Figure 9–35. It consists of the pattern 10101010 in one serial group of data bits followed by 01010101 in the next group, as shown in Figure 9–38. These patterns are generated on a repetitive basis by a special test-pattern generator. The basic test setup is shown in Figure 9–39.

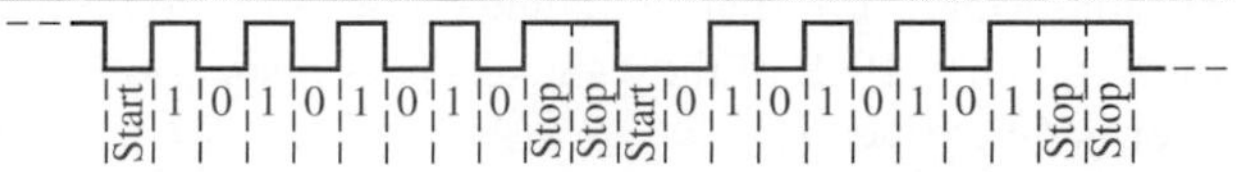

FIGURE 9–38

Sample test pattern.

FIGURE 9–39 Basic test setup for the serial-to-parallel data converter of Figure 9–34.

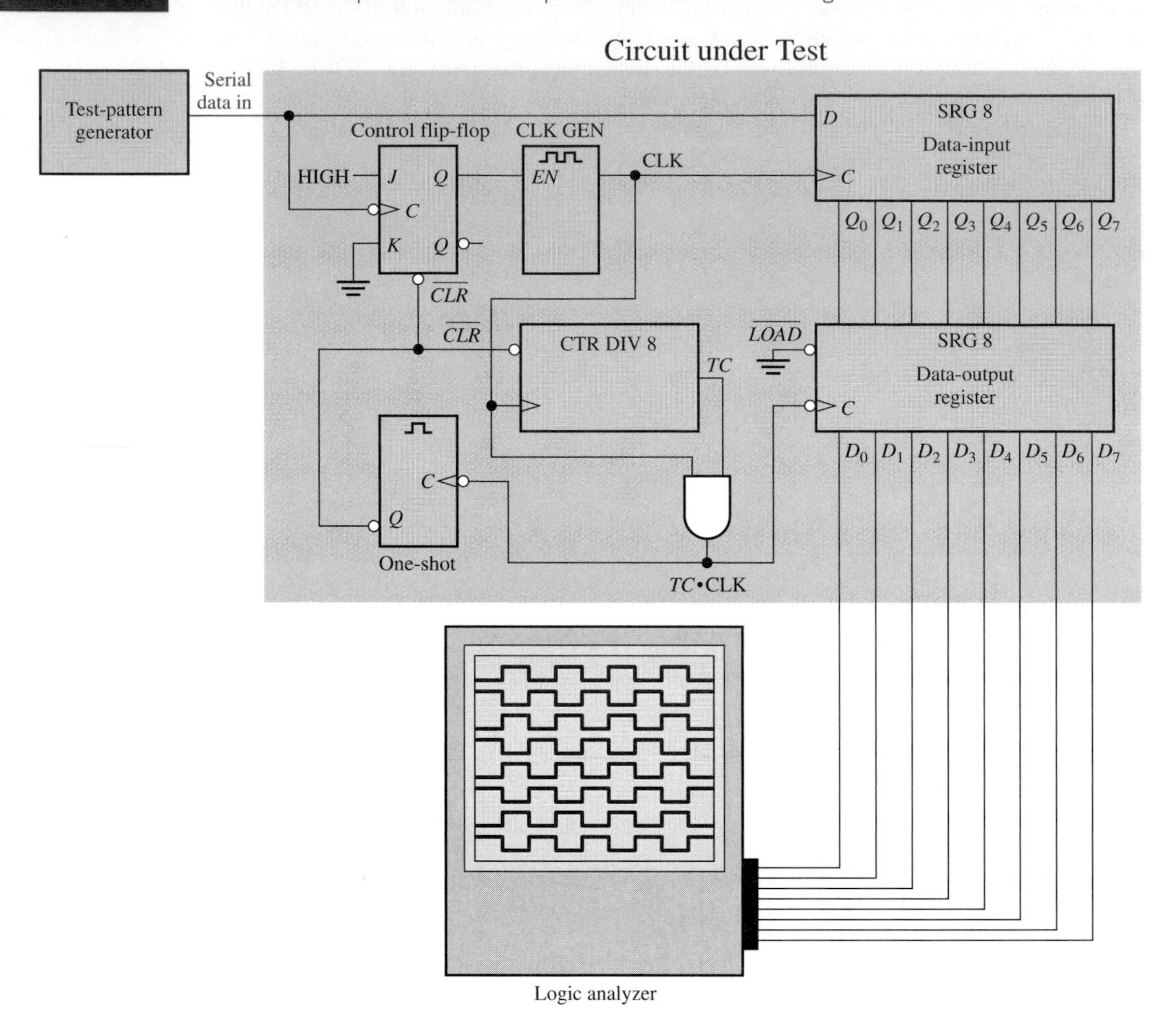

After both patterns have been run through the circuit under test, all the flip-flops in the data-input register and in the data-output register have resided in both SET and RESET states, the counter has gone through its sequence (once for each bit pattern), and all the other devices have been exercised.

To check for proper operation, each of the parallel data outputs is observed for an alternating pattern of 1s and 0s as the input test patterns are repetitively shifted into the data-input register and then loaded into the data-output register. The proper timing diagram is shown in Figure 9–40. The outputs can be observed in pairs with a dual-trace oscilloscope, or all eight outputs can be observed simultaneously with a logic analyzer configured for timing analysis.

If one or more outputs of the data-output register are incorrect, then you must back up to the outputs of the data-input register. If these outputs are correct, then the problem is associated with the data-output register. Check the inputs to the data-output register directly on the pins of the IC for an open input line. Check that power and ground are correct (look for the absence of noise on the ground line). Verify that the load input is a solid LOW and that there are clock pulses on the clock input of the correct amplitude. Make sure that the connection to the logic analyzer did not short two output lines together. If all of these checks pass inspection, then it is likely that the output register is defective.

SAFETY NOTE

Be aware of fire safety in the lab. Know the location of fire extinguishers. Electrical fires are never fought with water, which can conduct electricity. Use only a class-C extinguisher, which uses a nonconductive agent—typically CO_2. It is important to know what type of extinguisher you use for any fire. The wrong choice can make the fire worse.

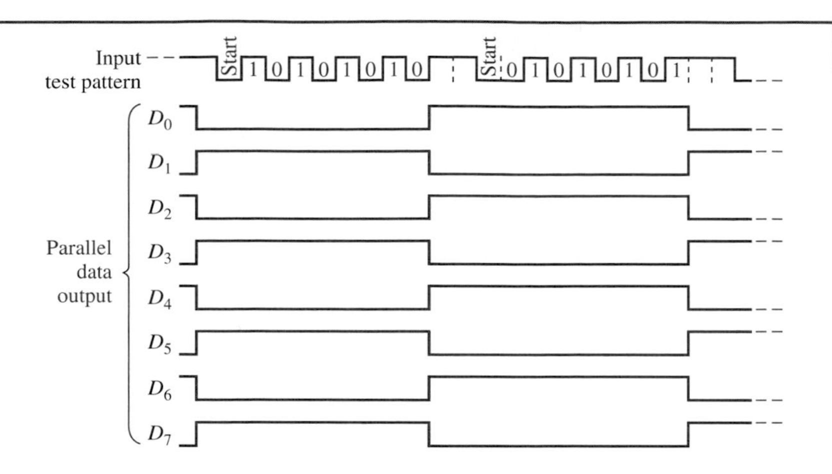

FIGURE 9–40

Proper outputs for the circuit under test in Figure 9–39. The input test pattern is shown.

If the data-input register outputs are also incorrect, the fault could be associated with the input register itself or with any of the other logic, and additional investigation is necessary to isolate the problem.

Review Questions

36. What should you do before testing a circuit?
37. What is the purpose of providing a test input to a sequential circuit?
38. Generally, when an output waveform is found to be incorrect, what is the next step to be taken?
39. What is meant by "exercising" a circuit?
40. What is a digital test pattern?

CHAPTER REVIEW

Key Terms

Bidirectional Having two directions. In a bidirectional shift register, the stored data can be shifted right or left.

Load To enter parallel data into a shift register.

Shift To move binary data from stage to stage within a shift register or other storage device or to move binary data into or out of the device.

Shift register Two or more flip-flops connected to temporarily store and move binary data from one flip-flop to another.

Shift register counter A type of shift register with the serial output connected back to the serial input that exhibits a special sequence of states.

Stage One storage element (flip-flop) in a shift register.

10 CHAPTER

PROGRAMMABLE LOGIC

CHAPTER OUTLINE

Study aids for this chapter are available at

http://www.prenhall.com/SOE

INTRODUCTION

The integrated circuits that you have studied in previous chapters contained various types of gates, latches, decoders, encoders, flip-flops, counters, and registers. These ICs are designed to provide fixed logic functions that can't be altered and are classified as either small-scale integration (SSI) or medium-scale integration (MSI). If you need a certain type of counter, for example, you pick a specific fixed-function IC device that meets your requirement.

In many applications, the programmable logic device (PLD) has replaced the fixed-function logic device. You can expect to see a continued growth in PLDs. However, fixed-function logic is still important and will be around for a long time but in more limited applications. One advantage of the PLD over fixed-function logic is that more complex logic circuits can be "stuffed" into a much smaller area on a chip with PLDs. A second advantage is that, with certain PLDs, logic circuits can be easily changed without rewiring or replacing components. A third advantage is that PLDs can be used to implement a custom logic circuit designed using a computer program. The implementation is accomplished faster than using many individual fixed-function ICs.

In this chapter, you will learn the basics of programmable logic and some of the main types of PLDs. Also, you will see how a PLD is programmed and will be introduced to VHDL, one of the most popular PLD programming languages.

KEY OBJECTIVES

A section number is given for each objective. After completing this chapter, you should be able to

- **10–1** Describe the basic concepts of a PLD and discuss some major types
- **10–2** Explain the structure of programmable array logic (PAL)
- **10–3** Explain the basics of generic array logic (GAL)
- **10–4** Discuss the fundamentals of CPLDs and FPGAs
- **10–5** Describe the process of programming a PLD
- **10–6** Discuss how VHDL is used to program a PLD to perform a specified logic function

LABORATORY EXPERIMENTS DIRECTORY

The following exercise is for this chapter.

- **Experiment 14**
  Programming for a PLD

KEY TERMS

- PLD
- SPLD
- CPLD
- FPGA
- Programmable arrays
- PAL
- GAL
- Target device
- Program
- VHDL
- Compiler

The programmable logic devices covered in this chapter are complex networks of logic circuits that are joined by a matrix of interconnecting paths. One of the more exciting prospects for future electronic systems is associated with life itself. DNA (deoxyribonucleic acid) contains the "blueprint" for replicating complex molecular networks. Using DNA, scientists have constructed a DNA "computer," which has been used to solve the "traveling salesman" problem: Find the shortest route between seven cities without retracing your steps. The problem has thousands of possible solutions, but only one that is the shortest path.

To solve the problem, a unique strand of DNA was made to represent each "city." Billions of copies of the strands were mixed together, producing all possible routes between the "cities." Then with a series of biochemical reactions, the shortest strand that had each city only once in the sequence was identified. This strand represented the shortest route between the "cities."

In the future, DNA might be used to build huge networks of logic circuits.

10–1 PROGRAMMABLE LOGIC DEVICES (PLD)

The **PLD** (programmable logic device) is a programmable integrated circuit into which any digital logic design can be programmed using a PLD programming language called a hardware description language (HDL).

In this section, you will learn the basic types of PLDs and PLD arrays.

Types of PLDs

Three major types of programmable logic devices are SPLD, CPLD, and FPGA. There may be two or more categories in each type.

FIGURE 10–1

Typical SPLD package.

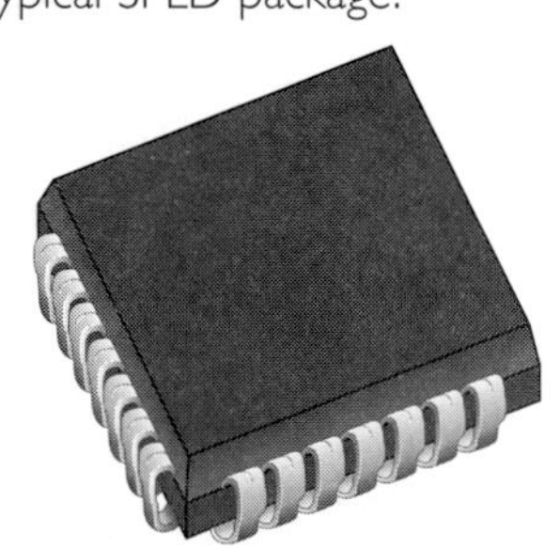

SPLD (Simple Programmable Logic Device)

The **SPLD** is the least complex form of PLD and was the first type available. Typically, one SPLD can replace several fixed-function SSI or MSI devices and their interconnections. A few categories of SPLD are listed here, but in this chapter we will cover only the PAL and the GAL. A typical SPLD package, such as shown in Figure 10–1, has 24 to 28 pins.

- PAL—programmable array logic
- GAL—generic array logic
- PLA—programmable logic array
- PROM—programmable read-only memory

CPLD (Complex Programmable Logic Device)

The CPLD has a much higher capacity than a SPLD so it can replace more complicated logic. Much more complex logic circuits can be programmed into them than into SPLDs. A typical **CPLD** is the equivalent of from two to sixty-four SPLDs.

The development of the CPLD followed the SPLD as advances in technology allowed higher-density chips to be implemented. The forms of CPLD vary in complexity and programming capability. CPLDs typically are in 44 to 160-pin packages. Typical packages are shown in Figure 10–2.

FIGURE 10–2

Typical CPLD packages.

FPGA (Field Programmable Gate Array)

FPGAs are different from SPLDs and CPLDs in their internal organization, and they have more logic capacity than CPLDs. The **FPGA** consists of an array of from sixty-four to several thousand logic gate groups that are usually called **logic blocks**. FPGAs come in package sizes ranging up to 1000 pins or more.

PLD Arrays

All PLDs, including SPLDs, CPLDs, and FPGAs, are made up of **programmable arrays**, which are logic gates with a grid of programmable interconnections. The interconnections form a matrix of rows and columns with an integrated programmable link at each cross point. A **fuse** is a one-time programmable (OTP) link in a PLD that is blown to eliminate selected variables from the output function. Another one-time programmable link is the **antifuse**; in this method of programming a PLD, a normally open contact is shorted by melting the antifuse material to form a connection. A third type of link is reprogrammable and will be covered in Section 10–3.

The Programmable AND Array

The programmable AND array consists of AND gates connected to an interconnection matrix with programmable fuse links at each cross point, as shown in Figure 10–3(a). Only twenty-four links are shown, but an actual array can have hundreds or thousands of programmable links. The AND array can be programmed by passing sufficient current through selected links to blow the fuses and eliminate the associated input from a particular gate. After programming, only one fuse is left intact in each row in order to connect the desired variable to the gate input, as illustrated in part (b). Notice the resulting output logic expressions for this particular AND array. Once a fuse is blown, it cannot be reconnected.

The Programmable OR Array

Although the programmable AND array is more commonly used, the programmable OR array is similar except for the type of gate. A simple illustration of an OR array with fuse links is shown in Figure 10–4. Notice the resulting output logic expressions.

FIGURE 10–3 Example of a simple programmable AND array.

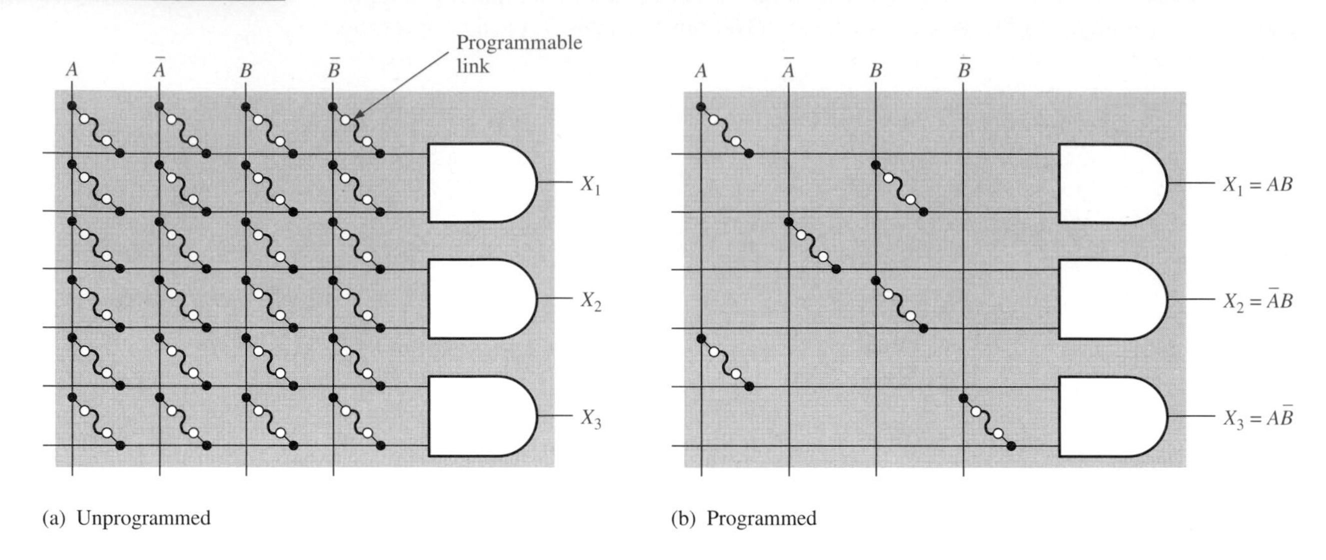

(a) Unprogrammed (b) Programmed

FIGURE 10–4 An example of a simple programmable OR array.

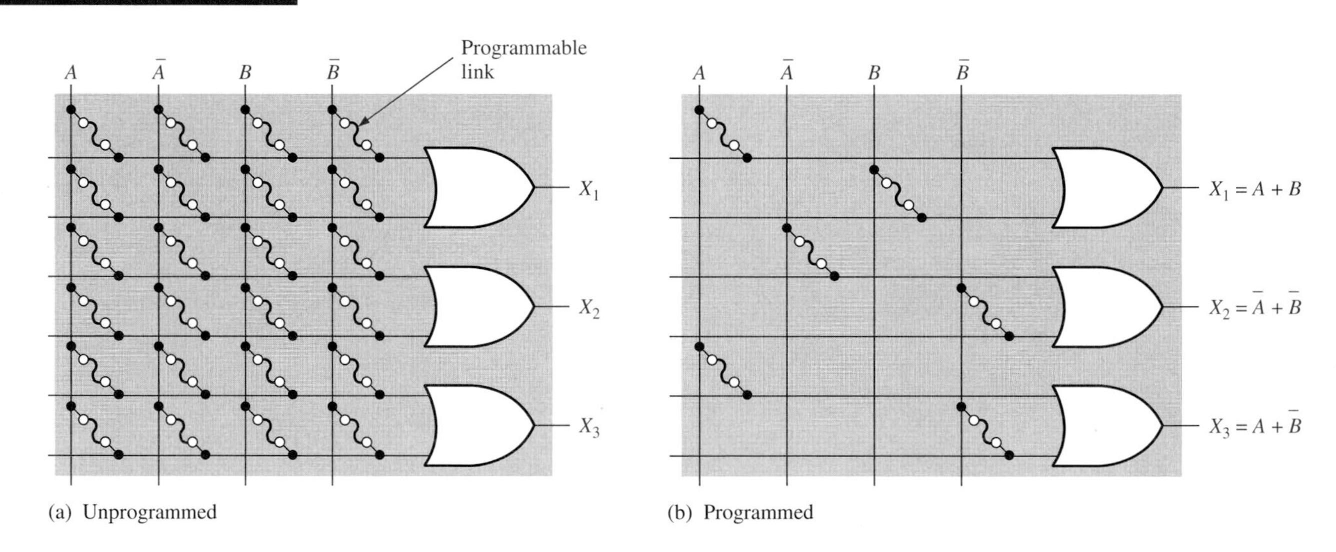

(a) Unprogrammed (b) Programmed

Array Programming

PLDs are programmed using a computer that is running a program based on a hardware description language such as VHDL. The PLD is plugged into a special fixture that is connected to the computer. When the program is run, signals are applied to the PLD pins that cause selected fuse links to be opened to break a connection or, in the case of the antifuse, the selected links are shorted to create a connection. Some PLDs can be programmed after they are mounted on a printed circuit board and do not require a special fixture. All PLDs are supported by manufacturers' development systems, which are software packages for device programming or by general-purpose development systems that can be used to program devices from several manufacturers. One advantage of the development systems is that combinational logic can often be reduced by software, freeing the designer from having to minimize Boolean expressions.

Review Questions

Answers are at the end of the chapter.

1. What does PLD stand for?
2. What are three types of PLDs?
3. What is represented by each of the following acronyms: SPLD, CPLD, and FPGA?
4. How are programmable fuse links used to program a PLD?
5. What are two types of PLD arrays?

PROGRAMMABLE ARRAY LOGIC (PAL) 10–2

The PAL and the GAL are the most common types of SPLDs used for logic implementation. Programmable array logic is used to produce specified combinational logic functions.

In this section, you will learn about the structure of the PAL.

PAL Operation

The **PAL** is an SPLD that is one-time programmable (OTP). A PAL consists of a programmable array of AND gates that connects to a fixed array of OR gates. This structure allows any sum-of-products (SOP) logic expression with a defined number of variables to be implemented. As you learned in Chapter 4, any logic function can be expressed in SOP form.

A greatly simplified array structure of a PAL is illustrated in Figure 10–5 for two input variables and one output. Actual PALs have many input and output variables and AND gates in their programmable array and many OR gates in their fixed array.

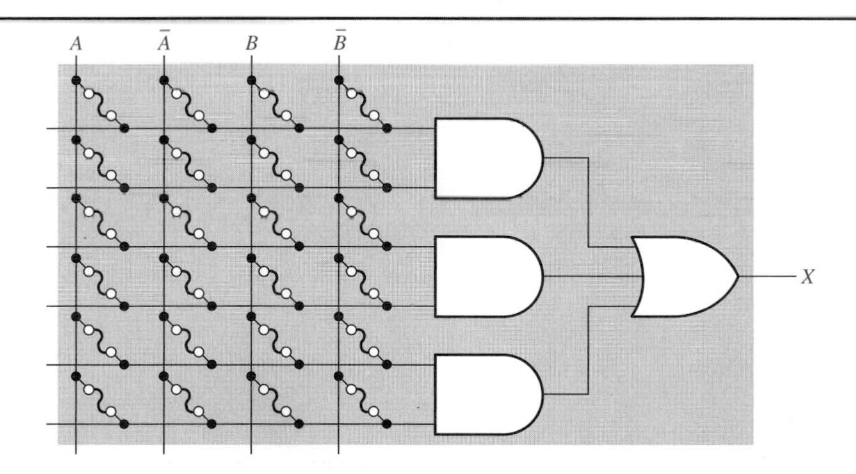

FIGURE 10–5

Simplified structure of a PAL with fuse cells.

As you have learned, a programmable array is essentially a grid or matrix of conductors that form rows and columns with a link at each cross point of a row and column. Each link between a row and column is called a **cell** and is the programmable element of a PAL. Each row in the matrix is connected to the input of an AND gate, and each column is connected to an input variable or the complement of an input variable. By properly programming the links to make or break connection between a row and column, various combinations of input variables or complements can be applied to the inputs of the AND gates to produce any desired product term.

Implementing a Sum-of-Products (SOP) Expression

Figure 10–6 shows how an SOP expression can be implemented with a PAL. The AND array is programmed so that the product term AB is produced by the top AND gate, the term $A\overline{B}$ is produced by the middle AND gate, and the term $\overline{A}\,\overline{B}$ is produced by the bottom AND gate. As you can see, the cells are left intact to connect the desired variables or their complements to the appropriate AND gate inputs. The cells are opened where a variable or its complement is not used in a given product term. There can be only one variable connected to each gate input. The AND gate array outputs are connected to the OR gate inputs. The final output from the OR gate is the SOP expression.

$$X = AB + A\overline{B} + \overline{A}\,\overline{B}$$

FIGURE 10–6

An SOP expression implemented in a PAL.

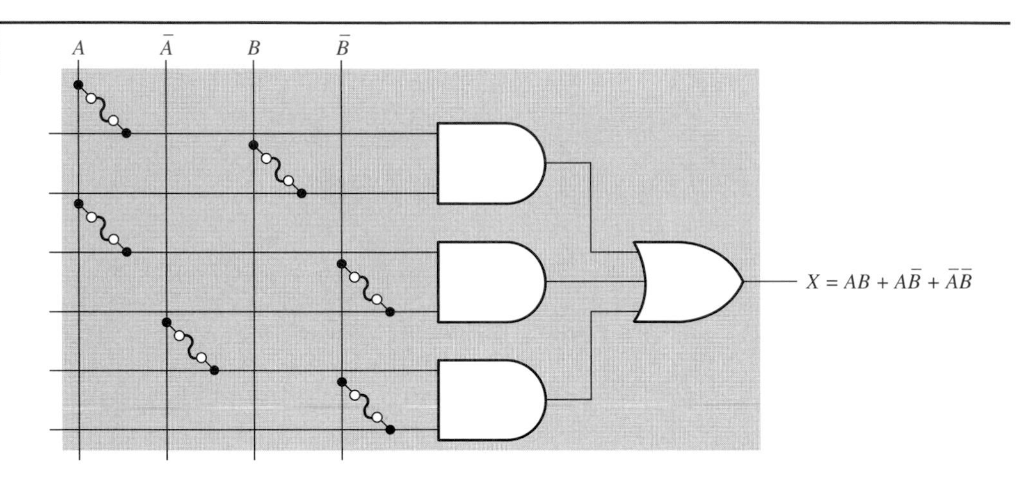

PAL Logic Diagram Symbols

As mentioned, actual PALs have many AND gates and many OR gates in addition to other circuitry and are capable of handling many input variables and their complements. Since PALs are relatively complicated, manufacturers have adopted a simplified notation for the logic diagrams to keep them from being overwhelmingly complex.

Input Buffers

The input variables to a PAL are buffered to prevent loading by the large number of AND gate inputs to which a variable or its complement may be connected. A buffer has both inverting and noninverting outputs to provide both the variable and its complement. A triangle represents a buffer circuit, as shown in the logic diagram in Figure 10–7 where the bubble output is the complement of the input variable.

AND Gate Inputs

A typical PAL AND array has a very large number of interconnecting lines, and each AND gate has multiple inputs. PAL logic diagrams show an AND gate that actually has several inputs by using an AND gate symbol with a single input line that represents all of its input lines, as shown in Figure 10–7. These multiple gate input lines are usually indicated by a slash with the number of lines, as shown in the figure for the case of two lines.

PAL Connections

To keep a logic diagram as simple as possible, the programmable links in an array are indicated by an X at the cross point if the cell is left intact and by the absence of an X if the cell is open, as indicated in Figure 10–7. Fixed connections use the standard dot notation.

FIGURE 10–7 Simplified diagram of a programmed PAL.

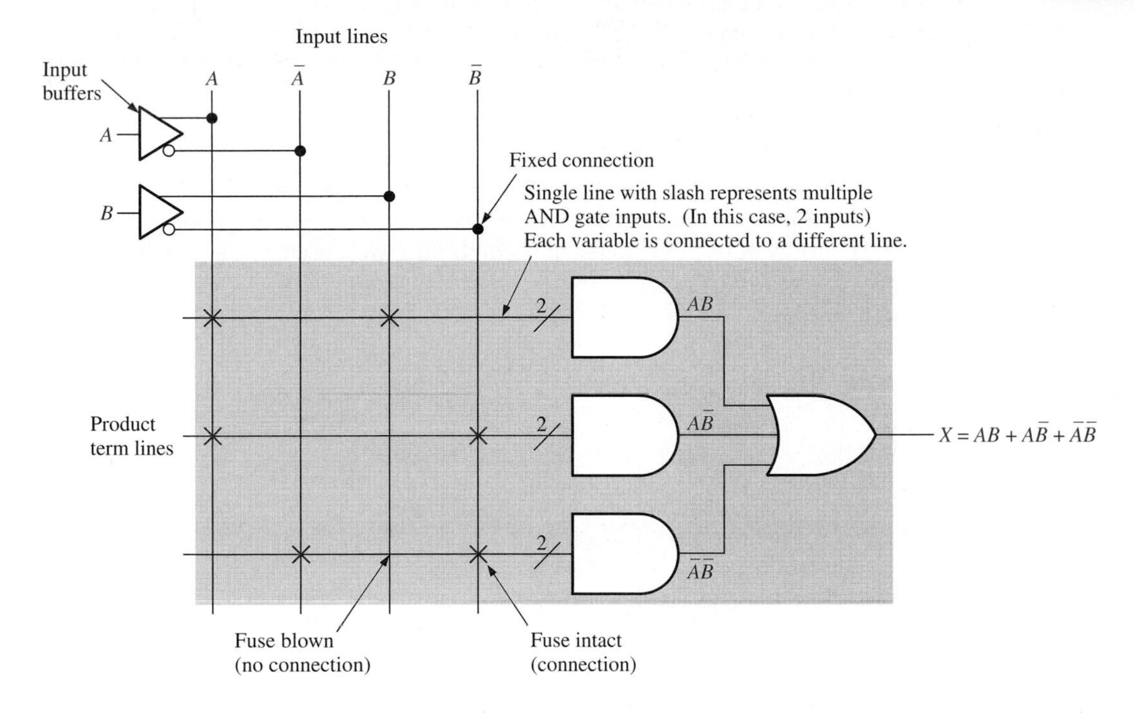

EXAMPLE 10–1

Problem

Show how a PAL is programmed for the following unreduced 3-variable logic function:

$$X = A\overline{B}C + \overline{A}B\overline{C} + \overline{A}\,\overline{B} + AC$$

Solution

The programmed array is shown in Figure 10–8. The intact programmable links are indicated by small red Xs. The absence of an X means that the fuse has been blown.

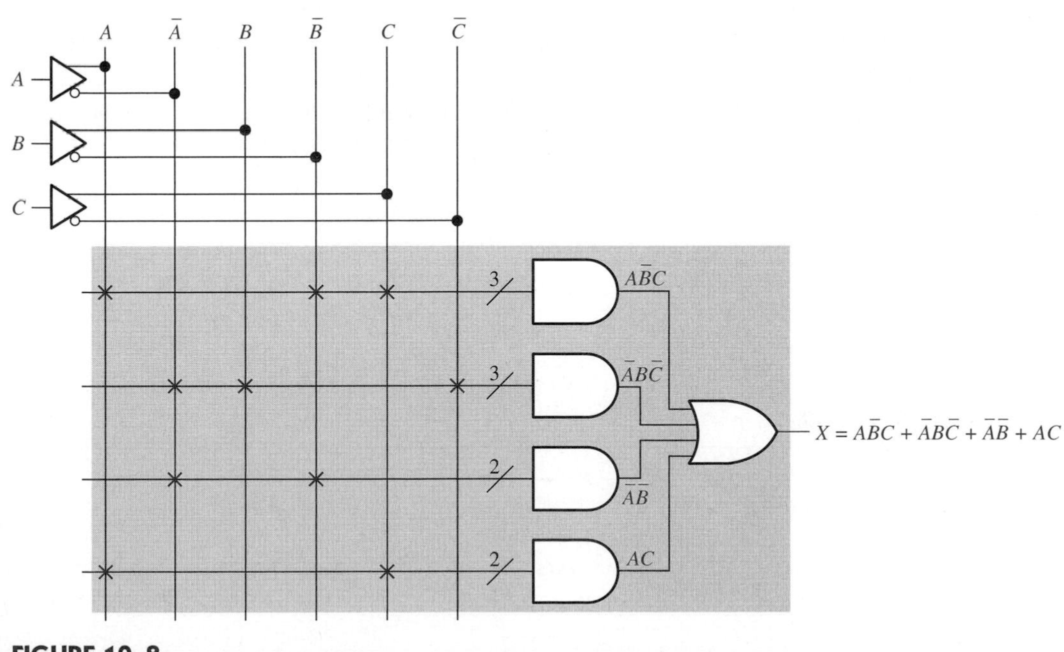

FIGURE 10–8

Question*
What is the expression for the output if the programmable links connecting input A to the top row and to the bottom row in Figure 10–8 are also blown?

Block Diagram of a PAL

A block diagram of a PAL is shown in Figure 10–9. The programmable AND array outputs go to the fixed OR array, and the output of each OR gate goes to its associated output logic. A typical PAL has eight or more inputs to its AND array and up to eight outputs from its output logic as indicated, where $n \geq 8$ and $m \leq 8$. Some PALs provide a combined input and output (I/O) pin that can be programmed as either an input or an output.

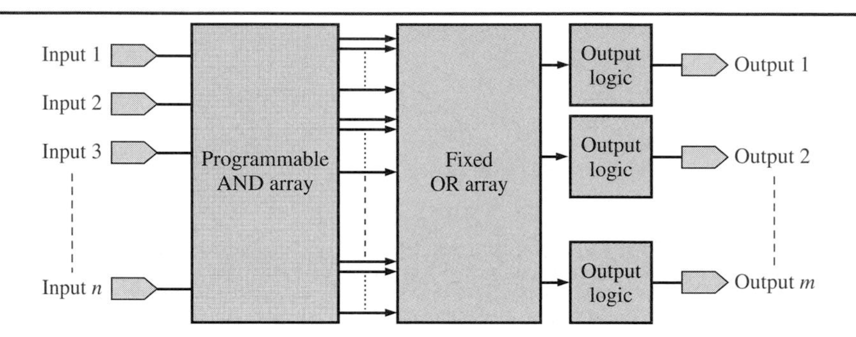

FIGURE 10–9
Block diagram of a PAL.

PAL Output Logic

Several basic types of PAL output logic allow you to configure the device for a specific application. In some types of PALs, the output logic provides only for combinational outputs, such as SOP functions. In other types, there are flip-flops in the output logic for providing what is known as registered output functions. Three typical types of PAL combinational output logic are output, input/output, and programmable polarity output.

- *Output* The combinational output logic type, shown in Figure 10–10(a), is used for an SOP function and is usually available as either an active-LOW or an active-HIGH output.
- *Input/output (I/O)* The combinational input/output logic type, shown in Figure 10–10(b), is used when the output function must be fed back into the array or when the Input/Output (I/O) pin is to be an input only.
- *Programmable polarity output* The programmable polarity output logic type, shown in Figure 10–10(c), is used to select either the output logic function or its complement by programming an exclusive-OR gate for inversion or noninversion. A programmable link from the exclusive-OR gate to ground is blown open for inversion or left intact for noninversion.

PAL16L8

The PAL16L8 is an example of a specific IC device. Its block diagram is shown in Figure 10–11. This particular PAL has ten dedicated inputs (I), two dedicated outputs (O), and six pins that can be used either as inputs or as outputs (I/O). Each of the outputs is active-LOW. In the part number PAL16L8, the sixteen indicates a total of sixteen available inputs, the

* *Answers are at the end of the chapter.*

FIGURE 10–10 Basic types of PAL combinational output logic.

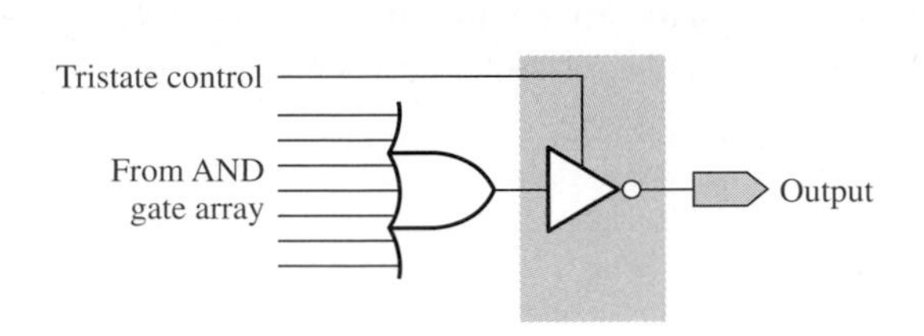

(a) Combinational output (active-LOW). An active-HIGH output would be shown without the bubble on the tristate gate symbol.

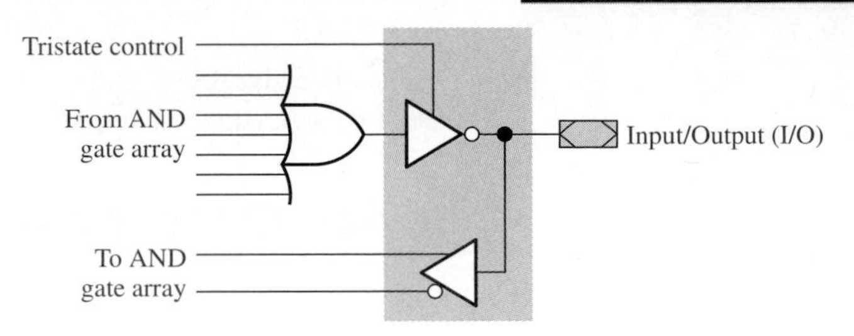

(b) Combinational input/output (active-LOW)

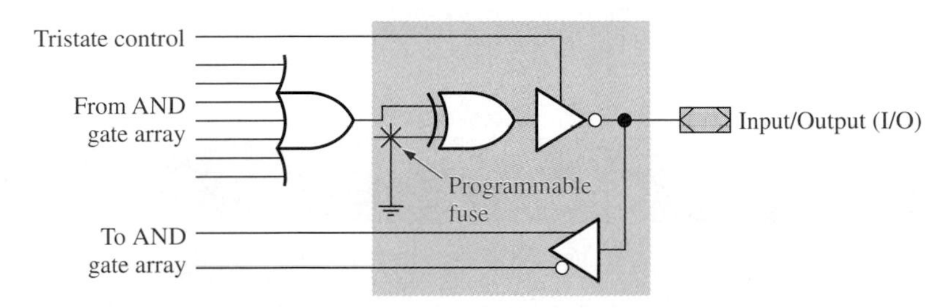

(c) Programmable polarity output

FIGURE 10–11 Block diagram of the PAL16L8.

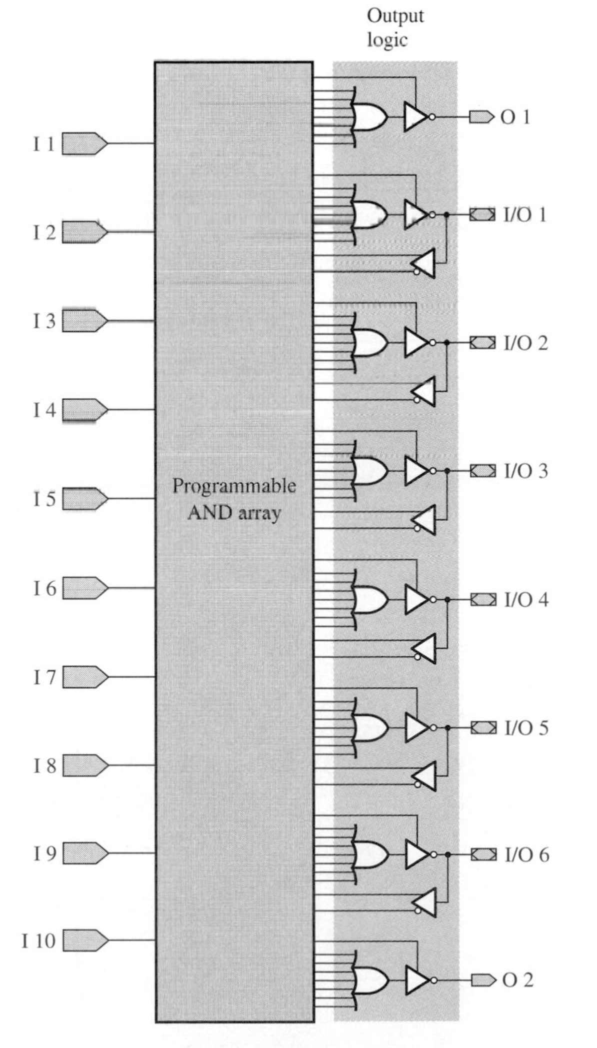

I = input, O = output, I/O = input/output

L indicates active-LOW outputs, and the 8 indicates a total of eight available outputs. In this case, two are dedicated outputs and six are programmable as either inputs or outputs.

Review Questions

6. What is a PAL?
7. Can any combinational logic function be implemented with a PAL?
8. What is the purpose of an input buffer?
9. What are three types of PAL combinational output logic?
10. What is meaning of the device numbering for the PAL16L8?

10–3 GENERIC ARRAY LOGIC (GAL)

GAL is a designation originally used by Lattice Semiconductor and later licensed to other manufacturers.

In this section, you will learn about basic GAL concepts.

The **GAL** is basically a PAL with a reprogrammable AND array, a fixed OR array, and programmable output logic. The PAL can be programmed only once, but the GAL can be programmed over and over.

GAL Operation

A simplified GAL structure is shown in Figure 10–12 with two input variables and their complements and one output. Actual GALs have many inputs and outputs. The reprogram-

FIGURE 10–12

Basic E^2CMOS array in a GAL.

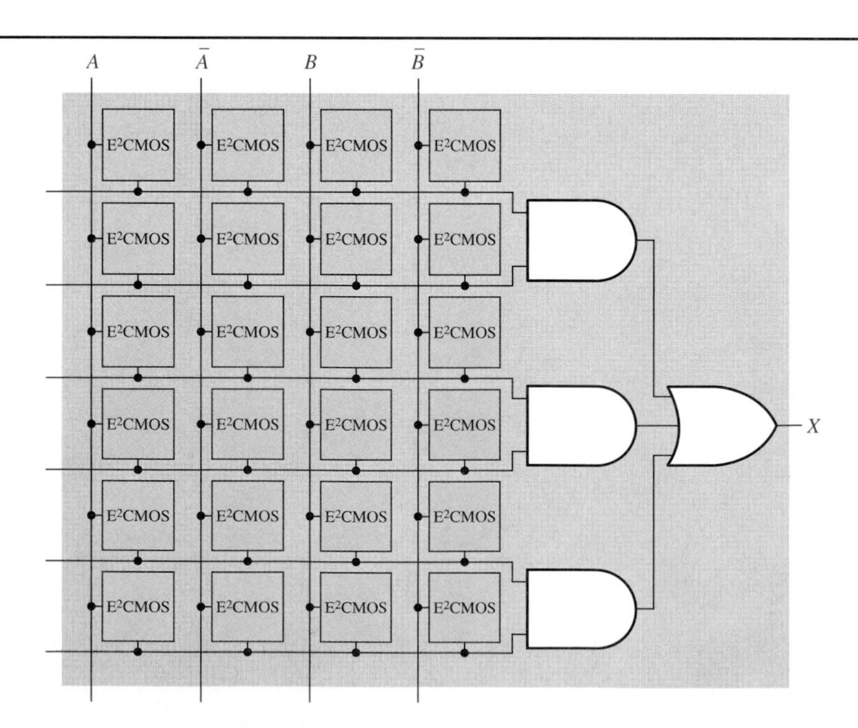

mable AND array connects to the fixed OR array, just as in a PAL. Instead of using a fuse as the programmable link, the GAL uses E^2CMOS cells. The designation E^2 means electrically erasable. These cells, shown as blocks in the figure, contain an electrically erasable CMOS circuit that is a type of field-effect transistor circuit that can be programmed to act like little switches.

Each row is connected to the input of an AND gate, and each column is connected to an input variable or its complement. By programming each E^2CMOS to be either *on* or *off,* any combination of input variables or complements can be applied to an AND gate to form any desired product term. A cell that is *on* effectively connects its corresponding row and column, and a cell that is *off* disconnects the row and column. A typical E^2CMOS cell can retain its programmed state for 20 years or more.

Implementing an SOP Expression

As an example, a simple GAL array is programmed as shown in Figure 10–13 so that the product term $\overline{A}B$ is produced by the top AND gate, the term AB is produced by the middle AND gate, and the term $\overline{A}\,\overline{B}$ is produced by the bottom AND gate. The E^2CMOS cells are *off* where a variable or its complement is not used in a given product term. The final output from the OR gate is an SOP expression.

$$X = \overline{A}B + AB + \overline{A}\,\overline{B}$$

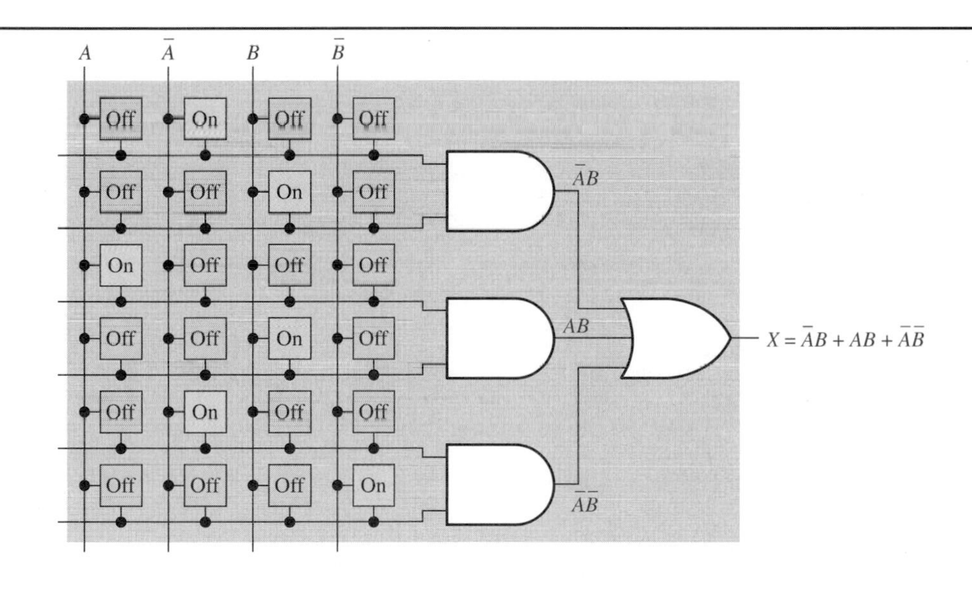

FIGURE 10–13

GAL implementation of a sum-of-products expression.

EXAMPLE 10–2

Problem

Show how a GAL is programmed for the following unreduced 3-variable logic function:

$$X = \overline{A}B\overline{C} + \overline{A}BC + BC + A\overline{B}$$

Solution

The programmed array using simplified notation is shown in Figure 10–14. Cells that are *on* are indicated by small red Xs. The absence of an X means that the cell is *off.*

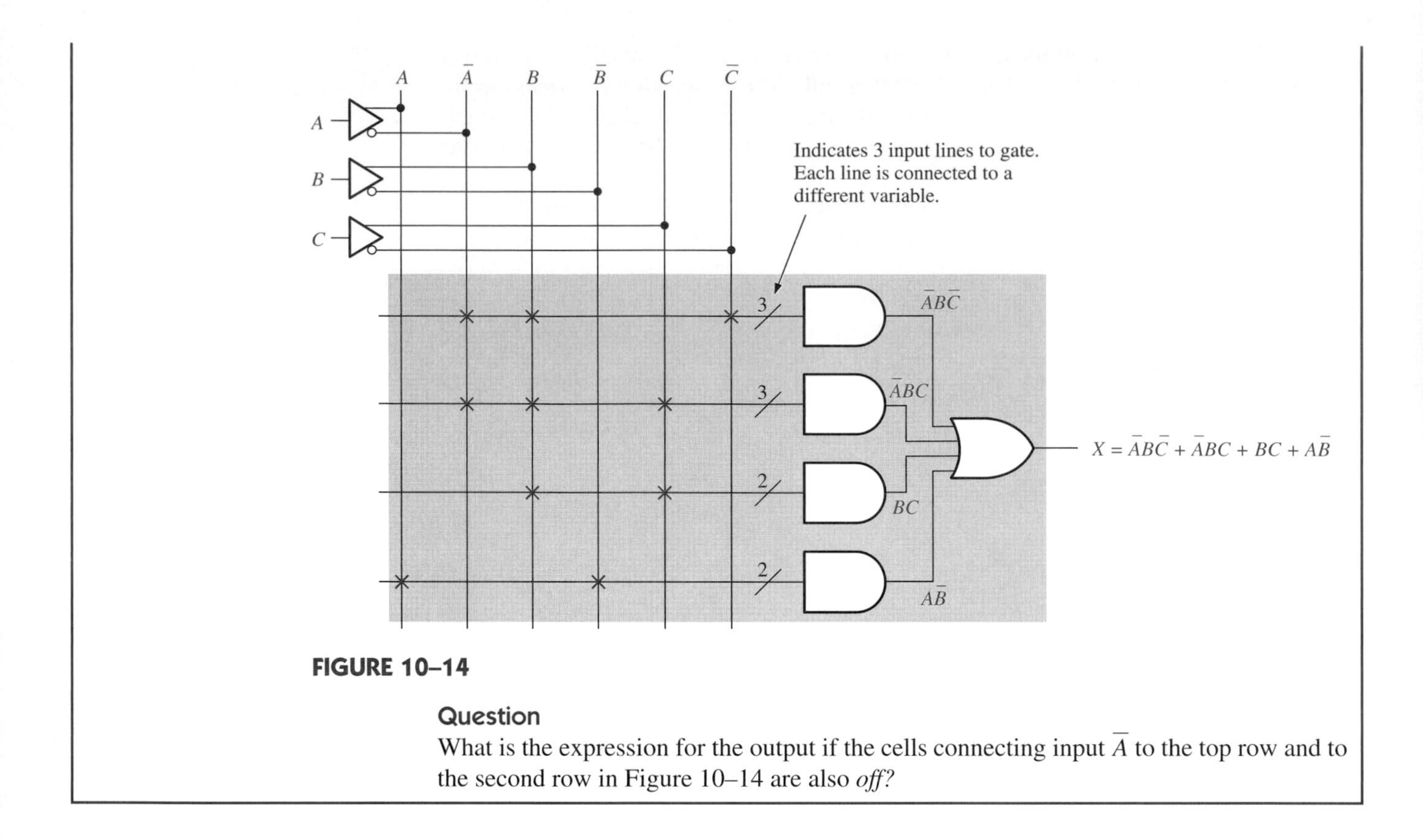

FIGURE 10–14

Question

What is the expression for the output if the cells connecting input $\bar{A}$ to the top row and to the second row in Figure 10–14 are also *off?*

Typical GALs

A block diagram of a GAL is shown in Figure 10–15. The AND array outputs go to output logic macrocells **(OLMCs)**, which contain the OR array and programmable output logic, which selects the direction for data. The typical GAL has eight or more inputs to its AND array and eight or more inputs/outputs (I/Os) from its OLMC. The OLMC is made up of logic circuits that can be programmed as either combinational logic or as registered logic, which contains flip-flops. The OLMC is more flexible than the fixed output logic in a PAL.

FIGURE 10–15

Block diagram of a GAL.

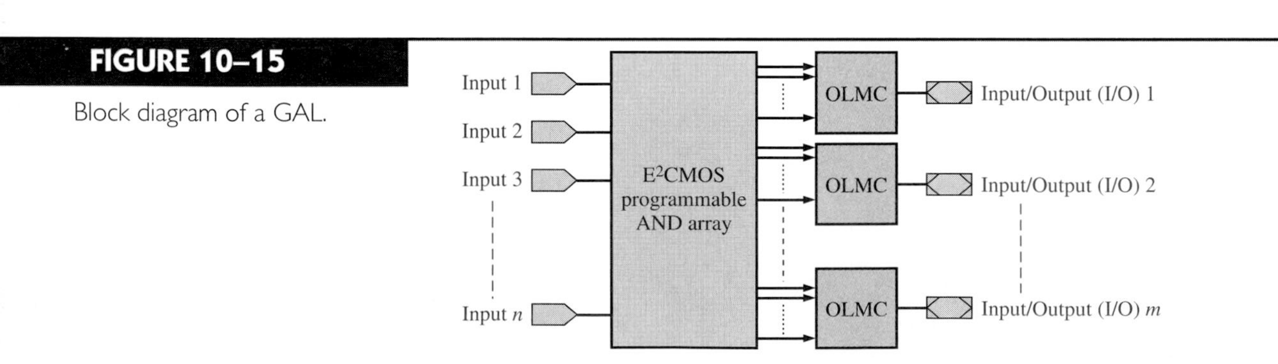

One specific GAL is the GAL16V8, which is comparable to the PAL16L8 except that it can be reprogrammed and it has a variable output configuration.

Another popular GAL is the GAL 22V10. Its block diagram is shown in Figure 10–16.

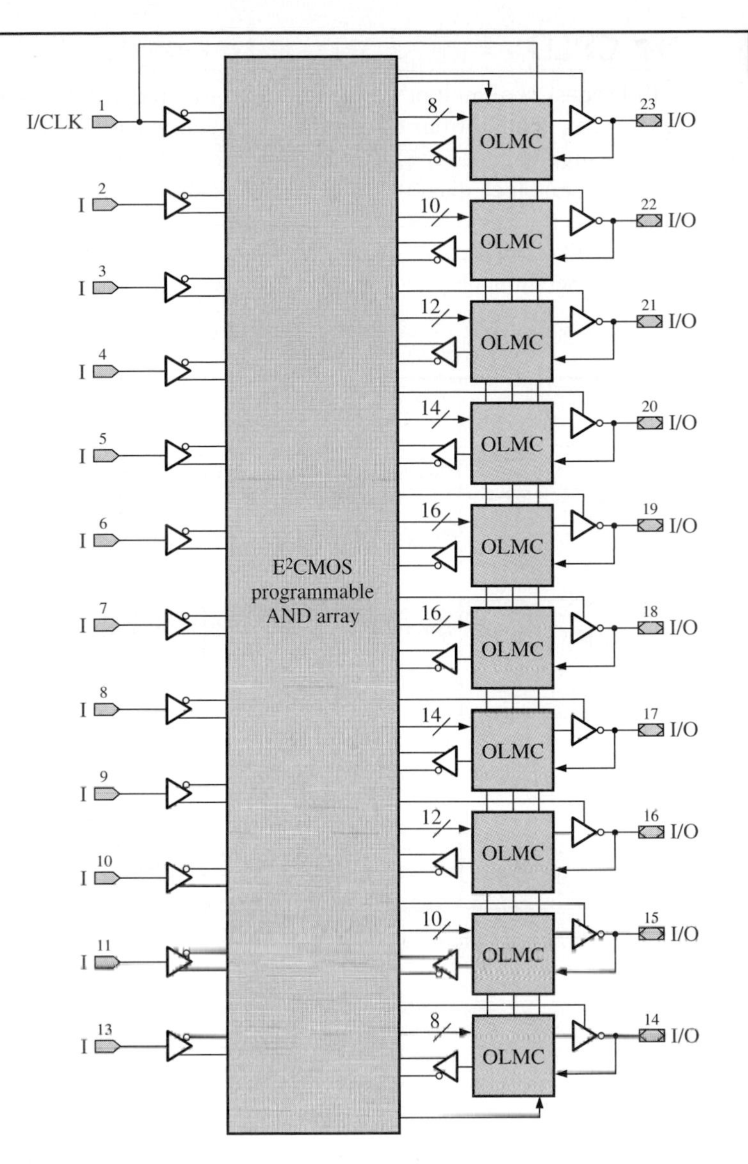

FIGURE 10–16

Block diagram of the GAL22V10.

Review Questions

11. What is a GAL?
12. What is the basic difference between a GAL and a PAL?
13. What makes the GAL reprogrammable?
14. What does OLMC stand for?
15. What does E^2CMOS stand for?

CPLDs AND FPGAs 10–4

You have learned about the PAL and the GAL, which are classified as SPLDs (Simplified Programmable Logic Devices).

In this section, you will learn about two other classes of PLD, the CPLD (Complex Programmable Logic Device) and the FPGA (Field Programmable Gate Array).

The CPLD

A CPLD consists of multiple SPLDs with programmable interconnections on a single chip. A CPLD can contain from two to sixty-four equivalent SPLDs, depending on the device type. The highest capacity CPLDs contain about 10,000 equivalent logic gates.

The general block diagram in Figure 10–17 shows how a typical CPLD is internally organized. Each logic array block (LAB) is essentially one SPLD containing several PAL/GAL-like arrays called **macrocells**. Each LAB can be interconnected with other LABs using the programmable inteconnect array (PIA) to implement large complex logic functions.

FIGURE 10–17

General block diagram of a CPLD.

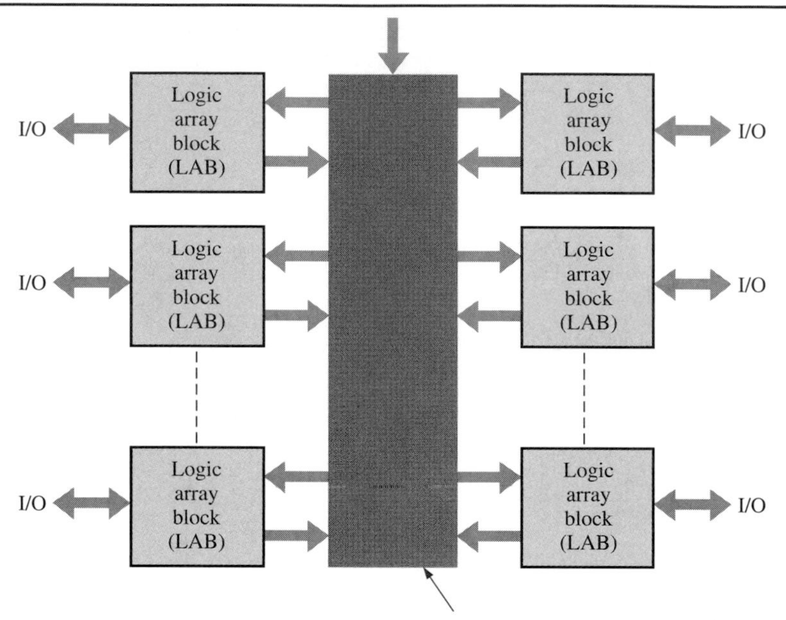

Macrocells

Each logic array block (LAB) in a CPLD contains several macrocells, as shown in Figure 10–18. The internal organization (architecture) of CPLDs varies from one manufacturer to another, but generally there are from 32 to several hundred macrocells in

FIGURE 10–18

A basic logic array block in a CPLD.

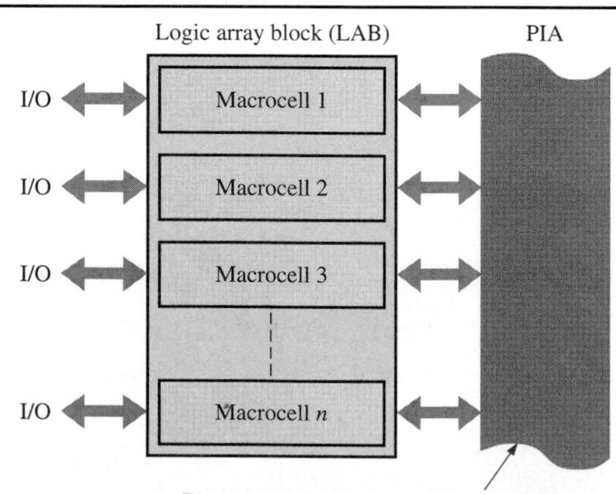

each LAB. Typically, a macrocell has a programmable AND array, a product-term selection matrix, an OR gate, and a programmable register section. A simplified logic diagram of a typical CPLD macrocell is shown in Figure 10–19.

FIGURE 10–19 A CPLD macrocell.

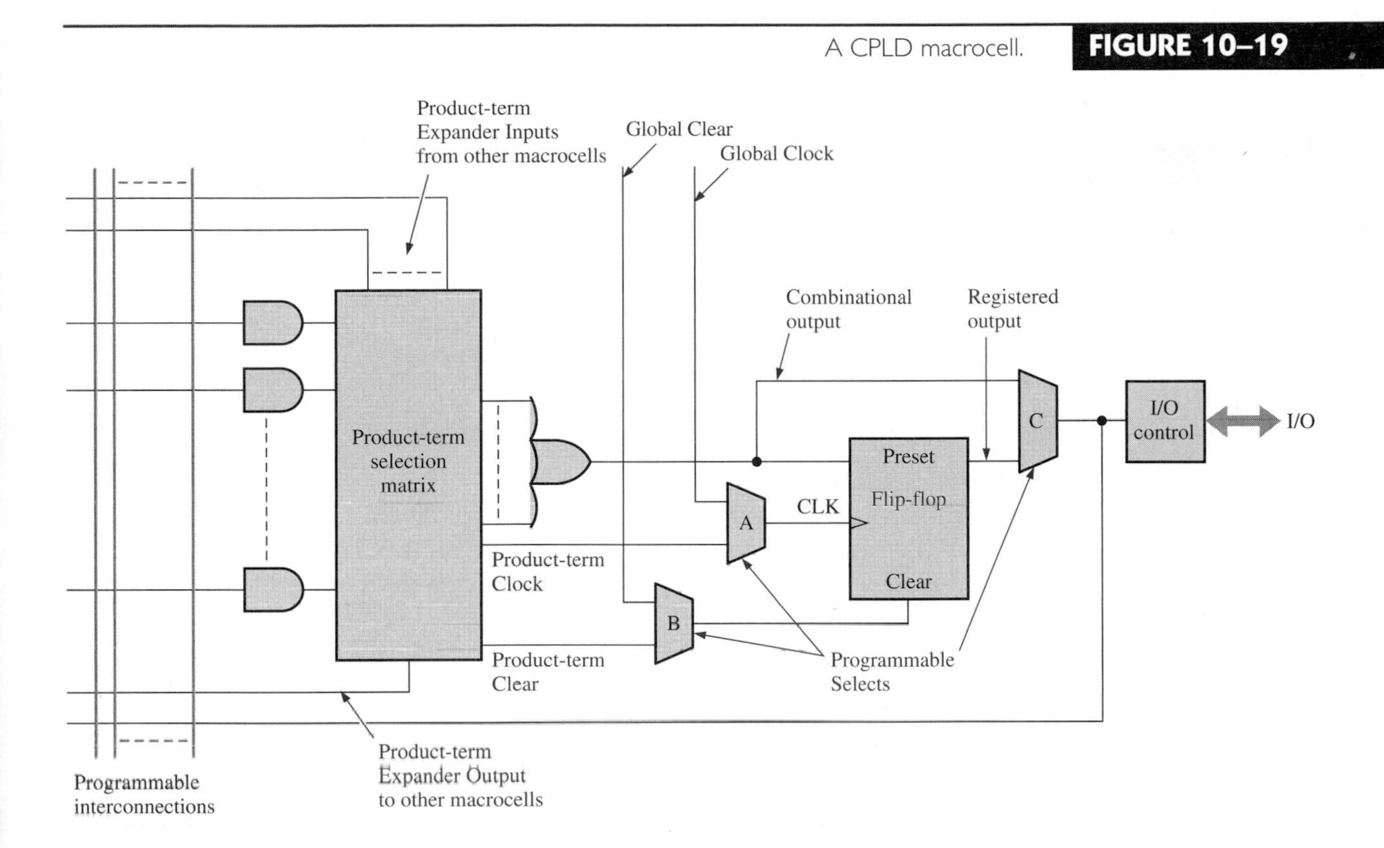

Each macrocell has a fixed number of AND gates that feed into a product-term selection matrix, where product terms can be selected and applied to the OR gate to create SOP logic functions. Additionally, product-term expander inputs from other macrocells allow more product terms to be selected in addition to those from the macrocell AND array. Also, a product-term expander output provides a path for any selected product term to other macrocells in the LAB or in other LABs via the programmable interconnection array (PIA).

The OR gate provides an SOP output through programmable select blocks to the I/O (input/output) or to the flip-flop. In this particular implementation, there are three programmable select blocks. These select blocks are essentially data selectors (multiplexers). One programmable select block A selects either a global clock or a product term to be used as the clock input for the flip-flop. The programmable select block B selects either a global clear or a product term to be used as the clear input of the flip-flop. The programmable select block C selects either the output of the OR gate or the output of the flip-flop and connects it to the I/O. The OR gate provides an SOP output, and the flip-flop provides a "registered" output. The flip-flops in a CPLD can be used for implementing shift register or counter logic.

Programmable Interconnect Array (PIA)

The programmable interconnect array (PIA) consists of conductors that run throughout the CPLD chip. Any macrocell can be connected to other macrocells within the same LAB, in other LABs, or to other I/Os using programming software.

Connections to the AND gates or other elements in a macrocell are accomplished by connecting a line in the PIA to an AND gate input or other macrocell line. Most CPLDs use what is known as E^2CMOS technology instead of fuse technology to make the connections. With E^2CMOS, a special type of transistor cell between two lines is programmed to the *on* state to form a connection or to the *off* state for no connection. The basic idea is illustrated in Figure 10–20.

FIGURE 10–20

Basic E^2CMOS interconnection technology.

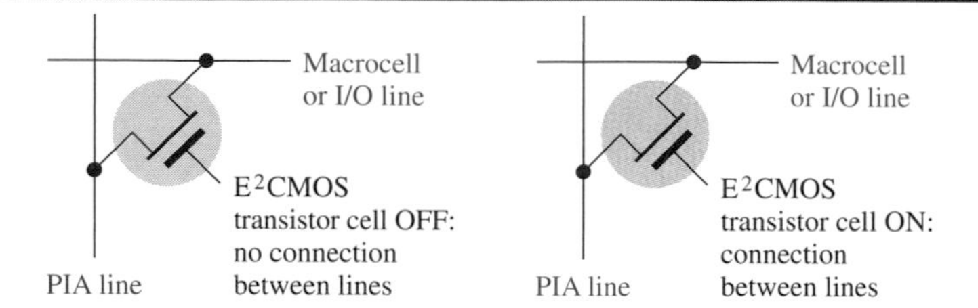

Specific CPLDs

Several companies, including Altera, Xilinx, Lattice, Cypress, and others manufacture CPLDs. Each company has its own approach to CPLD design, but all have one thing in common—they are based on PAL/GAL-like SOP logic arrays. Since we can't cover all of the devices that are available, two specific CPLD families will be introduced as representative examples: Altera's MAX 7000 family and Xilinx's XC9500 family.

Some CPLDs must be installed in a special fixture in order to be programmed; others, such as the MAX 7000S and all of the XC9500 family, are in-system programmable (ISP). **ISP** means that a CPLD can be programmed after it is mounted on a printed circuit board. We will discuss both of these programming methods in Section 10–5.

The MAX 7000 CPLDs The MAX 7000 family of CPLDs includes several variations ranging from 2 to 32 logic array blocks each with sixteen macrocells. These CPLDs are reprogrammable and can be programmed and erased up to 100 times.

Figure 10–21 shows the general block diagram for the MAX 7000 family. The logic array blocks (LABs) each have sixteen macrocells, and the LABs are linked together by the PIA. The PIA is fed by all I/Os and macrocells. The arrows with slashes indicate parallel lines and the associated number indicates the number of parallel lines. Where no numbers are shown by the slashes, the number of parallel lines vary depending on the particular device in the family. Each macrocell has the following inputs:

- 36 lines from the PIA
- Global control inputs (clock, clear, and enables that go to all macrocells)
- Inputs from I/Os

Combinational logic functions can be implemented in the logic array blocks as was previously discussed for the PAL and GAL. Each macrocell can be configured for either combinational logic or registered logic (using flip-flops).

The XC9500 CPLDs This family of CPLDs also includes several variations ranging from 2 to 16 logic array blocks (called function blocks in this case). Each function block has eighteen macrocells. These CPLDs are reprogrammable up to 10,000 times.

The general block diagram of the XC9500 family is shown in Figure 10–22. The function blocks are linked together by the programmable switch matrix. The switch matrix is similar to the PIA in the MAX 7000 devices and is fed by all I/Os and macrocells. The macrocells are similar to the general one shown in Figure 10–19.

FIGURE 10–21 General block diagram of the MAX 7000 family of CPLDs. Data sheets can be found at *www.altera.com*.

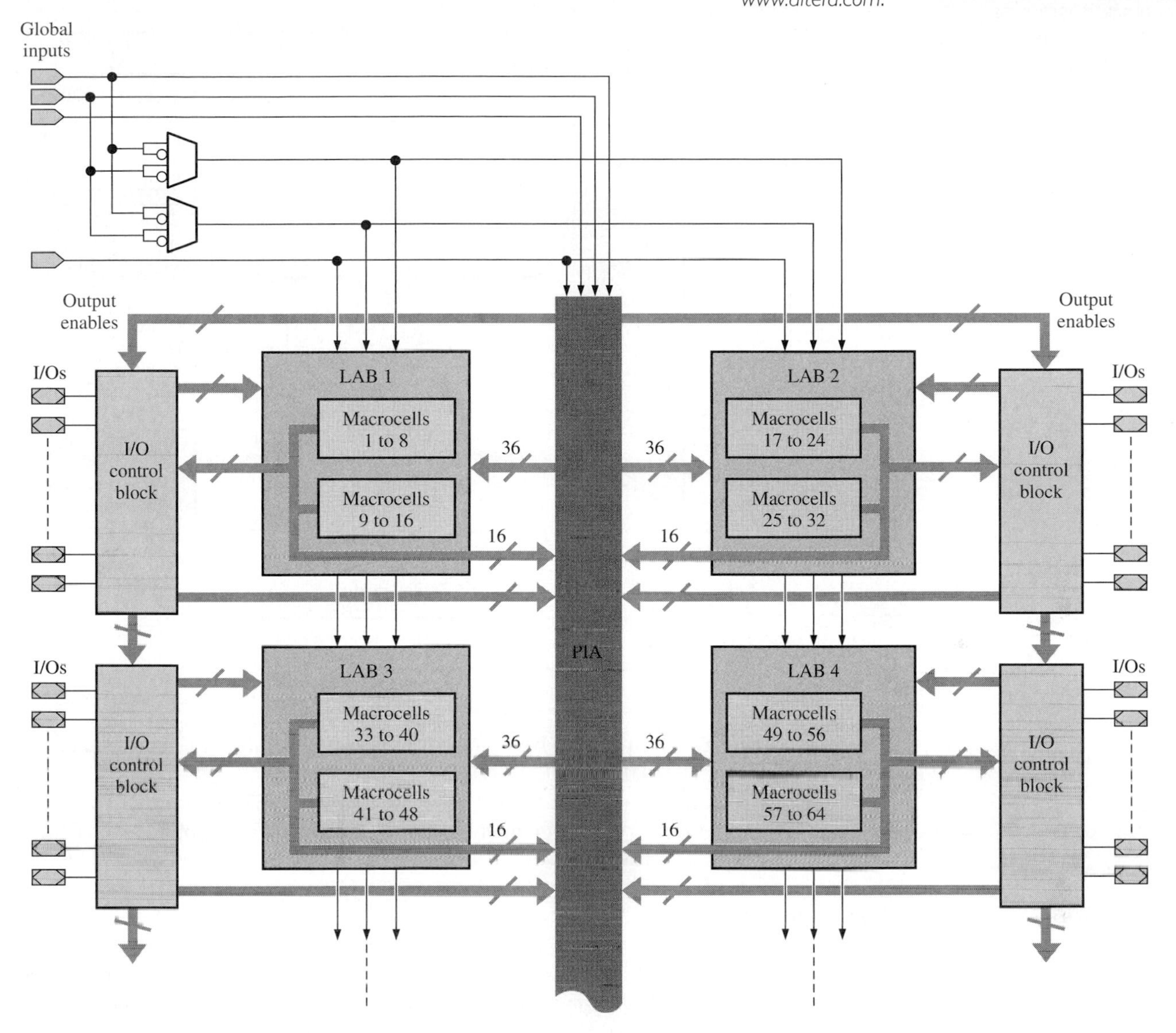

In-system programmability is achieved by the JTAG (Joint Test Action Group) controller and the in-system programmable controller. JTAG is an interface standard that allows connection to the CPLD mounted directly on a separate PC board.

The FPGA

The field programmable gate array (FPGA) is another major category of programmable logic device. There are basic differences between FPGAs and CPLDs. CPLDs are somewhat limited because they are essentially a matrix of interconnected SPLDs. FPGAs are more flexible because, instead of PAL/GAL-type logic, they have a very large number of logic blocks each containing a small array of gates and maybe a flip-flop. Also, FPGAs have many more interconnect points and often use different programmable interconnection technology than CPLDs, providing much more flexibility in routing the interconnections. The internal design (architecture) of FPGAs results in a logic capacity ranging to about 8 million equivalent gates compared to CPLDs, which range up to about 10,000 equivalent

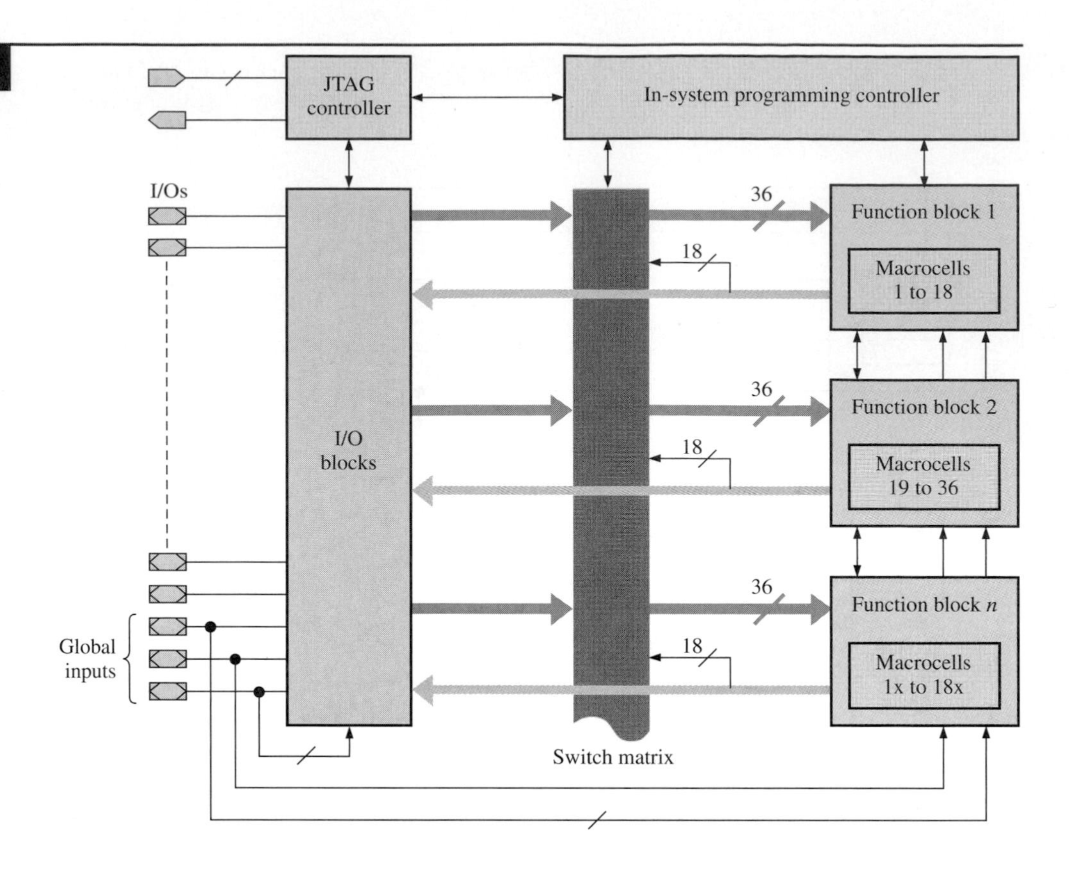

FIGURE 10–22 General block diagram of XC9500 CPLDs. Data sheets can be found at *www.xilinx.com.*

gates. Many FPGAs use a memory-based logic called a look-up table (LUT) instead of the AND-OR logic found in CPLDs.

A general block diagram of an FPGA is shown in Figure 10–23. It consists of an array of logic blocks surrounded by input/output blocks and laced with a grid of interconnection channels.

The Logic Block

Each logic block in an FPGA contains several logic elements, as shown in Figure 10–24. Basically, a logic element (LE) in many FPGAs contains a look-up table (LUT) or other means for generating either combinational logic or registered logic functions. The logic element also contains some associated logic and a flip-flop. The number of logic elements varies depending on the particular device.

The LUT

As mentioned before, the look-up table used in FPGAs is actually a memory device that can be programmed to perform logic functions (memories are covered in in Chapter 11). The LUT essentially replaces the AND/OR array logic in a CPLD. As an example of how an LUT can be used to produce a logic function, Figure 10–25 shows a simple diagram of an 8 bit by 1 bit (8×1) memory programmed to produce the SOP function $\overline{A}B\overline{C} + A\overline{B}C + ABC$. When any one of the three product terms appears on the LUT inputs, the corresponding memory cell storing a 1 is selected and the 1 (HIGH) appears on the output. For any product terms that are not part of the SOP function, the LUT output is 0 (LOW).

As another example, an 8×2 LUT is programmed as a full-adder in Figure 10–26. Recall that the sum and carry out expressions for a full-adder are as follows:

$$\Sigma = (A \oplus B) \oplus C_{in} = \overline{A}\,\overline{B}C_{in} + \overline{A}B\overline{C}_{in} + A\overline{B}\,\overline{C}_{in} + ABC_{in}$$

$$C_{out} = AB + (A \oplus B)C_{in} = \overline{A}BC_{in} + A\overline{B}C_{in} + AB\overline{C}_{in} + ABC_{in}$$

FIGURE 10–23 General block diagram of an FPGA.

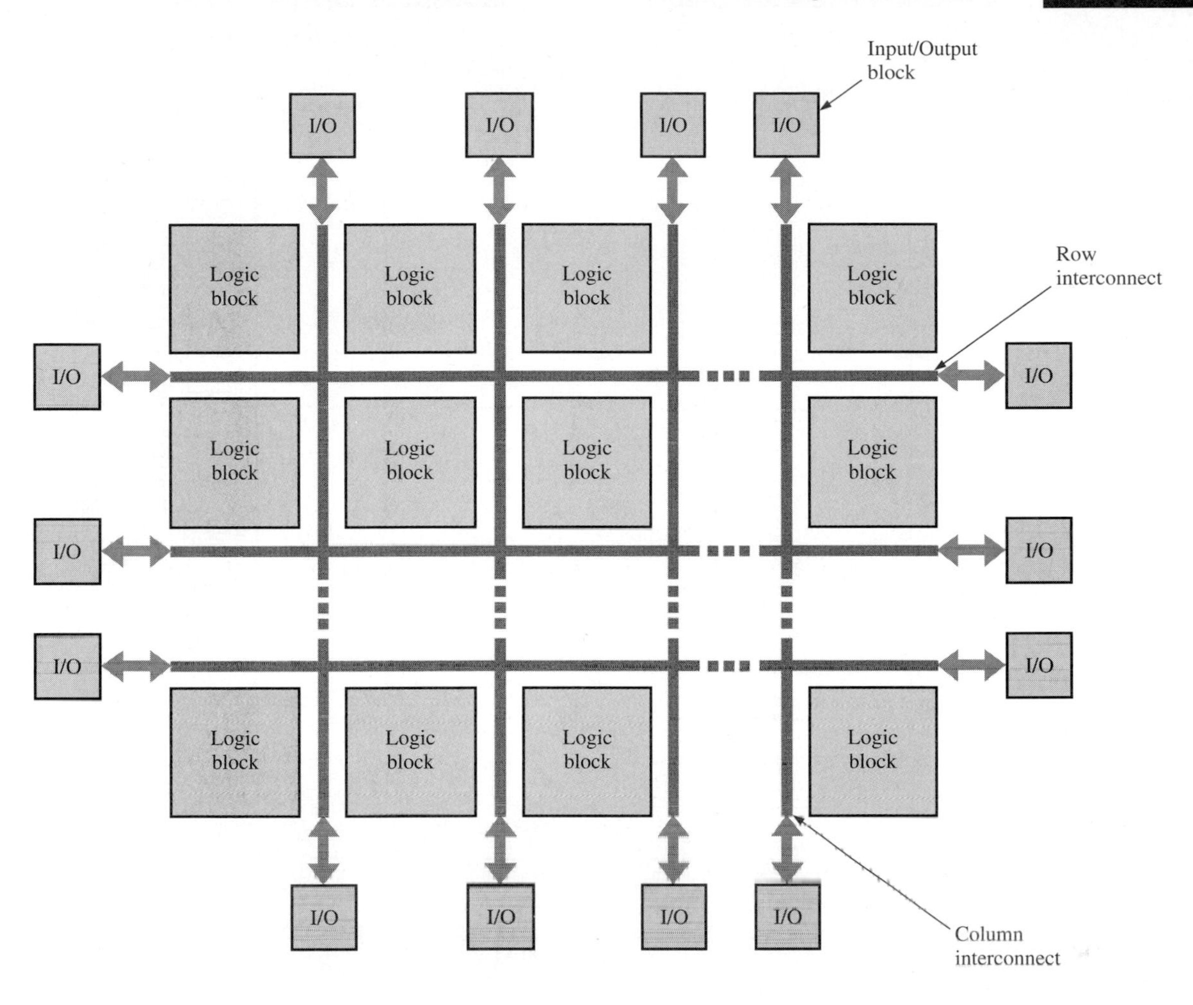

FIGURE 10–24 General logic block in an FPGA.

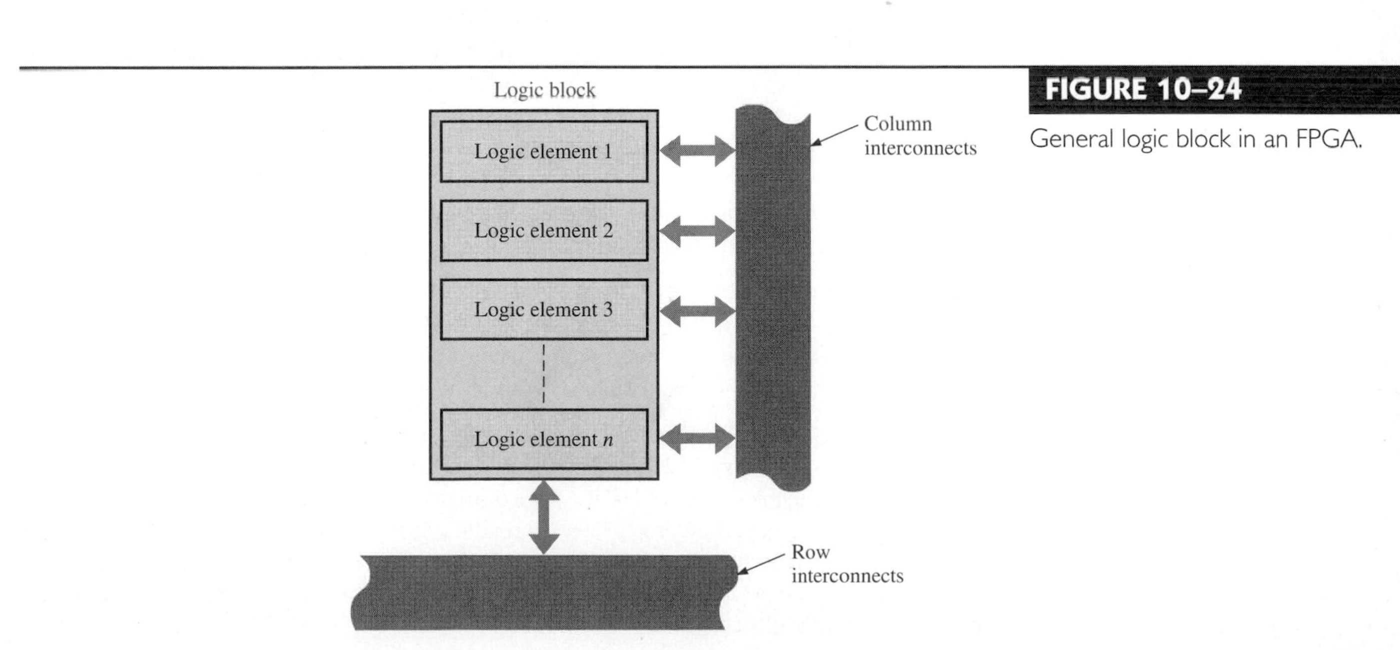

FIGURE 10–25

An LUT programmed to produce the SOP function shown. Each of the blue memory cells store a 1.

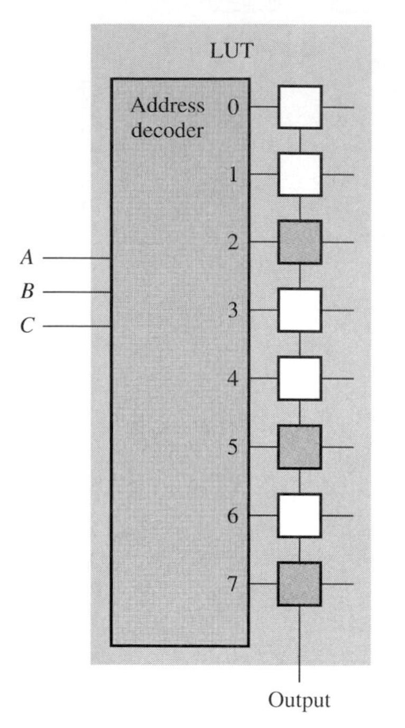

FIGURE 10–26

An LUT programmed as a full-adder.

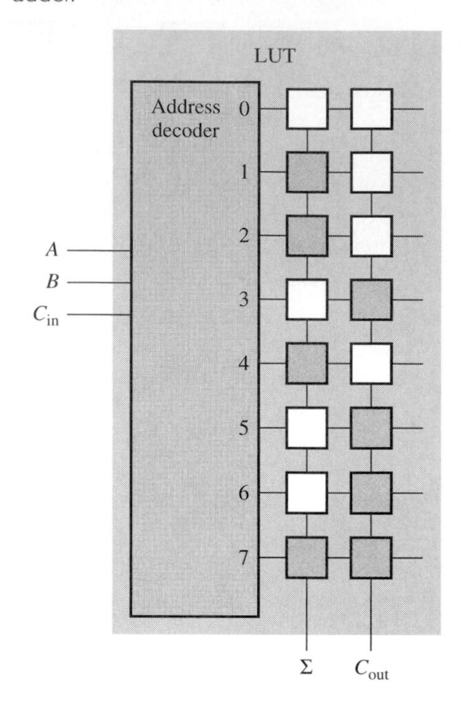

A Specific FPGA

The Altera FLEX 10K family is a widely used series of FPGAs that are in-system programmable. The logic blocks in the FLEX 10K devices are called logic array blocks (LABs), just as in the CPLDs. The number of LABs in this family range from 72 to 1520, depending on the particular device. Each LAB contains eight logic elements (LEs), so the highest capacity FPGA in this family has a total 12,160 logic elements. In addition to the LABs and logic elements, the FLEX 10K series contains blocks of random access memory (RAM) called embedded array blocks (EABs) for general-purpose use.

The LABs are arranged into rows and columns. The LABs and EABs are connected by row and column interconnections, and at the end of each row and column interconnect there is an input/output element (IOE).

The Logic Array Block (LAB) As mentioned, a logic array block in the FLEX 10K contains eight logic elements (LEs). In addition, a LAB also has a local interconnect that is separate from the row and column interconnects. The local interconnect allows interconnections among the eight LEs in an LAB without using the row and column interconnects.

The Logic Element (LE) The logic element is the smallest unit of logic in the FLEX 10K. Each LE contains a 4-input LUT and associated logic, as well as a flip-flop and associated logic. As discussed before, the LUT is a type of memory programmed to generate specified logic functions. This particular LUT can generate any 4-variable logic function because it has four inputs. In addition to the LUT, there is the carry chain and the cascade chain. The logic element can be connected to the local interconnect or the main interconnects (row and column) with the select logic.

Carry Chain The carry chain logic allows implementation of adders and other functions. For example, an 8-bit parallel adder can be implemented with the eight logic elements in a logic array block, as illustrated in Figure 10–27. Each LUT is programmed as an adder and generates a sum bit, and the carry chain logic produces the carry out bit. The input bits A and B and the initial carry inputs are connected via the main interconnects (row and column). The carry out from one LE is connected via the local interconnect to the next LE.

Any n-bit adder can be implemented by connecting the final carry out to the next LAB. For example, using two LABs you can implement a 16-bit adder.

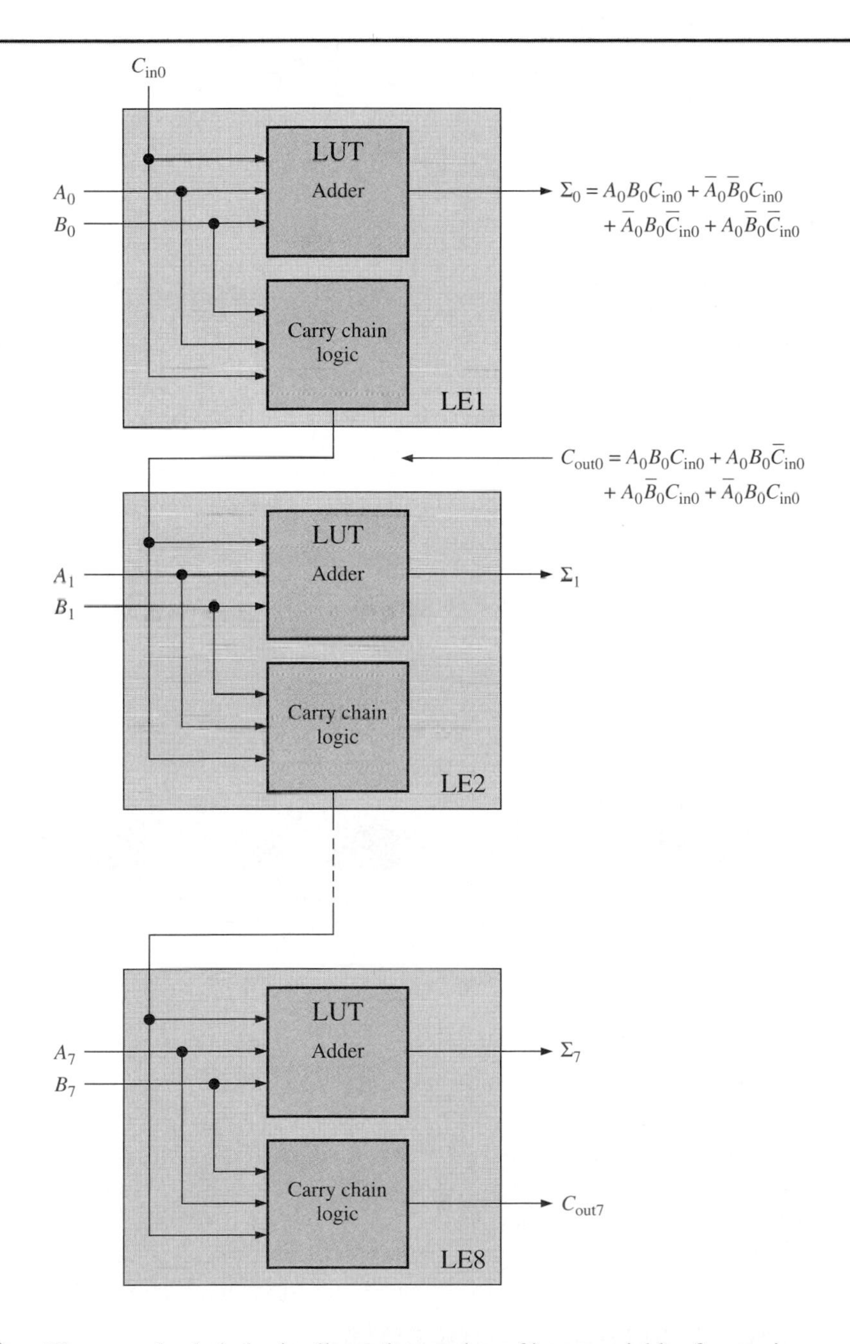

FIGURE 10–27

An illustration of using the eight LEs in a LAB to implement an 8-bit parallel adder.

Cascade Chain The cascade chain logic allows the number of input variables for any implemented logic function to be expanded from a minimum of four to a maximum of thirty-two in a single LAB. The cascade chain can be configured as either an AND function or an OR function. For example, Figure 10–28 shows the expansion of a product term to eight

FIGURE 10–28

Example of using LUTs programmed to generate product terms and an AND cascade chain to expand the number of variables in the product term.

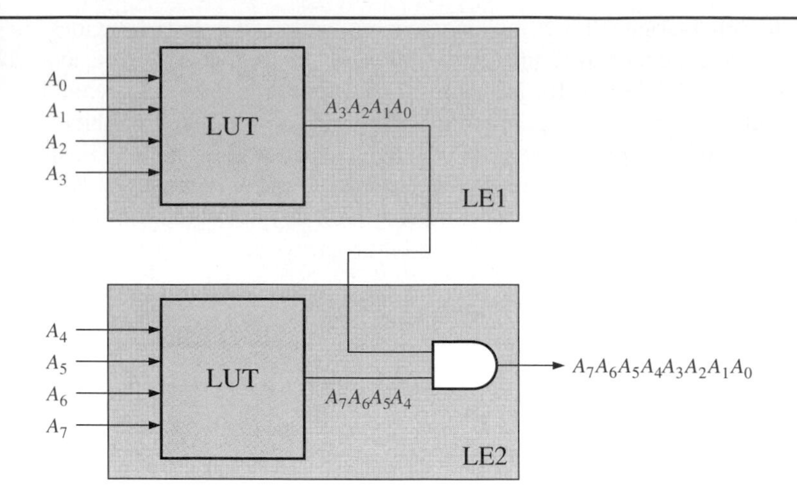

variables using two LEs in an AND cascade chain. Remember, one LE has only four inputs. In this example, additional LEs could be used in a cascade chain to produce a product term with any number of variables. Similarly, a sum term can be expanded using the cascade chain in an OR configuration.

Review Questions

16. What does CPLD stand for?
17. What components make up major elements in a CPLD?
18. What is a LAB and what does it consist of?
19. What does FPGA stand for?
20. What is an LUT?

10–5 PLD PROGRAMMING

In previous sections, three types of PLDs have been discussed. The PLD itself is a "blank slate." Software is used to put something intelligible onto the blank slate; that is, software is used to program a logic design into the PLD.

In this section, you will learn the general programming concepts and methods of programming PLDs.

PLDs must be programmed using a computer that is running special software called an **HDL** (hardware description language). The purpose of programming a PLD is to implement a logic design in the device. A PLD that is being programmed, whether it is an SPLD, a CPLD, or a FPGA, is generally called a **target device**.

Basic Software Concepts

Software is what runs a computer. Without software, a computer is just hardware, circuitry and physical components, that can do nothing. **Software** consists of programs that instruct a computer what to do in order to carry out a given set of tasks.

Algorithms

An **algorithm** is a description of how to accomplish a given task. It consists of a set of instructions that must be executed or carried out in a proper sequence in order to accomplish a task.

A food recipe is one way to illustrate the concept of an algorithm. For example, the following is a recipe for cornbread:

Step 1. Heat oven to 425°.
Step 2. Grease an 8-inch pan.
Step 3. Mix 1 egg, 1 ¼ cups of milk, ¼ cup of oil, and 2 cups of corn meal mix in large bowl.
Step 4. Pour the batter into the greased pan.
Step 5. Bake at 425° for 25 to 30 minutes.

Each of the steps in this algorithm is an instruction. Each step must be done in the proper sequence in order to get cornbread.

Programs

A **program** is an algorithm for a computer that consists of a list of instructions that the computer follows in order to achieve a specific result. Two broad categories of computer programs are system software and application software.

System software makes the computer accessible to the user by managing the various operations of the processor, the memory, and the files in order to run an application. An operating system, such as Windows, is an example of system software.

Application software is the software that you run on a computer to accomplish a certain result. As you may know, there are many types of application software for various purposes. These include word processing, accounting, publishing, computer-aided design (CAD), computer-aided engineering (CAE), databases for inventory control and scheduling, games, instruction, and many others. The particular application software for the purpose of programming PLDs is classified as CAE.

Programming Languages

A programming language is the mechanism that is used to construct a program. Basically the three levels of programming languages are high-level, assembly, and machine.

In the early days of computers, all programming was done in machine code, which consisted of the 1s and 0s that the computer could understand. However, machine code made it difficult for people to follow.

Assembly language was developed to avoid some of the tediousness and difficulty of writing a program in machine language using 1s and 0s. Assembly language uses words to represent certain binary codes and to form instructions. However, assembly language is "machine-dependent"; that is, different processors have different assembly languages.

High-level languages were developed to make programs easier to write and understand. English-like words are used along with other elements in high-level language to implement a specific function. Also, high-level languages are essentially "machine-independent" so that they can be used on any computer. Some examples of high-level computer languages are BASIC, C/C++, JAVA, and FORTRAN.

There are several hardware description languages (HDLs) available for programming PLDs. **VHDL** is one type of HDL used to program PLDs. The V in VHDL stands for VHSIC, an abbreviation for Very High Speed Integrated Circuit. VHDL is a standard language adopted by the IEEE (Institute of Electrical and Electronics Engineers) and is one of several HDLs that are available. Verilog, AHDL, ABEL, and CUPL are other examples of HDLs. Verilog is also an IEEE standard language and is similar to VHDL. AHDL, ABEL, and CUPL are proprietary languages owned by PLD manufacturers. VHDL is perhaps the most widely accepted of all the HDLs in industry and education.

HISTORICAL NOTE

Grace Hopper, a mathematician and pioneer programmer, developed considerable troubleshooting skills as a naval officer working with the Harvard Mark I computer in the 1940s. She found and documented in the Mark I's log the first real computer bug. It was a moth that had been trapped in one of the electromechanical relays inside the machine, causing the computer to malfunction. From then on, when asked if anything was being accomplished, those working on the computer would reply that they were "debugging" the system. The term stuck, and finding problems in a computer (or other electronic device), particularly the software, would always be known as debugging.

Using an HDL to Program a Digital Logic Design

As mentioned, an HDL is a high-level programming language that uses statements that are similar to the English language. These English-like statements are translated to a form that can be downloaded into a PLD. Using general-purpose high-level statements, HDLs can generally be read and understood even if the reader is not proficient in the language. HDL programs can be developed by following a general programming process as detailed in the following steps and illustrated in Figure 10–29.

1. Define and document what the program is going to do.
2. Determine a solution and document how the task is to be accomplished.
3. Create and document the program.
4. Test and debug the program and document any revisions.
5. Implement the design.

FIGURE 10–29 General HDL programming process.

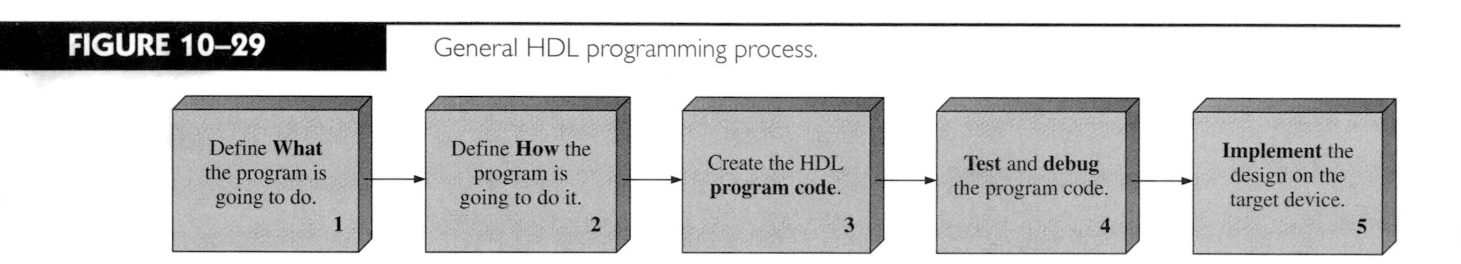

Once the program requirements have been defined and a solution has been determined, the HDL code can be written. (We will use the term *code* to mean program statements.) The ability to create code using general-purpose high-level program statements allows the programmer to concentrate on the concepts of the digital logic system operation. This permits a separation between the actual hardware being programmed and the code that defines what it is to do.

After completing the program, the code is compiled into an output file that can be tested and downloaded into the target device. A **compiler** is a program that can be thought of as a programming language translator. The high-level HDL statements are translated by the compiler to the 1s and 0s (machine code) required to program the selected target device. Because the application software (PLD development software) used to run an HDL program allows for the selection of the target device, the HDL code is said to be portable. Portable programs can be compiled or translated to an output file for a wide range of devices. After the HDL program is tested and debugged, it can be downloaded and implemented in the target device.

Development Software

As you have learned, the compiler translates the high-level HDL program statements into a form for targeting a specific hardware device, such as a CPLD. Every manufacturer of PLDs—including SPLDs, CPLDs, and FPGAs—provides development software for its line of products. Altera, Xilinx, Lattice, and Cypress are examples of companies that produce PLDs. One of Altera's development software packages for PLD programming is called MAX+Plus II, and one product from Xilinx is called Web-PACK Project Navigator.

The Altera and Xilinx development software packages perform similar functions in processing and synthesizing the HDL logic designs and translating the entered code to a form that can be downloaded to a target device.

Before an HDL program is compiled, the software allows the target device to be selected and the pin assignments made for the selected device. Some development software packages such as the one from Xilinx have graphical features that allow pin definition by dragging the signal name and dropping it onto the desired pin. Other development tools use list tables or external text data files.

The HDL program is entered, the target device is selected, and the device pin assignment is made before the program can be compiled. After the program is compiled, the development software provides for simulating the design to make sure it works properly. After simulation, the design is ready to be downloaded to the target device. The steps in using development software to program a PLD are shown in Figure 10–30.

FIGURE 10–30 Steps in applying development software to program a PLD.

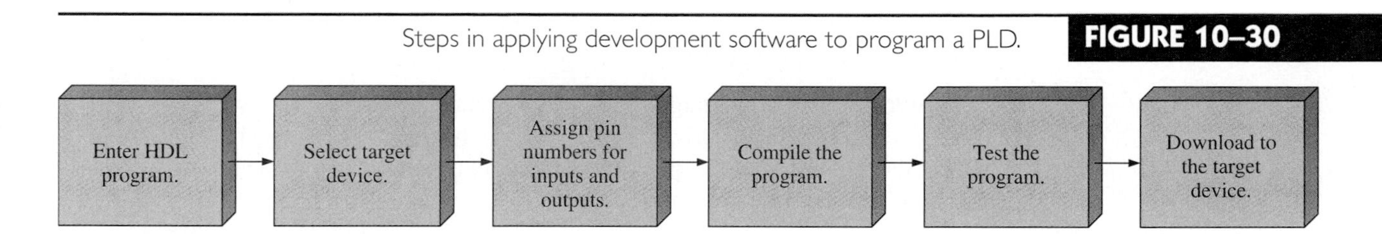

Programming Methods

From a hardware standpoint, programming generally falls into two major categories: the conventional method and the in-system programming (ISP) method. From a software standpoint, a logic design is programmed into a target device (CPLD, FPGA, or other).

Conventional Programming of a Target Device

CPLDs and FPGAs are conventionally programmed using programming software, a computer, and a device programming fixture (programmer) connected to the computer. The programming fixture has a socket that accepts the PLD package. With conventional programming, the device is programmed while plugged into the programming fixture, as illustrated in Figure 10–31. This programming configuration is for devices that have not yet been installed on a printed circuit board.

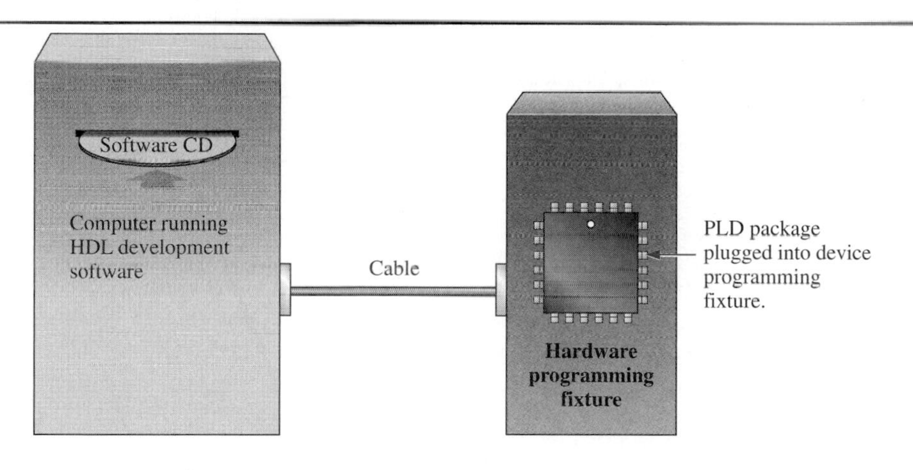

FIGURE 10–31 Typical configuration for conventional PLD programming.

As shown in Figure 10–31, the three components required to program a target device with HDL are

1. Development software (HDL compiler)
2. A computer that meets the software requirements
3. A software-driven device programming fixture

The computer must meet the software requirements of the development software and programming hardware. The *programming fixture* is a hardware device that accepts programming data from the computer and implements a specified logic design in the target device that is plugged into it.

Computer Any computer that meets the software and programmer specifications can be used. These specifications normally include the type of microprocessor on which the computer is based, the amount of memory, and the operating systems (DOS, Windows, and Macintosh operating system (OS) are examples).

Software The software packages for HDL programming are called logic compilers. Many HDL software packages are available from various vendors. These software packages fully support the HDL used and compile the logic design entered to a file that can be downloaded to the target device.

Programming Fixture The target device (PLD) is inserted into the hardware programming fixture, which has a software driver program that reads a data file compiled by the HDL software. The data file is converted to instructions for applying required voltages to specified target device pins in order to internally alter the device to implement the required logic.

In-System Programming (ISP) of a Target Device

Some PLDs have ISP capability that allows them to be programmed after they are installed on a circuit board. This method uses a standard 4-wire interface specified by IEEE Std. 1149.1, called the JTAG (Joint Test Action Group) standard. A JTAG cable is connected from the parallel port of a computer running the programming software to a socket on the PC board connected to the special JTAG pins of the target device. This configuration is shown in Figure 10–32.

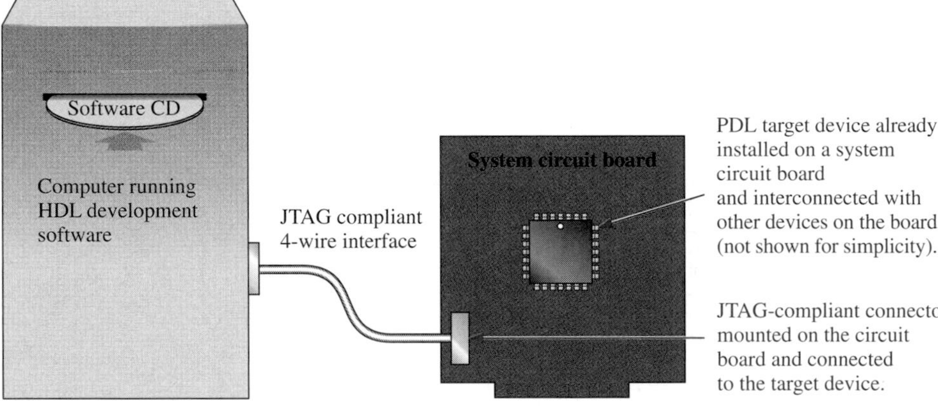

FIGURE 10–32 Typical configuration for in-system programming of a PLD.

In-system programmable devices provide an excellent way to change a circuit design or upgrade a system after it is already in use in the field without having to remove it from the circuit board on which it is installed. An upgrade disk can be used to reprogram a device directly on the circuit board via the computer's printer port or by modem.

The Programming Process

Regardless of the approach used to describe a digital circuit or system using an HDL, there is a general flow followed in the creation of any program. First, the requirements of the digital system are defined and refined to the point where they can be written as program code. The digital design is expressed in HDL and the software is loaded into the computer.

Entering the Design

The logic design is entered into the computer by creating an input or source file. Typically, some preliminary information is entered into the software development tool, such as user's name, company, date, and description of the design. Then, the target device type is assigned the input and output pin numbers. Finally, the logic of the digital system is entered. Any syntax errors or other errors made in entering data into the input file are indicated and must be corrected. Syntax is the prescribed format used in writing an HDL program.

Running the Software

The software compiler processes and translates the input file. The logic design is then checked using a set of test waveforms in a test simulation. This effectively "exercises" the design in software to determine if it works properly before the target device is actually programmed and the design committed to hardware. If any design flaws are discovered during software simulation, the design must be debugged and modified to correct the flaw. Once the design is finalized, the compiler generates an output file, which includes the final logic equations and device assignments to be programmed onto the target device.

Programming the Device

The output file is downloaded to the PLD via the JTAG interface. The output file tells the programming fixture or JTAG interface how to program the finalized logic onto the PLD target device.

Review Questions

21. What is a computer program?
22. What does HDL stand for?
23. What is required to program a target device?
24. What are the software packages for VHDL development called?
25. What is the difference between conventional programming and in-system programming of a target device?

INTRODUCTION TO VHDL 10–6

As you have learned, VHDL is one among many HDLs that are available for programming PLDs. Although others are used, VHDL is perhaps the most common because it has been established as a standard HDL by the IEEE (Institute of Electrical and Electronics Engineers) and is in widespread use. It would take a very thick textbook to cover VHDL in its entirety. Here, only a few fundamental ideas are presented.

In this section, you will learn how VHDL is used to program a PLD to perform a specified logic function.

VHDL was initiated in 1981 by the Department of Defense in an effort to create a hardware development language that was independent of any specific technology or design method. All rights to VHDL were transferred to the IEEE and the first standard version was published in 1987. A revision, was published in 1993 and is designated IEEE Std. 1076-1993. The "V" in VHDL stands for the acronym VHSIC (very high-speed integrated circuit).

Although there are many aspects to a typical VHDL program, all programs must have a minimum of two elements, the *entity* and the *architecture.* Every VHDL program contains at least one entity/architecture pair. These elements can be thought of as basic VHDL building blocks.

The Entity

Figure 10–33 shows the concept of an entity. The **entity** describes a given logic function, as viewed from the outside, by naming the function and defining the inputs and outputs and the type of data they handle. Each input and output to a logic function is called a **port**. In the figure, the ports are Input 1, Input 2, etc., and Output 1, Output 2, etc. An entity can have any number of inputs and outputs.

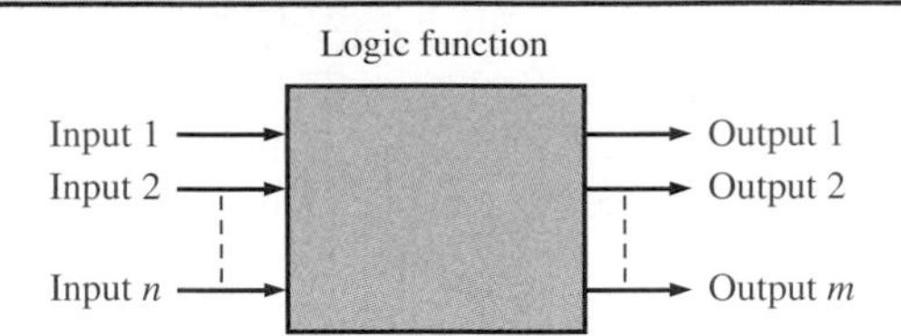

FIGURE 10–33

An entity describes the logic function in terms of its inputs and outputs.

Using an entity declaration, you can define any logic function from a simple gate to a complex system and provide the complete interface to connect the logic function to other circuits.

Every VHDL design must include at least one entity. The entity is where the interface of a logic function to the "outside world" is described. For example, if you wish to describe a logic gate using VHDL, you first must describe the input and output signals with an entity declaration. Entities require the use of identifiers for naming the entity itself as well as for describing the input and output signals.

Identifiers

User-defined names called *identifiers* are used to identify VHDL elements and units. VHDL supports two different types of identifiers, basic identifiers and extended identifiers, but we will focus only on the basic identifier. The rules for a VHDL basic identifier are as follows:

- A basic identifier consists only of lowercase or uppercase letters, numeric digits, and single underscores. You can use a mix of lower- and uppercase.
- A letter must be used as the first symbol in an identifier name.
- Basic identifiers must not contain spaces.
- VHDL keywords or reserved identifiers may not be used as identifiers. Keywords are also known as reserved words.

A few examples of basic identifier construction are A, BI, X, MyAND, ORgate, Decoder_1, INPUT1, input2, Input_A, Output_X, and OR_Out. You should always assign a meaningful name that is related to what is being identified. Identifiers are used in both the entities and the architectures covered in this section. Except for identifiers, VHDL ignores spaces and is also case insensitive.

The Entity Declaration

The general structure and syntax of a VHDL entity declaration is as follows. Identifiers are placed as indicated and the VHDL keywords are in boldface.

entity <entity identifier> **is**
  port (signal descriptions);
end entity <entity identifier>;

The first line in the entity declaration begins with the keyword **entity**, followed by an identifier assigned as the name of the entity and finally the keyword **is**.

The second line is the port statement that defines the input and output signals, called *ports*. Each input and output port is assigned an identifier.

The third line defines the end of the entity declaration using the keywords **end entity** and the entity name. Notice that there must be a semicolon after the port statement and the end statement.

The Port Statement

A port is an input or output signal. The VHDL keyword **port** is used to declare the input and output signal assignments. The port statement must specify the port identifier, the port direction, and the port data type for each signal.

Three directions can be assigned to a port: **in** for inputs, **out** for outputs, and **inout** for a port that can be both an input or an output, called a *bidirectional port.*

VHDL provides many different data types that essentially define the signal values. We will use the data type **bit**. A signal with this data type can have only two values, 1 or 0.

EXAMPLE 10–3

Problem

Write an entity declaration for a 2-input AND gate.

Solution

Step 1: Assign a name to the entity using an identifier of your choice. Since this is an AND gate, let's call it AND_Gate. This entity name is used in the first line.

Entity name

entity AND_Gate **is**

The VHDL keywords which, in this line, are **entity** and **is** are shown boldfaced. Many VHDL software editors automatically recognize keywords and other elements and usually display them in boldface or color, making the program easier to follow. Keywords can be either upper- or lowercase.

Step 2: Write the port statement and assign identifier names, directions, and data type to the two inputs and the output.

Input port names, direction, and data type

port (A, B: **in bit**; X: **out bit**);

Output port name, direction, and data type

The two inputs are called A and B. You could have used other identifiers such as IN_1 and IN_2 or A1 and A2. A, B designates the inputs followed by a colon, the port direction **in**, and the data type **bit**. The input data type is followed by a semicolon to separate the inputs from the output. The output is assigned the identifier X, which is followed by a colon, the port direction **out**, and the data type **bit**. The port assignments are enclosed in parentheses, and the port statement ends with a semicolon.

Step 3: The end statement indicates the end of the entity and includes the entity name followed by a semicolon.

end entity AND_Gate;

By putting all three lines together, you have the entity declaration for a 2-input AND gate.

```
entity AND_Gate is
   port (A, B: in bit; X: out bit);
end entity AND_Gate;
```

Question

How would an entity declaration for a 3-input AND gate be written?

The Architecture

For every entity there must be at least one architecture. While the entity declaration describes the external aspects of the logic function, the **architecture** declaration describes its internal operation. The architecture can describe the operation of the logic function in any one of three ways: structural description, data flow description, or behavioral description.

With the structural approach, a logic function is described in terms of the basic logic gates and their interconnections that make up the logic function. With the data flow approach, the logic function is described by how signals (data) flow through the logic gates that make up the circuit. Note that both the structural approach and the data flow approach deal with the internal details of the logic function.

The behaviorial approach is different from the structural or the data flow approaches because it does not deal with the type of gates, how they are interconnected, or how the data flow through them. With the behavioral approach, the architecture describes the logic function in terms of what happens on the outputs based on the current state and the inputs.

If a logic function is relatively simple, the structural or data flow approaches can be used in the architecture declaration. It is better to use the behavioral approach, if possible, when the design is so complex that it would be tedious to describe using the other approaches.

Figure 10–34 illustrates how an entity and an architecture are bound together in VHDL to describe an overall logic function. In simple terms, the entity describes the "exterior" as indicated by the blue area and the architecture describes the "interior" as indicated by the gray area.

FIGURE 10–34

Illustration of the relationship of the entity and the architecture in the description of a logic function.

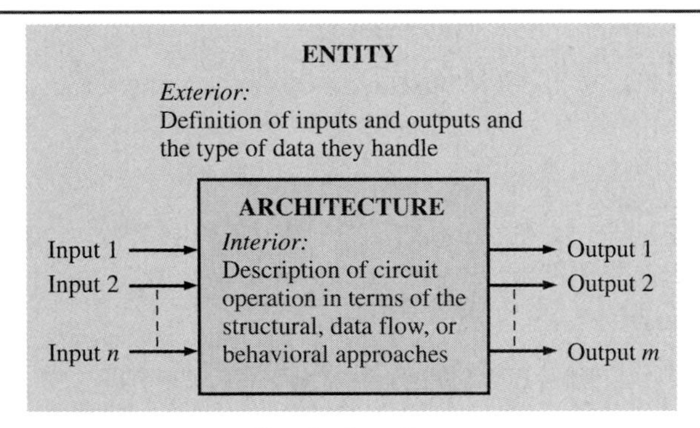

Each architecture declaration must be associated by name with an entity. The basic structure and syntax of an architecture declaration is as follows:

architecture <architecture name> **of** <entity name> **is**
begin
(The description of the logic function goes here.)
end architecture <architecture name>;

The first line in the architecture declaration begins with the keyword **architecture**, followed by an identifier assigned as the name of the architecture. The keyword **of** and the name of the associated entity follow the architecture name; the line ends with the keyword **is**. This binds the architecture to a specified entity.

The second line is simply the keyword **begin**. Following **begin** are the VHDL statements required to describe the logic function. There can be any number of lines of code in the functional description, depending on the type of logic function being described and its complexity.

The last line defines the end of the architecture declaration using the keywords **end architecture** and the architecture name. There must be a semicolon after the end statement.

Operators

Many types of operators in VHDL can be used in describing a logic function. We will introduce only two: the logic operator **and** and the signal assignment operator <=. These are used in Example 10–4.

EXAMPLE 10–4

Problem

Write an architecture declaration for the 2-input AND gate in Example 10–3.

Solution

Step 1: Assign a name to the architecture using an identifier of your choice. It must be different than the entity name. Since this is a description of the AND gate operation, let's call it ANDfunction. This architecture name is used in the first line. The entity to which the architecture is bound is also named.

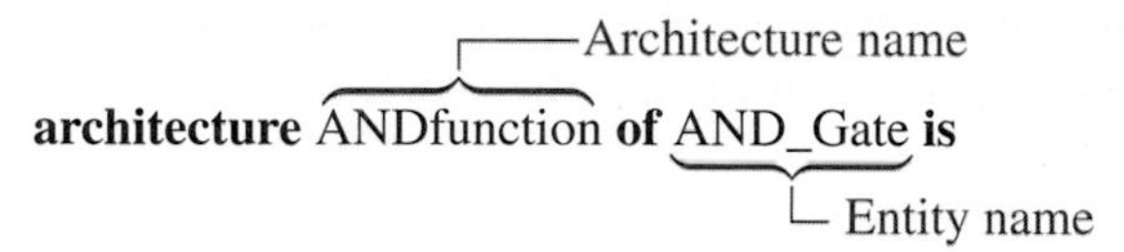

Step 2: Write the keyword **begin**, followed by the functional description of the AND gate.

begin ← Logical operator defines the relationship of A and B.
X <= A **and** B:
← Signal assignment operator assigns the value of A **and** B to the output X.

Step 3: The end statement indicates the end of the architecture and includes the architecture name followed by a semicolon.

end architecture ANDfunction;

By putting all three lines together, you have the architecture declaration for the 2-input AND gate that goes with the entity declaration in Example 10–3. Keep in mind that, although the functional description in this case is only one line, there can be many lines between the keyword **begin** and the end statement in an architecture.

```
architecture ANDfunction of AND_Gate is
begin
  X <= A and B;
end architecture ANDfunction;
```

Question

How would the architecture declaration for the 3-input AND gate in the Question in Example 10–3 be written?

The entity and architecture declarations from Examples 10–3 and 10–4 together form a simple VHDL program that completely defines the 2-input AND gate, as illustrated in Figure 10–35. As illustrated by the first line in the program in Figure 10–35, comments can be added to a VHDL program. A comment line is preceded by two hyphens --. Comment lines are not part of the program; but they can be used to explain or clarify certain parts or statements in the program and can be placed throughout a program.

FIGURE 10–35

VHDL program describing a 2-input AND gate.

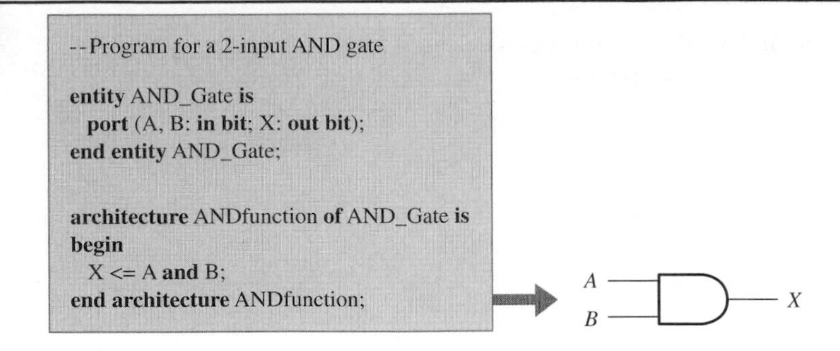

```
--Program for a 2-input AND gate

entity AND_Gate is
  port (A, B: in bit; X: out bit);
end entity AND_Gate;

architecture ANDfunction of AND_Gate is
begin
  X <= A and B;
end architecture ANDfunction;
```


Using VHDL logical operators, you can write a simple VHDL program for any type of basic gate with any number of inputs. The AND gate program in Figure 10–35 can easily be modified for any gate, as illustrated in Figure 10–36 for a 3-input OR gate. Notice that the entity and architecture names reflect the type of function that is being described. The three inputs are defined in the port statement, and the logical operator **or** is used in the architectural description.

FIGURE 10–36

VHDL program for a 3-input OR gate.

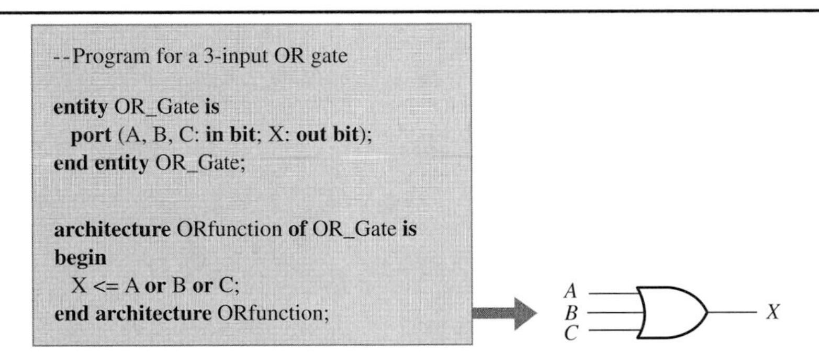

```
--Program for a 3-input OR gate

entity OR_Gate is
  port (A, B, C: in bit; X: out bit);
end entity OR_Gate;

architecture ORfunction of OR_Gate is
begin
  X <= A or B or C;
end architecture ORfunction;
```


EXAMPLE 10–5

Problem

Modern automobiles have protection systems built in for the safety of the driver. In this application, three conditions are required before the car can be started: *seat belt fastened* and *all doors closed* and *transmission in park*. Assume that a logic 0 is needed to engage the starter when all the conditions are met. Also, assume that each of the sensors produces a logic 1 when its condition is met. Develope a VHDL code.

Solution

This application can be implemented with a 4-input NAND gate, assuming a 2-door vehicle. The logic diagram is shown in Figure 10–37.

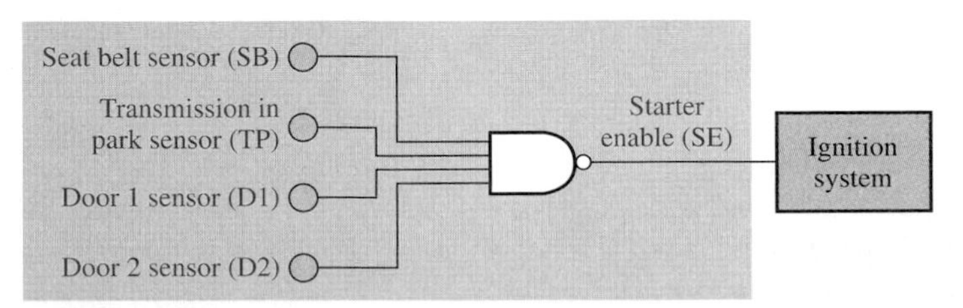

FIGURE 10–37

To develop the VHDL code necessary to implement the required logic gate shown in Figure 10–37 in the beige block, identifiers must be created for the entity and architecture names and for each input and output (**ports**). We will use SB for the seat belt sensor signal, TP for the transmission in park signal, D1 for door 1, D2 for door 2, and SE for starter enable. All of the inputs and the output are the data type **bit**.

```
--Program for automobile safety logic
entity AutoSafety is
   port (SB, TP, D1, D2: in bit; SE: out bit);
end entity AutoSafety;

architecture SafetyLogic of AutoSafety is
begin
   SE <= not(SB and TP and D1 and D2);
end architecture SafetyLogic;
```

Question

How would you modify the VHDL program for a 4-door automobile?

Packages

A VHDL program may contain several entity/architecture pairs as shown in block form in Figure 10–38. The *package* is a VHDL unit used to store various items that may be commonly used by any of the entity/architectures within the program. The *library* is another VHDL feature that stores items that can be accessed by several different programs. You may store packages in a user-defined library to make them more accessible and to keep them organized. Also, there are standard libraries available from PLD development software vendors that contain items commonly used by everyone who writes programs in VHDL.

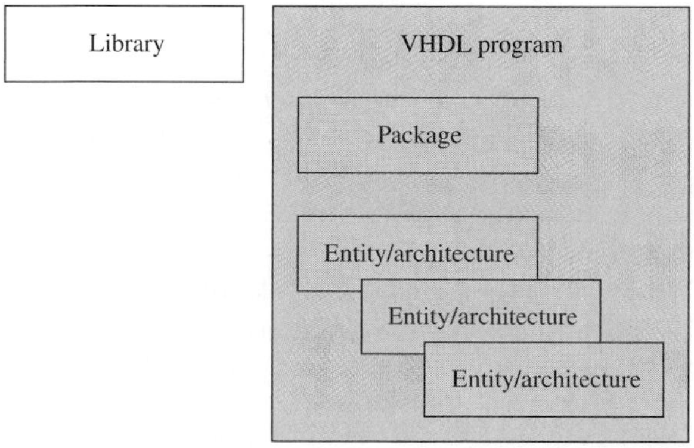

Library

VHDL program

Package

Entity/architecture

Entity/architecture

Entity/architecture

FIGURE 10–38

Representation of a VHDL program with three entity/architectures and a package. The program has access to the library and to common features described in the package.

VHDL Language Elements

VHDL has many different language-related elements. These are the "nuts and bolts" of the language that are used to construct the various types of statements in a program. A summary of the basic elements introduced in this section include keywords, basic identifiers, data objects, data types, literals, and operators.

- *Keywords* are reserved words in VHDL and are a type of identifier. Examples are **entity**, which defines an entity declaration, and **architecture**, which of course, defines an architecture declaration. Keywords are shown as black boldface words.
- *Basic identifiers* are used to name an entity or architecture as well as other items. You, as a programmer, can use any word as a basic identifier as long as you follow certain basic rules. Keywords cannot be used as basic identifiers.
- *Data objects* are used to hold a value or set of values. A data object can be a constant, variable, signal, or file.
- *Data types* are used to specify the type of data for data objects and other elements. The data type can be **bit** that has values of 1 and 0, **boolean** that has values of true and false, and **integer** that has integer values in a certain range, just to name a few.
- *Literals* are numeric values that can be characters, strings of characters, strings of bits, and decimal numbers to name a few.
- *Operators* are elements that perform certain operations or functions. There are many types of operators available in VHDL. The logical operators include **and**, **or**, and **not**, **nand**, **nor**, **xor**, and **xnor**. Relational operators include equality (=), inequality (/=), and a few others. The arithmetic operators allow you to add, subtract, multiply, and divide as well as do some more specialized operations.

SAFETY NOTE

Always be careful when you work with electricity. Electric shock can cause muscle spasms, weakness, shallow breathing, rapid pulse, severe burns, unconsciousness, or death. In a shock incident, the path that electric current takes through the body gets very hot. Burns occur all along that path, including the places on the skin where the current enters and leaves the body. It's not only power lines that are dangerous if you contact them. You can also be injured by a shock from an appliance, instrument, or power cord in your home or in the lab.

Boolean Algebra in VHDL Programming

A VHDL compiler converts a VHDL source program (the program that you write) into a data file that is downloaded into a programmable logic device (called the target device). The data contained in this data file must physically fit in the target device, thus making it important to always write the simplest code possible.

The basic rules of Boolean algebra that you have learned (see Appendix A for review) should be applied to any applicable VHDL code. Eliminating unnecessary gate logic allows you to create compact code that is easier to understand, especially when someone has to go back later and update or modify the program.

EXAMPLE 10–6

Problem

Write a VHDL program for the logic described by the following Boolean expression. Next, apply DeMorgan's theorems and Boolean rules to simplify the expression. Then write a program to reflect the simplified expression.

$$X = \overline{(AC + \overline{B\overline{C}} + D)} + \overline{\overline{BC}}$$

Solution

The VHDL program for the logic represented by the original expression follows.

```
--Program to implement the original Boolean expression
entity OriginalLogic is
  port (A, B, C, D: in bit; X: out bit);
end entity OriginalLogic;
architecture Expression1 of OriginalLogic is
begin
  X <= not((A and C) or not(B and not C) or D) or not(not(B and C));
end architecture Expression1;
```

Four inputs are described.

The original logic contains four inputs, 3 AND gates, 2 OR gates, and 5 inverters.

By selectively applying DeMorgan's theorem and the laws of Boolean algebra, you can reduce the Boolean expression to its simplest form.

$\overline{(AC + \overline{B\overline{C}} + D)} + \overline{\overline{BC}} = (\overline{AC})(\overline{\overline{B\overline{C}}})\overline{D} + \overline{\overline{BC}}$	Apply DeMorgan
$= (\overline{AC})(B\overline{C})\overline{D} + BC$	Cancel double complements
$= (\overline{A} + \overline{C})B\overline{C}\,\overline{D} + BC$	Apply DeMorgan and factor
$= \overline{A}B\overline{C}\,\overline{D} + B\overline{C}\,\overline{D} + BC$	Distributive law
$= B\overline{C}\,\overline{D}(1 + \overline{A}) + BC$	Factor
$= B\overline{C}\,\overline{D} + BC$	Rule: $1 + A = 1$

The VHDL program for the logic represented by the reduced expression follows.

```
--Program to implement the simplified Boolean expression
entity ReducedLogic is
  port (B, C, D: in bit; X: out bit);
end entity ReducedLogic;
architecture Expression 2 of ReducedLogic is
begin
  X <= (B and not C and not D) or (B and C);
end architecture Expression 2;
```

Three inputs are described.

The simplified logic contains three inputs, 3 AND gates, 1 OR gate, and 2 inverters.

As you can see, Boolean simplification is applicable to even simple VHDL programs.

Question

How would the VHDL architecture statement for the expression $X = (\overline{A} + B + C)D$ be written if the entity name is LogicCircuit?

Although we have discussed how to describe only simple combinational logic with VHDL, any logic, no matter how complex, can be implemented. Sequential logic such as flip-flops, counters, and registers can be implemented with VHDL too. For more advanced logic, VHDL has many features that are beyond the scope of this book. However, this brief introduction gives you a good start in becoming proficient in using this powerful language.

Levels of Approach to VHDL Design

In this section, you have seen one approach to implementing a logic function with VHDL. A given logic function can be described at three different levels. It can be described by a truth table or a state diagram, by a Boolean expression, or by its logic diagram (schematic).

The truth table and state diagram are the most abstract ways to describe a logic function. A Boolean expression is the next level of abstraction, and a schematic is the lowest level of abstraction. This concept is illustrated in Figure 10–39 for a simple logic circuit. VHDL provides three approaches for describing functions that correspond to the three levels.

- The behavioral approach is analogous to describing a logic function using a state diagram or truth table. However, this approach is the most complex; it is usually restricted to logic functions whose operations are time dependent and normally require some type of memory.
- The data flow approach is analogous to describing a logic function with a Boolean expression. The data flow approach specifies each of the logic gates and how the data flows through them. This is the approach that has been applied in this section.
- The structural approach is analogous to using a logic diagram or schematic to describe a logic function. It specifies the gates and how they are connected, rather than how signals (data) flow through them as in using Boolean expressions.

FIGURE 10–39

Illustration of the three levels for describing a logic function.

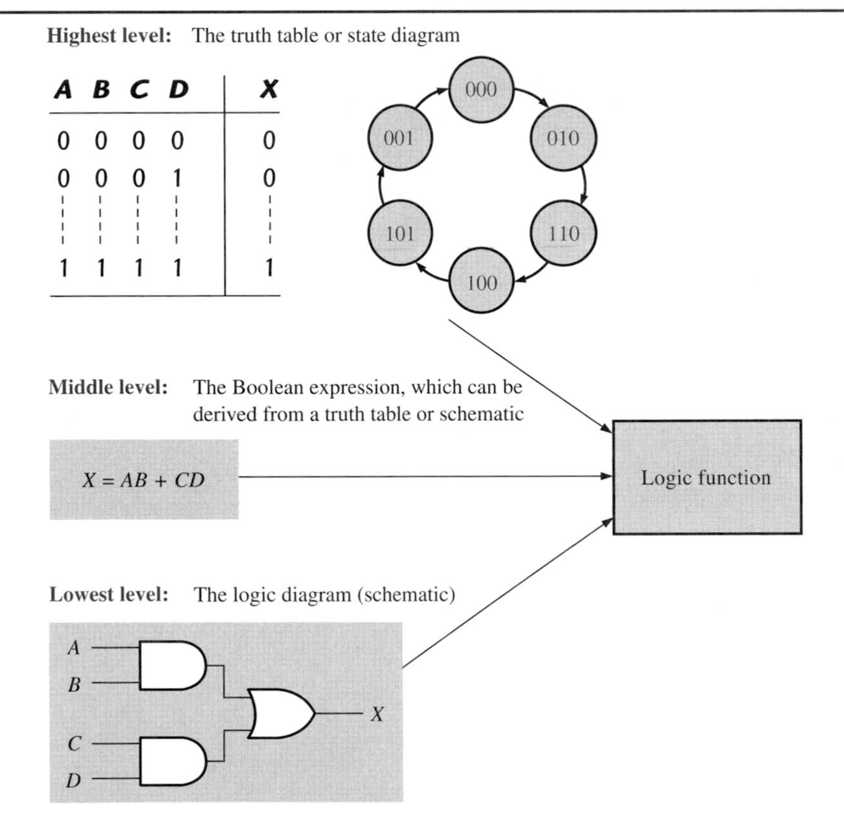

Review Questions

26. What does an entity do?
27. What does an architecture do?
28. What are the VHDL logic operators?
29. Where and how is an architecture bound to a specific entity?
30. What are the three approaches to describing a logic function in VHDL?

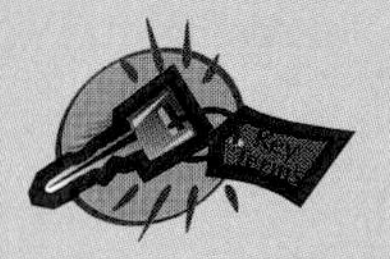

Key Terms

Compiler A program that functions as a programming language translator. A compiler translates high-level language to machine code.

CPLD Complex programmable logic device; a type of PLD that can contain the equivalent of from two to sixty-four SPLDs.

FPGA Field programmable gate array; a type of PLD that generally has much more complexity than a CPLD and a different organization.

GAL Generic array logic; a type of SPLD that is essentially a reprogrammable PAL.

PAL Programmable array logic; a type of SPLD that is generally one-time programmable.

PLD Programmable logic device; an integrated circuit that can be programmed with any specified logic function.

Program A list of instructions that a computer follows in order to achieve a specific result.

Programmable arrays An array of logic gates with a grid of programmable interconnections.

SPLD Simple programmable logic device; the least complex form of PLD. The PAL and the GAL are types of SPLDs.

Target device A PLD that is being programmed.

VHDL A standard hardware description language for programming PLDs.

Important Facts

- ❑ A PLD is an integrated circuit that can be programmed to implement logic functions.
- ❑ There are three basic categories of PLDs: SPLDs, CPLDs, and FPGAs.
- ❑ An SPLD can be used to replace several fixed-function logic ICs.
- ❑ All PLDs are supported by manufacturer's development systems that are software packages for programming the devices.
- ❑ A PAL is an SPLD that consists of a one-time programmable array of AND gates that connects to a fixed array of OR gates.
- ❑ A GAL is a PAL that is reprogrammable.
- ❑ A CPLD contains several PAL/GAL-type arrays.
- ❑ An FPGA can contain much more implemented logic than a CPLD.
- ❑ Computer programs are written using a programming language.
- ❑ There are three levels of programming languages: high-level, assembly, and machine.
- ❑ High-level languages use English-like words and other symbols to make a program easier to write and to understand. High-level languages are "machine-independent," so they can be used on any computer.
- ❑ Assembly languages use English-like words but are "machine dependent." Each type of microprocessor has its own assembly language.
- ❑ Machine languages are binary codes that represent instructions.
- ❑ A PLD is programmed using a type of programming language known as a hardware description language (HDL).

16. What is one widely used standard language for programming PLDs?
17. What is a target device?
18. How does ISP differ from conventional programming?
19. What are the two essential parts of a VHDL program?
20. What is a VHDL package used for?

PROBLEMS

Basic Problems

Answers to odd-numbered problems are at the end of the book.

1. Name each of the types of CPLD packages shown in Figure 10–40.

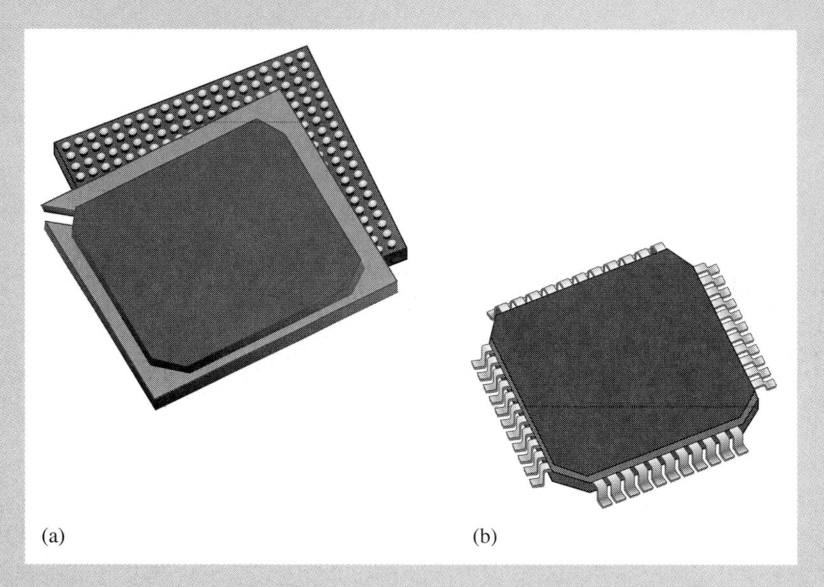

FIGURE 10–40

2. Name four types of SPLDs.
3. In the simple programmed AND array in Figure 10–41, determine the logic expressions at each gate output.

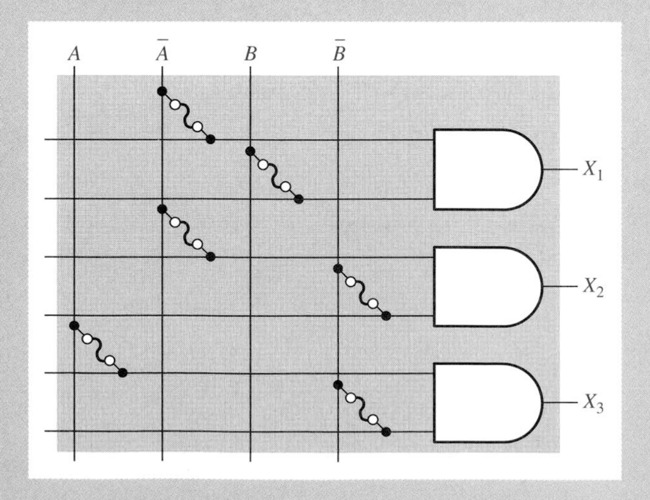

FIGURE 10–41

4. Determine the output expression for the simple PAL array in Figure 10–42. The Xs represent where connections are made (intact fusible links).

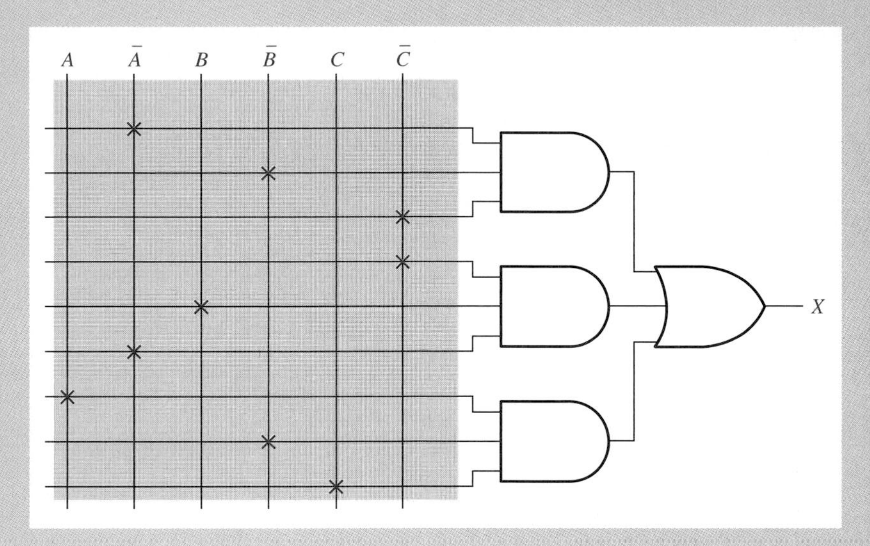

FIGURE 10–42

5. What does the V in VHDL stand for?
6. List the steps in a general HDL programming procedure.
7. Name three components required to program a target device (PLD).
8. List six VHDL elements.
9. Which of the following VHDL entity declarations is correct?

```
(a) entity AND_Gate is
        port (A, B: in bit: X; out bit)
    end entity AND_Gate
(b) entity AND_Gate:
        port (A, B: input; X: output);
    end entity AND_Gate;
(c) entity AND_Gate is
        port (A, B: in bit; X: out bit);
    end entity AND_Gate;
```

10. What type of logic gate with how many inputs does the following VHDL entity/architecture describe?

```
entity Logic is
  port (A, B, C, D: in bit; X: out bit);
end entity Logic;
architecture Gate of Logic is
begin
  X <= A or B or C or D;
end architecture Gate;
```

Basic-Plus Problems

11. Determine by row and column number for which fuses must be retained in the programmable AND array of Figure 10–43 to implement the following expressions: $X_1 = \overline{A}BC, X_2 = AB\overline{C}, X_3 = \overline{A}B\overline{C}$.

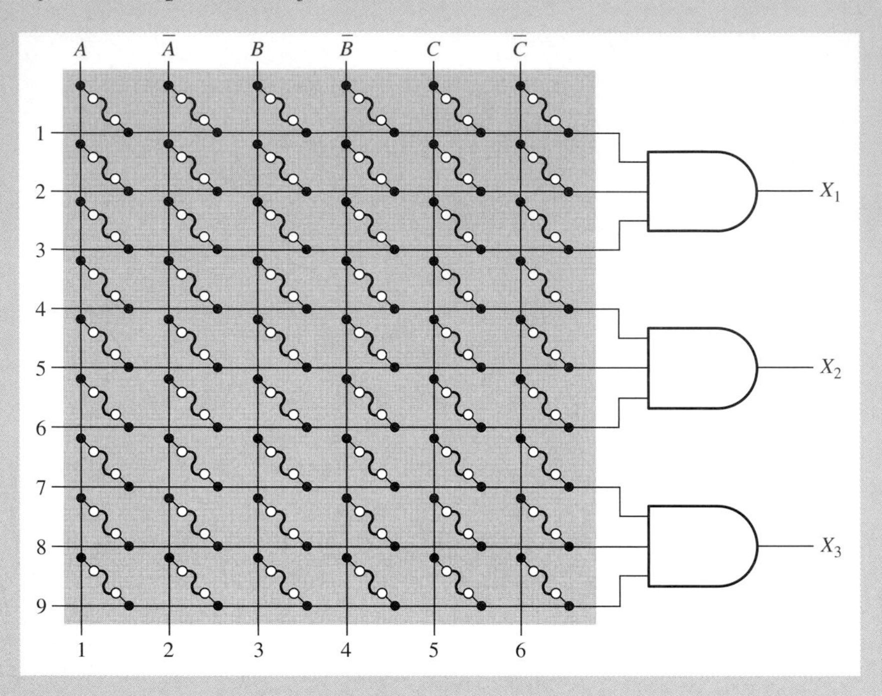

FIGURE 10–43

12. Show how the PAL array represented in Figure 10–44 should be programmed in order to implement the following SOP expression. Place an X to indicate a connection point (intact fuse). $X = A\overline{B}C + \overline{A}B\overline{C} + ABC$

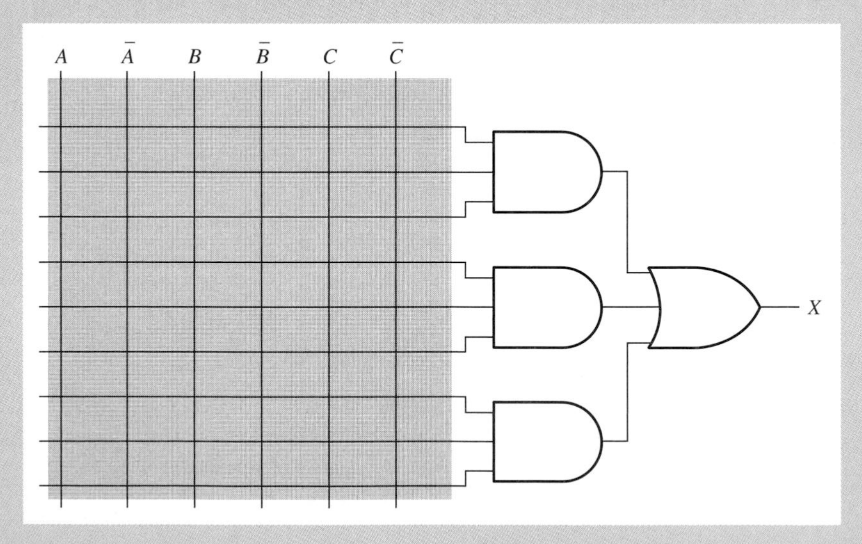

FIGURE 10–44

13. Show how the PAL array represented in Figure 10–44 should be programmed in order to implement the following SOP expression. If possible, reduce the expression using a Karnaugh map or Boolean algebra to fit the PAL array. Place an X to indicate a connection point (intact fuse): $X = A\overline{B}C + \overline{A}\,\overline{B}C + A\overline{B}\,\overline{C} + \overline{A}BC$.

14. Write an entity and architecture for a 3-input AND gate. Use A, B, and C as input identifiers and X as an output identifier.
15. Write an entity and architecture for a 4-input NAND gate. Use A, B, C, and D as input identifiers and X as an output identifier.
16. Write a VHDL program to describe an exclusive-OR gate. Choose your own indentifiers.
17. Show the logic circuit that is described by the following VHDL program.

```
entity Logic_Circuit is
  port (A, B, C, D: in bit; X: out bit);
end entity Logic_Circuit;
architecture Design of Logic_Circuit is
begin
  X <= (A or not B) nand (C nand D) or (A and B and C) and not (C or D);
end architecture Design;
```

18. Write a VHDL program to describe the following logic expression:

$$X = A\overline{B}C + \overline{A}B\overline{C}D + \overline{B}CD + AD$$

19. Write a VHDL program to implement the logic circuit in Figure 10–45.

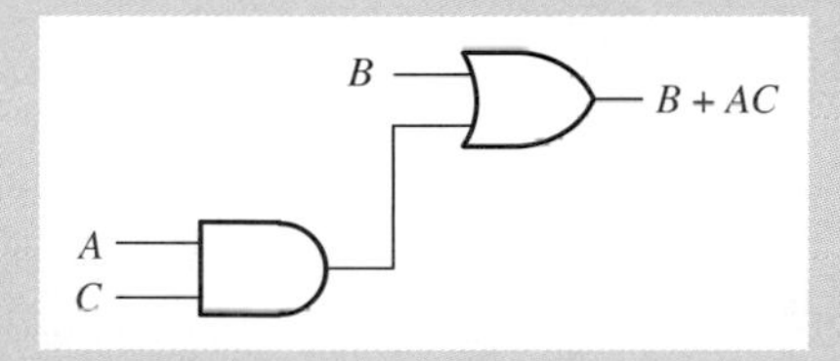

FIGURE 10–45

20. Write a VHDL program for the circuit shown (not simplified) in Figure 10–46.

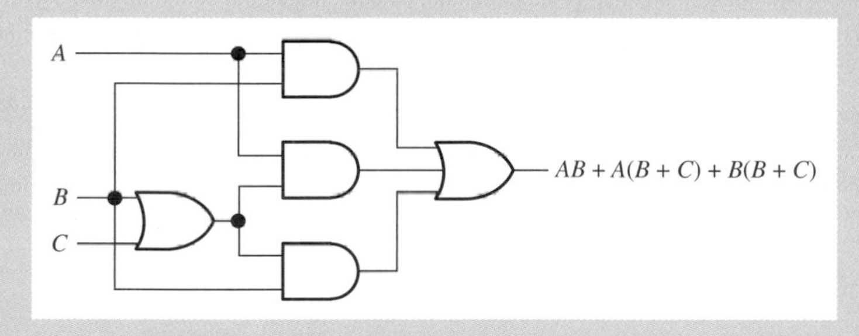

FIGURE 10–46

21. Write a VHDL program for the circuit shown (not simplified) in Figure 10–47.

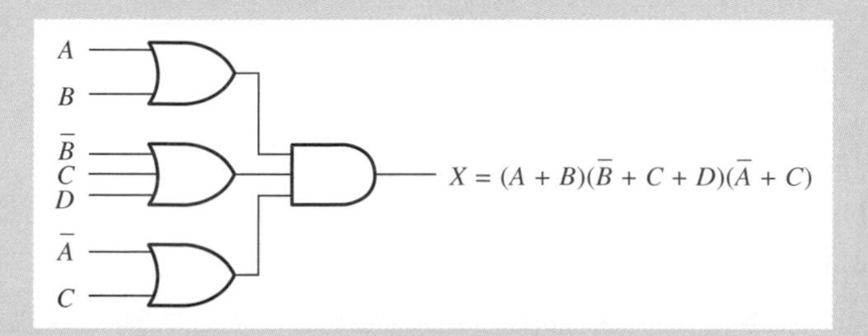

FIGURE 10–47

22. Digital imaging has created many new opportunities for using programmable logic devices. Use the Internet as a resource to investigate some of the applications for using PLDs in digital imaging and summarize your findings in a short report. You might start with manufacturers of PLDs such as Altera and Xilinx.

ANSWERS

Example Questions

10–1: $X = \overline{B}C + \overline{A}B\overline{C} + \overline{A}\,\overline{B} + C$

10–2: $X = B\overline{C} + BC + \cancel{BC} + A\overline{B}$

10–3:

```
entity AND3 is
   port (A, B, C: in bit; X: out bit);
end entity AND3;
```

10–4:

```
architecture ANDfunction of AND3 is
begin
   X <= A and B and C;
end architecture ANDfunction;
```

10–5:

```
entity AutoSafety is
   port (SB, TP, D1, D2, D3, D4: in bit; SE: out bit);
end entity AutoSafety;
architecture SafetyLogic of AutoSafety is
begin
   SE <= not(SB and TP and D1 and D2 and D3 and D4);
end architecture SafetyLogic;
```

10–6:

```
architecture Expression of LogicCircuit is
begin
   X <= (not A or B or C) and D;
end architecture Expression;
```

Review Questions

1. PLD: Programmable Logic Device
2. SPLD, CPLD, and FPGA
3. SPLD: Simple Programmable Logic Device, CPLD: Complex Programmable Logic Device, FPGA: Field-Programmable Gate Array
4. The fuse links are left intact to make a connection and blown open for no connection.
5. AND array; OR array
6. PAL: Programmable Array Logic; a type of SPLD
7. Yes, unless it is limited by the capacity of the PAL
8. Input buffers prevent loading by AND gate inputs.
9. Combinational output, combinational input/output, and programmable polarity output

10. The 16 indicates that 16 inputs are available, the L indicates active-LOW outputs, and the 8 indicates a total of 8 outputs are available.
11. GAL: Generic Array Logic; a type of SPLD
12. A GAL is reprogrammable: a PAL is not.
13. GALs use reprogrammable E^2CMOS cells for interconnection instead of fusible links.
14. OLMC: Output Logic Macrocell
15. E^2CMOS: Electrically Erasable Complementary Metal-Oxide Semiconductor.
16. CPLD: Complex Programmable Logic Device
17. A CPLD consists of several PAL/GAL-type logic blocks and programmable interconnections.
18. LAB: Logic Array Block; it is equivalent to a single SPLD.
19. FPGA: Field Programmable Gate Array
20. LUT: Look-Up Table; A type of memory programmed to produce logic functions.
21. An algorithm for a computer consisting of a list of instructions that the computer follows.
22. HDL: Hardware Description Language
23. Computer, HDL software, programming fixture
24. Development software
25. In conventional PLD programming, the target device is inserted into a programming fixture. For in-system programming the target device remains mounted on its system printed circuit board.
26. An entity describes the inputs and outputs of a logic function.
27. An architecture describes the operation of a logic function.
28. **not, and, or, nand, nor, xor, xnor**
29. In the architecture statement the entity name follows the keyword **of**.
30. Structural approach, data flow approach, and behavioral approach

Chapter Checkup

1. (b)	2. (a)	3. (c)	4. (b)	5. (b)
6. (c)	7. (a)	8. (a)	9. (c)	10. (b)
11. (b)	12. (a)	13. (c)	14. (b)	15. (c)

11

CHAPTER

COMPUTER BASICS

CHAPTER OUTLINE

Study aids for this chapter are available at

http://www.prenhall.com/SOE

INTRODUCTION

This chapter introduces basic computer concepts. It is not a comprehensive or in-depth coverage but is meant only to familiarize you with computer terminology and principal components and to show you a little about what goes on inside a computer.

A computer system is comprised of two major aspects, hardware and software. The hardware is the physical part of the computer, such as the integrated circuits (chips) that make up the microprocessor, memory, and other functions. Hardware is also the motherboard, the hard disk, the connectors, and the input and output devices. The software is information stored in the computer that provides instructions to enable the computer to perform specified tasks. Software is what makes a computer run.

KEY OBJECTIVES

A section number is given for each objective. After completing this chapter, you should be able to

- **11–1** Describe a basic computer system
- **11–2** Describe the elements in a micoprocessor
- **11–3** Explain basic memory concepts
- **11–4** Discuss static and dynamic RAMs
- **11–5** Discuss read-only memories
- **11–6** Describe various types of storage devices used in computers
- **11–7** Explain some programming concepts and the levels of programming languages

LABORATORY EXPERIMENTS DIRECTORY

The following exercises are for this chapter.

- ◆ **Experiment 15**
  Semiconductor Memory
- ◆ **Experiment 16**
  Introduction to the Intel Processors

KEY TERMS

- CPU
- RAM
- ROM
- Cache
- Hard disk
- Bus
- Software
- Microprocessor
- ALU
- Byte
- Address
- Flash memory
- Machine language
- High-level language
- Assembly language
- Compiler

Early computers were relatively slow primarily because they were sequential; that is, they could do only one thing at a time. For example, those computers could "look" at a picture pixel by pixel. By contrast, the human brain can instantly perceive an entire picture and simultaneously relate it to every image that it knows. This limitation of computers has led to the development of parallel supercomputers that can do many things at one time.

Simulation of the atmosphere for predicting weather is a good example of the application of parallel computing. Various numbers representing temperature, pressure, wind, and humidity are viewed in a 3-dimensional "cube" of air. The supercomputer processes millions of these "cubes" using parallel processing to create a simulation of a broad area of the atmosphere.

In contrast, the prediction of the positions of the planets far into the future is basically a sequential process. To predict a future position, the position between now and then must be determined at each intermediate time along the way. This can only be done serially.

11–1 THE BASIC COMPUTER SYSTEM

Special-purpose computers control various functions in automobiles or appliances, control manufacturing processes in industry, provide games for entertainment, and are used in navigation systems such as GPS (Global Positioning System), to name a few areas. However, the most familiar type of computer is the general-purpose computer that can be programmed to do many different types of things.

In this section, you will learn what makes up a typical general-purpose computer system, such as your desktop or laptop.

You know what a computer looks like from the outside. In a desktop system, as shown in Figure 11–1(a), you generally see a tower unit that contains the computer itself, the monitor, the keyboard, the mouse, and maybe some speakers. In a laptop model, of course you see the monitor and the keyboard when you open it up. The desktop computer itself is contained in the "tower," as shown by the cutaway view in Figure 11–1(b).

FIGURE 11–1 Typical general-purpose desktop computer.

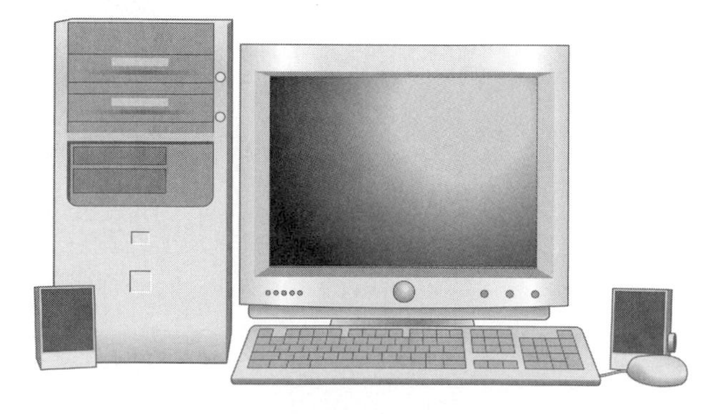

(a)

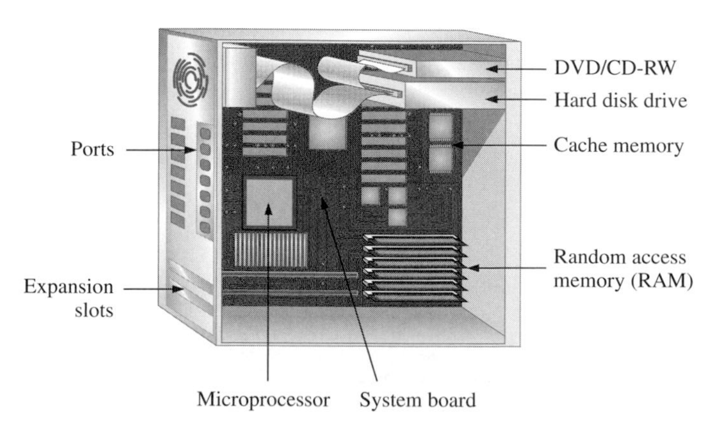

(b)

Parts of a Computer

The block diagram in Figure 11–2 shows the main elements in a typical computer and how they are interconnected. For a computer to accomplish a given task, it must communicate with the "outside world" by interfacing with people, sensing devices, or devices to be controlled in some way. To do this, there is a keyboard for data entry, a mouse, a video monitor, a printer, a modem, and a CD drive in most basic systems. These are called **peripherals**.

Let's briefly look at each of the computer hardware elements.

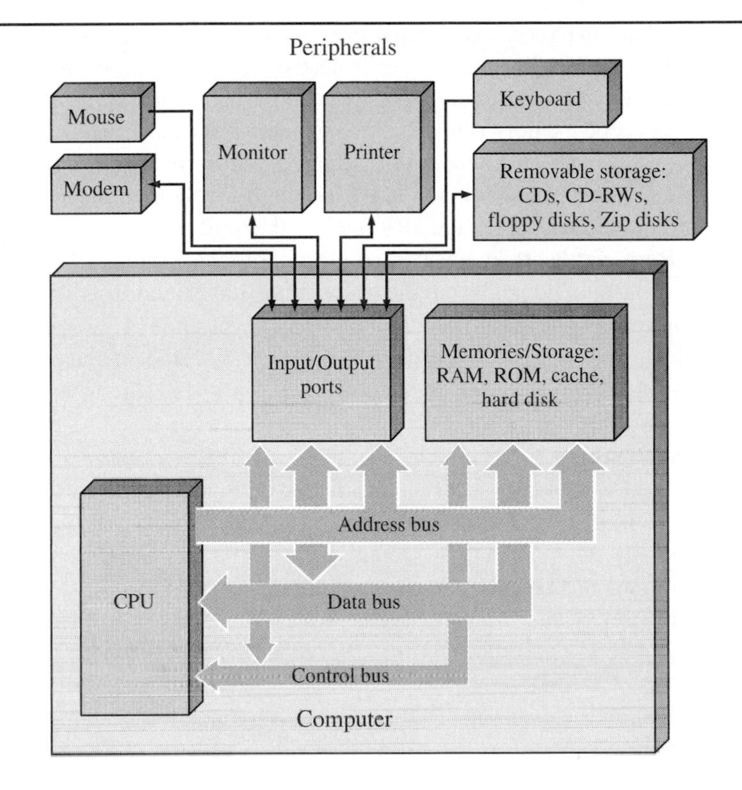

FIGURE 11–2

Basic block diagram of a typical computer system including common peripherals. The computer itself is shown within the gray block.

Central Processing Unit (CPU)

The **CPU** is the "brain" of the computer; it oversees everything that the computer does. The CPU is a microprocessor with associated circuits that control the running of the computer software programs. Basically, the CPU obtains (fetches) each program instruction from memory and carries out (executes) the instruction.

After completing one instruction, the CPU moves on to the next one and in most cases can operate on more than one instruction at the same time. This "fetch and execute" process is repeated until all of the instructions in a specific program have been executed. For example, an application program may require the sum of a series of numbers. The instructions to add the numbers are stored in the form of binary codes that direct the CPU to fetch a series of numbers from memory, add them, and store the sum back in memory.

Memories and Storage

Several types of memories are used in a typical computer. The **RAM** (random-access memory) stores binary data and programs temporarily during processing. Data are numbers and other information, and programs are lists of instructions. Data can be written into and read out of a RAM at any time. The RAM is volatile, meaning that the information is lost if power is turned off or fails. Therefore, any data or program that needs to be saved should be moved to nonvolatile memory (such as a CD or hard disk) before power is removed.

The **ROM** (read-only memory) stores a permanent system program called the **BIOS** (Basic Input/Output System) and certain locations of system programs in memory. The ROM is nonvolatile, which means it retains what is stored, even when the power is *off*. As

HISTORICAL NOTE

John von Neumann (1903–1957) was an Hungarian-born American mathematician. He and Albert Einstein were on the first faculty at the Institute for Advanced Study at Princeton. Dr. von Neumann also worked on the Manhattan Project. He is known as the father of the modern computer. In a paper written in 1945, he proposed the stored-program concept and described what is now known as the von Neumann architecture, which consists of five parts: an arithmetic logic unit (ALU), a control unit, a memory that stores both instructions and data, some form of input/output, and buses that provide interconnections among the parts. Most computers today are based on the von Neumann architecture with the ALU and control unit being parts of the central processing unit (CPU).

the name implies, the programs and data in ROM cannot be altered. Sometimes it is referred to as "firmware" because it is permanent software for a given system.

The BIOS is the lowest level of the computer's operating system. It contains instructions that tell the CPU what to do when power is first applied; the first instruction executed is in the BIOS. It controls the computer's basic start-up functions that include a self-test and a disk self-loader to bring up the rest of the operating system. In addition, the BIOS stores locations of system programs that handle certain requests from peripherals called **interrupts**, which cause the current processing to be temporarily stopped.

The **cache** memory is a small RAM that is used to store a limited amount of frequently used data that can be accessed much faster than the main RAM. The cache stores "close at hand" information that will be used again instead of having to retrieve it from farther away in the main memory. Most microprocessors have internal cache memory called level-1, or simply L1. External cache memory is in a separate memory chip and is referred to as level-2, or L2.

The **hard disk** is the major storage medium in a computer because it can store large amounts of data and is nonvolatile. The high-level operating systems as well as applications software and data files are all stored on the hard disk.

Removable storage is part of most computer systems. The most common types of removable storage media are the CDs, floppy disks, and Zip disks (magnetic storage media). Floppy disks have limited storage capability of about 1.4 MB, (megabyte). CDs are available as CD-ROMs (Compact Disk Read Only Memory) and as CD-RWs (Rewritable) and can store huge amounts of data (typically 650 MB). Zip drives typically store 250 MB.

Input/Output Ports

Generally, the computer sends data to a peripheral device through an output port and receives information through an input port. Ports can be configured in software to be either an input or output port. The keyboard, mouse, video monitor, printer, and other peripherals communicate to the CPU through individual ports. Ports are generally classified as either serial ports, with a single data line, or parallel ports, with multiple data lines.

Buses

Peripherals are connected to the computer ports with standard interface buses. A **bus** can be thought of as a highway for digital signals that consists of a set of physical connections, as well as electrical specifications for the signals. Examples of serial buses are FireWire and USB (Universal Serial Bus). The most common parallel bus is simply called the *parallel bus,* which connects to a port commonly referred to as the printer port (although this port can be used by other peripherals.) Another example of a parallel bus, for connecting lab instruments to a computer, is called the General Purpose Interface Bus (GPIB).

The three basic types of internal buses that interconnect the CPU with memory and storage and with input and output ports are the address bus, data bus, and control bus. These buses are usually lumped into what is called the *local bus.* The address bus is used by the CPU to specify memory locations or addresses and to select ports. The data bus is used to transfer program instructions and data between the CPU, memories, and ports. The control bus is used for transferring control signals to and from the CPU.

Computer Software

The other major aspect of a computer system is the **software**. The software makes the hardware perform. The two major categories of software used in computers are system software and applications software.

System Software

The system software is called the operating system of a computer and allows the user to interface with the computer. The most common operating systems used in desktop and laptop computers are Windows, MacOS, and UNIX. Many other operating systems are used in special-purpose computers and in mainframe computers.

HISTORICAL NOTE

Charles Babbage (1791–1871) was an Englishman who is considered to be a pioneer of the computer. Although his achievements are documented in several different fields, his greatest achievement is considered to be his detailed plans for "calculating engines," which were basically punched-card controlled calculators. The concepts employed in these "calculating engines" included many features that later reappeared in the modern stored-program computer such as punched-card control, separate storage, internal registers, multiplier/divider, array processing, and a range of peripherals.

HISTORICAL NOTE

Dr. John V. Atanasoff was the first person to apply electronic processing to digital computing. With the help of graduate student Clifford E. Berry, he constructed the first computer that used electronics to perform switching and arithmetic functions in 1939. Their computer, called the ABC (for Atanasoff-Berry Computer), was also the first electronic computer to separate the processing of data from memory and use memory refresh, a capacitor drum memory, clocked control, and vector processing, among other innovations. It was completed in 1942, but further work was abandoned because of World War II. In November 1990, Dr. Atanasoff received the National Medal of Technology from President Bush in recognition of his achievements.

System software performs two basic functions. It manages all the hardware and software in a computer. For example, the operating system manages and allots space on the hard disk. It also provides a consistent interface between applications software and hardware. This allows an applications program to work on various computers that may differ in hardware details.

The operating system on your computer allows you to have several programs running at the same time. This is called multitasking. For example, you can be using the word processor while downloading something from the Internet and printing an e-mail message.

Applications Software

You use applications software to accomplish a specific job or task. Table 11–1 lists several types of application software.

TABLE 11–1

Applications software.

Application	Function	Examples
Word processing	Prepare text documents and letters	Microsoft Word, WordPerfect
Drawing	Prepare technical drawings and pictures	CorelDraw, Freehand, Illustrator
Spreadsheet	Manipulate numbers and words in an array	Excel, Lotus 123
Desktop publishing	Prepare newsletters, flyers, books, and other printed material	Quark XPress, Pagemaker
Photography	Manipulate digital pictures, add special effects to pictures	Photoshop, Image Expert
Accounting	Tax preparation, bookkeeping	Quickbooks, Turbotax, MYOB
Presentations	Prepare slide shows and technical presentations	PowerPoint, Harvard Graphics
Data management	Manipulate large databases	Filemaker, Access
Multimedia	Digital video editing, produce moving images in presentations	Premier, Dreamweaver, After Effects
Speech recognition	Converts speech to text	NaturallySpeaking
Website preparation	Tools to create web pages and websites on the Internet	FrontPage, Acrobat
Circuit simulation	Create and test electronic circuits	Multisim

Sequence of Operation

When you first turn on your computer, this is what happens:

1. BIOS from ROM is loaded into RAM and a self-test is performed to check all major components and memory. Also, the BIOS provides information about storage, boot sequence, and the like.
2. The operating system (such as Windows) on the hard disk is loaded into RAM.
3. Application programs (such as Microsoft Word) are stored on the hard disk. When you select one, it is loaded into RAM. Sometimes, only portions are loaded as needed.
4. Files required by the application are loaded from the hard disk into RAM.
5. When a file is saved and the application is closed, the file is written back to the hard disk and both the application and the file are removed from RAM.

The Internet

The Internet is the world's largest computer network, connecting many millions of computers together so that each can communicate with any of the others. It can be thought of as a large collection of interconnected networks, where many smaller networks are linked together in a massive array known as the World Wide Web. Many companies and organizations such as schools have local area networks (LANs) that are part of the interconnected network of computers on the Internet.

The Internet has its roots in research done in the late 1960s by the research arm of the Department of Defense, known then as ARPA (Advanced Research Projects Administration). The military, national labs, and universities were the first users to link computers into networks. Although the original networks that led to the Internet are no longer in service, many of the concepts developed in these forerunner systems were adopted for use in today's Internet. There is no one controlling computer; rather any computer can communicate with any other through a series of protocols (standards) for communicating, even among computers of different types.

The most common way to connect to the Internet is through servers, called hosts, provided by Internet service providers (ISPs). An ISP communicates with other servers by sending "packets" of information with data formatted using a Transmission Control Protocol (TCP) or an Internet protocol (IP) that identifies the sending and receiving address as well as control information that allows the packets to be routed. Many other protocols are used within the Internet for other tasks. Other ways of connecting to the Internet are through LANs or commercial on-line services.

Some of the important applications for the Internet are

- Information search and retrieval
- Education
- Products and services
- Electronic mail service
- News

Information Search and Retrieval Many companies and organizations have set up large data files of information that are available to anyone who wants them. For example, in the electronics industry, the specifications for virtually all current electronic devices are available from the manufacturer. In some cases, the information is available as software that can be downloaded directly to the user's computer. Special tools known as *search engines* (such as Google) are available to help you locate specific information.

Education Formal education is available through various organizations including many colleges and universities, where on-line classes can be taken for credit. Even when classes are not available on-line, descriptions about classes and schools are readily available. Informal education is also available and includes training or just looking up information of interest at museums or libraries. Some sites provide animation to illustrate concepts. Many research organizations, including government, have photographs and descriptions of their latest research.

Products and Services The Internet is a convenient source for consumers to shop electronically. Consumers can research manufacturer's sites to obtain specific information about products or services of interest. Many retail businesses compete for your dollar, and competition generally means prices are lower. The Internet also made it possible for special interest products to have a wider audience.

Electronic Mail (e-mail) Electronic mail has revolutionized communication because of its speed and convenience; you can exchange e-mail with people anywhere in the world. The sender and receiver each have a unique e-mail address that works something like your street address. E-mail is an excellent way to send pictures or other information by including an attachment to the basic e-mail (but be aware of possible problems opening an attachment from an unknown sender).

News Of course, one of the attributes of the Internet is speed. News and related information can be quickly disseminated. Many news organizations keep archive files of older news stories so you can look them up days later. All of the information that is available in a newspaper, from the front page to the comics, is available on the Internet.

Review Questions

Answers are at the end of the chapter.

1. What are the major elements or blocks in a computer?
2. What is the difference between RAM and ROM?
3. What are five applications of the Internet?
4. What are peripherals?
5. What is the difference between computer hardware and computer software?

MICROPROCESSORS 11–2

The **microprocessor** is a digital integrated circuit that can be programmed with a series of instructions to perform various operations on data. A microprocessor is the CPU of a computer. It can do arithmetic and logic operations, move data from one place to another, and make decisions based on certain instructions.

In this section, you will learn about the basic elements of a microprocessor.

Basic Elements

A microprocessor consists of several units, each designed for a specific job. The specific units, their design and organization, are called the architecture (do not confuse the term with the VHDL element). The architecture determines the instruction set and the process for executing those instructions. Four basic units that are common to all microprocessors are the arithmetic logic unit (ALU), the instruction decoder, the register array, and the control unit, as shown in Figure 11–3.

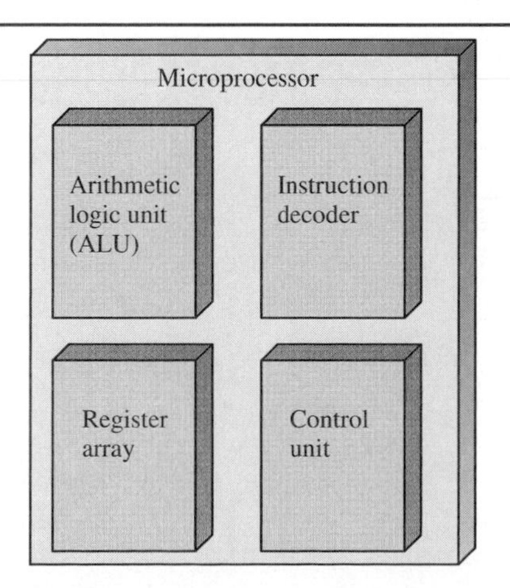

FIGURE 11–3

Arithmetic Logic Unit

The **ALU** is the key processing element of the microprocessor. It is directed by the control unit to perform arithmetic operations (addition, subtraction, multiplication, and division) and logic operations (NOT, AND, OR, and exclusive-OR), as well as many other types of operations. Data for the ALU are obtained from the register array.

Instruction Decoder

The instruction decoder can be considered as part of the ALU, although we are treating it as a separate function in this discussion because the instructions and the decoding of them are key to a microprocessor's operation. The microprocessor accomplishes a given task as directed by programs that consist of lists of instructions stored in memory. The instruction decoder takes each binary instruction in the order in which it appears in memory and decodes it.

Register Array

The register array is a collection of registers that are contained within the microprocessor. During the execution of a program, data and memory addresses are temporarily stored in registers that make up this array. The ALU can access the registers very quickly, making the program run more efficiently. Some registers are classed as general-purpose, meaning they can be used for any purpose dictated by the program. Other registers have specific capabilities and functions and cannot be used as general-purpose registers. Still others are called program invisible registers, used only by the microprocessor and not available to the programmer.

Control Unit

The control unit is "in charge" of the processing of instructions once they are decoded. It provides the timing and control signals for getting data into and out of the microprocessor and for synchronizing the execution of instructions.

Basic Operations

As mentioned, a microprocessor is capable of executing a set of machine language instructions. Every microprocessor has a specific instruction set from which programs can be written. Based on the instructions, a microprocessor does the following three basic operations:

1. Performs arithmetic and logic operations
2. Moves data to and from memory locations and to and from input/output ports
3. Makes decisions and jumps to a new set of instructions based on those decisions

Microprocessor Buses

The three buses mentioned earlier are the connections for microprocessors to allow data, addresses, and instructions to be moved.

The Address Bus

The **address bus** is a "one-way street" over which the microprocessor sends an address code to a memory or other external device. The size or width of the address bus is specified by the number of conductive paths or bits. Early microprocessors had sixteen address lines that could select 65,536 (2^{16}) unique locations in memory. The more bits there are in the address, the higher the number of memory locations that can be accessed. The number of address bits has advanced to the point where the Pentium 4 has 36 address bits and can access over 68 G (68,000,000,000) memory locations.

The Data Bus

The **data bus** is a "two-way street" on which data or instruction codes are transferred into the microprocessor or the result of an operation or computation is sent out. The original microprocessors had 8-bit data buses. Today's microprocessors have up to 64-bit data buses.

The Control Bus

The **control bus** is used by the microprocessor to coordinate its operations and to communicate with external devices. The control bus has signals that enable either a memory or an input/output operation at the proper time to read or write data. Control bus lines are also used to insert special wait states for slower devices and prevent bus contention, a condition that can occur if two or more devices try to communicate at the same time.

Technological Progress

The first microprocessor, the Intel 4004, was introduced in 1971. Basically, all it could do was add and subtract only 4 bits at a time. In 1974, the Intel 8080 became the first microprocessor to be used as the CPU in a computer. The 8080 chip had 6,000 transistors, an 8-bit data bus, and it ran at a clock frequency of 2 MHz. The 8080 could perform about 0.64 million instructions per second (MIPS). The Intel family has evolved from the 8080 through several different processors to the Pentium 4. This latest microprocessor (it may not be at the time you are reading this) has about 42,000,000 transistors on the chip and a 64-bit data bus. It runs at clock frequencies of up to over 3 GHz and it can do approximately 1,700 MIPS. The instruction sets have also changed drastically, but the Pentium 4 can execute any instruction code that ran on the 8086, the 1979 device that came after the 8080.

The number of transistors available has a tremendous impact on the performance and the types of things that a microprocessor can do. For example, the large number of transistors on a chip has made a technology called *pipelining* possible. Basically, pipelining allows more than one instruction to be in the process of execution at one time. Also, modern microprocessors have multiple instruction decoders, each with its own pipeline. This allows several streams of instructions to be processed simultaneously.

Microprocessor Programming

All microprocessors work with an instruction set that implements the basic operations previously discussed. The Pentium, for example, has hundreds of variations of its instruction set divided into seven basic groups.

- Data transfer
- Arithmetic and logic
- Bit manipulation
- Loops and jumps
- Strings
- Subroutines and interrupts
- Control

Each instruction consists of a group of bits (1s and 0s) that is decoded by the microprocessor before being executed. These binary code instructions are called machine language and are all that the microprocessor recognizes. The first computers were programmed by actually writing instructions in binary code, which was a tedious job and prone to error. This primitive method of programming in binary code has evolved to a higher form where coded instructions are represented by English-like words to form what is known as assembly language. This will be discussed further in Section 11–7.

SAFETY NOTE

Five rules should be observed when using hand or power tools:

- Keep all tools in good condition with regular maintenance.
- Use the right tool for the job.
- Examine each tool for damage before use and do not use damaged tools.
- Operate tools according to the manufacturers' instructions.
- Use the right personal protective equipment.

Review Questions

6. What are the four basic elements in a microprocessor?
7. What are the three types of buses in a microprocessor?
8. What function does a microprocessor perform in a computer?
9. What are the three basic operations that a microprocessor performs?
10. What is pipelining?

11–3 SEMICONDUCTOR MEMORY ORGANIZATION

Memories are a crucial part of a computer and are used to store binary data. Generally, when we refer to computer *memory*, we mean the semiconductor memories such as RAM and ROM. The term *storage* is generally used to refer the internal hard disk and the removable storage media such as CD-ROM and CD-RW that are used for large amounts of data (mass storage).

In this section, you will learn about semiconductor memories.

Semiconductor RAM memories generally use either latches or capacitors as the basic memory element and are classified as volatile because data is lost when the power is off. Semiconductor ROM memories generally use fuse, antifuse, or E^2CMOS technology (just as in PLDs) for their basic memory element and are classified as nonvolatile because data is retained when power is off. Another type of memory called flash memory is often used for nonvolatile storage.

Units of Binary Data: Bits, Bytes, Nibbles, and Words

As a rule, memories store data in units that have from one to eight bits. The smallest unit of binary data, as you know, is the bit. In many applications, data are handled in an 8-bit unit called a **byte** or in multiples of 8-bit units. The byte can be split into two 4-bit units that are called *nibbles*. A complete unit of information is called a **word** and generally consists of one or more bytes. In many computer languages, a word is defined as two bytes. Some memories store data in 9-bit groups; a 9-bit group consists of a byte plus a parity bit.

The Basic Semiconductor Memory Array

Each storage element in a memory can retain either a 1 or a 0 and is called a **cell**. Memories are made up of arrays of cells, as shown in Figure 11–4 using 64 cells for illustration. Each block in the memory array represents one storage cell, and its location can be identified by specifying a row and a column.

The 64-cell array can be organized in several ways based on units of data. Figure 11–4(a) shows an 8×8 array, which can be viewed as either a 64-bit memory or an 8-byte memory. Part (b) shows a 16×4 array, which is a 16-nibble memory, and part (c) shows a 64×1 array which is a 64-bit memory. A memory is identified by the number of words it can store times the word size. For example, a $16k \times 8$ memory can store 16,384 words of eight bits each. The inconsistency here is common in memory terminology. The actual number of words is always a power of 2, which, in this case, is $2^{14} = 16{,}384$. However, it is common practice to state the number to the nearest thousand, in this case, 16k.

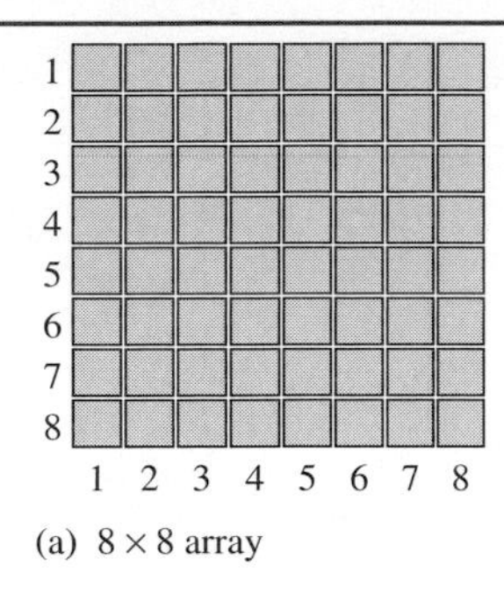

(a) 8 × 8 array

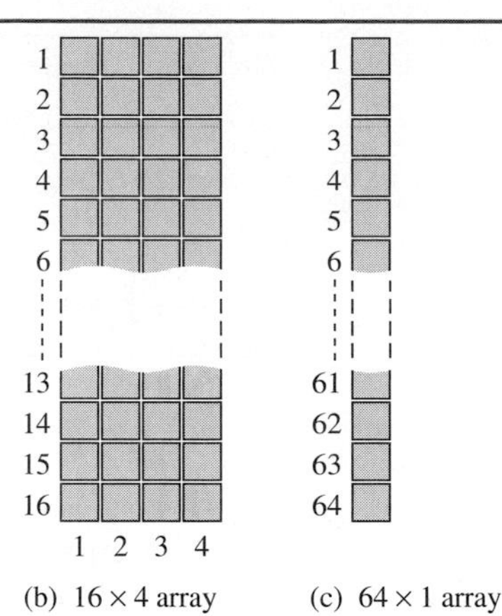

(b) 16 × 4 array (c) 64 × 1 array

FIGURE 11–4

A 64-cell memory array organized in three different ways.

Memory Address and Capacity

The location of a unit of data in a memory array is called its **address**. For example, in Figure 11–5(a), the address of a bit in the array is specified by the row and column as shown. In Figure 11–5(b), the address of a byte is specified only by the row. So, as you can see, the address depends on how the memory is organized into units of data. Personal computers have random-access memories organized in bytes. This means that the smallest group of bits that can be addressed is eight.

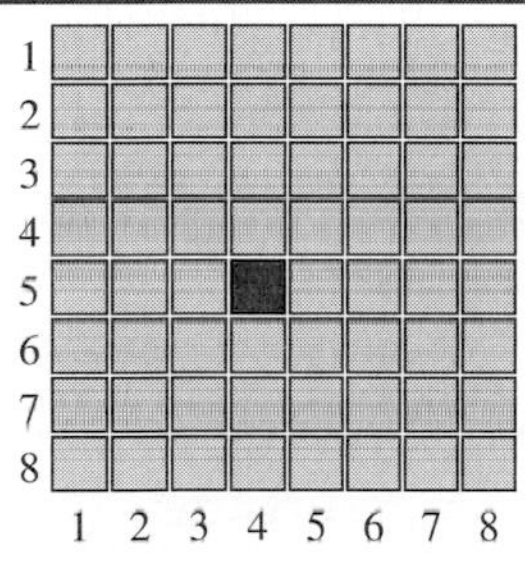

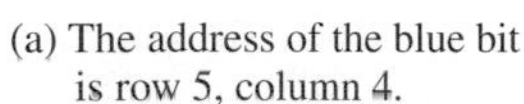

(a) The address of the blue bit is row 5, column 4.

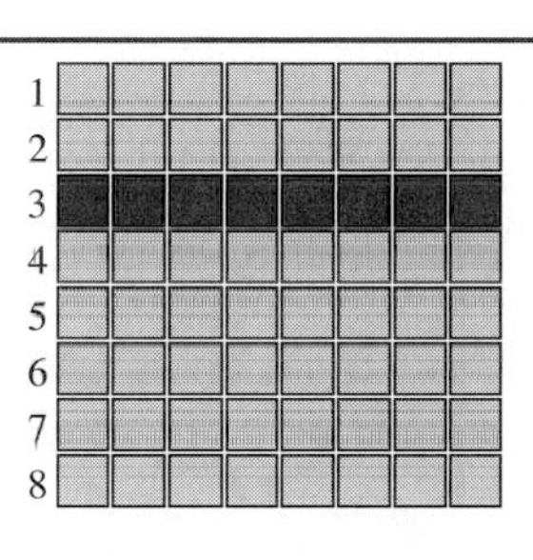

(b) The address of the blue byte is row 3.

FIGURE 11–5

Examples of memory address.

The **capacity** of a memory is the total number of data units that can be stored. For example, in the bit-organized memory array in Figure 11–5(a), the capacity is 64 bits. In the byte-organized memory array in Figure 11–5(b), the capacity is 8 bytes, which is also 64 bits. Computer memories typically have 256 MB or more of internal memory.

Basic Memory Operations

Since a memory stores binary data, data must be put into the memory and data must be copied from the memory when needed. The **write** operation puts data into a specified address in the memory, and the **read** operation copies data out of a specified address in the memory. The addressing operation, which is part of both the write and the read operations, selects the specified memory address.

Data units go into the memory during a write operation and come out of the memory during a read operation on a data bus. As indicated in Figure 11–6, the data bus is bidirectional, which means that data can go in either direction (into the memory or out of the memory). In this case of a byte-organized memory, the data bus has at least eight lines so that all eight bits in a selected address are transferred in parallel. For a write or a read operation, an

FIGURE 11–6

Block diagram of a memory showing address bus, address decoder, bidirectional data bus, and read/write inputs.

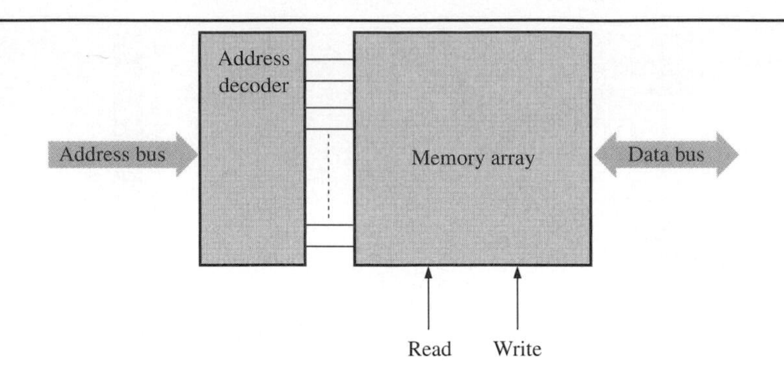

address is selected by placing a binary code representing the desired address on an address bus. The address code is decoded internally, and the appropriate address is selected. The number of lines in the address bus depends on the capacity of the memory. For example, a 15-bit address code can select 32,768 locations (2^{15}) in the memory, a 16-bit address code can select 65,536 locations (2^{16}) in the memory, and so on. In computers a 32-bit address bus is common; it can select 4,294,967,296 locations (2^{32}), expressed as 4G.

The Write Operation

A simplified write operation is illustrated in Figure 11–7. To store a byte of data in the memory, a code held in the address register is placed on the address bus. Once the address code is on the bus, the address decoder decodes the address and selects the specified location in the memory. The memory then gets a write command, and the data byte held in the data register is placed on the data bus and stored in the selected memory address, thus completing the write operation. When a new data byte is written into a memory address, the current data byte stored at that address is overwritten (replaced with a new data byte).

FIGURE 11–7

Illustration of a simplified write operation.

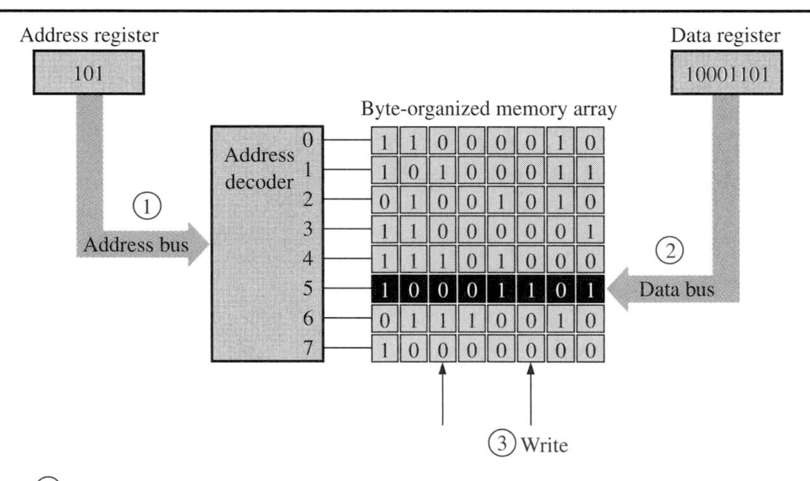

① Address code 101 is placed on the address bus and address 5 is selected.
② Data byte is placed on the data bus.
③ Write command causes the data byte to be stored in address 5, replacing previous data.

The Read Operation

A simplified read operation is illustrated in Figure 11–8. Again, a code held in the address register is placed on the address bus. Once the address code is on the bus, the address decoder decodes the address and selects the specified location in the memory. The memory then gets a read command, and a "copy" of the data byte that is stored in the selected memory address is placed on the data bus and loaded into the data register, thus completing the read operation. When a data byte is read from a memory address, it also remains stored at that address. This is called *nondestructive read.*

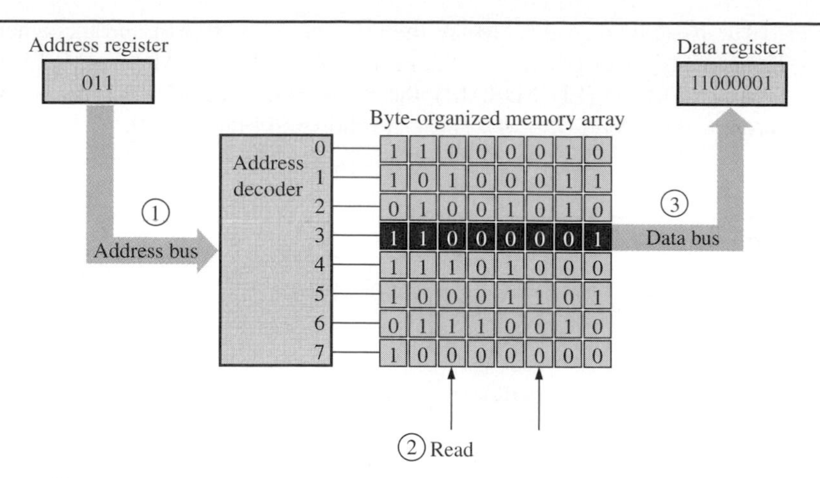

① Address code 011 is placed on the address bus and address 3 is selected.

② Read command is applied.

③ The contents of address 3 is placed on the data bus and shifted into data register. The contents of address 3 is not destroyed by the read operation.

FIGURE 11–8 Illustration of a simplified read operation.

Review Questions

11. What is a memory address?
12. What is the bit capacity of a memory that can store 256 bytes of data?
13. What is a write operation?
14. What is a read operation?
15. How is a given unit of data located in a memory?

RAM 11–4

RAMs are read/write memories in which data can be written into or read from any selected address in any sequence (thus the name *random access*). When a unit of data is written into a selected address in the RAM, the unit of data previously stored at that address is replaced by the new data. When a unit of data is read from a selected address in the RAM, the unit of data remains at the address and is not destroyed by the read operation. Effectively, the unit of data is copied while leaving the original unit of data intact. RAMs are typically used for short-term data storage because stored data cannot be retained when the power is turned off.

In this section, you will learn about several basic types of RAM.

The two categories of RAM are the static RAM (SRAM) and the dynamic RAM (DRAM). Static RAMs use latches or flip-flops as storage elements and can therefore store data indefinitely as long as dc power is applied. Dynamic RAMs use capacitors as storage elements and cannot retain data very long without the capacitors being recharged by a process called *refreshing*. Both SRAMs and DRAMs will lose stored data when dc power is removed and, therefore, are classified as volatile memories.

Data can be read much faster from SRAMs than from DRAMs. However, DRAMs can store much more data than SRAMs for a given physical size and cost because the DRAM cell is much simpler, and more cells can be crammed into a given area than in the SRAM.

The basic types of SRAM are the asynchronous SRAM and the synchronous SRAM. The basic types of DRAM are the Fast Page Mode DRAM (FPM DRAM), the Extended Data Out DRAM (EDO DRAM), the Burst EDO DRAM (BEDO DRAM), and the synchronous DRAM (SDRAM). These are shown in Figure 11–9.

FIGURE 11–9 The RAM family.

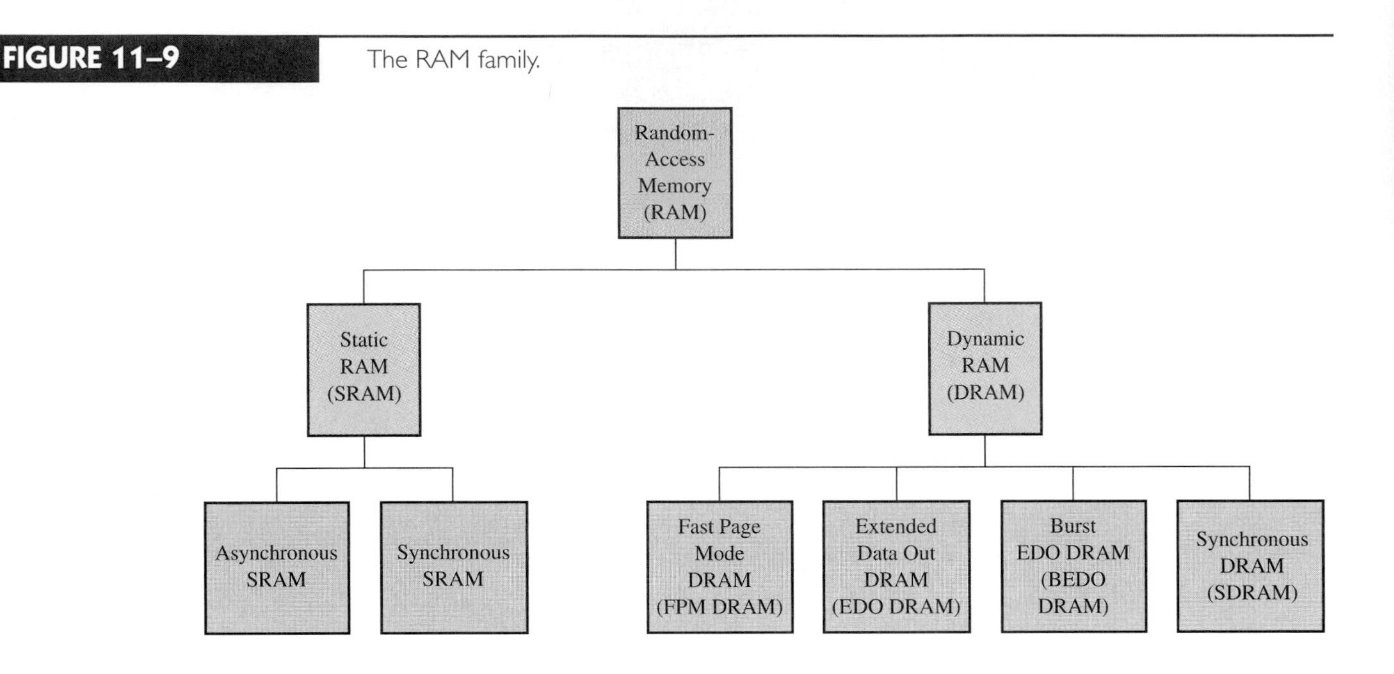

Static RAMs (SRAMs)

Storage Cell All static RAMs are characterized by latch or flip-flop memory cells that are typically implemented in integrated circuits. As long as dc power is applied to a static memory cell, it can retain a 1 or 0 state indefinitely. If power is removed, the stored data bit is lost.

Static Memory Cell Array The storage cells in a SRAM are organized in rows and columns, as illustrated in Figure 11–10 for the case of an $n \times 4$ array. All the cells in a row share the same Row Select line. Each set of Data In and Data Out lines go to each cell in a given column and are connected to a single data line that serves as both an input and output (Data I/O) through the data input and data output buffers. For simplicity, think of each cell as a gated D latch like the one you studied in Chapter 7.

To store a data unit, in this case 4 bits (called a nibble), into a given row of cells in the memory array, the Row Select line is taken to its active state and four data bits are placed on the Data I/O lines. The Write line is then taken to its active state, which causes each data bit to be stored in the selected cell in the associated column. To read a data unit, the Read line is taken to its active state, which causes the four data bits stored in the selected row to appear on the Data I/O lines.

Asynchronous SRAM

An asynchronous SRAM is one in which the operation is not synchronized with a system clock. To illustrate the general organization of a SRAM, a 32k $\times$ 8 bit memory is used. A logic symbol for this memory is shown in Figure 11–11.

In the READ mode, the eight data bits that are stored in a selected address appear on the data output lines. In the WRITE mode, the eight data bits that are applied to the data input lines are stored at a selected address. The data input and data output (I/O_0 through I/O_7) share the same lines. During READ, they act as output lines (O_0 through O_7) and during WRITE they act as input lines (I_0 through I_7).

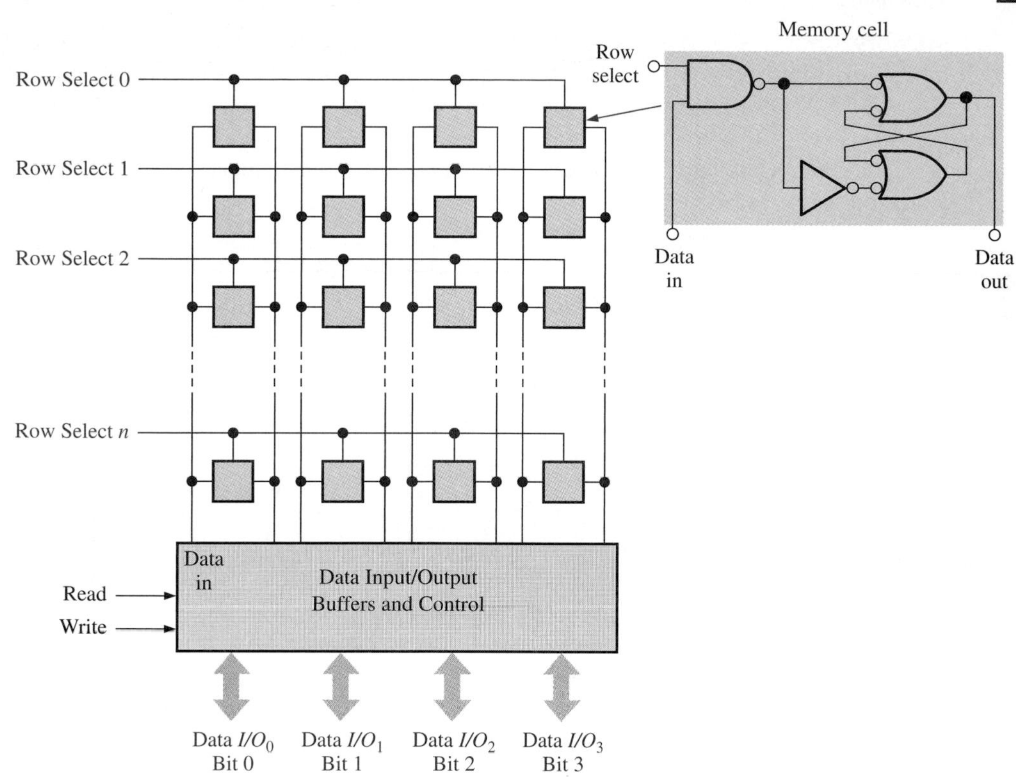

FIGURE 11–10 Basic SRAM array.

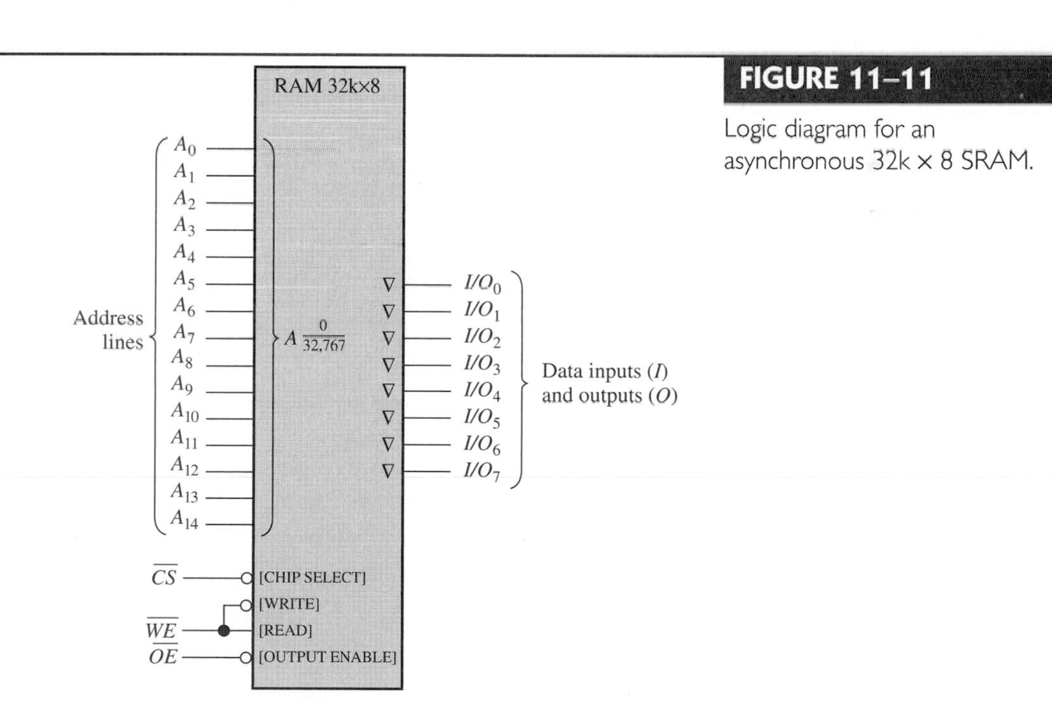

FIGURE 11–11 Logic diagram for an asynchronous 32k × 8 SRAM.

Tristate Outputs and Buses Tristate buffers in a memory allow the data lines to act as either input or output lines and connect the memory to the data bus in a computer. These buffers have three output states: HIGH (1), LOW (0), and HIGH-Z (open). Tristate outputs are indicated on logic symbols by a small inverted triangle (∇), as shown in Figure

11–11, and are used for compatibility with bus structures found in microprocessor-based systems.

Memory Array SRAM chips can be organized in single bits, nibbles (4 bits), bytes (8 bits), or multiple bytes (16, 24, 32, and 64 bits). Figure 11–12 shows the organization of a 32k × 8 SRAM for illustration. The memory cell array is arranged in 256 rows and 128 columns, each with 8 bits, as shown in part (a). There are actually $2^{15} = 32{,}768$ addresses and each address contains 8 bits. The capacity of this example memory is 32,768 bytes, typically expressed as 32 kB (kilobyte).

FIGURE 11–12 Basic organization of an asynchronous 32k × 8 SRAM.

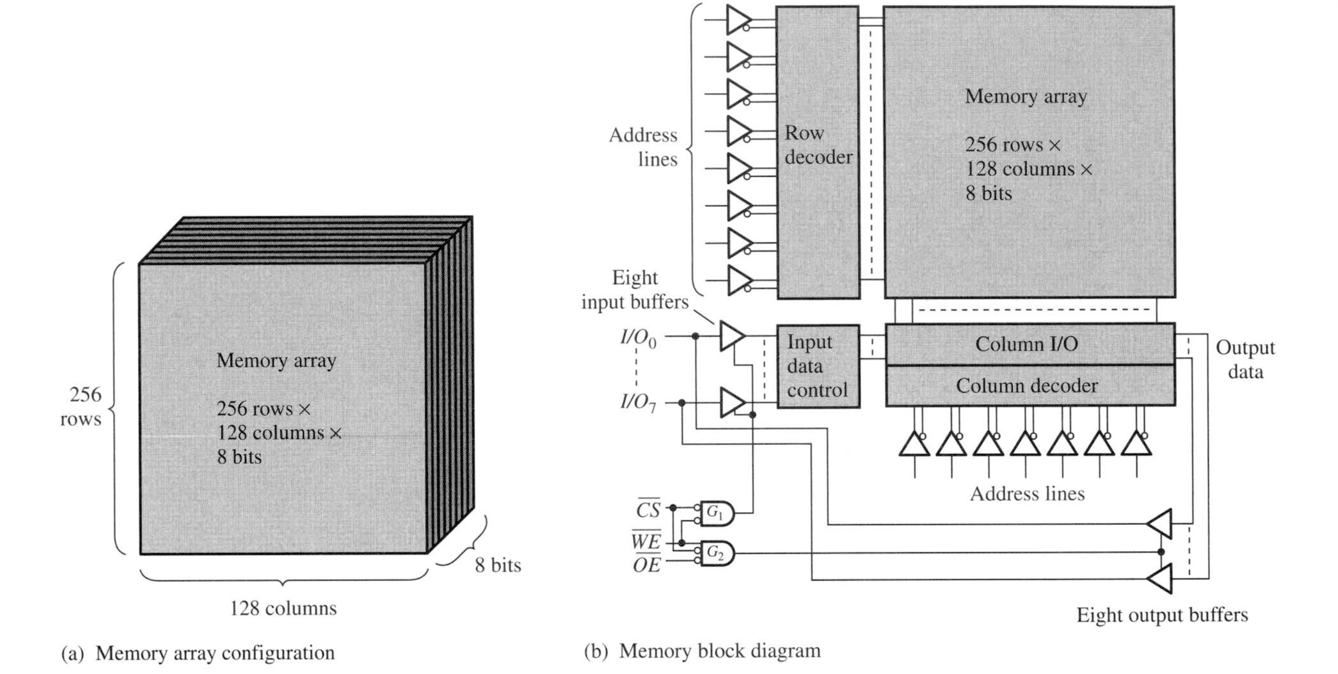

The SRAM in Figure 11–12(b) works as follows. First, the chip select, $\overline{CS}$, must be LOW for the memory to operate. Eight of the fifteen address lines are decoded by the row decoder to select one of the 256 rows. Seven of the fifteen address lines are decoded by the column decoder to select one of the 128 8-bit columns.

READ In the READ mode, the write enable input, $\overline{WE}$, is HIGH (disabled) and the output enable, $\overline{OE}$, is LOW. The input tristate buffers are disabled by gate G_1, and the column output tristate buffers are enabled by gate G_2. Therefore, the eight data bits from the selected address are routed through the column I/O to the data lines (I/O_0 though I/O_7), which are acting as data output lines.

WRITE In the WRITE mode, $\overline{WE}$ is LOW and $\overline{OE}$ is HIGH. The input buffers are enabled by gate G_1, and the output buffers are disabled by gate G_2. Therefore, the eight input data bits on the data lines are routed through the input data control and the column I/O to the selected address and stored.

Synchronous SRAM

Unlike the asynchronous SRAM, a synchronous SRAM is synchronized with the system clock. For example, in a computer system, the synchronous SRAM operates with the same clock signal that operates the microprocessor so that the microprocessor and memory are synchronized for faster operation.

The fundamental concept of the synchronous feature of a SRAM can be shown with Figure 11–13, which is a simplified block diagram of a 32k × 8 memory for purposes of illustration. The synchronous SRAM is similar to the asynchronous SRAM in terms of the memory array, address decoder, and read/write and enable inputs. The basic difference is that the synchronous SRAM uses clocked registers to synchronize all inputs with the system clock. The address, the read/write input, the chip enable, and the input data are all latched into their respective registers on an active clock pulse edge. Once this information is latched, the memory operation is in sync with the clock.

FIGURE 11–13 A basic block diagram of a synchronous SRAM.

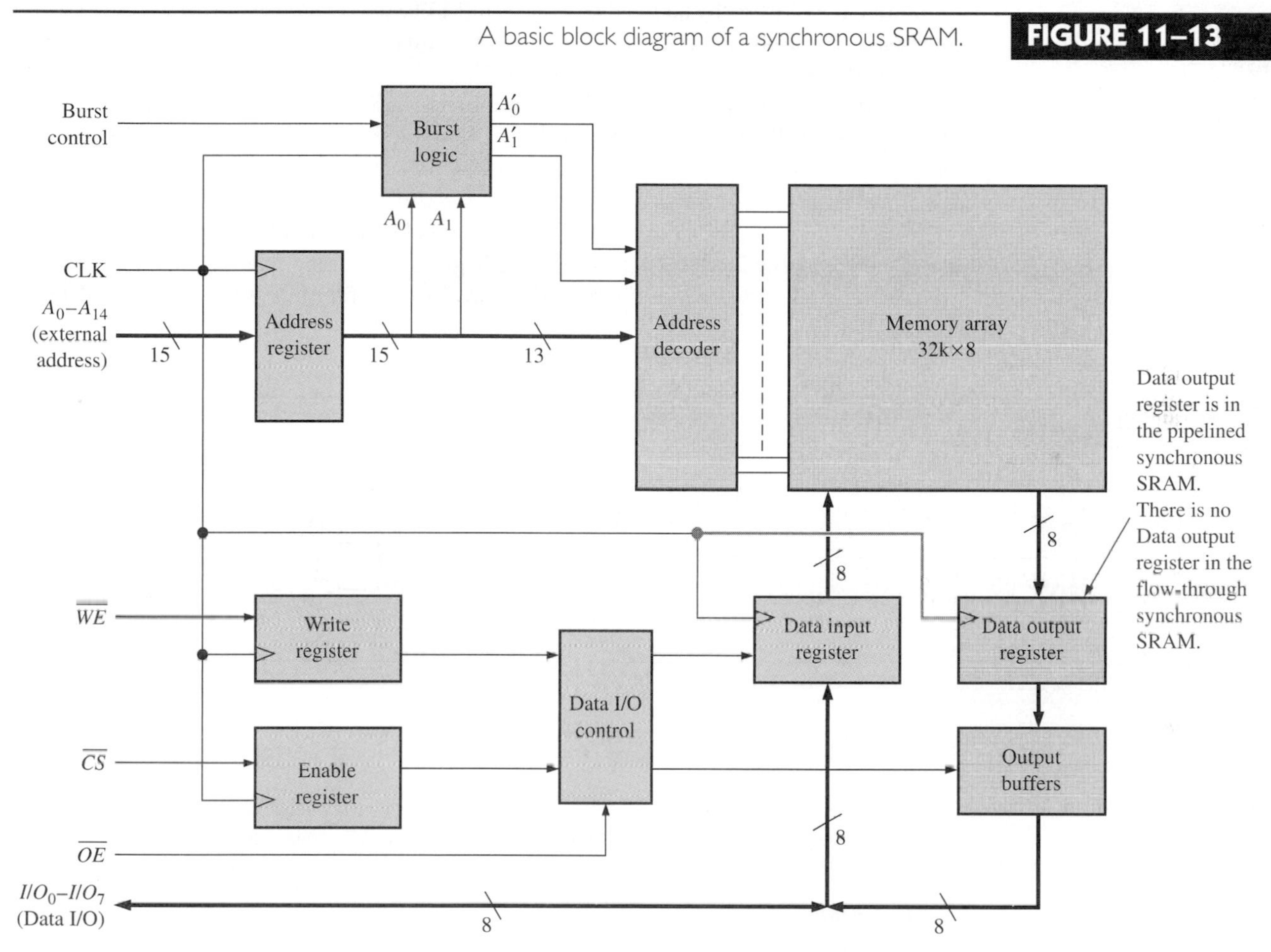

Data output register is in the pipelined synchronous SRAM. There is no Data output register in the flow-through synchronous SRAM.

For the purpose of simplification, a notation for multiple parallel lines or bus lines is shown in Figure 11–13, as an alternative to drawing each line separately. This notation was also used for PLDs. A set of parallel lines can be indicated by a single heavy line with a slash and the number of separate lines in the set. For example, the following notation represents a set of 8 parallel lines:

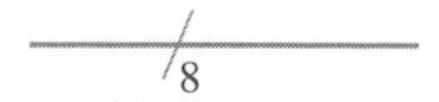

The address bits A_0 through A_{14} are latched into the Address register on the positive edge of a clock pulse. On the same clock pulse, the state of the write enable $(\overline{WE})$ line and chip select $(\overline{CS})$ are latched into the Write register and the Enable register respectively. These are one-bit registers or simply flip-flops. Also, on the same clock pulse the input data are latched into the Data input register for a Write operation, and data in a selected memory address are latched into the Data output register for a Read operation, as determined by the Data I/O control based on inputs from the Write register, Enable register, and the Output enable $(\overline{OE})$.

Two basic types of synchronous SRAM are the flow-through and the pipelined. The *flow-through* synchronous SRAM does not have a Data output register, so the output data flow asynchronously to the data I/O lines through the output buffers. The *pipelined* synchronous SRAM has a Data output register, as shown in Figure 11–13, so the output data are synchronously placed on the data I/O lines.

The Burst Feature As shown in Figure 11–13, synchronous SRAMs normally have an address burst feature, which allows the memory to read or write at up to four locations using a single address. When an external address is latched in the address register, the two lowest-order address bits. A_0 and A_1, are applied to the burst logic. This produces a sequence of four internal addresses by adding 00, 01, 10, and 11 to the two lowest-order address bits on successive clock pulses. The sequence always begins with the base address, which is the external address held in the address register.

The address burst logic in a typical synchronous SRAM consists of a binary counter and exclusive-OR gates, as shown in Figure 11–14. For 2-bit burst logic, the internal burst address sequence is formed by the base address bits A_2–A_{14} plus the two burst address bits A'_1 and A'_0.

To begin the burst sequence, the counter is in its 00 state and the two lowest-order address bits are applied to the inputs of the XOR gates. Assuming that A_0 and A_1 are both 0, the internal address sequence in terms of its two lowest-order bits is 00, 01, 10, and 11.

FIGURE 11–14

Address burst logic.

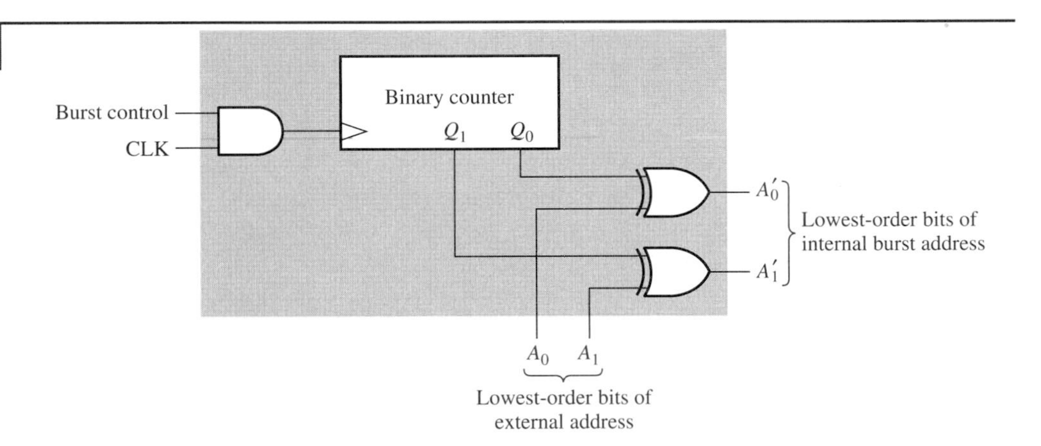

Cache Memory

Cache memories were briefly discussed in Section11–1. One of the major applications of SRAMs is in cache memories in computers. *Cache memory* is a relatively small, high-speed memory that stores the most recently used instructions or data from the larger but slower main memory. Cache memory can also use dynamic RAM (DRAM), which is covered next. Typically, SRAM is several times faster than DRAM. Overall, a cache memory gets stored information to the microprocessor much faster than if only high-capacity DRAM is used. Cache memory is basically a cost-effective method of improving system performance without having to resort to the expense of making all of the memory faster.

The concept of cache memory is based on the idea that computer programs tend to get instructions or data from one area of main memory before moving to another area. Basically, the cache controller "guesses" which area of the slow dynamic memory the CPU (central-processing unit) will need next and moves it to the cache memory so that it is ready when needed. If the cache controller guesses right, the data are immediately available to the microprocessor. If the cache controller guesses wrong, the CPU must go to the main memory and wait much longer for the correct instructions or data. Fortunately, the cache controller is right most of the time.

There are many analogies that can be used to describe a cache memory, but comparing it to a home refrigerator is perhaps the most effective. A home refrigerator can be thought of as a "cache" for certain food items while the supermarket is the main memory where all foods are kept. Each time you want something to eat or drink, you can go to the refrigerator (cache) first to see if the item you want is there. If it is, you save a lot of time. If it is not there, then you have to spend extra time to get it from the supermarket (main memory).

A first-level cache (L1 cache) is usually integrated into the microprocessor chip and has a very limited storage capacity. L1 cache is also known as *primary cache.* A second-level cache (L2 cache) is a separate memory chip or set of chips external to the processor and usually has a larger storage capacity than an L1 cache. L2 cache is also known as *secondary cache.* Some systems may have higher-level caches (L3, L4, etc.), but L1 and L2 are the most common.

Dynamic RAMs (DRAMs)

The DRAM memory cell typically consists of one transistor and a capacitor and is much simpler than the SRAM cell. This allows much greater densities in DRAMs and results in greater bit capacities for a given chip area, although much slower access time.

Because charge stored in a capacitor will leak off, the DRAM cell requires a frequent refresh operation to preserve the stored data bit. This requirement results in more complex circuitry than in a SRAM. Several features common to most DRAMs are now discussed using a generic 1M × 1 bit DRAM as an example.

Address Multiplexing DRAMs use a technique called *address multiplexing* to reduce the number of address lines. Figure 11–15 shows the block diagram of a 1,048,576-bit

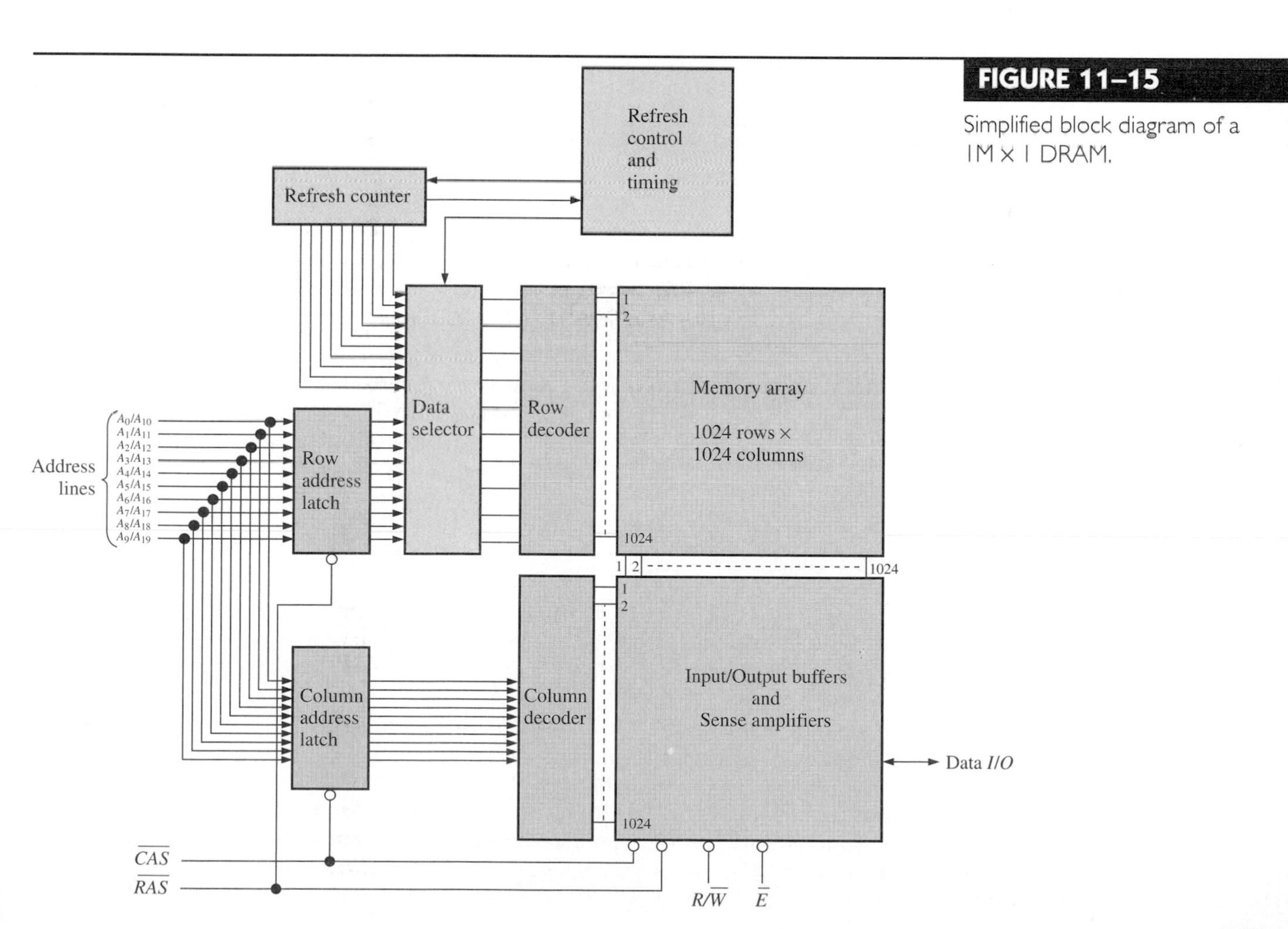

FIGURE 11–15

Simplified block diagram of a 1M × 1 DRAM.

(1 Mbit) DRAM with a 1M × 1 organization. We will focus on the beige blocks to illustrate address multiplexing. The blue blocks represent the refresh logic.

The ten address lines are time multiplexed at the beginning of a memory cycle by the row address select ($\overline{RAS}$) and the column address select ($\overline{CAS}$) into two separate 10-bit address fields. First, the 10-bit row address is latched into the row address latch. Next, the 10-bit column address is latched into the column address latch. The row address and the column address are decoded to select one of the 1,048,576 addresses ($2^{20} = 1,048,576$) in the memory array.

Read and Write Cycles At the beginning of each read or write memory cycle, $\overline{RAS}$ and $\overline{CAS}$ go active (LOW) to multiplex the row and column addresses into the latches and decoders. For a read cycle, the $R/\overline{W}$ input is HIGH. For a write cycle, the $R/\overline{W}$ input is LOW.

Fast Page Mode In the normal read or write cycle described previously, the row address for a particular memory location is first loaded by an active-LOW $\overline{RAS}$ and then the column address for that location is loaded by an active-LOW $\overline{CAS}$. The next location is selected by another $\overline{RAS}$ followed by a $\overline{CAS}$, and so on.

A "page" is a section of memory available at a single row address and consists of all the columns in a row. Fast page mode allows fast successive read or write operations at each column address in a selected row. A row address is first loaded by $\overline{RAS}$ going LOW and remaining LOW while $\overline{CAS}$ is toggled between HIGH and LOW. A single row address is selected and remains selected while $\overline{RAS}$ is active. Each successive $\overline{CAS}$ selects another column in the selected row. So, after a fast page mode cycle, all of the addresses in the selected row have been read from or written into, depending on $R/\overline{W}$. For example, a fast page mode cycle for the DRAM in Figure 11–15 requires $\overline{CAS}$ to go active 1024 times for each row selected by $\overline{RAS}$.

Refresh Cycles As you know, DRAMs are based on capacitor charge storage for each bit in the memory array. This charge degrades (leaks off) with time and temperature, so each bit must be periodically refreshed (recharged) to maintain the correct bit state. Typically, a DRAM must be refreshed every 8 ms to 16 ms, although for some devices the refresh period can exceed 100 ms.

A read operation automatically refreshes all the addresses in the selected row. However, in typical applications, you cannot always predict how often there will be a read cycle and so you cannot depend on a read cycle to occur frequently enough to prevent data loss. Therefore, special refresh cycles must be implemented in DRAM systems.

Burst refresh and *distributed refresh* are the two basic refresh modes for refresh operations. In burst refresh, all rows in the memory array are refreshed consecutively each refresh period. For a memory with a refresh period of 8 ms, a burst refresh of all rows occurs once every 8 ms. The normal read and write operations are suspended during a burst refresh cycle. In distributed refresh, each row is refreshed at intervals interspersed between normal read or write cycles. The memory in Figure 11–15 has 1024 rows. As an example, for an 8 ms refresh period, each row must be refreshed every 8 ms/1024 = 7.8 μs when distributed refresh is used.

Types of DRAMs

FPM DRAM Fast page mode operation was described earlier. This type of DRAM traditionally has been the most common and has been the type used in computers until the development of the EDO DRAM. Recall that a page in memory is all of the column addresses contained within one row address.

The basic idea of the FPM DRAM is based on the probability that the next several memory addresses to be accessed are in the same row (on the same page). Fortunately, this happens a large percentage of the time. FPM saves time over pure random accessing because in FPM the row address is specified only once for access to several successive column addresses whereas for pure random accessing, a row address is specified for each column address.

EDO DRAM The Extended Data Output DRAM, sometimes called *hyper page mode DRAM,* is similar to the FPM DRAM. The key difference is that the $\overline{CAS}$ signal in the EDO DRAM does not disable the output data when it goes to its nonasserted state because the valid data from the current address can be held until $\overline{CAS}$ is asserted again. This means that the next column address can be accessed before the external system accepts the current valid data. The idea is to speed up the access time.

BEDO DRAM The Burst Extended Data Output DRAM is an EDO DRAM with address burst capability. Recall from the discussion of the synchronous burst SRAM that the address burst feature allows up to four addresses to be internally generated from a single external address, which saves some access time. This same concept applies to the BEDO DRAM.

SDRAM Faster DRAMs are needed to keep up with the ever-increasing speed of microprocessors. The Synchronous DRAM is one way to accomplish this. Like the synchronous SRAM discussed earlier, the operation of the SDRAM is synchronized with the system clock, which also runs the microprocessor in a computer system. The same basic ideas described in relation to the synchronous SRAM also apply to the SDRAM.

This synchronized operation makes the SDRAM totally different from the other DRAM types, which are asynchronous. With asynchronous memories, the microprocessor must wait for the DRAM to complete its internal operations. However, with synchronous operation, the DRAM latches addresses, data, and control information from the processor under control of the system clock. This allows the processor to handle other tasks while the memory read or write operations are in progress.

Memory Modules

RAMs are commonly supplied as single in-line memory modules (SIMMs), dual in-line memory modules (DIMMs), or rambus in-line memory modules (RIMM). SIMMs and DIMMs are small circuit boards on which memory chips (ICs) are mounted with the inputs and outputs connected to an edge connector on the bottom of the board. DIMMs are generally faster, but they can only be installed in machines that are designed for them.

Two classifications of SIMMs are the 30-pin and the 72-pin. These are illustrated in Figure 11–16. Although available memory capacities for SIMMs vary anywhere from 256 kB to 32 MB, the key difference in the two pin configurations is the size of the data path. Generally, 30-pin SIMMs are designed for 8-bit data buses, and more SIMMs are required for handling larger buses. The 72-pin SIMMs can accommodate a 32-bit data bus, so for 64-bit data buses a pair of SIMMs is required.

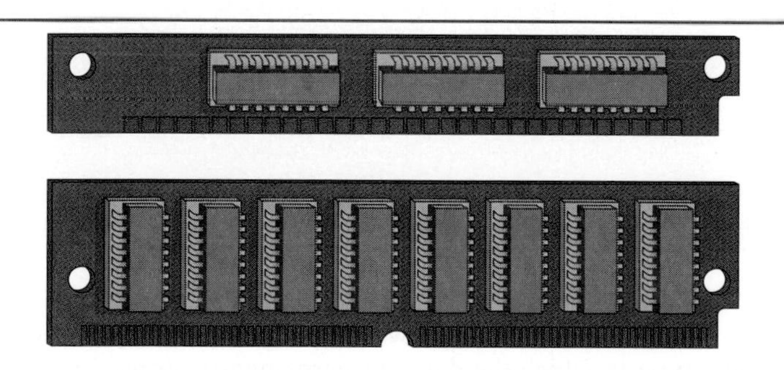

FIGURE 11–16

30-pin and 72-pin SIMMs.

DIMMs look similar to SIMMs but provide an increase in memory density with only a relatively slight increase in physical size. The key difference is that DIMMs distribute the input and output pins on both sides of the printed circuit (PC) board, whereas SIMMs use only one side. Common DIMM configurations provide up to 168-pins that accommodate both 32-bit and 64-bit data paths. Generally, DIMM capacities range from 4 MB to 512 MB.

As you can see, the example ROM in Figure 11–19 is organized into 16 addresses, each of which stores 8 data bits. Thus, it is a 16 × 8 (16-by-8) ROM, and its total capacity is 128 bits or 16 bytes.

Internal ROM Organization

Most IC ROMs have a more complex internal organization than that in the basic simplified example just presented. To illustrate how an IC ROM is structured, let's use a 1024-bit device with a 256 × 4 organization. The logic symbol is shown in Figure 11–20. When any one of 256 binary codes (eight bits) is applied to the address lines, four data bits appear on the outputs if the chip enable inputs are LOW. (256 addresses require eight address lines.)

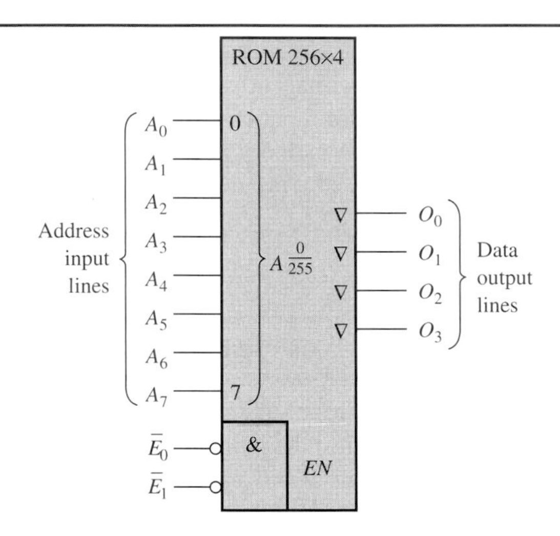

FIGURE 11–20

A 256 × 4 ROM logic symbol. The $A\frac{0}{255}$ designator means that the 8-bit address code selects addresses 0 through 255. Recall that the symbol ∇ indicates a tristate output.

Although the 256 × 4 organization of this device implies that there are 256 rows and 4 columns in the memory array, this is not actually the case. The memory cell array is actually a 32 × 32 matrix (32 rows and 32 columns), as shown in the block diagram in Figure 11–21.

The ROM in Figure 11–21 works as follows: Five of the eight address lines (A_0 through A_4) are decoded by the row decoder (often called the *Y* decoder) to select one of the 32 rows. Three of the eight address lines (A_5 through A_7) are decoded by the column decoder (often called the *X* decoder) to select four of the 32 columns. Actually, the column decoder consists of four 1-of-8 decoders (data selectors), as shown in Figure 11–21.

The result of this structure is that when an 8-bit address code (A_0 through A_7) is applied, a 4-bit data word appears on the data outputs when the chip enable lines ($\overline{E}_0$ and $\overline{E}_1$) are LOW to enable the output buffers. This type of internal organization (architecture) is typical of IC ROMs of various capacities.

Review Questions

21. What is a UV EPROM?
22. What is an E^2PROM?
23. What is the bit storage capacity of a ROM with a 512 × 8 organization?
24. What are the types of read-only memories?
25. How many address lines are required for a 2048-bit memory organized as a 256 × 8 memory?

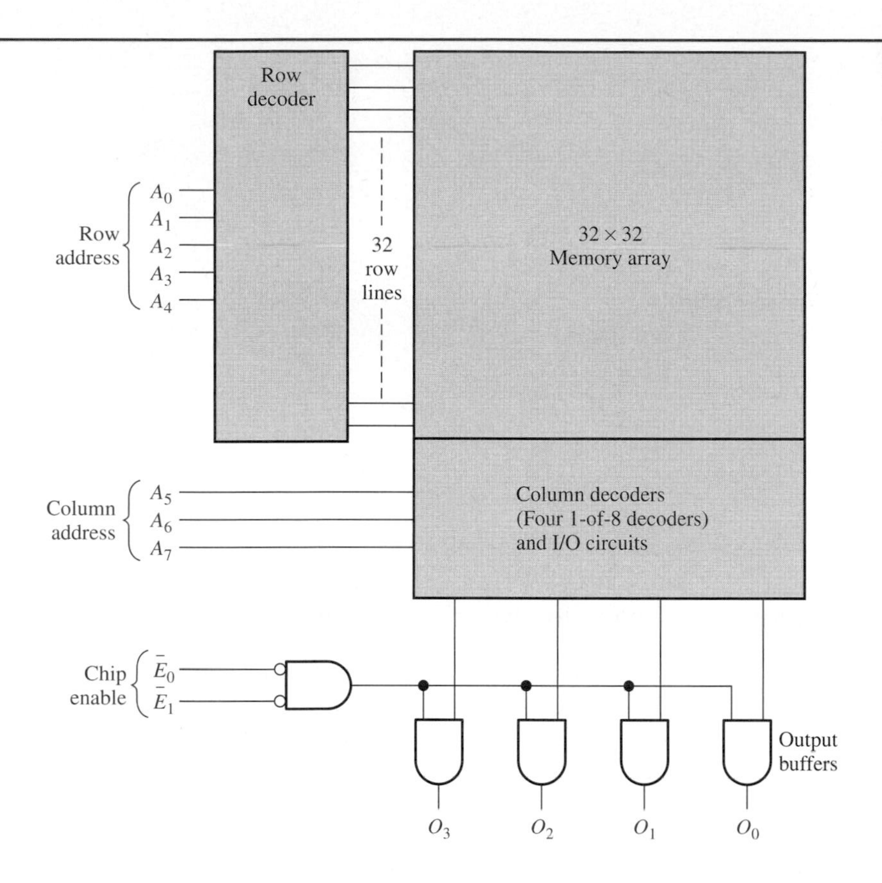

FIGURE 11–21

A typical ROM. This particular example is a 1024-bit ROM with a 256 × 4 organization based on a 32 × 32 array.

MASS STORAGE DEVICES 11–6

Mass storage media are important in computer applications where they are used for non-volatile storage of large amounts of data.

In this section, you will learn the basics of magnetic disks, magneto-optical disks, optical disks, and flash memory.

Magnetic Storage

Hard Disks

Computers use hard disks as the internal mass storage media. Hard disks are rigid "platters" made of aluminum alloy or a mixture of glass and ceramic covered with a magnetic coating. Hard disk drives mainly come in two diameter sizes, 5.25 in. and 3.5 in. although 2.5 in. and 1.75 in. are also available. A hard disk drive is hermetically sealed to keep the disks dust-free.

Typically, two or more platters are stacked on top of each other on a common shaft or spindle that turns the assembly at several thousand rpm. There is a separation between each disk to allow for a magnetic read/write head that is mounted on the end of an actuator arm, as shown in Figure 11–22. There is a read/write head for both sides of each disk since data are recorded on both sides of the disk surface. The drive actuator arm synchronizes all the read/write heads to keep them in perfect alignment as they "fly" across the disk surface with a separation of only a fraction of a millimeter from the disk. A small dust particle could cause a head to "crash," causing damage to the disk surface.

FIGURE 11–22

A hard disk drive.

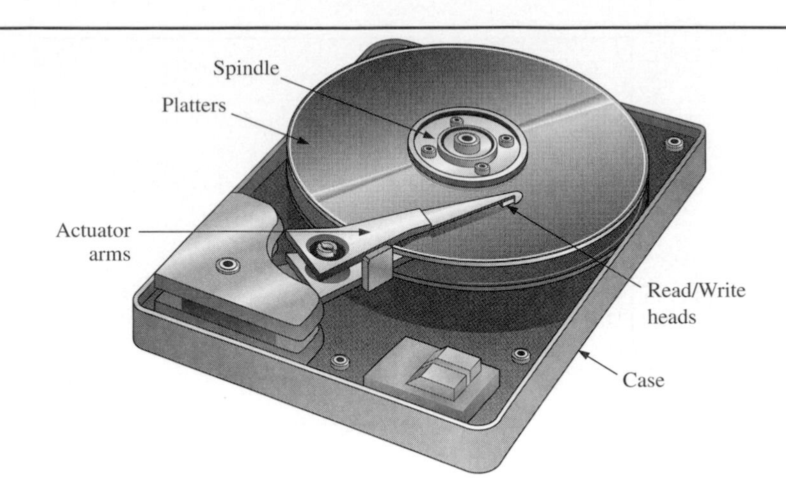

Basic Read/Write Head Principles The hard drive is a random-access device because it can retrieve stored data anywhere on the disk in any order. A simplified diagram of the magnetic surface read/write operation is shown in Figure 11–23. The direction or polarization of the magnetic domains on the disk surface is controlled by the direction of the magnetic flux lines (magnetic field) produced by the write head according to the direction of a current pulse in the winding. This magnetic flux magnetizes a small spot on the disk surface in the direction of the magnetic field. A magnetized spot of one polarity represents a binary 1, and one of the opposite polarity represents a binary 0. Once a spot on the disk surface is magnetized, it remains until written over with an opposite magnetic field.

When the magnetic surface passes a read head, the magnetized spots produce magnetic fields in the read head, which induce voltage pulses in the winding. The polarity of these pulses depends on the direction of the magnetized spot and indicates whether the stored bit is a 1 or a 0. The read and write heads are usually combined in a single unit.

FIGURE 11–23

Simplified read/write head operation.

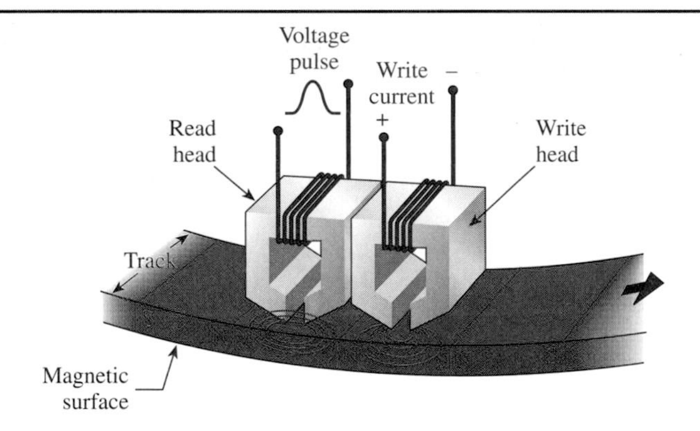

Hard Disk Format A hard disk is organized or formatted into tracks and sectors, as shown in Figure 11–24(a). Each track is divided into a number of sectors, and each track and sector has a physical address that is used by the operating system to locate a particular data record. Hard disks typically have from a few hundred to thousands of tracks. As you can see in the figure, there is a constant number of tracks/sector, with outer sectors using more surface area than the inner sectors. The arrangement of tracks and sectors on a disk is known as the *format.*

A hard disk stack is illustrated in Figure 11–24(b). Hard disk drives differ in the number of platters in a stack, but there is always a minimum of two. All of the same corresponding tracks on each platter are collectively known as a cylinder, as indicated.

Hard Disk Performance Several basic parameters determine the performance of a given hard disk drive. A *seek* operation is the movement of the read/write head to the de-

Hard disk organization and formatting. **FIGURE 11–24**

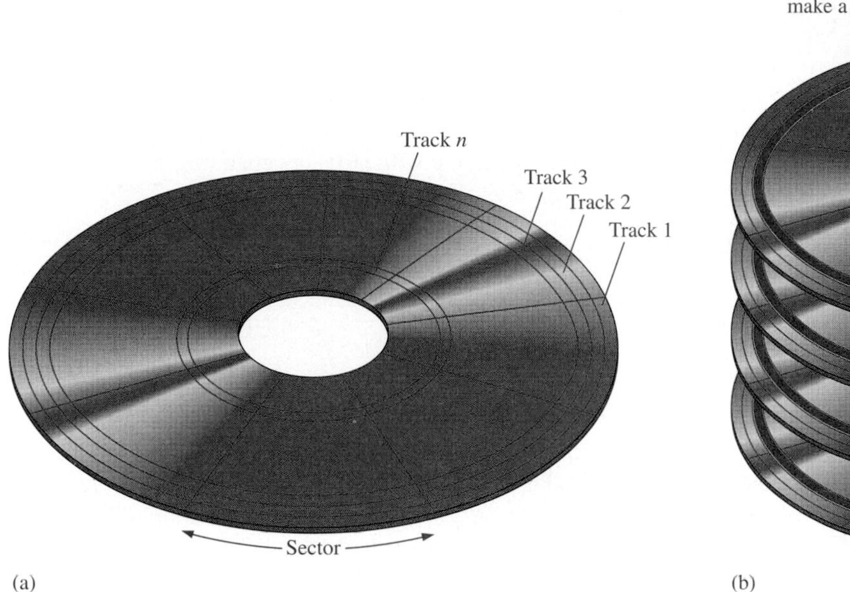

sired track. The *seek time* is the average time for this operation to be performed. Typically, hard disk drives have an average seek time of several milliseconds, depending on the particular drive.

The *latency period* is the time it takes for the desired sector to spin under the head once the head is positioned over the desired track. A worst case is when the desired sector is just past the head position and spinning away from it. The sector must rotate almost a full revolution back to the head position. Average latency period assumes that the disk must make half of a revolution. Obviously, the latency period depends on the constant rotational speed of the disk. Disk rotation speeds are different for different disk drives but typically are 3600 rpm, 4500 rpm, 5400 rpm, and 7200 rpm. Some disk drives rotate at 10,033 rpm and have an average latency period of less than 3 ms.

The sum of the average seek time and the average latency period is the *access time* for the disk drive.

HISTORICAL NOTE

Bill Gates is the richest man in the world. He formed Microsoft in 1976 at the age of 21. With his close friend Paul Allen, he developed a version of the computer language BASIC for the early Altair PC. Later, Gates licensed a computer operating system, called DOS, and offered it to IBM. When IBM rejected the offer, he bought it himself and the rest is history. Almost half of the revenue from computer software worldwide goes to Microsoft.

Floppy Disks

The floppy disk is an older technology and derives its name because it is made of a flexible polyester material with a magnetic coating on both sides. The early floppy disks were 5.25 inches in diameter and were packaged in a semiflexible jacket. Current floppy disks or diskettes are 3.5 inches in diameter and are encased in a rigid plastic jacket. A spring-loaded shutter covers the access window and remains closed until the disk is inserted into a disk drive. A metal hub has one hole to center the disk and another for spinning it within the protective jacket. Obviously, floppy disks are removable disks, whereas hard disks are not. Floppy disks are formatted into tracks and sectors similar to hard disks except for the number of tracks and sectors. The high-density 1.44 MB floppies have 80 tracks per side with 18 sectors. With the advent of other types of removable disks such as Zip, the floppy has serious competition, but its low cost may continue to make the floppy disk useful for smaller storage applications.

Zip™

The Zip drive is one type of removable magnetic storage device that has replaced the limited-capacity floppy. Like the floppy disk, the Zip disk cartridge is a flexible disk housed in a rigid case about the same size as that of the floppy disk but thicker. The typical Zip drive is much faster than the floppy drive because it has a 3000 rpm spin rate compared to the floppy's 300 rpm. The Zip drive has a storage capacity of up to 250 MB, over 173 times more than the 1.44 MB floppy.

Jaz™

Another type of removable magnetic storage device is the Jaz drive, which is similar to a hard disk drive except that two platters are housed in a removable cartridge protected by a dust-proof shutter. The Jaz cartridges are available with storage capacities of 1 or 2 GB.

Magneto-Optical Storage

As the name implies, magneto-optical (MO) storage devices use a combination of magnetic and optical (laser) technologies. A magneto-optical disk is formatted into tracks and sectors similar to magnetic disks.

The basic difference between a purely magnetic disk and an MO disk is that the magnetic coating used on the MO disk requires heat to alter the magnetic polarization. Therefore, the MO is extremely stable at ambient temperature, making data unchangeable. To write a data bit, a high-power laser beam is focused on a tiny spot on the disk, and the temperature of that tiny spot is raised above a temperature level called the Curie point (about 200°C). Once heated, the magnetic particles at that spot can easily have their direction (polarization) changed by a magnetic field generated by the write head. Information is read from the disk with a less-powerful laser than used for writing, making use of the Kerr effect where the polarity of the reflected laser light is altered depending on the orientation of the magnetic particles. Spots of one polarity represent 0s and spots of the opposite polarity represent 1s. Basic MO operation is shown in Figure 11–25, which represents a small cross-sectional area of a disk.

Optical Storage

CD-ROM

The basic Compact Disk-Read-Only Memory is a 120 mm diameter disk with a sandwich of three coatings: a polycarbonate plastic on the bottom, a thin aluminum sheet for reflectivity, and a top coating of lacquer for protection. The CD-ROM disk is formatted in a single spiral track with sequential 2 kB sectors and has a capacity of 680 MB. Data is prerecorded at the factory in the form of minute indentations called *pits* and the flat area surrounding the pits called *lands.* The pits are stamped into the plastic layer and cannot be erased.

A CD player reads data from the spiral track with a low-power infrared laser, as illustrated in Figure 11–26. The data are in the form of pits and lands as shown. Laser light reflected from a pit is 180° out-of-phase with the light reflected from the lands. As the disk rotates, the narrow laser beam strikes the series of pits and lands of varying lengths, and a photodiode detects the difference in the reflected light. The result is a series of 1s and 0s corresponding to the configuration of pits and lands along the track.

WORM

Write Once/Read Many (WORM) is a type of optical storage that can be written onto one time after which the data cannot be erased but can be read many times. To write data, a low-power laser is used to burn microscopic pits on the disk surface. 1s and 0s are represented by the burned and nonburned areas.

CD-R

This is essentially a type of WORM. The difference is that the CD-Recordable allows multiple write sessions to different areas of the disk. The CD-R disk has a spiral track like the CD-ROM, but instead of mechanically pressing indentations on the disk to

Basic principle of a magneto-optical disk. **FIGURE 11–25**

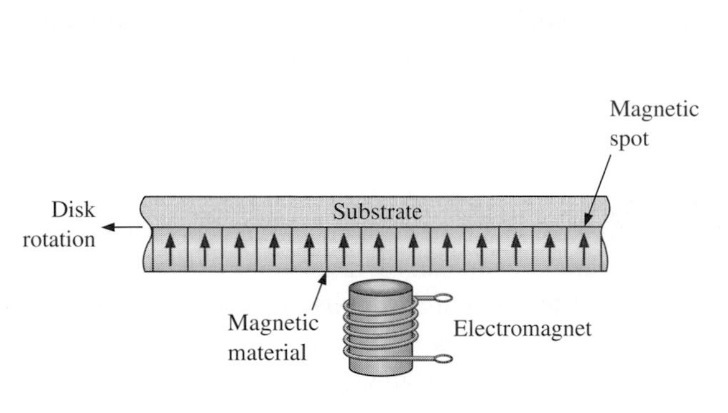

(a) Unrecorded disk

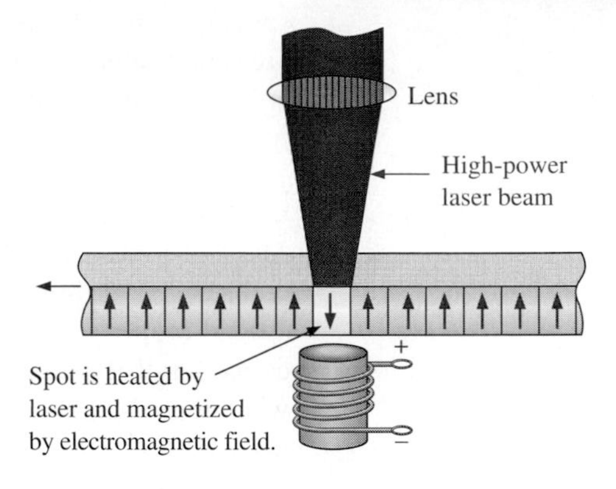

(b) Writing: A high-power laser beam heats the spot, causing the magnetic particles to align with the electromagnetic field.

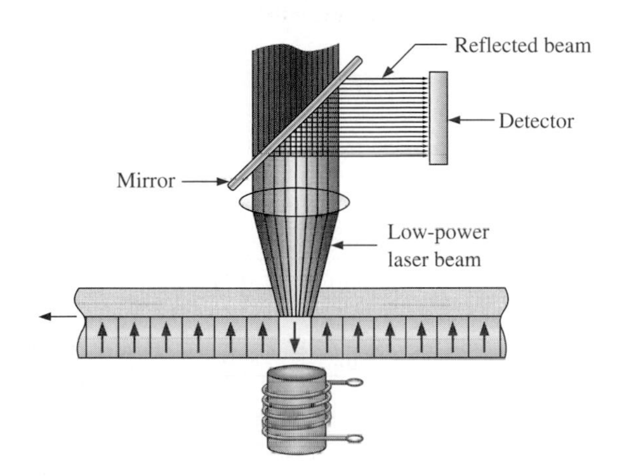

(c) Reading: A low-power laser beam reflects off of the reversed-polarity magnetic particles and its polarization shifts. If the particles are not reversed, the polarization of the reflected beam is unchanged.

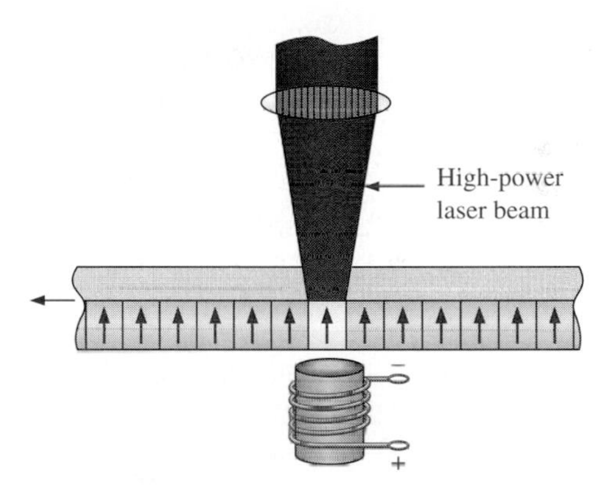

(d) Erasing: The electromagnetic field is reversed as the high-power laser beam heats the spot, causing the magnetic particles to be restored to the original polarity.

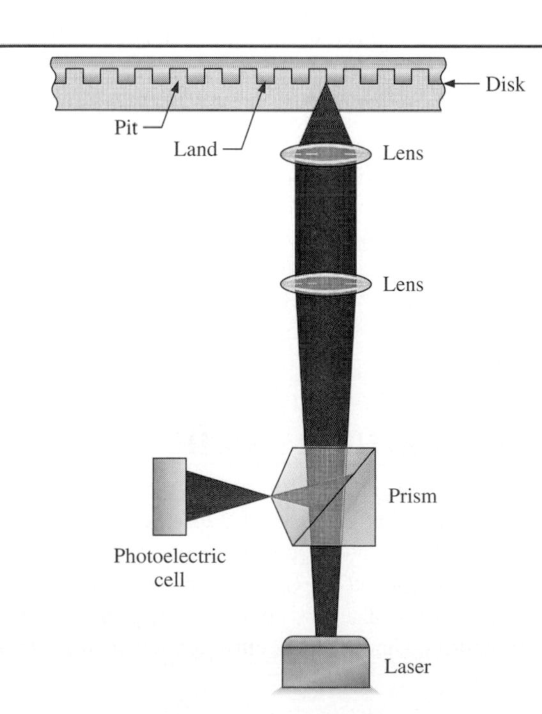

FIGURE 11–26 Basic operation of reading data from a CD-ROM.

represent data, the CD-R uses a laser to burn microscopic spots into an organic dye surface. When heated beyond a critical temperature with a laser during read, the burned spots change color and reflect less light than the nonburned areas. Therefore, 1s and 0s are represented on a CD-R by burned and nonburned areas, whereas on a CD-ROM they are represented by pits and lands. Like the CD-ROM, the data cannot be erased once it is written.

CD-RW

The CD-Rewritable disk can be used to read and write data. Instead of the dye-based recording layer in the CD-R, the CD-RW commonly uses a crystalline compound with a special property. When it is heated to a certain temperature, it becomes crystalline when it cools; but if it is heated to a certain higher temperature, it melts and becomes amorphous when it cools. To write data, the focused laser beam heats the material to the melting temperature resulting in an amorphous state. The resulting amorphous areas reflect less light than the crystalline areas, allowing the read operation to detect 1s and 0s. The data can be erased or overwritten by heating the amorphous areas to a temperature above the crystallization temperature but lower than the melting temperature that causes the amorphous material to revert to a crystalline state.

DVD-ROM

Originally DVD was an abbreviation for Digital Video Disk but eventually came to represent *Digital Versatile Disk.* Like the CD-ROM, DVD-ROM data are prestored on the disk. However, the pit size is smaller than for the CD-ROM, allowing more data to be stored on a track. The major difference between CD-ROM and DVD-ROM is that the CD is single-sided, while the DVD has data on both sides. Also, in addition to double-sided DVDs, there are also multiple-layer disks that use semitransparent data layers placed over the main data layers, providing storage capacities of tens of gigabytes (GB). To access all the layers, the laser beam requires refocusing going from one layer to the other.

Flash Memories

The ideal memory has high storage capacity, nonvolatility, in-system read and write capability, comparatively fast operation, and cost effectiveness. The traditional memory technologies such as ROM, PROM, EPROM, EEPROM, SRAM, and DRAM individually exhibit one or more of these characteristics, but none of these technologies has all of them except the flash memory. The flash memory is generally considered a solid-state storage device.

Flash memories are high-density read/write memories (high-density translates into large bit storage capacity) that are nonvolatile, which means that data can be stored indefinitely without power. They are sometimes used in place of floppy or small-capacity hard disk drives in portable computers.

High-density means that a large number of cells can be packed into a given surface area on a chip; that is, the higher the density, the more bits that can be stored on a given size chip. This high density is achieved in flash memories with a storage cell that consists of a single floating-gate MOS transistor. A data bit is stored as charge or the absence of charge on the floating gate depending if a 0 or a 1 is stored.

Flash Memory Cell

A single-transistor cell in a flash memory is represented in Figure 11–27. The stacked gate MOS transistor consists of a control gate and a floating gate in addition to the drain and source. The floating gate stores electrons (charge) as a result of a sufficient voltage applied to the control gate. *A 0 is stored when there is more charge and a 1 is stored when there is less or no charge.* The amount of charge present on the floating gate determines if the transistor will turn on and conduct current when a control voltage is applied during a read operation.

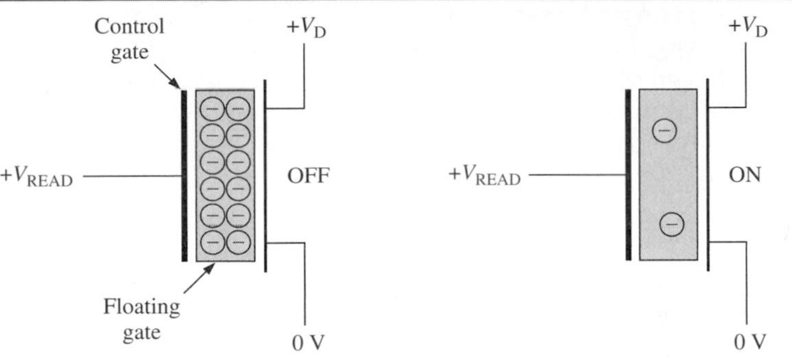

FIGURE 11–27

The read operation of a flash cell in an array.

During an erase operation, charge is removed from all the memory cells. A sufficient positive voltage is applied to the transistor source with respect to the control gate. This is opposite in polarity to that used in programming. This voltage attracts electrons from the floating gate and depletes it of charge. A flash memory is always erased prior to being reprogrammed.

The memory stick is a storage medium that uses flash memory technology in a physical configuration smaller than a stick of chewing gum. Memory sticks are typically available in 4 MB, 8 MB, 16 MB, 32 MB, 64 MB, and 128 MB capacities and as a kit with a PC card adaptor. Because of its compact design, it is ideal for use in small digital electronics products, such as laptop computers and digital cameras.

Review Questions

26. What are the major types of magnetic storage?
27. Generally, how is a magnetic disk organized?
28. How are data written on and read from a magneto-optical disk?
29. What are the types of optical storage?
30. What type of storage cell does a flash memory use?

COMPUTER PROGRAMMING 11–7

Assembly language is a way to express machine language in English-like terms, so there is a one-to-one correspondence. Assembly language has limited applications and is not portable from one processor to another, so most computer programs are written in high-level languages such as C, C++, JAVA, BASIC, COBOL, and FORTRAN. High-level languages are portable and therefore can be used in different computers. High-level languages must be converted to the machine language for a specific microprocessor by a process called *compiling*.

In this section, you will learn some programming concepts and the levels of programming languages.

Levels of Programming Languages

A hierarchy diagram of computer programming languages relative to the computer hardware is shown in Figure 11–28. At the lowest level is the computer hardware (CPU, memory, disk drive, input/output). Next is the **machine language** that the hardware understands

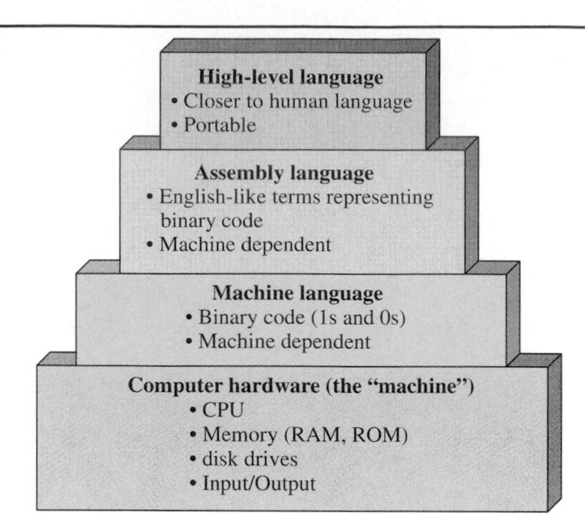

FIGURE 11–28

Hierarchy of programming languages relative to computer hardware.

because it is written with 1s and 0s (remember, a logic gate can recognize only a LOW (0) or a HIGH (1). At the level above machine language is assembly language where the 1s and 0s are represented by English-like words. Assembly languages are considered low-level because they are closely related to machine language and are machine dependent, which means a given assembly language can only be used on a specific microprocessor.

At the level above assembly language is **high-level language**, which is closer to human language and further from machine language. An advantage of high-level language over assembly language is that it is portable, which means that a program can run on a variety of computers. Also, high-level language is easier to read, write, and maintain than assembly language. Most system software (e.g., Windows and Unix), and application software (e.g., word processors and spreadsheets) are written with high-level languages.

Assembly Language

To avoid having to write out long strings of 1s and 0s to represent microprocessor instructions, English-like terms called mnemonics or **op-codes** are used. Each type of microprocessor has its own set of mnemonic instructions that represent binary codes for the instructions. All of the mnemonic instructions for a given microprocessor are called the instruction set. **Assembly language** uses the instruction set to create programs for the microprocessor; and because an assembly language is directly related to the machine language (binary code instructions), it is classified as a low-level language. Assembly language is one step removed from machine language.

Assembly language and the corresponding machine language that it represents is specific to the type of microprocessor or microprocessor family. Assembly language is not portable; that is, you cannot run an assembly language program written for one type of microprocessor on another type of microprocessor. For example, an assembly program for the Motorola processors will not work on the Intel processors. Even within a given family different microprocessors may have different instruction sets.

An **assembler** is a program that converts an assembly language program to machine language that is recognized by the microprocessor. Also, programs called **cross-assemblers** translate an assembly language program for one type of microprocessor to an assembly language for another type of microprocessor.

Assembly language is rarely used to create large application programs. However, assembly language is often used in a subroutine (a small program within a larger program) that can be called from a high-level language program. Assembly language is useful in subroutine applications because it usually runs faster and has none of the restrictions of a high-level language. Assembly language is also used in machine control, such as for industrial processes. Another area for assembly language is in video game programming.

Software Model of the Intel Processors for Assembly Language

Assembly language programmers need to have an understanding of the internal structure of the target microprocessor for which the program is written. The brief discussion that follows is intended to give you an idea of how a microprocessor executes instructions. To learn assembly language programming is a whole course in itself.

Microprocessors have *revolutionized* computing—hardly too strong a word when you look at the history of microprocessors. As Intel introduced newer microprocessors, capabilities and speed increased very dramatically from the earliest 16-bit processors to new 64-bit processors. In addition to internal processor improvements, many other improvements to computers occurred (such as the communication links to the outside world, the size and speed of hard drives and other memory, as well as external peripheral improvements). Despite all of these changes to the speed and capability of computers, the designers of new processors maintained compatibility with all assembly language software written for earlier Intel processors; that is, the newest Pentium could still run the same programs written for any the processors that preceded it.

To understand an assembly language program, you need to be able to visualize the internal structure of the processor for which it is written. Internally, all processors use registers to store information. A **register** is an internal memory location within the processor but registers also have special capabilities. Special names are assigned to registers to indicate their function. One very important register, called the *accumulator,* is always involved in arithmetic operations. Other registers have other capabilities such as holding address information of locations in RAM where data is held.

Normally the assembly language programmer is not concerned with minute details of the operation of the processor, and frequently the smaller 16-bit version of the operation is satisfactory for writing a very fast, efficient program. The 16-bit registers are a subset of the full register set within newer processors. The 16-bit registers are illustrated in the yellow areas of Figure 11–29. These are the only registers involved in programs running in real

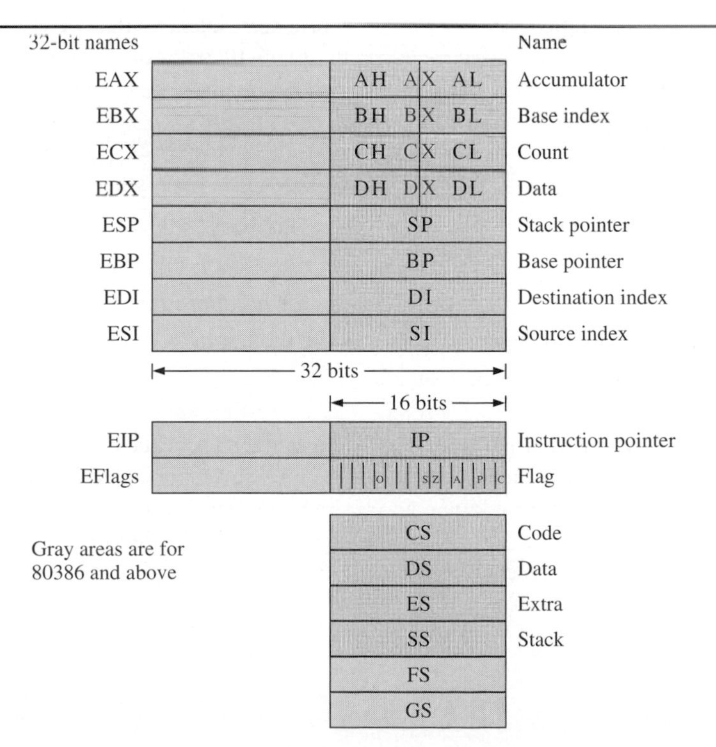

FIGURE 11–29

Registers in a microprocessor.

mode. *Real mode* is any operation that allows the processor to only access the first 1 MB of memory.

The registers shown in Figure 11–29 are divided into three groups. The larger 32-bit version names have an E written in front, which stands for *extended.* Our focus here is on the 16-bit registers, which will be used in real-mode programming. In the first group are four *data registers* (AX, BX, CX, and DX) that can be addressed either as 16 bits (using the X in the name) or by bytes (using either an H or an L in the name). Notice that the accumulator (AX) is in this group. Some instructions in assembly language are designed to work directly with the accumulator or one of the other three data registers. The remaining four registers in this group are designed to "point" to different locations in memory by holding part of the address information for a memory location. These registers are the stack pointer (SP), the base pointer (BP), and two index registers (DI and SI).

The second group has two registers — the instruction pointer and the flag register. The *instruction pointer* holds the address of the next instruction to be executed. The *flag register* has independent one-bit indicators that show the result of certain operations in the microprocessor. For example, one flag is called the zero flag (ZF), which is set by an arithmetic operation that results in a zero answer. Another flag is the carry flag (CF), which is set by an arithmetic operation that generates a carry. A programmer can check flags and change what the program does next as a consequence of their setting.

The third group of registers is called the *segment registers.* These registers hold the other part of a memory address and include the CS, DS, ES and SS registers. Intel developed a method for forming an address by combining the address in the segment register with one of the pointer registers. The method simplifies the overall processor operation. Keep in mind that segment registers just are part of the address information. Usually programmers refer to an address in two parts: a *segment* and an *offset.* The segment address is combined with the offset address that is stored in one of the pointer registers. For example, an address might be written 10AF:258E. The address will be stored in two parts; the 10AF will be stored in a segment register, and the 258E will be stored in a pointer register.

You may find it instructive to see how a programmer writes instructions to produce a result. The instructions will use several of the registers discussed previously to complete a task.

Example of an Assembly Language Program

For a simple assembly language program, let's say that we want the computer to add a list of numbers from the memory and place the sum of the numbers back into the memory. A zero is used as the last number in the list to indicate the end of the list of numbers. The steps required to accomplish this task are as follows:

1. Clear a register (in the microprocessor) for the total or sum of the numbers.
2. Point to the first number in the memory (RAM).
3. Check to see if the number is zero. If it is zero, all the numbers have been added.
4. If the number is not zero, add the number in the memory to the total in the register.
5. Point to the next number in the memory.
6. Repeat steps 3, 4, and 5.

A flow chart is often used to diagram the sequence of steps in a computer program. Figure 11–30 shows the flow chart for the program represented by the six steps listed above.

The assembly language program implements the addition problem shown in the flow chart in Figure 11–30. Two of the registers in the microprocessor are named ax and bx. The comments preceeded by a semicolon are not recognized by the microprocessor; they are for explanation only.

FIGURE 11–30

Flow chart for adding a list of numbers.

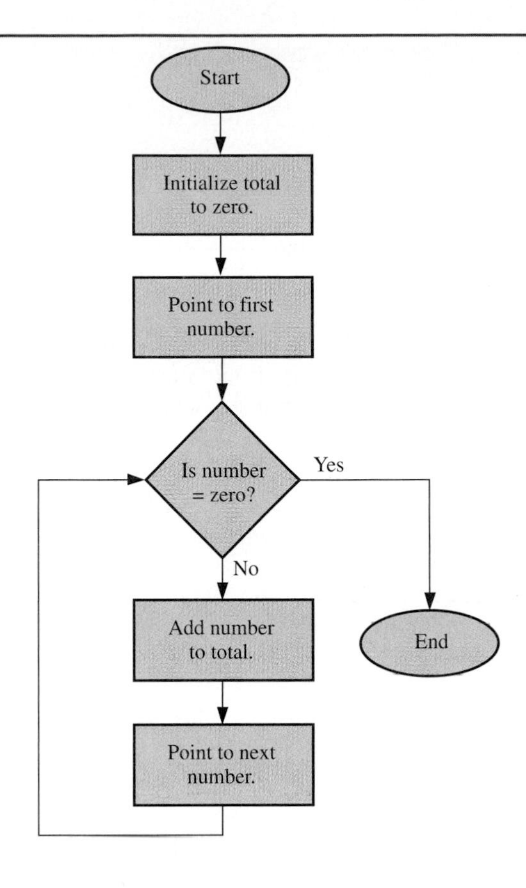


```
      mov ax, 0               ; Replaces the contents of the ax register with zero.
                              ; Register ax will store the total of the addition.
      mov bx, 50H             ; Places memory address hexadecimal 50 into the bx register.
next: cmp word ptr [bx], 0    ; Compares the number stored in the memory location pointed to by the bx register to zero.
      jz done                 ; If the number in the memory location is zero, jump to "done".
      add ax, [bx]            ; Add the number in the memory location pointed to by the bx register to the number in the ax
                              ; register and place the sum into the ax register.
      add bx, 02              ; 2 is added to the address in the bx register. Two addresses are required to store each number
                              ; which is two bytes long.
      jmp next                ; Loop back to "next" and repeat the process.
done: mov [bx], ax            ; Replace the zero last number in the memory location pointed to by the bx register with the
                              ; total in the ax register.
      nop                     ; No operation, this indicates the end of the program.
```

Conversion of a Program to Machine Language

All programs written in either an assembly language or a high-level language must be converted into machine language in order for a particular computer to recognize the program instructions.

Assemblers

An assembler translates and converts a program written in assembly language into machine code, as indicated in Figure 11–31. The term **source program** is often used to refer

FIGURE 11–31

Assembly to machine conversion using an assembler.

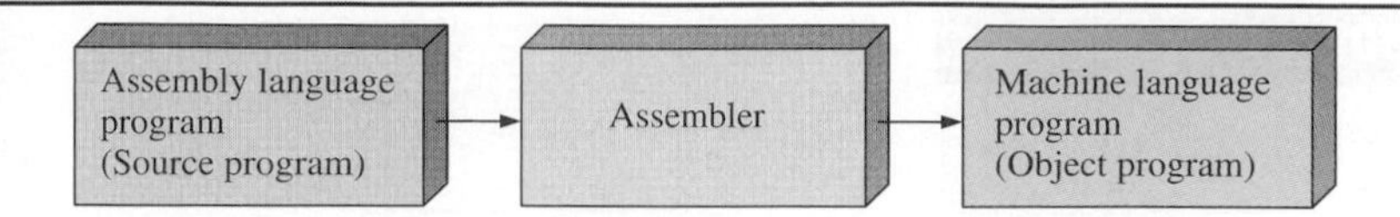

to a program written in either assembly or high-level language. The term **object program** refers to a machine language translation of a source program.

Compilers

A **compiler** is a program that compiles or translates a program written in a high-level language and converts it into machine code, as shown in Figure 11–32. The compiler examines the entire source program and collects and reorganizes the instructions. Every high-level language comes with a specific compiler for a specific computer, making the high-level language indepenent of the computer on which it is used. Some high-level languages are translated using what is called an *interpreter* that translates each line of program code to machine language.

FIGURE 11–32

High-level to machine conversion with a compiler.

As previously mentioned, all high-level languages, such as C, C++, FORTRAN, and COBOL, will run on any computer. A given high-level language is valid for any computer, but the compiler that goes with it is specific to a particular type of CPU. This is illustrated in Figure 11–33, where the same high-level language program (written in C++ in this case) is converted by different machine-specific compilers.

FIGURE 11–33

Machine independence of a program written in a high-level language.

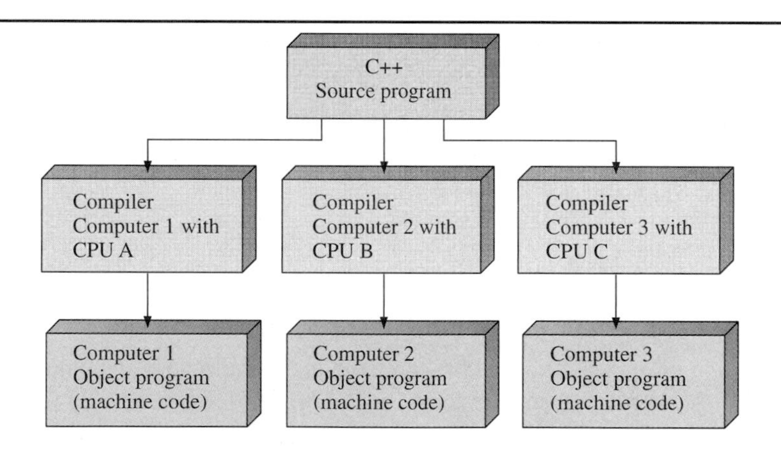

The Process of Programming

The basic steps to take when you write a computer program, regardless of the particular programming language that you use, are as follows:

1. Determine and specify the problem that is to be solved or task that is to be done.
2. Create an algorithm; that is, develop a series of steps to accomplish the task.
3. Express the steps using a particular programming language and enter them on the software text editor.
4. Compile (or assemble) and run the program.

A simple program will show an example of high-level programming. The following C++ program implements the same addition problem defined by the flow chart in Figure 11–30 and implemented using assembly language.

```
int total = 0;                  //Initialize the total to 0.
while (*number != 0x00)         //Loop while the value is not found. The asterisk
                                //preceding the pointer identifier number says that the
                                //contents of the memory location pointed to by the
                                //identifier number are being evaluated.
{
     total = total + *number;   //Accumulative summation of total.
     number++;                  // Increment pointer to next number in memory.
}
```

This C++ program is equivalent to the assembly program that adds a series of numbers and produces a total value.

```
int total = 0;
while (*number != 0x00)
{
     total = total + *number;
     number++;
}
```

C++

Equivalent

```
        mov ax, 0
        mov bx, 50H
next:   cmp word ptr [bx], 0
        jz done
        add ax, [bx]
        add bx, 02
        jmp next
done:   mov [bx], ax
        nop
```

Assembly

Review Questions

31. How is an assembly program converted to machine language?
32. How is a high-level language program converted to machine language?
33. Which language level is closest to the computer hardware?
34. Which language level is closer to human speech?
35. What does machine language consist of?

CHAPTER REVIEW

Key Terms

Address A location in memory.

ALU The part of a microprocessor that performs arithmetic operations, logic operations, and comparison operations on binary data.

Assembly language A programming language that uses English-like words and has a one-to-one correspondence to machine language.

12

CHAPTER

DIGITAL SIGNAL PROCESSING

CHAPTER OUTLINE

Study aids for this chapter are available at

http://www.prenhall.com/SOE

INTRODUCTION

Digital signal processing is a powerful technology that is widely used in many applications, such as automotive, consumer, graphics/imaging, industrial, instrumentation, medical, military, telecommunications, and voice/speech applications. Digital signal processing incorporates mathematics, software programming, and processing hardware to manipulate analog signals. For example, digital signal processing can be used to enhance images, compress data for efficient transmission and storage, recognize and generate speech, and clean up noisy or deteriorated audio.

This chapter provides a brief look at digital signal processing. To completely cover the topic in the depth necessary to have a detailed understanding would take much more than a single chapter. Entire books are available on the subject. Much information, including data sheets, on the TMS320 family of DSPs is available at the Texas Instruments Website (*www.ti.com*). Information about other DSPs can be found on the Motorola Website (*www.motorola.com*) and the Analog Devices Website (*www.analogdevices.com*).

- Analog-to-digital converter (ADC)
- DSP
- Digital-to-analog converter (DAC)
- Sampling
- Nyquist frequency
- Aliasing
- Quantization
- Pipeline
- Fetch
- Decode
- Execute

KEY OBJECTIVES

A section number is given for each objective. After completing this chapter, you should be able to

12–1 List the essential elements in a digital signal processing system

12–2 Explain how analog signals are converted to digital form

12–3 State the purpose of analog-to-digital conversion and how several types of ADCs work

12–4 Explain the basic concepts of a digital signal processor (DSP)

12–5 Explain how digital signals are converted to analog form

LABORATORY EXPERIMENTS DIRECTORY

The following laboratory exercise is for this chapter.

- **Experiment 17**
  Analog-to-Digital Conversion

Researchers have devised many methods to look for signals from an extraterrestrial civilization. Some use radio telescopes to look for radio signals; others use optical telescopes to look for light signals. Astronomers occasionally detect artificial signals produced by our own noisy civilization. Today, Earth satellites produce signals that really are coming from space, and researchers have to be careful not to be fooled by them. In 1967, astronomers in England discovered strange radio signals from space that sounded like the ticking of an old-fashioned clock. They were definitely produced by something outside the solar system because the signals didn't move among the stars. They were unlike anything ever seen in astronomy and seemed to be artificial. The scientists even nicknamed them "LGM," for Little Green Men. When more scientists studied the signals, they concluded that they were neutron stars. We now call these objects pulsars because of the pulsating radio signals we detect from them.

12–1 DIGITAL SIGNAL PROCESSING BASICS

Digital signal processing converts signals that naturally occur in analog form, such as sound, video, and information from sensors, to digital form and uses digital techniques to enhance and modify analog signal data for various applications.

In this section, you will learn the essential elements in a digital signal processing system.

A digital signal processing system first translates a continuously varying analog signal into a series of discrete levels. This series of levels follows the variations of the analog signal and resembles a staircase, as illustrated for the case of a sine wave in Figure 12–1. The process of changing the original analog signal to a "stairstep" approximation is accomplished by a sample-and-hold circuit.

FIGURE 12–1

An original analog signal (sine wave) and its "stairstep" approximation.

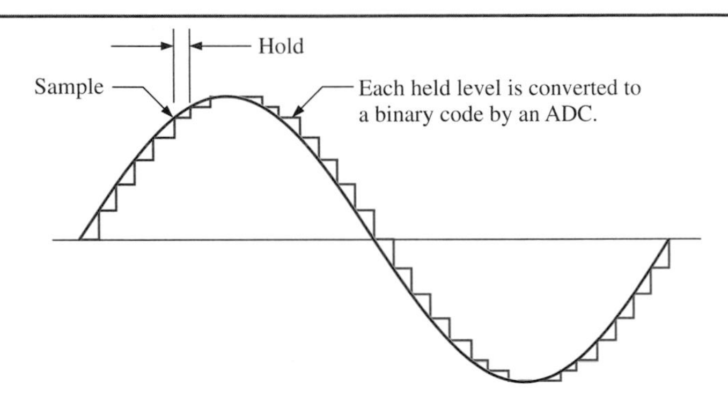

Next, the "stairstep" approximation is quantized into binary codes that represent each discrete step on the "stairsteps" by a process called analog-to-digital (A/D) conversion. The circuit that performs A/D conversion is an **analog-to-digital converter (ADC)**.

Once the analog signal has been converted to a binary coded form, it is applied to a **DSP** (digital signal processor). The DSP can perform various operations on the incoming data, such as removing unwanted interference, increasing the amplitude of some signal frequencies and reducing others, encoding the data for secure transmissions, and detecting and correcting errors in transmitted codes. DSPs make possible, among many other things, the cleanup of sound recordings, the removal of echos from communications lines, the enhancement of images from CT scans for better medical diagnosis, and the scrambling of cellular phone conversations for privacy.

After a DSP processes a signal, the signal can be converted back to a much improved version of the original analog signal. This is accomplished by a **digital-to-analog converter (DAC)**. Figure 12–2 shows a basic block diagram of a typical digital signal processing system.

FIGURE 12–2 Basic block diagram of a typical digital signal processing system.

Anti-aliasing filter → Sample-and-hold circuit → ADC → DSP → DAC → Reconstruction filter

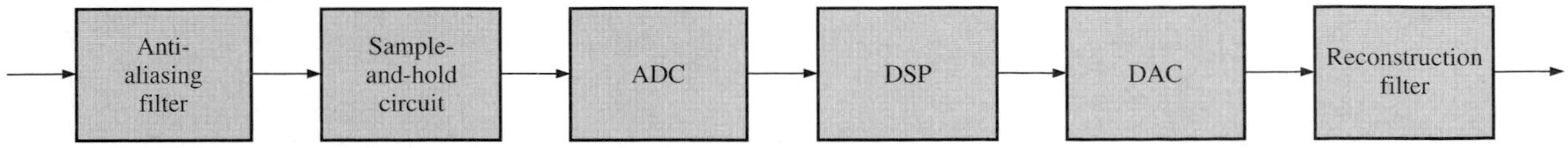

DSPs are actually a specialized type of microprocessor but are different from general-purpose microprocessors in a couple of significant ways. Typically, microprocessors are designed for general-purpose functions and operate with large software packages. DSPs are used for special-purpose applications; they are very fast number crunchers that must work in real time by processing information as it happens using specialized algorithms (programs). The analog-to-digital converter (ADC) in a system must take samples of the incoming analog data often enough to catch all the relevant fluctuations in the signal amplitude, and the DSP must keep pace with the sampling rate of the ADC by doing its calculations as fast as the sampled data are received. Once the digital data are processed by the DSP, they go to the digital-to-analog converter (DAC) for conversion back to analog form.

Review Questions

Answers are at the end of the chapter.

1. What does DSP stand for?
2. What does ADC stand for?
3. What does DAC stand for?
4. An analog signal is changed to a binary coded form by what circuit?
5. A binary coded signal is changed to analog form by what circuit?

CONVERTING ANALOG SIGNALS TO DIGITAL 12–2

In order to process signals using digital techniques, the incoming analog signal must be converted into digital form.

In this section, you will learn how analog signals are converted to digital form.

Filtering and Sampling

The first two blocks in the system diagram of Figure 12–2 are the anti-aliasing filter and the sample-and-hold circuit. The sample-and-hold function does two operations, the first of which is sampling. **Sampling** is the process of taking a sufficient number of discrete values at points on a waveform that will define the shape of the waveform. The more samples you take, the more accurately you can define a waveform. Sampling converts an analog signal into a series of impulses, each representing the amplitude of the signal at a given instant in time. Figure 12–3 illustrates the process of sampling.

When an analog signal is to be sampled, there are certain criteria that must be met in order to accurately represent the original signal. All analog signals (except a pure sine wave) contain a spectrum of component frequencies called *harmonics*. The harmonics of an analog signal are sine waves of different frequencies and amplitudes. When the harmonics of a given periodic waveform are added, the result is the original signal. Before a signal can be sampled, it must be passed through a low-pass filter (anti-aliasing filter) to eliminate harmonic frequencies above a certain value as determined by the Nyquist frequency.

The Sampling Theorem

Notice in Figure 12–3 that there are two input waveforms. One is the analog signal and the other is the sampling pulse waveform. The **Nyquist frequency**, f_{Nyquist}, specifies the limit of the analog signal frequency to be sampled and defines it to be one-half of the sampling frequency. The sampling theorem requires that, in order to properly represent an analog signal, the sampling frequency, f_{sample}, must be greater than twice the Nyquist frequency.

$$f_{\text{sample}} > 2f_{\text{Nyquist}}$$

HISTORICAL NOTE

Harry Nyquist (1889–1976) was an American physicist and electrical engineer. During his 37 years with AT&T and Bell Labs, he received 138 U.S. patents and published twelve technical articles. He wrote a landmark paper "Certain Topics in Telegraph Transmission Theory," published in 1928 that described the criteria for what we know today as sampled data systems. Nyquist taught us that for periodic functions, if you sampled at a rate that was at least twice as fast as the signal of interest, then no information (data) would be lost upon reconstruction.

FIGURE 12–3

Illustration of the sampling process.

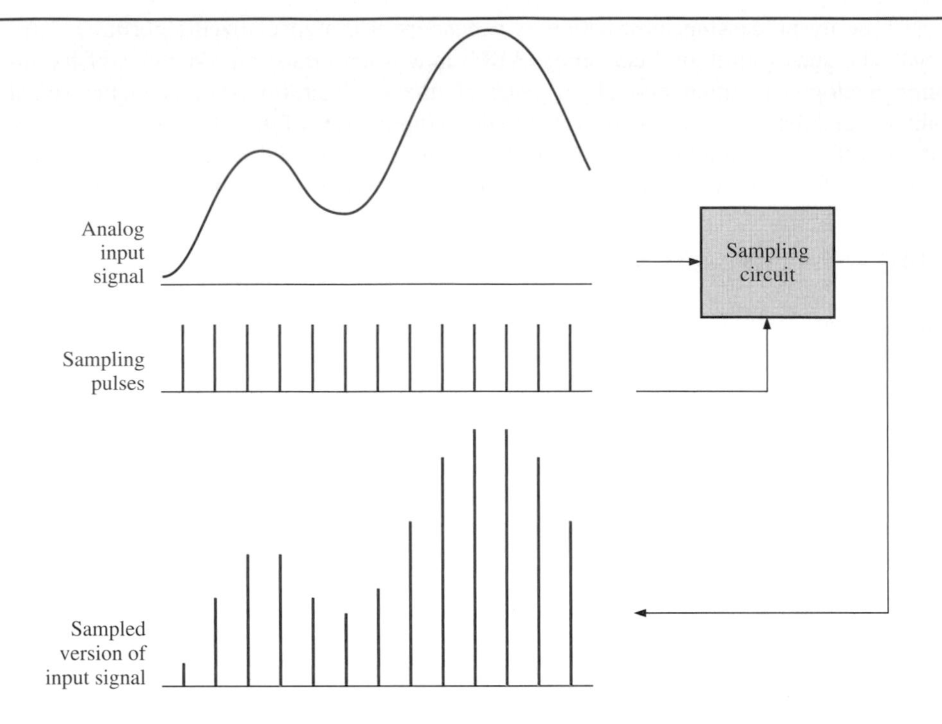

To intuitively understand the sampling theorem, a simple "bouncing-ball" analogy may be helpful. Although it is not a perfect representation of the sampling of electrical signals, it does serve to illustrate the basic idea. If a ball is photographed (sampled) at one instant during a single bounce, as illustrated in Figure 12–4(a), you cannot tell anything about the path of the ball except that it is off the floor. You can't tell whether it is going up or down or the distance of its bounce. If you take photos at two equally-spaced instants during one bounce, as shown in part (b), you can obtain only a minimum amount of information about its movement and nothing about the distance of the bounce. In this particular case, you know only that the ball is in the air at the times the two photos were taken and that the maximum height of the bounce is at least equal to the height shown in each photo. If you take four photos, as shown in part (c), then the path that the ball follows during a bounce begins to emerge. The more photos (samples) that you take, the more accurately you can determine the path of the ball as it bounces.

FIGURE 12–4 "Bouncing ball" analogy of sampling theory.

(a) One sample of a ball during a single bounce

(b) Two samples of a ball during a single bounce. This is the absolute minimum required to tell anything about its movement, but generally insufficient to describe its path.

(c) Four samples of a ball during a single bounce form a rough picture of the path of the bounce.

The Need for Filtering

Low-pass filtering is necessary to remove all frequency components (harmonics) of the analog signal that exceed the Nyquist frequency. If there are any frequency components in the analog signal that exceed the Nyquist frequency, an unwanted condition known as

aliasing will occur. An alias is a signal produced when the sampling frequency is not at least twice the Nyquist frequency. An alias signal has a frequency that is less than the highest frequency in the analog signal being sampled and therefore falls within the spectrum or frequency band of the input analog signal, causing distortion. Such a signal is actually "posing" as part of the analog signal when it really isn't, thus the term *alias.*

Another way to view aliasing is by considering that the sampling pulses produce a spectrum of harmonic frequencies above and below the sample frequency, as shown in Figure 12–5. If the analog signal contains frequencies above the Nyquist frequency, these frequencies overlap into the spectrum of the sample waveform as shown and interference occurs. The lower-frequency components of the sampling waveform become mixed in with the frequency spectra of the analog waveform, resulting in an aliasing error.

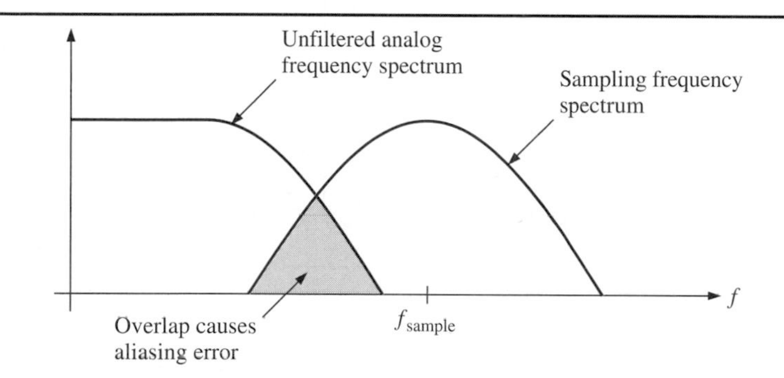

FIGURE 12–5

Aliasing occurs when $f_{sample} < 2f_{Nyquist}$.

A low-pass anti-aliasing filter must be used to limit the frequency spectrum of the analog signal for a given sample frequency. To avoid an aliasing error, the filter must at least eliminate all analog frequencies above the minimum frequency in the sampling spectrum, as illustrated in Figure 12–6. Aliasing can also be avoided by sufficiently increasing the sampling frequency. However, the maximum sampling frequency is usually limited by the performance of the analog-to-digital converter (ADC) that follows it.

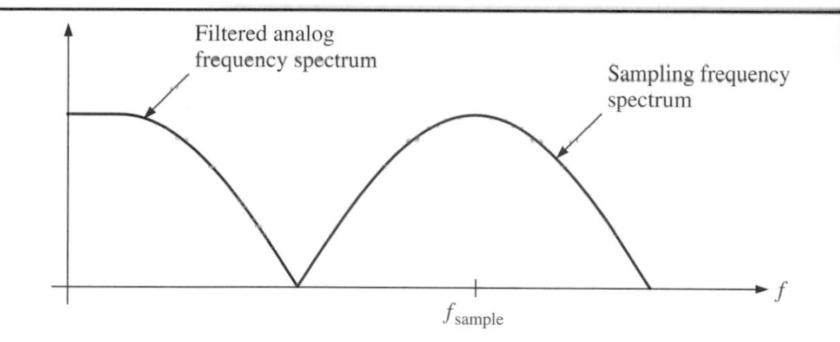

FIGURE 12–6

After low-pass filtering, the frequency spectra of the analog and the sampling signals do not overlap, thus eliminating aliasing error.

An Application

An example of the application of sampling is in digital audio equipment. The sampling rates used are 32 kHz, 44.1 kHz, or 48 kHz (the number of samples per second). The 48 kHz rate is the most common, but the 44.1 kHz rate is used for audio CDs and prerecorded tapes. According to the Nyquist rate, the sampling frequency must be at least twice the audio signal. Therefore, the CD sampling rate of 44.1 kHz captures frequencies up to about 22 kHz, which exceeds the 20 kHz specification that is common for most audio equipment.

Many applications do not require a wide frequency range to obtain reproduced sound that is acceptable. For example, human speech contains some frequencies near 10 kHz and, therefore, requires a sampling rate of at least 20 kHz. However, if only frequencies up to 4 kHz (ideally requiring an 8 kHz sampling rate) are reproduced, voice is very understandable. On the other hand, if a sound signal is not sampled at a high enough rate, the effect of aliasing will become noticeable with background noise and distortion.

Holding the Sampled Value

The holding operation is part of the sample-and-hold block shown in Figure 12–2. After filtering and sampling, the sampled level must be held constant until the next sample occurs. This is necessary for the ADC to have time to process the sampled value. This sample-and-hold operation results in a "stairstep" waveform that approximates the analog input waveform, as shown in Figure 12–7.

FIGURE 12–7 The sample-and-hold operation.

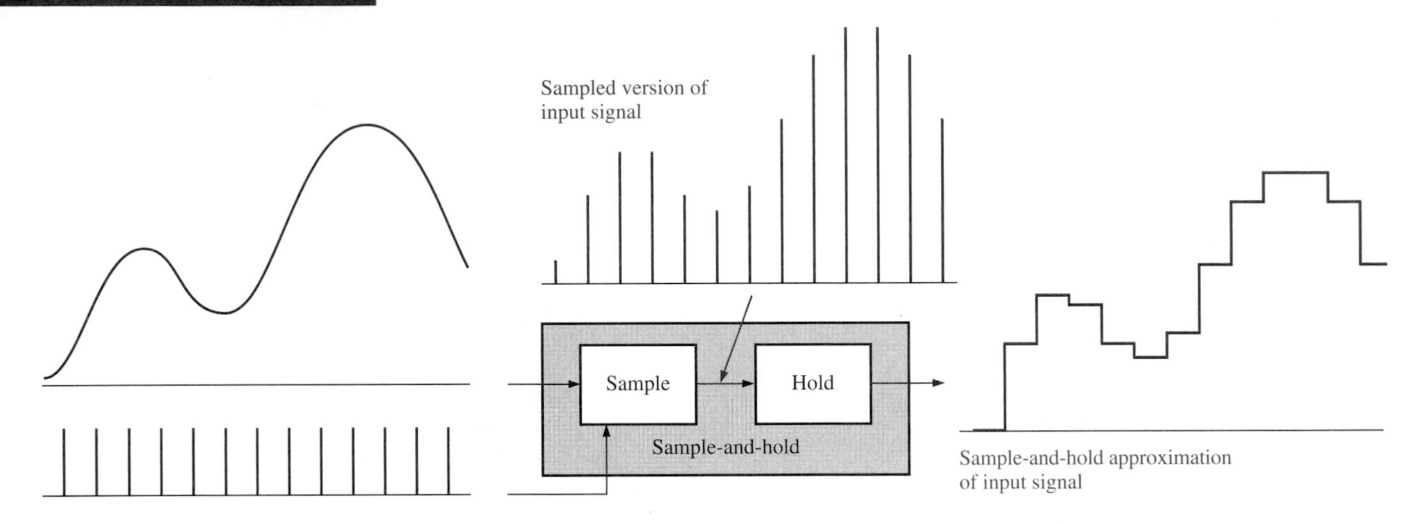

Analog-to-Digital Conversion

Analog-to-digital conversion is the process of converting the output of the sample-and-hold circuit to a series of binary codes that represent the amplitude of the analog input at each of the sample times. The sample-and-hold process keeps the amplitude of the analog input signal constant between sample pulses; therefore, the analog-to-digital conversion can be done using a constant value rather than having the analog signal change during a conversion interval, which is the time between sample pulses. Figure 12–8 illustrates the basic function of an analog-to-digital (ADC) converter. The sample intervals are indicated by dashed lines.

FIGURE 12–8 Basic function of an analog-to-digital (ADC) converter (The binary codes and number of bits are arbitrarily chosen for illustration only). The ADC output waveform that represents the binary codes is also shown.

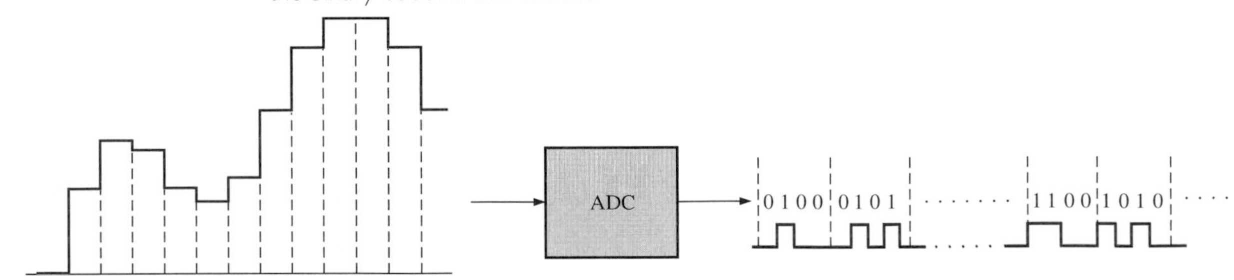

Quantization

The process of converting an analog value to a code is called **quantization**. During the quantization process, the ADC converts each sampled value of the analog signal to a binary code. The more bits that are used to represent a sampled value, the more accurate is the representation.

To illustrate, let's quantize a reproduction of the analog waveform into four levels. As shown in Figure 12–9, two bits are required. Note that each quantization level is represented by a 2-bit code on the vertical axis, and each sample interval is numbered along the horizontal axis. The quantization process is summarized in Table 12–1.

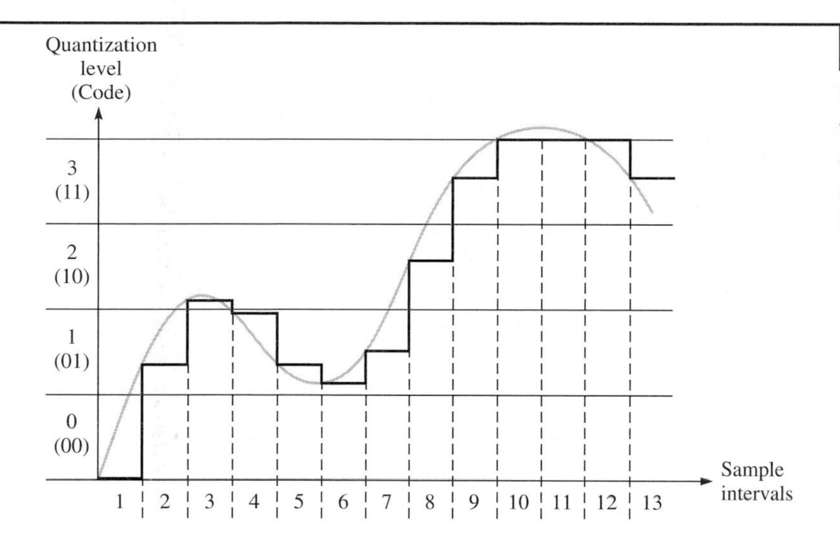

FIGURE 12–9

Sample-and-hold output waveform with four quantization levels. The original analog waveform is shown in light gray for reference.

TABLE 12–1

Two-bit quantization for the waveform in Figure 12–9.

Sample interval	Quantization level	Code
1	0	00
2	1	01
3	2	10
4	1	01
5	1	01
6	1	01
7	1	01
8	2	10
9	3	11
10	3	11
11	3	11
12	3	11
13	3	11

If the resulting 2-bit digital codes are used to reconstruct the original waveform, which is done by digital-to-analog converts (DAC), you would get the waveform shown in Figure 12–10. As you can see, quite a bit of accuracy is lost using only two bits to represent the sampled values.

Now, let's see how more bits will improve the accuracy. Figure 12–11 shows the same waveform with sixteen quantization levels (4 bits). The 4-bit quantization process is summarized in Table 12–2.

If the resulting 4-bit digital codes are used to reconstruct the original waveform, you would get the waveform shown in Figure 12–12. As you can see, the result is much more like the original waveform than for the case of four quantization levels in Figure 12–10. This shows that greater accuracy is achieved with more quantization bits. Most integrated circuit ADCs use from 8 to 24 bits, and the sample-and-hold function is sometimes contained on the ADC chip. Several types of ADCs are introduced in the next section.

HISTORICAL NOTE

In 1960, a high-performance ADC that could sample 50,000 samples per second cost about $8000, used several hundred watts of power, and occupied the space of a large oscilloscope. Today, a higher-resolution and faster ADC sells for a few dollars, uses less than 3 mW of power, and is housed in a single integrated circuit the size of a pencil eraser. Advances like this have affected the way data is acquired and processed.

FIGURE 12–10

The reconstructed waveform in Figure 12–9 using four quantization levels (2 bits). The original analog waveform is shown in light gray for reference.

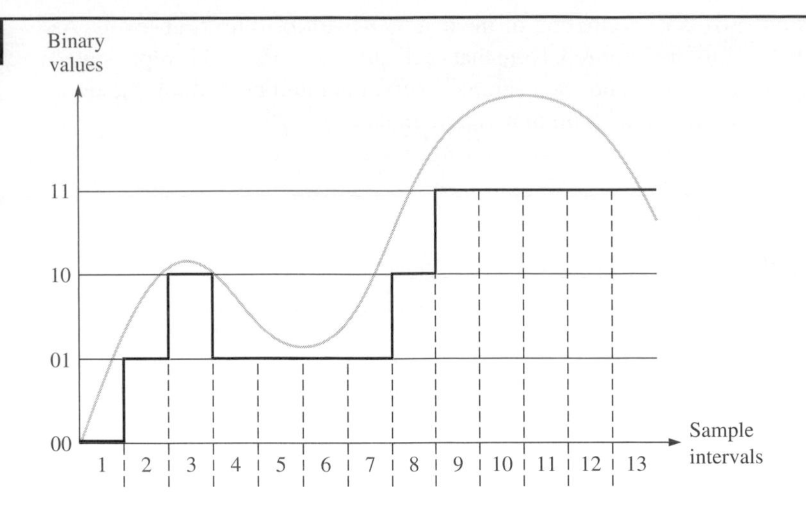

FIGURE 12–11

Sample-and-hold output waveform with sixteen quantization levels. The original analog waveform is shown in light gray for reference.

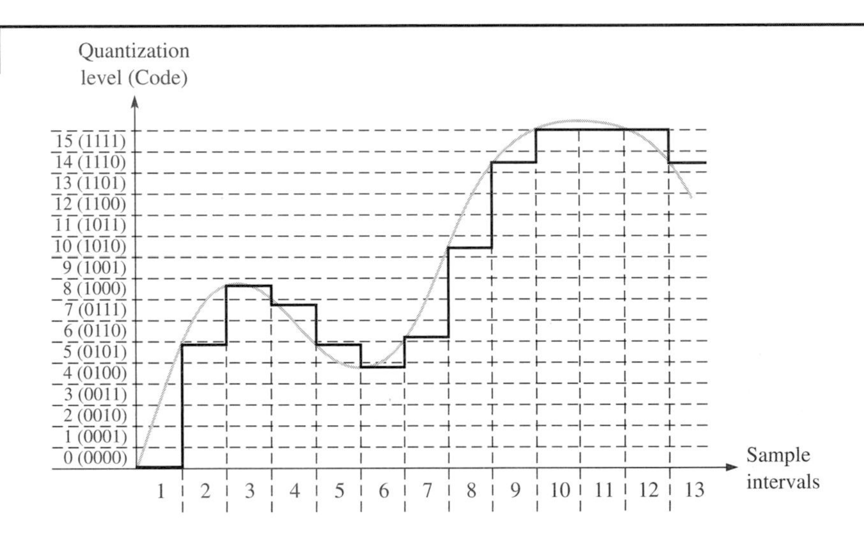

TABLE 12–2

Four-bit quantization for the waveform in Figure 12–11.

Sample interval	Quantization level	Code
1	0	0000
2	5	0101
3	8	1000
4	7	0111
5	5	0101
6	4	0100
7	6	0110
8	10	1010
9	14	1110
10	15	1111
11	15	1111
12	15	1111
13	14	1110

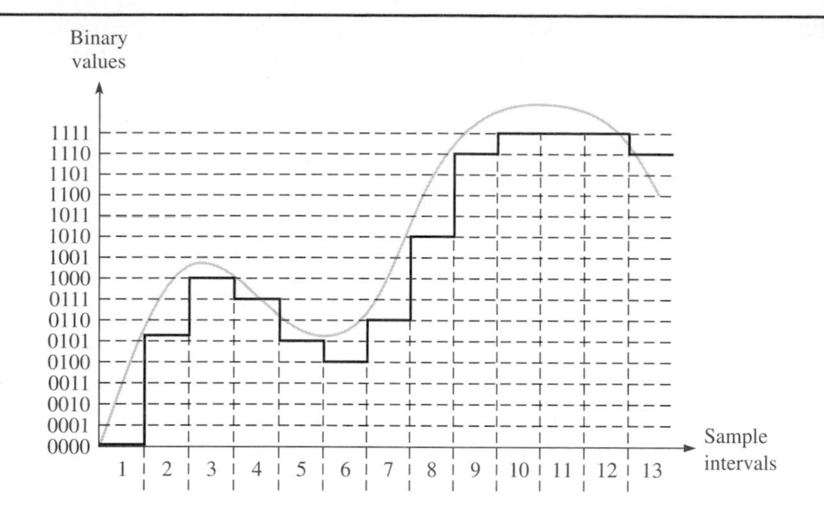

FIGURE 12–12

The reconstructed waveform in Figure 12–11 using sixteen quantization levels (4 bits). The original analog waveform is shown in light gray for reference.

Review Questions

6. What does sampling mean?
7. Why must you hold a sampled value?
8. If the highest frequency component in an analog signal is 20 kHz, what is the minimum sample frequency?
9. What does quantization mean?
10. What determines the accuracy of the quantization process?

ANALOG-TO-DIGITAL CONVERSION METHODS 12–3

As you have seen, analog-to-digital conversion is the process by which an analog quantity is converted to digital form. It is necessary when measured quantities must be in digital form for processing or for display or storage.

In this section, you will learn the purpose of analog-to-digital conversion and how several types of ADCs work.

A Quick Look at an Operational Amplifier

Before getting into analog-to-digital converters (ADCs), let's look briefly at an element that is common to most types of ADC and digital-to-analog converters (DACs). This element is the operational amplifier, or op-amp for short. This coverage of the op-amp is very abbreviated. For a complete understanding of the op-amp and its operation, refer to a text that provides a more thorough coverage.

An *op-amp* is a linear amplifier that has two inputs (inverting and noninverting) and one output. It has a very high voltage gain and a very high input resistance, as well as a very low output resistance. The op-amp symbol is shown in Figure 12–13(a). When used as an inverting amplifier, the op-amp is configured as shown in part (b). The feedback resistor, R_f, and the input resistor, R_{in}, control the voltage gain according to the following formula:

$$\frac{V_{out}}{V_{in}} = -\frac{R_f}{R_{in}}$$

where V_{out}/V_{in} is the closed-loop voltage gain (closed loop refers to the feedback from output to input provided by R_f). The negative sign indicates inversion.

FIGURE 12–13

The operational amplifier (op-amp).

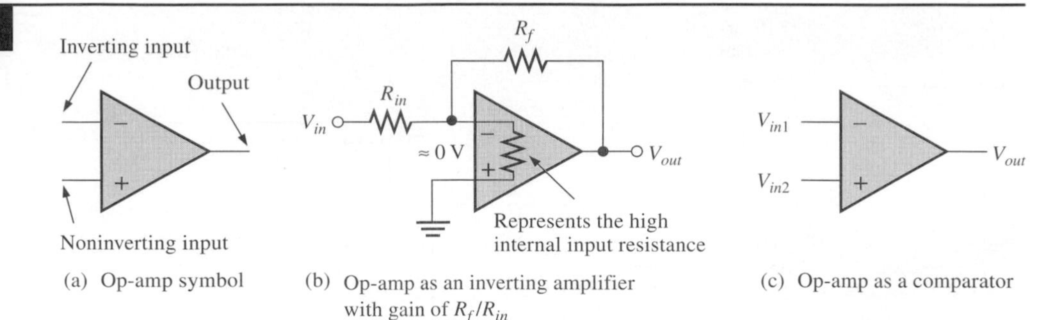

In the inverting amplifier configuration, the inverting input of the op-amp is approximately at ground potential (0 V) because feedback and the extremely high open-loop gain make the differential voltage between the two inputs extremely small. Since the noninverting input is grounded, the inverting input is at approximately 0 V, which is called *virtual ground.*

When the op-amp is used as a comparator, as shown in Figure 12–13(c), two voltages are applied to the inputs. When these input voltages differ by a very small amount, the op-amp is driven into one of its two saturated output states, either HIGH or LOW, depending on which input voltage is greater.

Flash (Simultaneous) Analog-to-Digital Converter

The flash (do not confuse with flash memory) method utilizes comparators that compare reference voltages with the analog input voltage. When the input voltage exceeds the reference voltage for a given comparator, a HIGH is generated. Figure 12–14 shows a 3-bit converter

FIGURE 12–14

A 3-bit flash ADC.

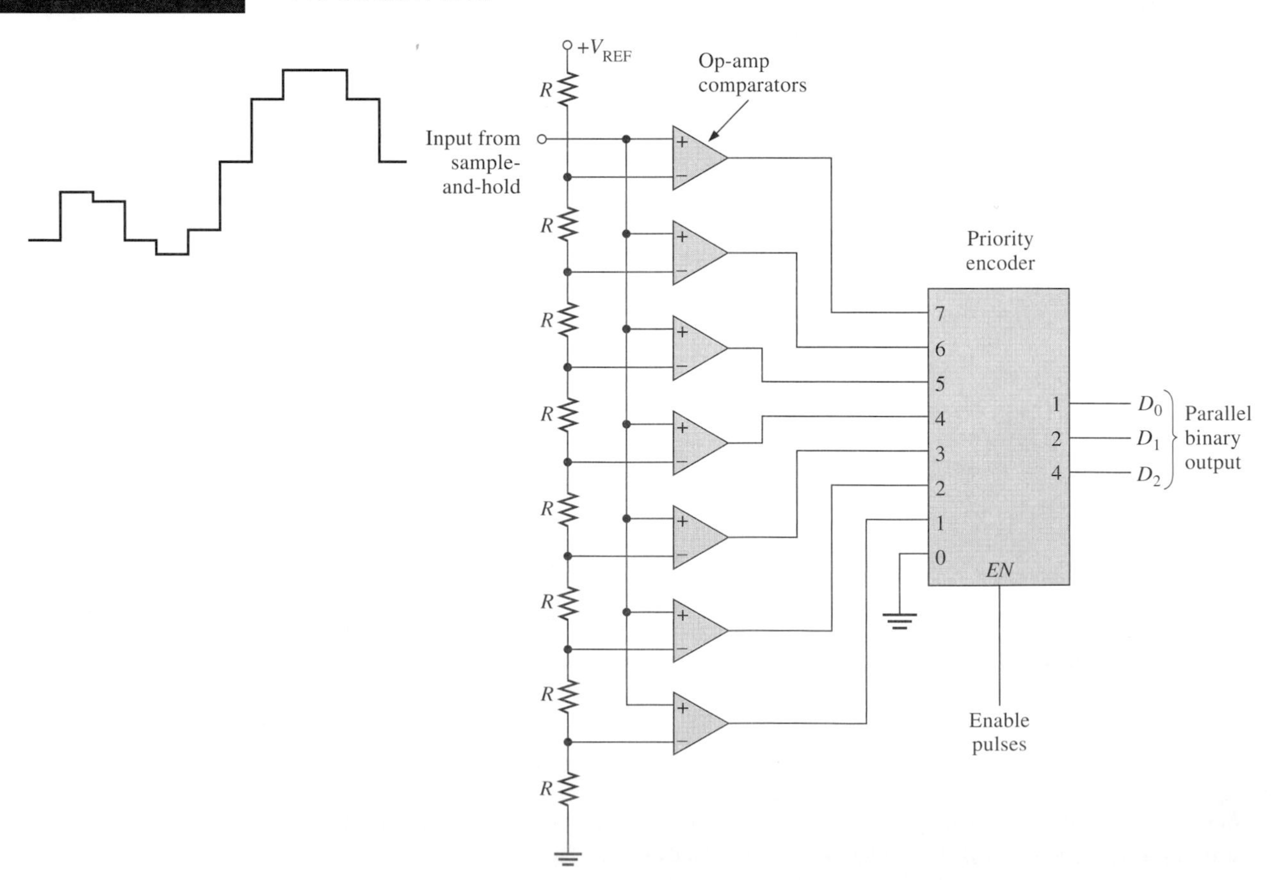

that uses seven comparator circuits; a comparator is not needed for the all-0s condition. A 4-bit converter of this type requires fifteen comparators. In general, $2^n - 1$ comparators are required for conversion to an n-bit binary code. The number of bits used in an ADC is its resolution. The large number of comparators necessary for a reasonable-sized binary number is one of the disadvantages of the flash ADC. Its chief advantage is that it provides a fast conversion time because of a high throughput, measured in samples per second (sps).

The reference voltage for each comparator is set by the resistive voltage-divider circuit. The output of each comparator is connected to an input of the priority encoder. The encoder is enabled by a pulse on the *EN* input, and a 3-bit code representing the value of the input appears on the encoder's outputs. The binary code is determined by the highest-order input having a HIGH level.

The frequency of the enable pulses and the number of bits in the binary code determine the accuracy with which the sequence of binary codes represents the input of the ADC. There should be one enable pulse for each sampled level of the input signal.

EXAMPLE 12–1

Problem

Determine the binary code output of the 3-bit flash ADC in Figure 12–14 for the input signal in Figure 12–15 and the encoder enable pulses shown. $V_{REF} = +8$ V.

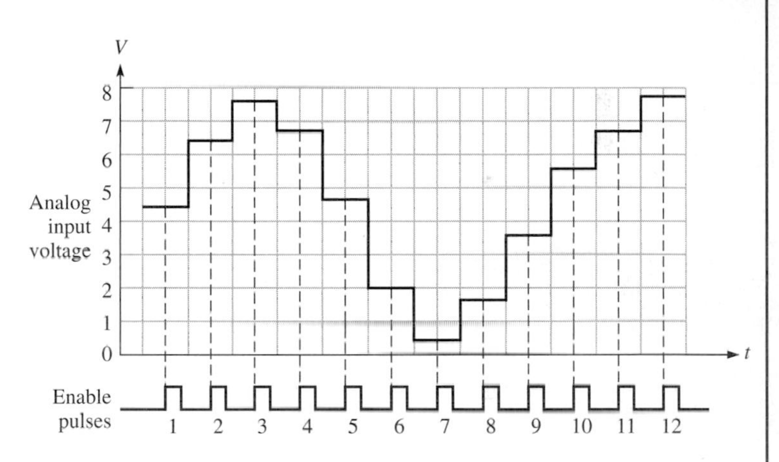

FIGURE 12–15
Sampling of values on a waveform for conversion to binary code.

Solution

The resulting digital output sequence is listed as follows and shown in the waveform diagram of Figure 12–16 in relation to the enable pulses:

100, 110, 111, 110, 100, 010, 000, 001, 011, 101, 110, 111

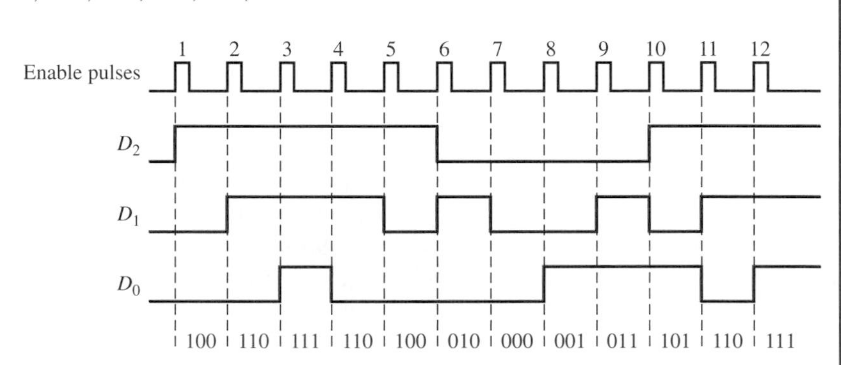

FIGURE 12–16
Resulting digital outputs for sample-and-hold values. Output D_0 is the LSB of the 3-bit binary code.

Question*

If the enable pulse frequency in Figure 12–15 were halved, what binary numbers are represented by the resulting digital output sequence for 6 pulses? Is any information lost?

* *Answers are at the end of the chapter.*

Dual-Slope Analog-to-Digital Converter

A dual-slope ADC is common in digital voltmeters and other types of measurement instruments. A ramp generator (integrator) is used to produce the dual-slope characteristic. A block diagram of a dual-slope ADC is shown in Figure 12–17.

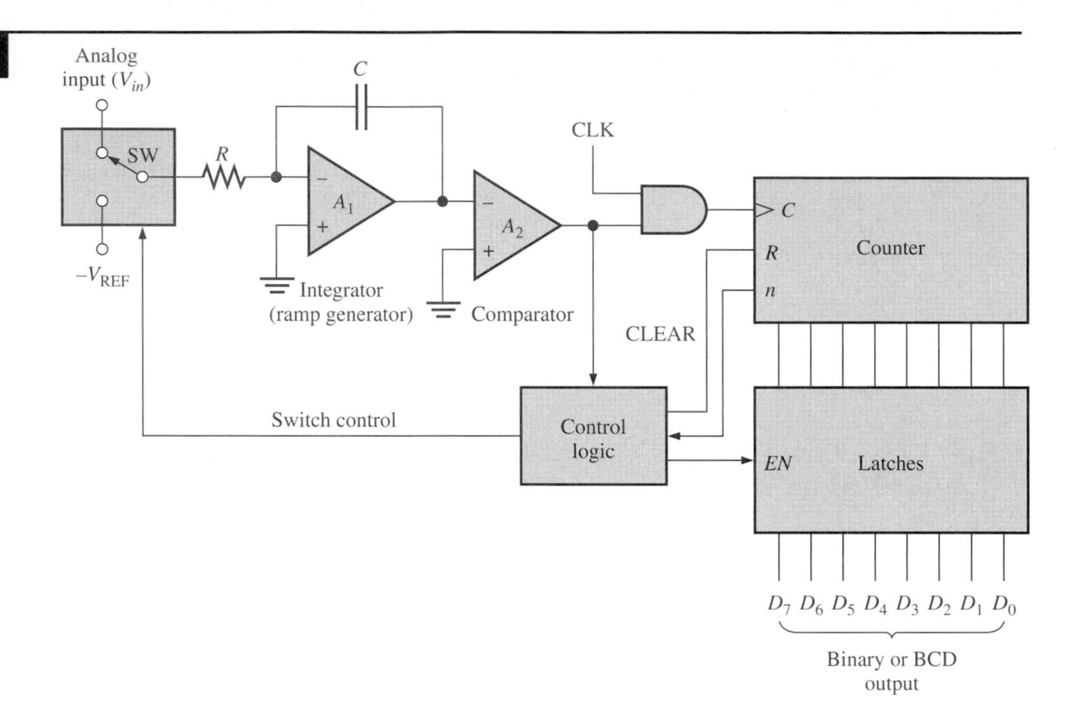

FIGURE 12–17 Basic dual-slope ADC.

Figure 12–18 illustrates dual-slope conversion. Start by assuming that the counter is reset and the output of the integrator is zero. Now assume that a positive input voltage is applied to the input through the switch (SW) as selected by the control logic. Since the inverting input of A_1 is at virtual ground, and assuming that V_{in} is constant for a period of time, there will be constant current through the input resistor R and therefore through the capacitor C. Capacitor C will charge linearly because the current is constant, and as a result, there will be a negative-going linear voltage ramp on the output of A_1, as illustrated in Figure 12–18(a).

When the counter reaches a specified count, it will be reset, and the control logic will switch the negative reference voltage ($-V_{REF}$) to the input of A_1, as shown in Figure 12–18(b). At this point the capacitor is charged to a negative voltage ($-V$) proportional to the input analog voltage.

Now the capacitor discharges linearly because of the constant current from the $-V_{REF}$, as shown in Figure 12–18(c). This linear discharge produces a positive-going ramp on the A_1 output, starting at $-V$ and having a constant slope that is independent of the charge voltage. As the capacitor discharges, the counter advances from its RESET state. The time it takes the capacitor to discharge to zero depends on the initial voltage $-V$ (proportional to V_{in}) because the discharge rate (slope) is constant. When the integrator (A_1) output voltage reaches zero, the comparator (A_2) switches to the LOW state and disables the clock to the counter. The binary count is latched, thus completing one conversion cycle. The binary count is proportional to V_{in} because the time it takes the capacitor to discharge depends only on $-V$, and the counter records this interval of time.

FIGURE 12–18 Illustration of dual-slope conversion.

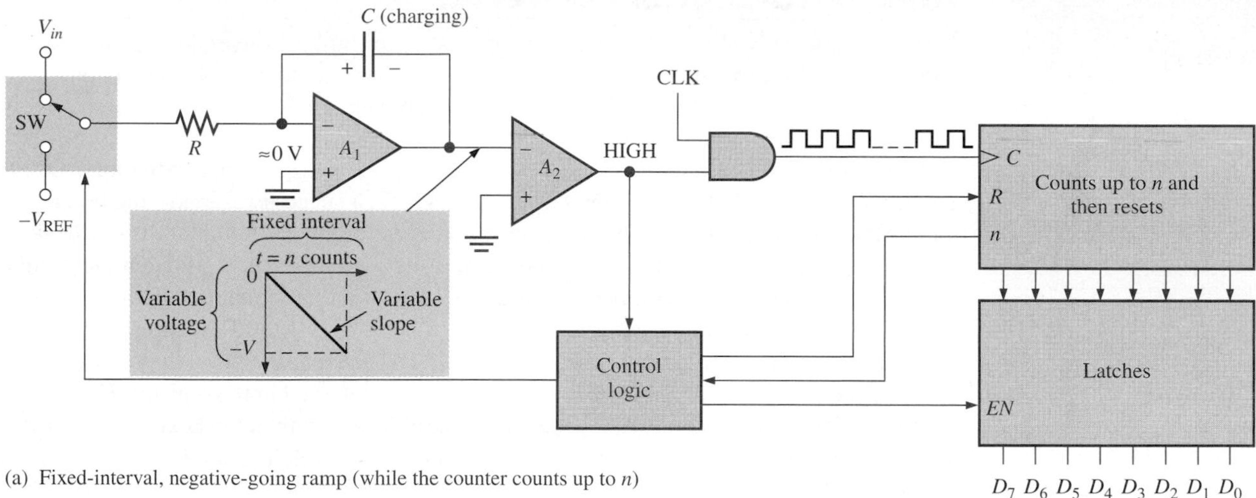

(a) Fixed-interval, negative-going ramp (while the counter counts up to n)

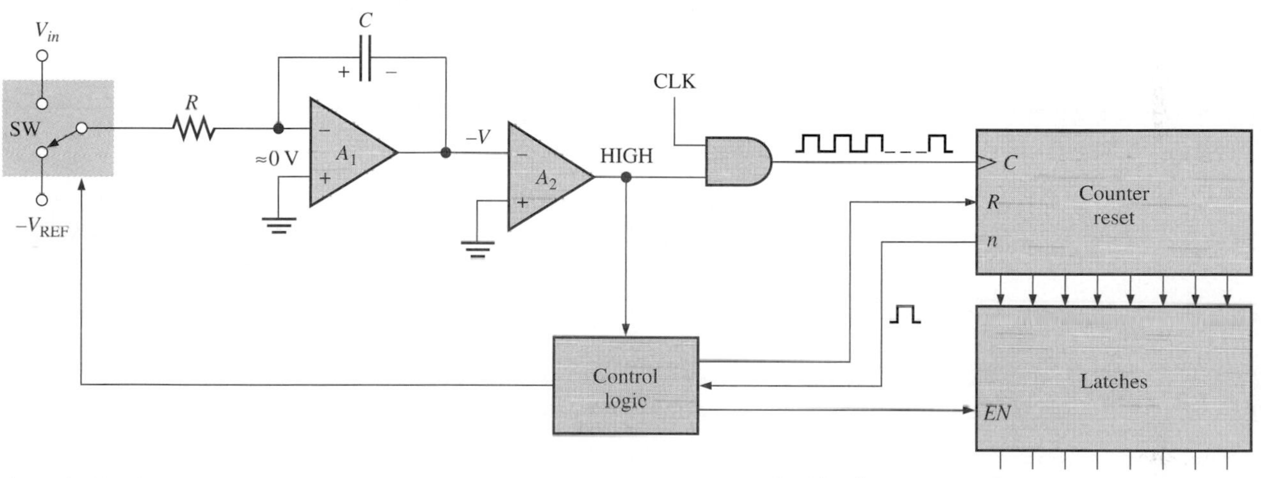

(b) End of fixed-interval when the counter sends a pulse to control logic to switch SW to the $-V_{REF}$ input

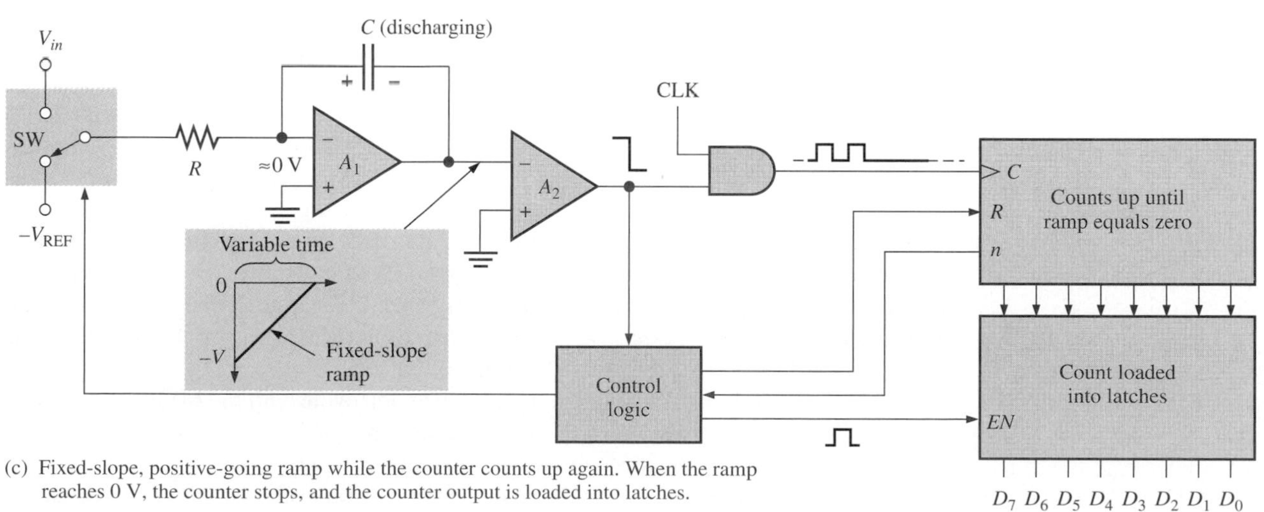

(c) Fixed-slope, positive-going ramp while the counter counts up again. When the ramp reaches 0 V, the counter stops, and the counter output is loaded into latches.

Successive-Approximation Analog-to-Digital Converter

One of the most widely used methods of analog-to-digital conversion is successive-approximation. It has a much faster conversion time than the dual-slope conversion, but it is slower than the flash method. It also has a fixed conversion time that is the same for any value of the analog input.

Figure 12–19 shows a basic block diagram of a 4-bit successive approximation ADC. It consists of a DAC (DACs are covered in Section 12–5), a successive-approximation register (SAR), and a comparator. The basic operation is as follows: The input bits of the DAC are enabled (made equal to a 1) one at a time, starting with the most significant bit (MSB). As each bit is enabled, the comparator produces an output that indicates whether the input signal voltage is greater or less than the output of the DAC. If the DAC output is greater than the input signal, the comparator's output is LOW, causing the bit in the register to reset. If the output is less than the input signal, the 1 bit is retained in the register. The system does this with the MSB first, then the next most significant bit, then the next, and so on. After all the bits of the DAC have been tried, the conversion cycle is complete.

FIGURE 12–19

Successive-approximation ADC.

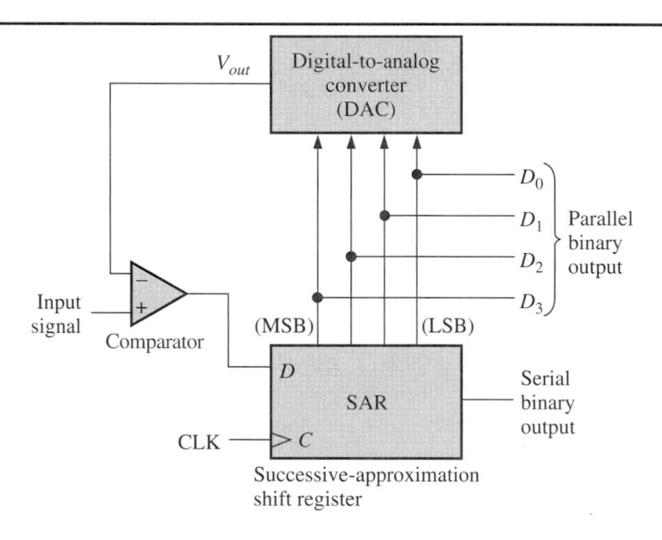

In order to better understand the operation of the successive-approximation ADC, let's take a specific example of a 4-bit conversion. Figure 12–20 illustrates the step-by-step conversion of a constant input voltage (5.1 V in this case). Let's assume that the DAC has the following output characteristic: V_{out} = 8 V for the 2^3 bit (MSB), V_{out} = 4 V for the 2^2 bit, V_{out} = 2 V for the 2^1 bit, and V_{out} = 1 V for the 2^0 bit (LSB).

Figure 12–20(a) shows the first step in the conversion cycle with the MSB = 1. The output of the DAC is 8 V. Since this is greater than the input of 5.1 V, the output of the comparator is LOW, causing the MSB in the SAR to be reset to a 0.

Figure 12–20(b) shows the second step in the conversion cycle with the 2^2 bit equal to a 1. The output of the DAC is 4 V. Since this is less than the input of 5.1 V, the output of the comparator switches to a HIGH, causing this bit to be retained in the SAR.

Figure 12–20(c) shows the third step in the conversion cycle with the 2^1 bit equal to a 1. The output of the DAC is 6 V because there is a 1 on the 2^2 bit input and on the 2^1 bit in-

FIGURE 12–20 Illustration of the successive-approximation conversion process.

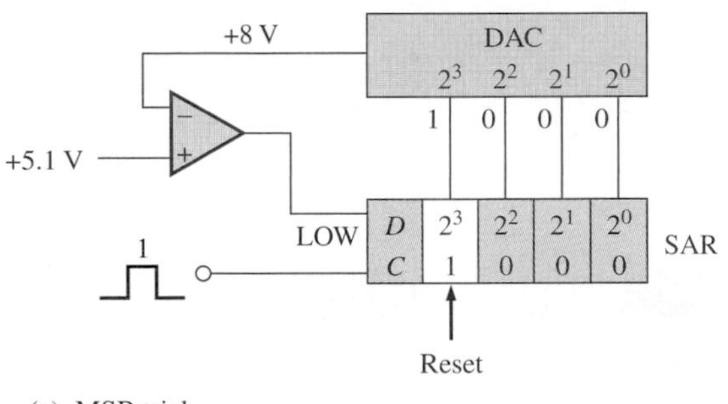

(a) MSB trial

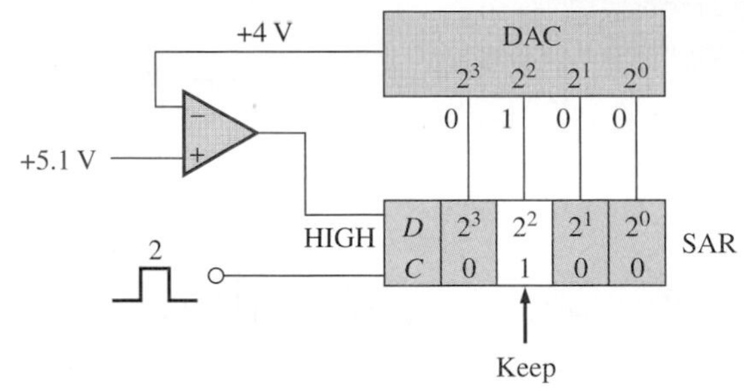

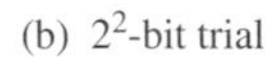

(b) 2^2-bit trial

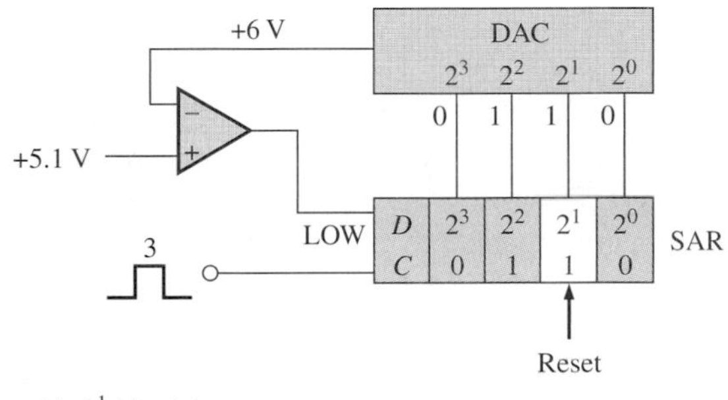

(c) 2^1-bit trial

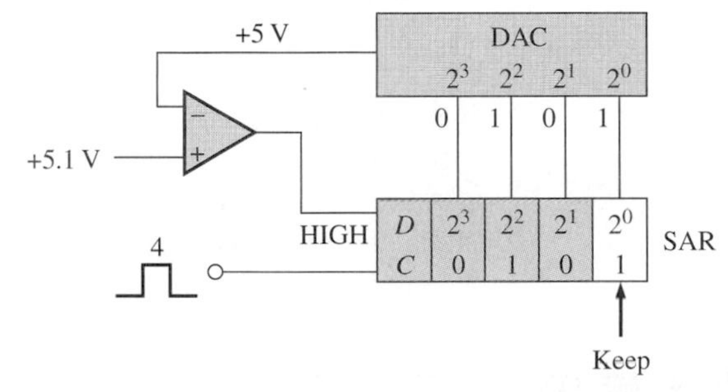

(d) LSB trial (conversion complete)

put; 4 V + 2 V = 6 V. Since this is greater than the input of 5.1 V, the output of the comparator switches to a LOW, causing this bit to be reset to a 0.

Figure 12–20(d) shows the fourth and final step in the conversion cycle with the 2^0 bit equal to a 1. The output of the DAC is 5 V because there is a 1 on the 2^2 bit input and on the 2^0 bit input; 4 V + 1 V = 5 V.

The four bits have all been tried, thus completing the conversion cycle. At this point the binary code in the register is 0101, which is approximately the binary value of the input of 5.1 V. Additional bits will produce an even more accurate result. Another conversion cycle now begins, and the basic process is repeated. The SAR is cleared at the beginning of each cycle.

SAFETY NOTE

Iron or steel hand tools may produce sparks that can be an ignition source around flammable substances. If this hazard exists, use spark-resistant tools made of nonferrous materials.

Sigma-Delta Analog-to-Digital Converter

Sigma-delta is a widely used method of analog-to-digital conversion, particularly in telecommunications using audio signals. The method is based on *delta modulation* where the *difference* between two successive samples (increase or decrease) is quantized; other ADC methods were based on the absolute value of a sample. Delta modulation is a 1-bit quantization method.

The output of a delta modulator is a single-bit data stream where the relative number of 1s and 0s indicates the level or amplitude of the input signal. The number of 1s over a given number of clock cycles establishes the signal amplitude during that interval. A maximum number of 1s corresponds to the maximum positive input voltage. A number of 1s equal to one-half the maximum corresponds to an input voltage of zero. No 1s (all 0s) corresponds to the maximum negative input voltage. This is illustrated in a simplified way in Figure 12–21. For example, assume that 4096 1s occur during the interval when the input signal is a positive maximum. Since zero is the midpoint of the dynamic range of the input signal, 2048 1s occur during the interval when the input signal is zero. There are no 1s during the interval when the input signal is a negative maximum. For signal levels in between, the number of 1s is proportional to the level.

FIGURE 12–21

A simplified illustration of sigma-delta analog-to-digital conversion.

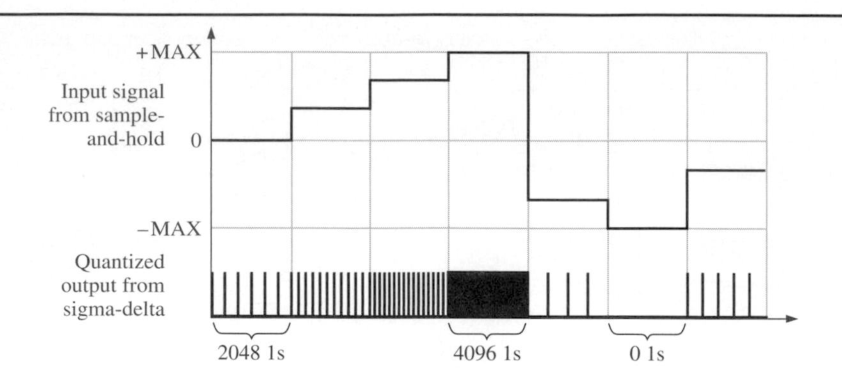

The Sigma-Delta ADC Functional Block Diagram

The basic block diagram in Figure 12–22 accomplishes the conversion illustrated in Figure 12–21. The analog input signal and the analog signal from the converted quantized bit stream from the DAC in the feedback loop are applied to the summation (Σ) point. The difference (Δ) signal out of the Σ is integrated, and the 1-bit ADC increases or decreases the number of 1s depending on the difference signal. This action attempts to keep the quantized signal that is fed back equal to the incoming analog signal. The 1-bit quantizer is essentially a comparator followed by a latch.

FIGURE 12–22

Partial functional block diagram of a sigma-delta ADC.

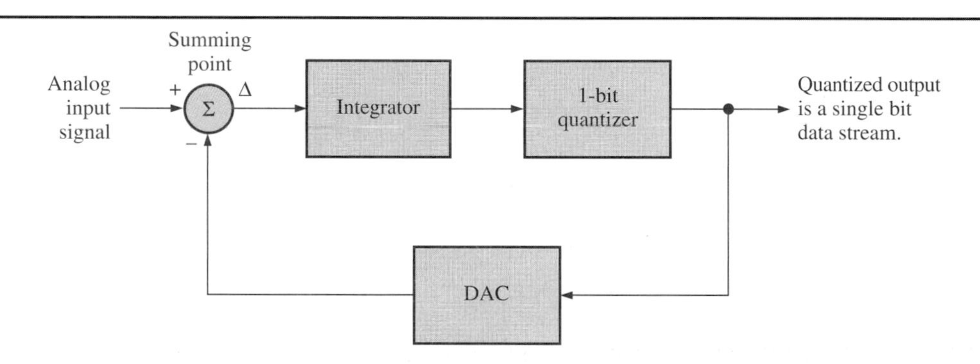

To complete the sigma-delta conversion process using one particular approach, the single bit data stream is converted to a series of binary codes, as shown in Figure 12–23. The counter counts the 1s in the quantized data stream for successive intervals and is cleared at the beginning of each interval. The code in the counter then represents the amplitude of the analog input signal for each interval. These codes are shifted out into the latch for temporary storage. What comes out of the latch is a series of *n*-bit codes, which completely represent the analog signal.

FIGURE 12–23 One type of sigma-delta ADC.

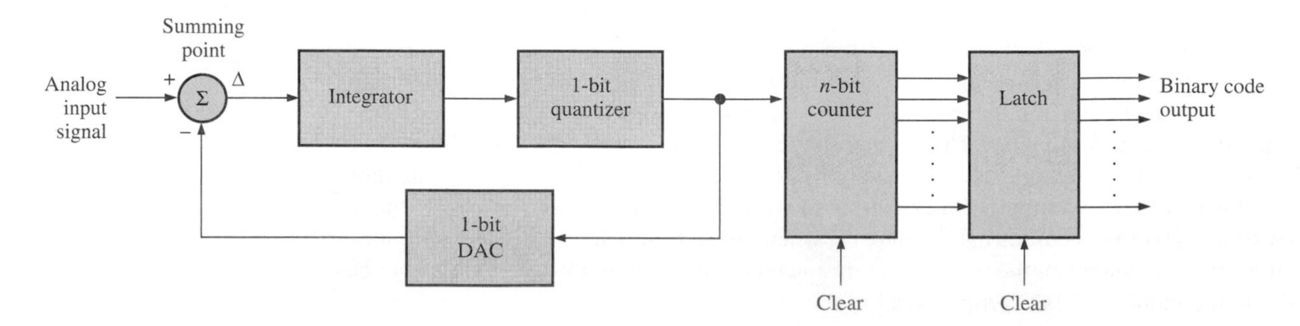

Testing Analog-to-Digital Converters

One method for testing ADCs is shown in Figure 12–24. A DAC is used as part of the test setup to convert the ADC output back to analog form for comparison with the test input.

A test input in the form of a linear ramp is applied to the input of the ADC. The resulting binary output sequence is then applied to the DAC test unit and converted to a stairstep ramp. The input and output ramps are compared for any deviation.

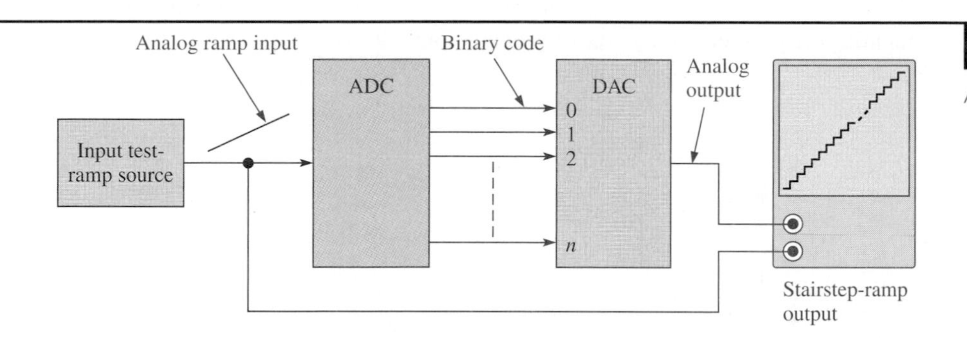

FIGURE 12–24

A method for testing ADCs.

Analog-to-Digital Conversion Errors

Again, a 4-bit conversion is used to illustrate the principles. Let's assume that the test input is an ideal linear ramp.

Missing Code

The stairstep output in Figure 12–25(a) indicates that the binary code 1001 does not appear on the output of the ADC. Notice that the 1000 value stays for two intervals and then the output jumps to the 1010 value.

In a flash ADC, for example, a failure of one of the comparators can cause a missing-code error.

FIGURE 12–25

Illustrations of analog-to-digital conversion errors.

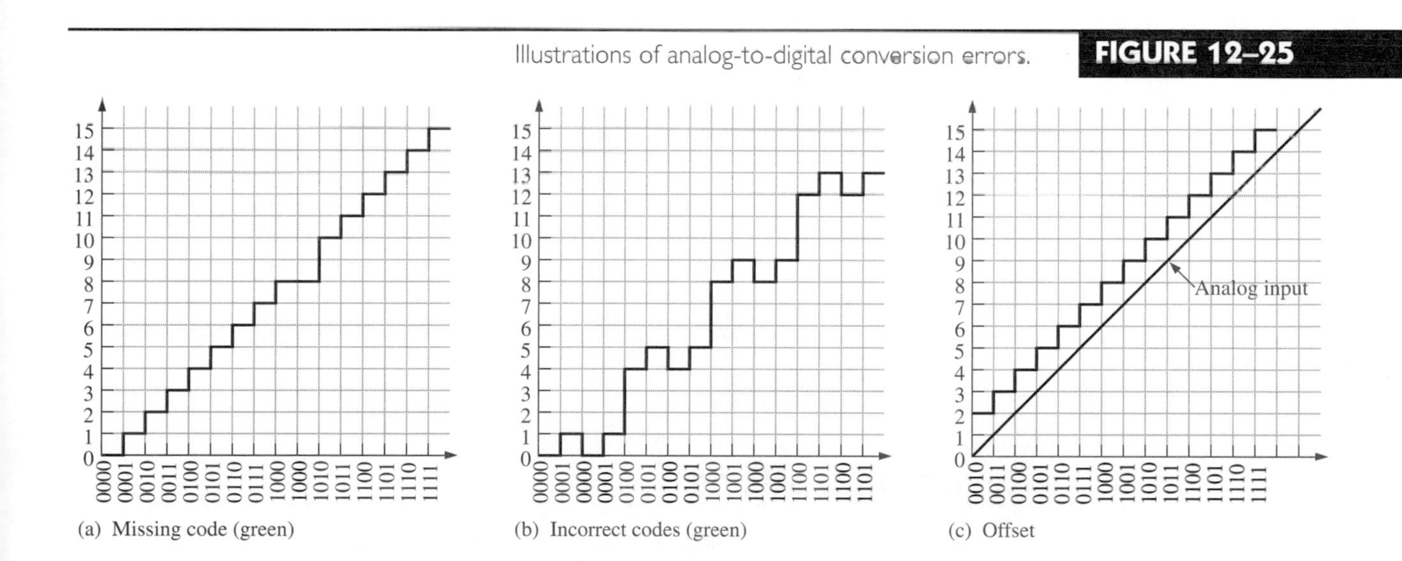

(a) Missing code (green) (b) Incorrect codes (green) (c) Offset

Incorrect Codes

The stairstep output in Figure 12–25(b) indicates that several of the binary code words coming out of the ADC are incorrect. Analysis indicates that the 2^1-bit line is stuck in the LOW (0) state in this particular case.

Offset

Offset conditions are shown in 12–25(c). In this situation the ADC interprets the analog input voltage as greater than its actual value.

EXAMPLE 12–2

Problem

A 4-bit flash ADC is shown in Figure 12–26(a). It is tested with a setup like the one in Figure 12–24. The resulting reconstructed analog output is shown in Figure 12–26(b). Identify the problem and the most probable fault.

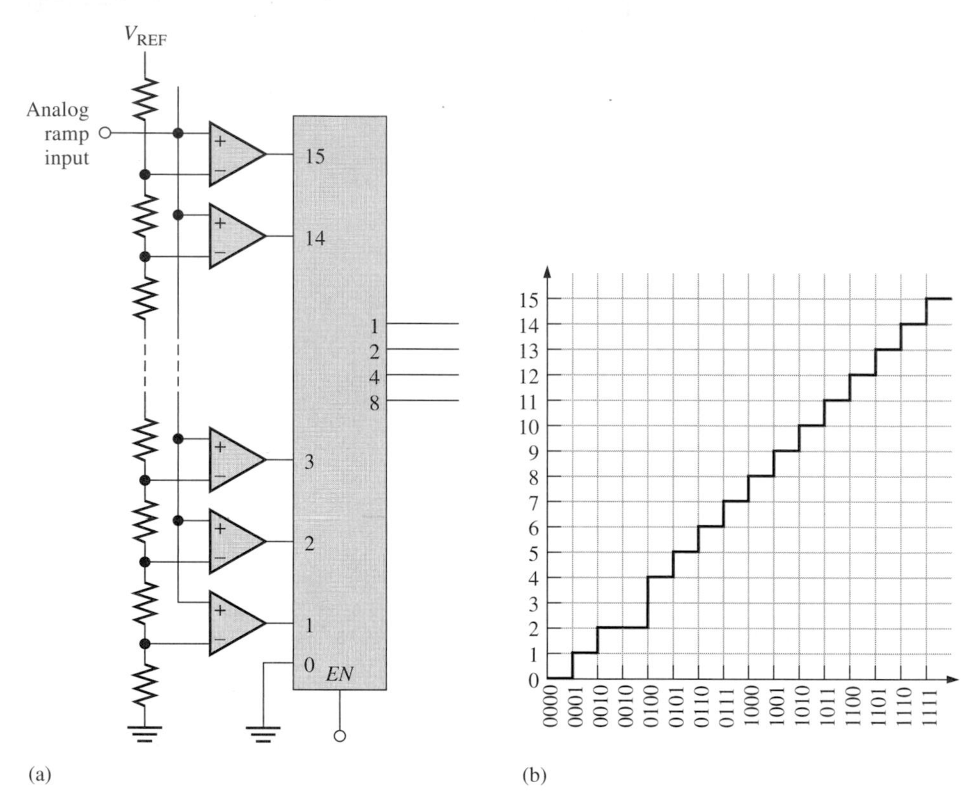

FIGURE 12–26

Solution

The binary code 0011 is missing from the ADC output, as indicated by the missing step. Most likely, the output of comparator 3 is stuck in its inactive state (LOW).

Question

Reconstruct the analog output in a test setup like in Figure 12–24 if the ADC in Figure 12–26(a) has comparator 8 stuck in the HIGH output state.

Review Questions

11. What is the fastest method of analog-to-digital conversion?
12. Which analog-to-digital conversion method produces a single-bit data stream?
13. Does the successive-approximation converter have a fixed conversion time?
14. What is an SAR?
15. What are two types of output errors in an ADC?

THE DIGITAL SIGNAL PROCESSOR (DSP) 12–4

Essentially, a digital signal processor (DSP) is a special type of microprocessor that processes data in real time. Its applications focus on the processing of digital data that represents analog signals. A DSP, like a microprocessor, has a central processing unit (CPU) and memory units in addition to many interfacing functions. Every time you use your cellular telephone, you are using a DSP, and this is only one example of its many applications.

In this section, you will learn the basic concepts of a digital signal processor.

The digital signal processor (DSP) is the heart of a digital signal processing system. It takes its input from an ADC and produces an output that goes to a DAC, as shown in Figure 12–27. As you have learned, the ADC changes an analog waveform into data in the form of a series of binary codes that are then applied to the DSP for processing. After being processed by the DSP, the data go to a DAC for conversion back to analog form.

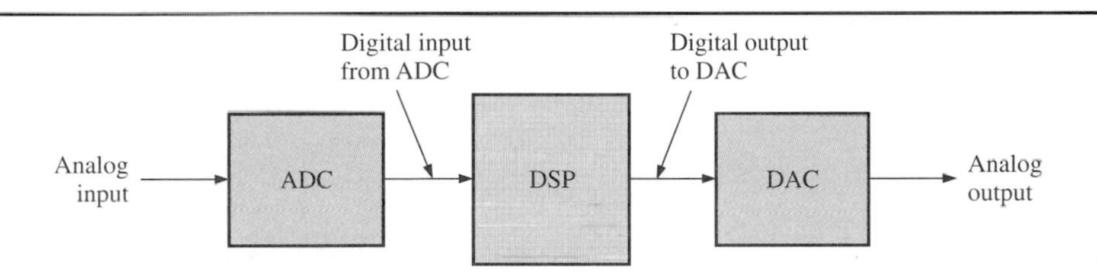

FIGURE 12–27

The DSP has a digital input and produces a digital output.

DSP Programming

DSPs are typically programmed in either assembly language or in C. Because programs written in assembly language can usually execute faster and because speed is critical in most DSP applications, assembly language is used much more in DSPs than in general-purpose microprocessors. Also, DSP programs are usually much shorter than traditional microprocessor programs because of their very specialized applications where much redundancy is used. In general, the instruction sets for DSPs tend to be smaller than for microprocessors.

DSP Applications

The DSP, unlike the general-purpose microprocessor, must typically process data in real time; that is, as it happens. Many applications in which DSPs are used cannot tolerate any noticeable delays, requiring the DSP to be extremely fast. The DSP manipulates data for a variety of applications too numerous to mention, but some of the areas where it has had a major impact are as follows.

Telecommunications

The field of telecommunications involves transferring all types of information from one location to another, including telephone conversations, television signals, and digital data. Among other functions, the DSP facilitates multiplexing many signals onto one transmission channel because information in digital form is relatively easy to multiplex and demultiplex.

At the transmitting end of a telecommunications system, DSPs are used to compress digitized voice signals for conservation of bandwidth. Compression is the process of reducing the data rate. Generally, a voice signal is converted to digital form at 8000 sps (samples per second), based on a Nyquist frequency of 4 kHz. If 8 bits are used to encode each sample, the data rate is 64 kbits/s. In general, reducing (compressing) the data rate from 64 kbits/s to 32 kbits/s results in no loss of sound quality. When the data are compressed to 8 kbits/s, the sound quality is reduced noticeably. When compressed to the minimum of 2 kbits/s, the sound is greatly distorted but still usable for some applications where only word recognition and not quality is important. At the receiving end of a telecommunications system, the DSP decompresses the data to restore the signal to its original form.

Echoes, a problem in many long distance telephone connections, occur when a portion of a voice signal is returned with a delay. For shorter distances, this delay is barely noticeable; but as the distance between the transmitter and the receiver increases, so does the delay time of the echo. DSPs are used to effectively cancel the annoying echo, which results in a clear, undisturbed voice signal.

Music Processing

The DSP is used in the music industry to provide filtering, signal addition and subtraction, and signal editing in music preparation and recording. Also, another application of the DSP is to add artificial echo and reverberation, which are usually minimized by the acoustics of a sound studio, in order to simulate ideal listening environments from concert halls to small rooms. DSPs are also used in electronic music synthesizers.

Speech Generation and Recognition

DSPs are used in speech generation and recognition to enhance the quality of man/machine communication. The most common method used to produce computer-generated speech is digital recording. In digital recording, the human voice is digitized and stored, usually in a compressed form. During playback the stored voice data are uncompressed and converted back into the original analog form. Approximately an hour of speech can be stored using about 3 MB of memory.

Speech recognition is much more difficult to accomplish than speech generation. Even with today's computers, speech recognition is very limited and, with a few exceptions, the results are only moderately successful. The DSP is used to isolate and analyze each word in the incoming voice signal. Certain parameters are identified in each word and compared with previous examples of the spoken word to create the closest match. Most systems are limited to a few hundred words at best. Also, significant pauses between words are usually required and the system must be "trained" for a given individual's voice. Speech recognition is an area of tremendous research effort and will eventually be applied in many commercial applications.

Radar

In *r*adio *d*etection *a*nd *r*anging (radar) applications, DSPs provide more accurate determination of distance using data compression techniques. They also can decrease noise using filtering techniques, thereby increasing the range, and optimize the ability of the radar system to identity specific types of targets. DSPs are also used in similar ways in sonar systems.

Image Processing

The DSP is used in image-processing applications such as the computed tomography (CT) and magnetic resonance imaging (MRI), which are widely used in the medical field for looking inside the human body. In CT, X-rays are passed through a section of the body from many directions. The resulting signals are converted to digital form and stored. This stored information is used to produce calculated images that appear to be slices through the human body that show great detail and permit better diagnosis.

Instead of X-rays, MRI uses magnetic fields in conjunction with radio waves to probe inside the human body. MRI produces images, just as CT, and provides excellent discrimination between different types of tissue as well as information such as blood flow through arteries. MRI depends entirely on digital signal processing methods.

In applications such as video telephones, digital television, and other media that provide moving pictures, the DSP uses image compression to reduce the number of bits needed, making these systems commercially feasible.

Filtering

DSPs are commonly used to implement digital filters for the purposes of separating signals that have been combined with other signals or with interference and noise and for restoring signals that are distorted. Although analog filters are quite adequate for some applications, the digital filter is generally much superior in terms of the performance that can be achieved. One drawback to digital filters is that the execute time required produces a delay from the time the analog signal is applied until the time the output appears. Analog filters present no delay problems because as soon as the input occurs, the response appears on the output. Analog filters are also less expensive than digital filters. Regardless of this, the overall performance of the digital filter is far superior in many applications.

Basic DSP Architecture

As mentioned before, a DSP is basically a specialized microprocessor optimized for speed in order to process data in real time. Many DSPs are based on what is known as the Harvard architecture, which consists of a central processing unit (CPU) and two memories, one for data and the other for the program, as shown by the block diagram in Figure 12–28.

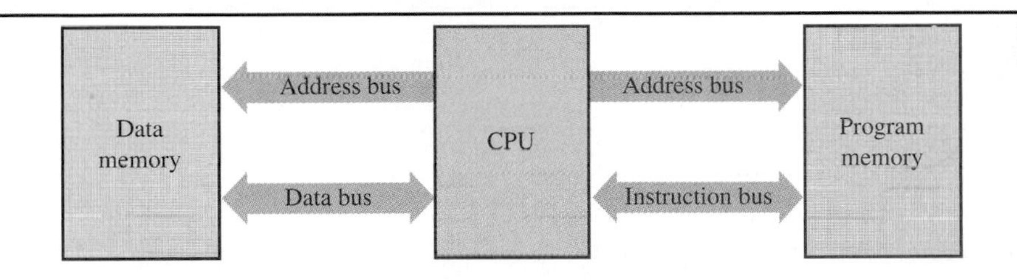

FIGURE 12–28

Many DSPs use the Harvard architecture (two memories).

Specific DSPs—The TMS320C6000 Series

DSPs are manufactured by several companies including Texas Instruments, Motorola, and Analog Devices. DSPs are available for both fixed-point and floating-point processing. All floating-point DSPs can also handle numbers in fixed-point format. Fixed-point DSPs are less expensive than the floating-point versions and, generally, can operate faster. The details of DSP architecture can vary significantly, even within the same family. Let's look briefly at one particular DSP series as an example of how a DSP is generally organized.

Examples of DSPs available in the TMS320C6000 series include the TMS320C62xx, the TMS320C64xx, and the TMS320C67xx, which are part of Texas Instrument's TMS320 family of devices. A general block diagram for these devices is shown in Figure 12–29.

The DSPs have a central processing unit (CPU) also known as the *DSP core,* that contains 64 general-purpose 32-bit registers in the C64xx and 32 general-purpose 32-bit registers in the C62xx and the C67xx. The C67xx can handle floating-point operations, whereas the C62xx and C64xx are fixed-point devices.

Each DSP has eight functional units that contain two 16-bit multipliers and six arithmetic logic units (ALUs). The performance of the three DSPs in the C6000 series in terms of **MIPS** (Million Instructions Per Second), **MFLOPS** (Million Floating-point Operations Per Second), and **MMACS** (Million Multiply/Accumulates per Second) is shown in Table 12–3.

Data Paths in the CPU

In the CPU, the program fetch, instruction dispatch, and instruction decode sections can provide eight 32-bit instructions to the functional units during every clock cycle. The CPU is split into two data paths, and instruction processing occurs in both data paths A and B. Each data path contains half of the general-purpose registers. (16 in the C62xx and C67xx or 32 in the C64xx) and four functional units. The control register and logic are used to configure and control the various processor operations.

FIGURE 12–29 General block diagram of the TMS320C6000 series DSP.

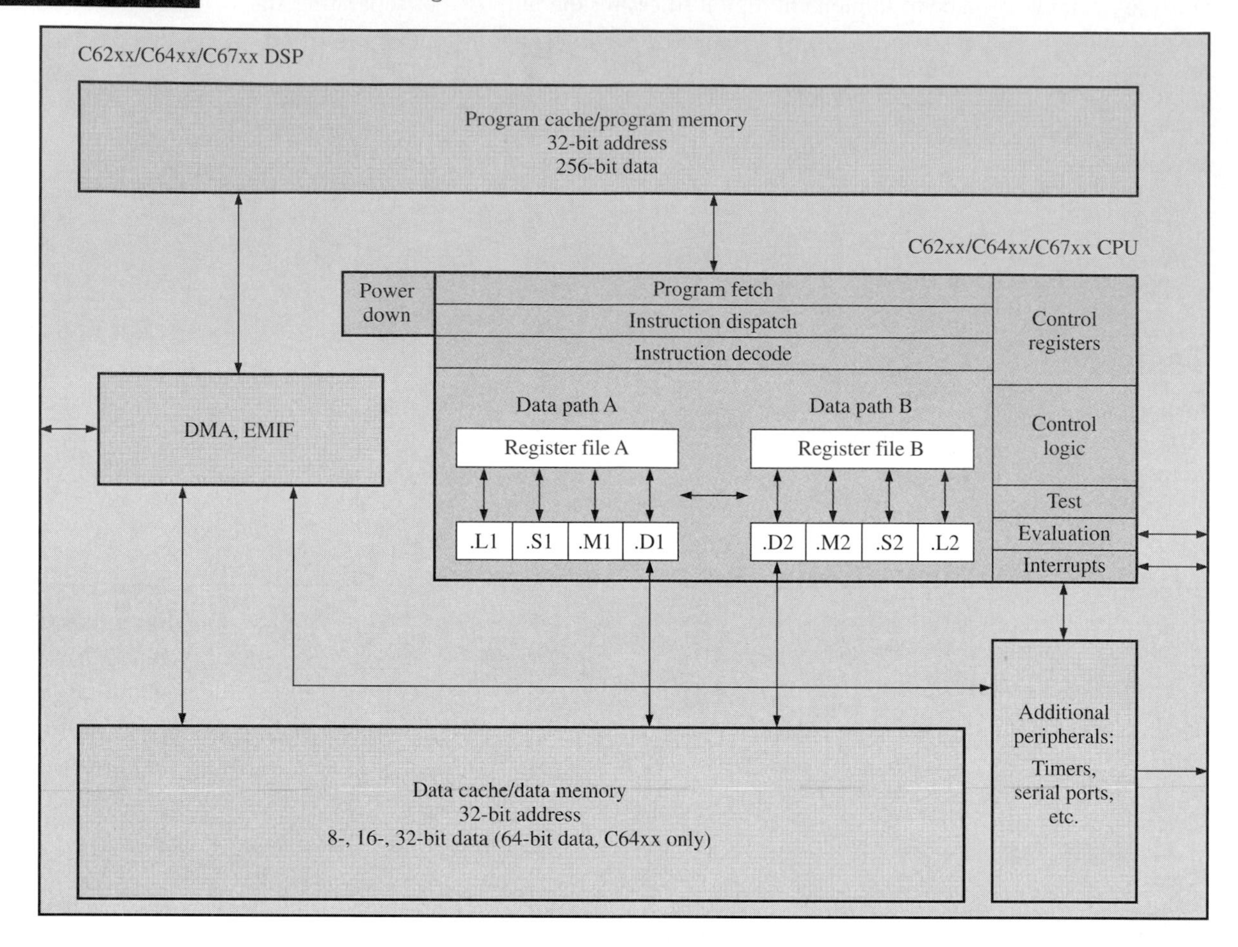

TABLE 12–3 TMSC6000 series DSP data processing performance.

DSP	Type	Application	Processing speed	Multiply/ accumulate speed
C62xx	Fixed-point	General-purpose	1200–2400 MIPS	300–600 MMACS
C64xx	Fixed-point	Special-purpose	3200–4800 MIPS	1600–2400 MMACS
C67xx	Floating-point	General-purpose	600–1000 MFLOPS	200–333 MMACS

Functional Units

Each data path has four functional units. The M units (labeled .M1 and .M2 in Figure 12–29) are dedicated multipliers. The L units (labeled .L1 and .L2) perform arithmetic, logic, and miscellaneous operations. The S units (labeled .S1 and .S2) perform compare, shift, and miscellaneous arithmetic operations. The D units (labeled .D1 and .D2) perform load, store, and miscellaneous operations.

Pipeline

A **pipeline** allows multiple instructions to be processed simultaneously. A pipeline operation consists of three processes: *fetch, decode, execute.* Eight instructions at a time are first fetched from the program memory; they are then decoded, and finally they are executed.

During **fetch**, the eight instructions (called a packet) are taken from memory in four phases, as shown in Figure 12–30.

- *Program address generate (PG).* The program address is generated by the CPU.
- *Program address send (PS).* The program address is sent to the memory.
- *Program access ready wait (PW).* A memory read operation occurs.
- *Program fetch packet receive (PR).* The CPU receives the packet of instructions.

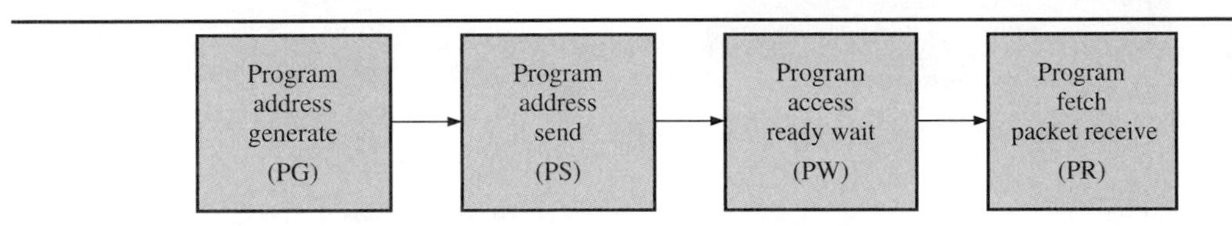

FIGURE 12–30
The four fetch phases of the pipeline operation.

Two phases make up the instruction **decode** process of pipeline operation, as shown in Figure 12–31. The instruction dispatch (DP) phase is where the instruction packets are split into execute packets and assigned to the appropriate functional units. The instruction decode (DC) phase is where the instructions are decoded.

FIGURE 12–31
The two decode phases of the pipeline operation.

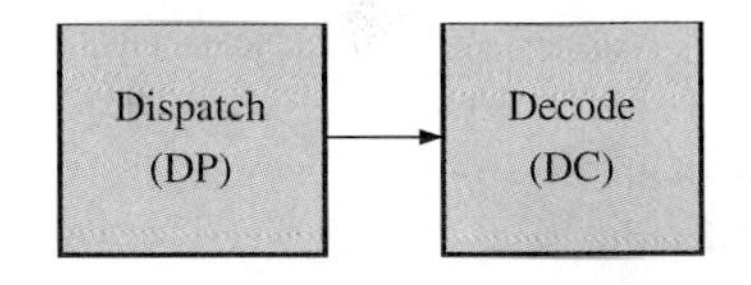

The **execute** process of the pipeline operation is where the instructions from the decode stage are carried out. The execute process has a maximum of five phases (E1 through E5), as shown in Figure 12–32. All instructions do not use all five phases. The number of phases used during execution depends on the type of instruction. Part of the execution of an instruction requires getting data from the data memory.

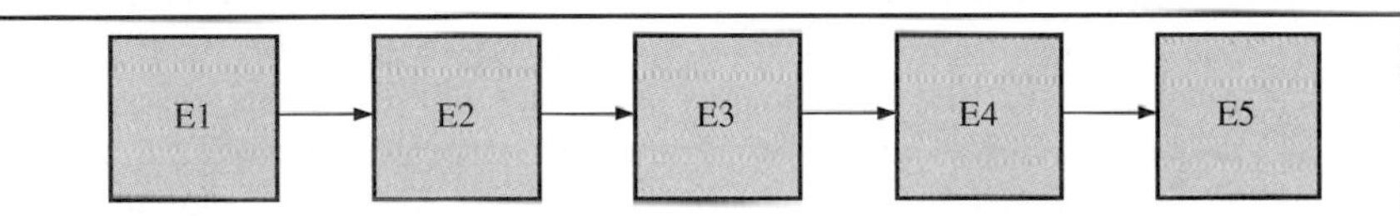

FIGURE 12–32
The five execute phases of pipeline operation.

Internal DSP Memory and Interfaces

As you can see in Figure 12–29, there are two internal memories, one for data and one for program. Notice that the program memory is organized in 256 bit packets (eight 32-bit instructions) so there are 64 kB of capacity. The data memory also has a capacity of 64 kB and can be accessed in 8, 16, 32, or 64-bit word lengths, depending on the specific device in the series. Both internal memories are accessed with a 32-bit address. The DMA (Direct Memory Access) is used to transfer data without going through the CPU. The EMIF (External Memory Interface) is used to support external memories when required in an application. Additional interface is provided for serial I/O ports and other external devices.

Timers

There are two general-purpose timers in the DSP that can be used for timed events, counting, pulse generation, CPU interrupts, and more.

Packaging

These particular processors are available in 352-pin ball grid array (BGA) packages, as shown in Figure 12–33, and are implemented with CMOS technology.

FIGURE 12–33

A 352-pin BGA package.

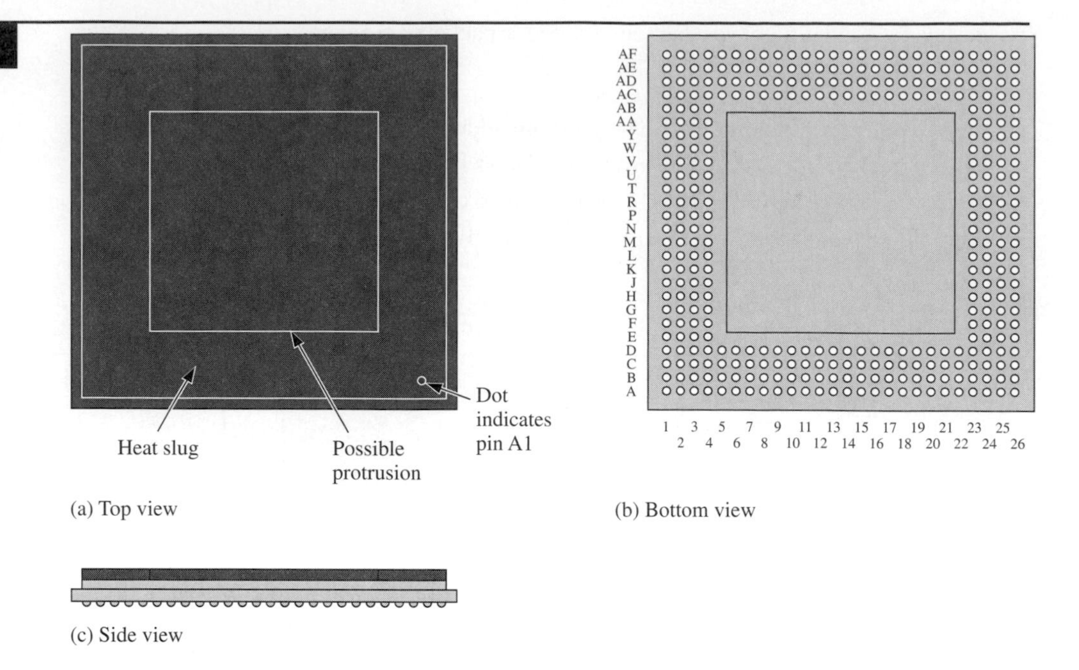

The DSP in a Cellular Telephone

The digital cellular telephone is an example of how a DSP can be used. Figure 12–34 shows a simplified block diagram of a digital cell phone. The voice *codec* (codec is the abbreviation for coder/decoder) contains, among other functions, the ADC and DAC necessary to convert between the analog voice signal and a digital voice format. Sigma-delta conversion is typically used in most cell phone applications. For transmission, the voice signal from the microphone is converted to digital form by the ADC in the codec and then it goes to the DSP for processing. From the DSP, the digital signal goes to the rf (radio frequency) section where it is modulated and changed to the radio frequency for transmission. An incoming rf signal containing voice data is picked up by the antenna, demodulated, and changed to a digital signal. It is then applied to the DSP for processing, after which the digital signal goes to the codec for conversion back to the original voice signal by the DAC. It is then amplified and applied to the speaker.

FIGURE 12–34

Simplified block diagram of a digital cellular phone.

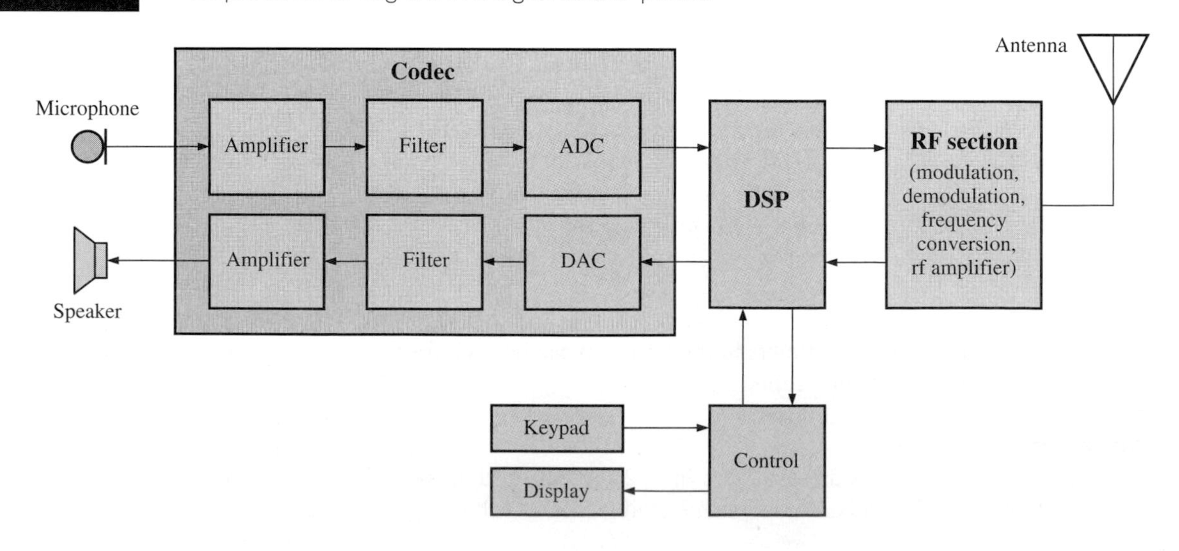

Functions Performed by the DSP

In a cellular phone application, the DSP performs many functions to improve and facilitate the reception and transmission of a voice signal. Some of these DSP functions are as follows:

- *Speech compression.* The rate of the digital voice signal is reduced significantly for transmission in order to meet the bandwidth requirements.
- *Speech decompression.* The rate of the received digital voice signal is returned to its original rate in order to properly reproduce the analog voice signal.
- *Protocol handling.* The cell phone communicates with the nearest base in order to establish the location of the cell phone, allocates time and frequency slots, and arranges handover to another base station as the phone moves into another cell.
- *Error detection and correction.* During transmission, error detection and correction codes are generated and, during reception, detect and correct errors induced in the rf channel by noise or interference.
- *Encryption.* Converts the digital voice signal to a form for secure transmission and converts it back to original form during reception.

Review Questions

16. What is meant by the Harvard architecture?
17. What is a DSP core?
18. What does each of the following mean? (a) MIPS (b) MFLOPS (c) MMACS
19. Basically, what does pipelining accomplish?
20. What are the three stages of pipeline operation?

DIGITAL-TO-ANALOG CONVERSION METHODS 12–5

Digital-to-analog conversion is an important part of the digital processing system. Once the digital data have been processed by the DSP, they are converted back to analog form.

In this section, you will learn how digital signals are converted to analog form.

Binary-Weighted-Input Digital-to-Analog Converter

One method of digital-to-analog conversion uses a resistor network with resistance values that represent the binary weights of the input bits of the digital code. Figure 12–35 shows a 4-bit DAC of this type. Each of the input resistors will either have current or have no cur-

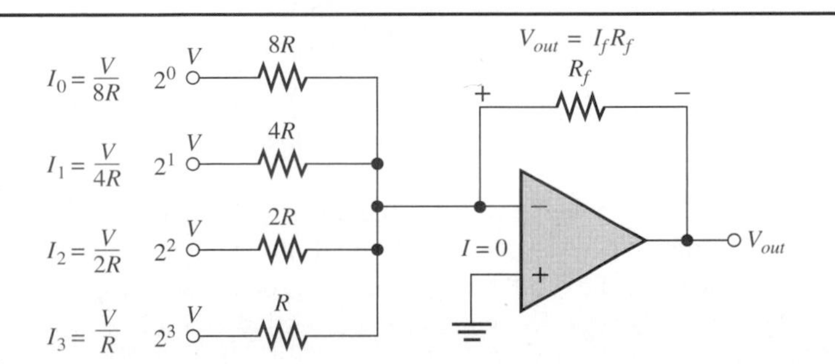

FIGURE 12–35

A 4-bit DAC with binary-weighted inputs.

rent, depending on the input voltage level. If the input voltage is zero (binary 0), the current is also zero. If the input voltage is HIGH (binary 1), the amount of current depends on the input resistor value and is different for each input resistor.

Since there is practically no current into the op-amp inverting (−) input, all of the input currents I_0 through I_3, sum together and go through R_f. Since the inverting input is at 0 V (virtual ground), the drop across R_f is equal to the output voltage, so $V_{out} = I_f R_f$.

The values of the input resistors are chosen to be inversely proportional to the binary weights of the corresponding input bits. The lowest-value resistor (R) corresponds to the highest binary-weighted input (2^3). The other resistors are multiples of R (that is, $2R$, $4R$, and $8R$) and correspond to the binary weights 2^2, 2^1, and 2^0, respectively. The input currents are also proportional to the binary weights. Thus, the output voltage is proportional to the sum of the binary weights because the sum of the input currents is through R_f.

Disadvantages of this type of DAC are the number of different resistor values and the fact that the voltage levels must be exactly the same for all inputs. For example, an 8-bit converter requires eight resistors, ranging from some value of R to $128R$ in binary-weighted steps. This range of resistors requires tolerances of one part in 255 (less than 0.5%) to accurately convert the input, making this type of DAC very difficult to mass-produce.

EXAMPLE 12–3

Problem

Determine the output of the DAC in Figure 12–36(a) if the waveforms representing a sequence of 4-bit numbers in Figure 12–36(b) are applied to the inputs. Input D_0 is the least significant bit (LSB).

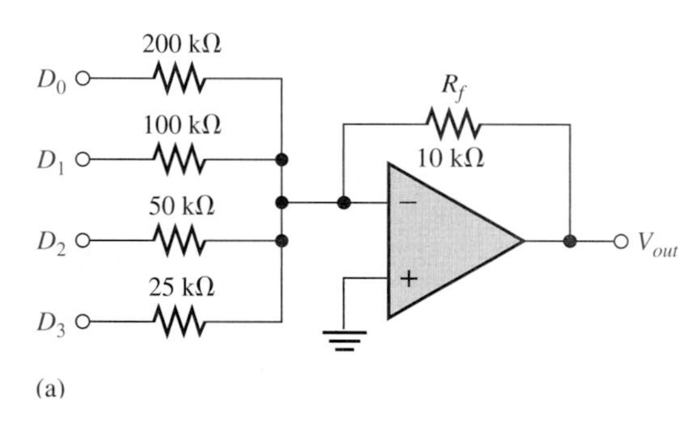

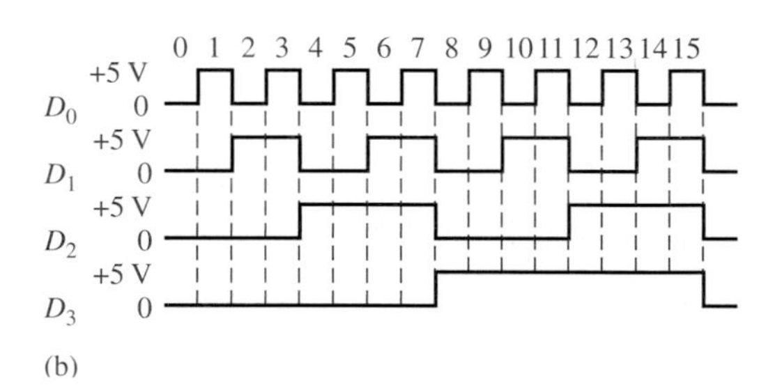

FIGURE 12–36

Solution

First, determine the current for each of the weighted inputs. Since the inverting (−) input of the op-amp is at 0 V (virtual ground) and a binary 1 corresponds to + 5 V, the current through any of the input resistors is 5 V divided by the resistance value.

$$I_0 = \frac{5\text{ V}}{200\text{ k}\Omega} = 0.025\text{ mA}$$

$$I_1 = \frac{5\text{ V}}{100\text{ k}\Omega} = 0.05\text{ mA}$$

$$I_2 = \frac{5\text{ V}}{50\text{ k}\Omega} = 0.1\text{ mA}$$

$$I_3 = \frac{5\text{ V}}{25\text{ k}\Omega} = 0.2\text{ mA}$$

Almost no current goes into the inverting op-amp input because of its extremely high impedance. Therefore, assume that all of the current goes through the feedback resistor R_f. Since one end of R_f is at 0 V (virtual ground), the drop across R_f equals the output voltage, which is negative with respect to virtual ground.

$$V_{out(D0)} = (10\ \text{k}\Omega)(-0.025\ \text{mA}) = -0.25\ \text{V}$$
$$V_{out(D1)} = (10\ \text{k}\Omega)(-0.05\ \text{mA}) = -0.5\ \text{V}$$
$$V_{out(D2)} = (10\ \text{k}\Omega)(-0.1\ \text{mA}) = -1\ \text{V}$$
$$V_{out(D3)} = (10\ \text{k}\Omega)(-0.2\ \text{mA}) = -2\ \text{V}$$

From Figure 12–36(b), the first binary input code is 0000, and the code then progresses in a count sequence to 1111. Each successive binary code increases the output voltage by -0.25 V, so for this particular straight binary sequence on the inputs, the output is a stairstep waveform going from 0 V to -3.75 V in -0.25 V steps. This is shown in Figure 12–37.

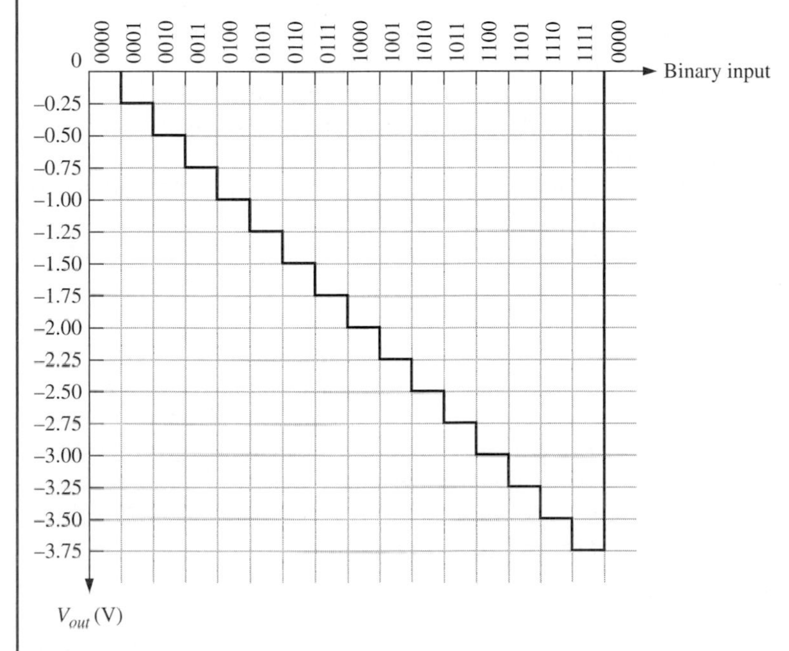

FIGURE 12–37
Output of the DAC in Figure 12–36.

Question
If the input waveforms to the DAC in Figure 12–36 are reversed (D_3 to D_0, D_2 to D_1, D_1 to D_2, D_0 to D_3), what is the output?

The R/2R Ladder Digital-to-Analog Converter

Another method of digital-to-analog conversion is the *R*/2*R* ladder, as shown in Figure 12–38 for four bits. It overcomes one of the problems in the binary-weighted-input DAC in that it requires only two resistor values.

Start by assuming that the D_3 input is HIGH (+5 V) and the others are LOW (ground, 0 V). This condition represents the binary number 1000. A circuit analysis will show that this reduces to the equivalent form shown in Figure 12–39(a). Essentially no current goes through the 2*R* equivalent resistance because the inverting input is at virtual ground. Thus, the current ($I = 5$ V/2*R*) through R_7 is the same as through R_f, and the output voltage is $-$ 5 V. The operational amplifier keeps the inverting ($-$) input near zero volts ($\approx$0 V) because of negative feedback. Because of the high input resistance of the op-amp, all current is through R_f.

FIGURE 12–38

An $R/2R$ ladder DAC.

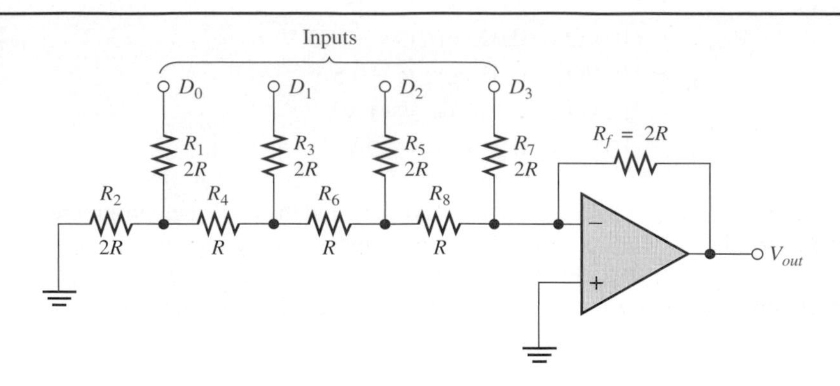

Figure 12–39(b) shows the equivalent circuit when the D_2 input is at +5 V and the others are at ground. This condition represents 0100. If we thevenize* looking from R_8, we get 2.5 V in series with R, as shown. This results in a current through R_f of $I = 2.5\text{ V}/2R$, which gives an output voltage of −2.5 V. Keep in mind that there is no current at the op-amp inverting input and that there is no current through the equivalent resistance R_{EQ} because it has 0 V across it, due to the virtual ground.

Figure 12–39(c) shows the equivalent circuit when the D_1 input is at +5 V and the others are at ground. This condition represents 0010. Again thevenizing looking from R_8, you get 1.25 V in series with R as shown. This results in a current through R_f of $I = 1.25\text{ V}/2R$, which gives an output voltage of −1.25 V.

In part (d) of Figure 12–39, the equivalent circuit representing the case where D_0 is at +5 V and the other inputs are at ground is shown. This condition represents 0001. Thevenizing from R_8 gives an equivalent of 0.625 V in series with R as shown. The resulting current through R_f is $I = 0.625\text{ V}/2R$, which gives an output voltage of −0.625 V.

Notice that each successively lower-weighted input produces an output voltage that is halved, so that the output voltage is proportional to the binary weight of the input bits.

Performance Characteristics of Digital-to-Analog Converters

The performance characteristics of a DAC include resolution, accuracy, linearity, monotonicity, and settling time, each of which is discussed in the following list:

- *Resolution.* The resolution of a DAC is the reciprocal of the number of discrete steps in the output. This, of course, is dependent on the number of input bits. For example, a 4-bit DAC has a resolution of one part in $2^4 - 1$ (one part in fifteen). Expressed as a percentage, this is $(1/15)100\% = 6.67\%$. The total number of discrete steps equals $2^n - 1$, where n is the number of bits. Resolution can also be expressed as the number of bits that are converted.
- *Accuracy.* Accuracy is derived from a comparison of the actual output of a DAC with the expected output. It is expressed as a percentage of a full-scale, or maximum, output voltage. For example, if a converter has a full-scale output of 10 V and the accuracy is ±0.1%, then the maximum error for any output voltage is (10 V)(0.001) = 10 mV. Ideally, the accuracy should be no worse than ±1/2 of a least significant bit. For an 8-bit converter, the least significant bit is 0.39% of full scale. The accuracy should be approximately ±0.2%.
- *Linearity.* A linear error is a deviation from the ideal straight-line output of a DAC. A special case is an offset error, which is the amount of output voltage when the input bits are all zeros.

* Recall from your dc/ac studies that Thevenin's theorem states that any circuit can be reduced to an equivalent voltage source in series with an equivalent resistance.

FIGURE 12–39 Analysis of the $R/2R$ ladder DAC.

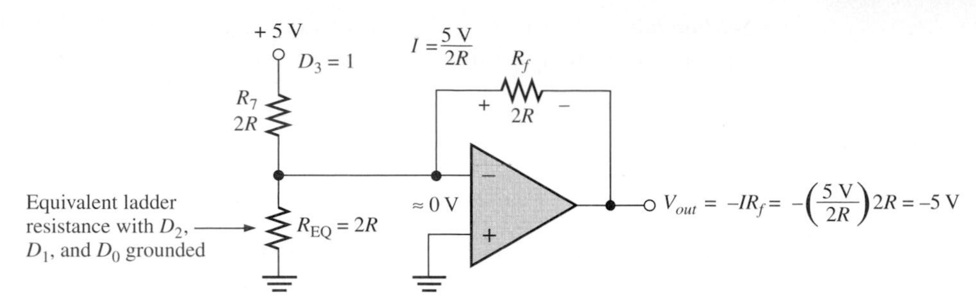

(a) Equivalent circuit for $D_3 = 1$, $D_2 = 0$, $D_1 = 0$, $D_0 = 0$

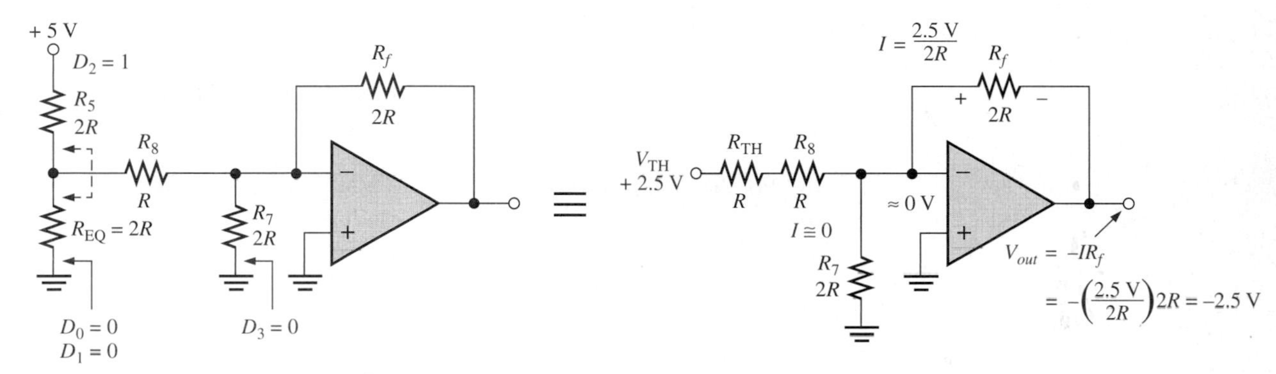

(b) Equivalent circuit for $D_3 = 0$, $D_2 = 1$, $D_1 = 0$, $D_0 = 0$

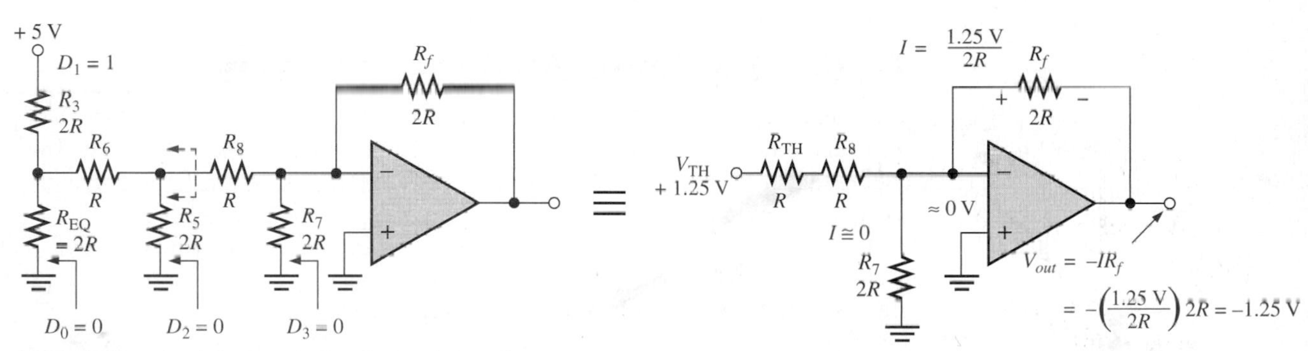

(c) Equivalent circuit for $D_3 = 0$, $D_2 = 0$, $D_1 = 1$, $D_0 = 0$

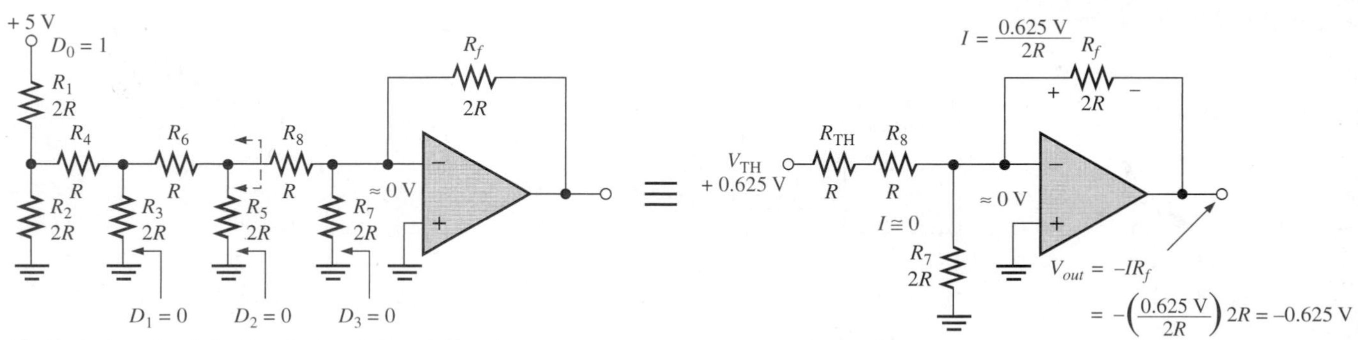

(d) Equivalent circuit for $D_3 = 0$, $D_2 = 0$, $D_1 = 0$, $D_0 = 1$

- *Monotonicity.* A DAC is monotonic if it does not take any reverse steps when it is sequenced over its entire range of input bits.
- *Settling time.* Settling time is normally defined as the time it takes a DAC to settle within ±1/2 LSB of its final value when a change occurs in the input code.

EXAMPLE 12–4

Problem

Determine the resolution, expressed as a percentage, of the following:

(a) an 8-bit DAC (b) a 12-bit DAC

Solution

(a) For the 8-bit converter,

$$\frac{1}{2^8 - 1} \times 100\% = \frac{1}{255} \times 100\% = \mathbf{0.392\%}$$

(b) For the 12-bit converter,

$$\frac{1}{2^{12} - 1} \times 100\% = \frac{1}{4095} \times 100\% = \mathbf{0.0244\%}$$

Question

What is the resolution of a 16-bit DAC?

Testing Digital-to-Analog Converters

The concept of DAC testing is illustrated in Figure 12–40. In this basic method, a sequence of binary codes is applied to the inputs, and the resulting output is observed. The binary code sequence extends over the full range of values from 0 to $2^n - 1$ in ascending order, where n is the number of bits.

The ideal output is a straight-line stairstep as indicated. As the number of bits in the binary code is increased, the resolution is improved. That is, the number of discrete steps increases, and the output approaches a straight-line linear ramp.

FIGURE 12–40

Basic test setup for a DAC.

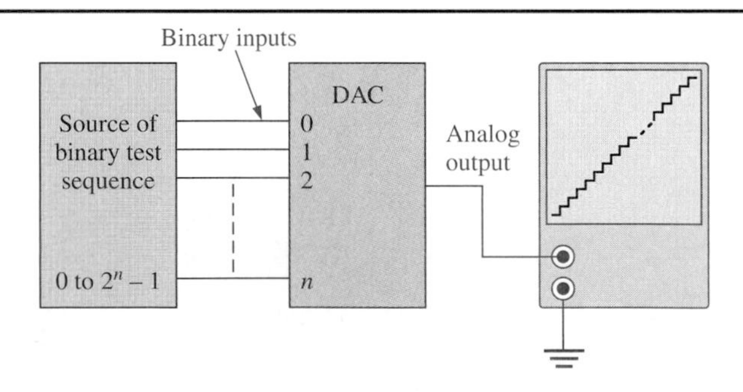

Digital-to-Analog Conversion Errors

Several digital-to-analog conversion errors to be checked for are shown in Figure 12–41, which uses a 4-bit conversion for illustration purposes. A 4-bit conversion produces fifteen discrete steps. Each graph in the figure includes an ideal stairstep ramp for comparison with the faulty outputs.

FIGURE 12–41

Illustrations of several digital-to-analog conversion errors.

(a) Nonmonotonic output (green)

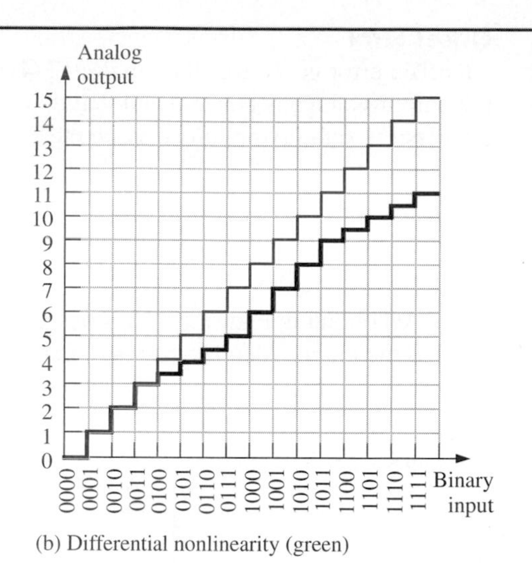

(b) Differential nonlinearity (green)

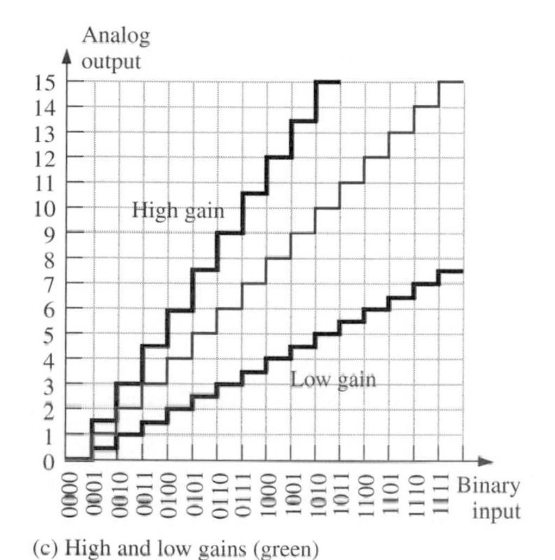

(c) High and low gains (green)

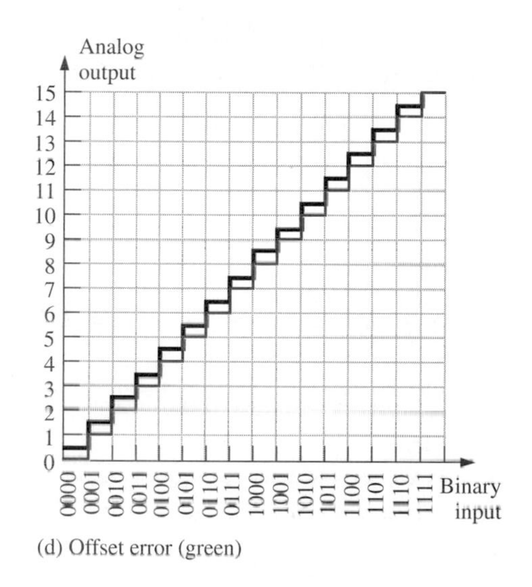

(d) Offset error (green)

Nonmonotonicity

The step reversals in Figure 12–41(a) indicate nonmonotonic performance, which is a form of nonlinearity. In this particular case, the error occurs because the 2^1 bit in the binary code is interpreted as a constant 0. That is, a short is causing the bit input line to be stuck LOW.

Differential Nonlinearity

Figure 12–41(b) illustrates differential nonlinearity in which the step amplitude is less than it should be for certain input codes. This particular output could be caused by the 2^2 bit having an insufficient weight, perhaps because of a faulty input resistor. We could also see steps with amplitudes greater than normal if a particular binary weight were greater than it should be.

Low or High Gain

Output errors caused by low or high gain are illustrated in Figure 12–41(c). In the case of low gain, all of the step amplitudes are less than ideal. In the case of high gain, all of the step amplitudes are greater than ideal. This situation may be caused by an incorrect feedback resistor value in the op-amp circuit.

Offset Error

An offset error is illustrated in Figure 12–41(d). Notice that when the binary input is 0000, the output voltage is nonzero and that this amount of offset is the same for all steps in the conversion. A faulty op-amp may be the culprit in this situation.

EXAMPLE 12–5

Problem

The DAC output in Figure 12–42 is observed when a straight 4-bit binary sequence is applied to the inputs. Identify the type of error, and suggest an approach to isolate the fault.

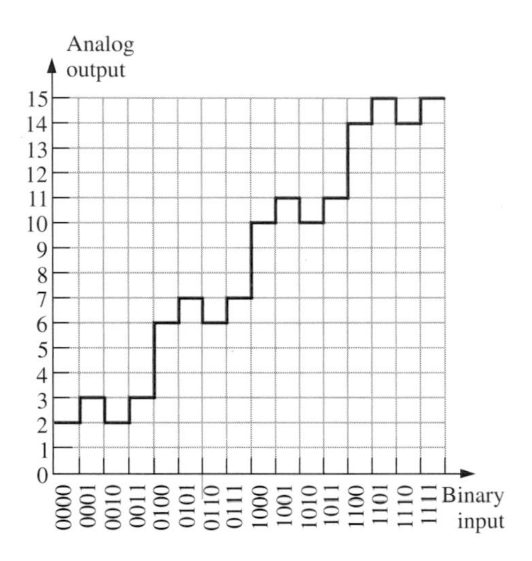

FIGURE 12–42

Solution

The DAC in this case is nonmonotonic. Analysis of the output reveals that the device is converting the following sequence, rather than the actual binary sequence applied to the inputs.

0010, 0011, 0010, 0011, 0110, 0111, 0110, 0111, 1010, 1011, 1010, 1011, 1110, 1111, 1110, 1111

Apparently, the 2^1 bit is stuck in the HIGH (1) state. To find the problem, first monitor the bit input pin to the device. If it is changing states, the fault is internal to the DAC and it should be replaced. If the external pin is not changing states and is always HIGH, check for an external short to $+V$ that may be caused by a solder bridge somewhere on the circuit board.

Question

What is the output of a DAC when a straight 4-bit binary sequence is applied to the inputs and the 2^0 bit is stuck HIGH?

The Reconstruction Filter

The output of the DAC is a "stairstep" approximation of the original analog signal after it has been processed by the DSP. The purpose of the low-pass reconstruction filter (sometimes called a postfilter) is to smooth out the DAC output by eliminating the higher frequency content that results from the fast transitions of the "stairsteps," as roughly illustrated in Figure 12–43.

FIGURE 12–43

The reconstruction filter smooths the output of the DAC.

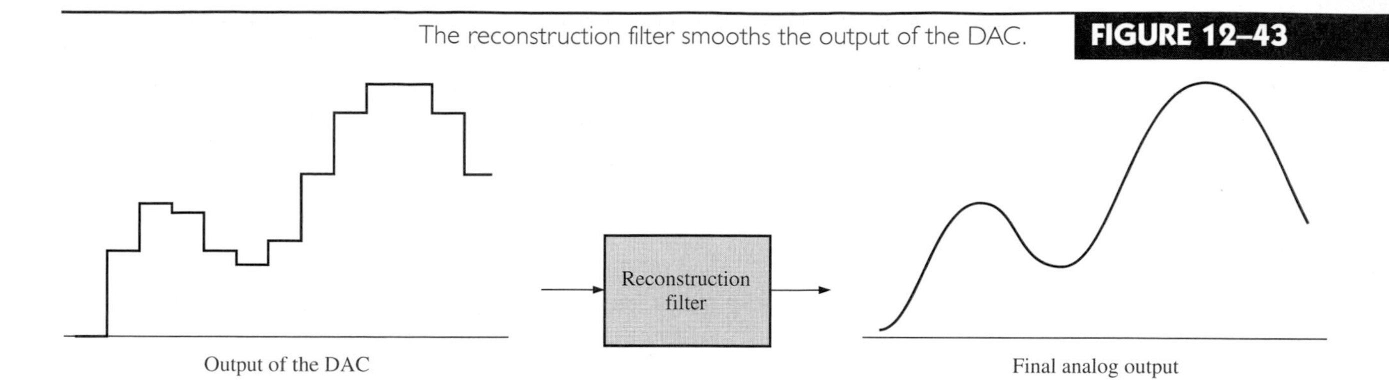

Review Questions

21. What are two types of DAC?
22. What is the disadvantage of the DAC with binary weighted inputs?
23. What is the resolution of a 4-bit DAC?
24. How do you detect nonmonotonic behavior in a DAC?
25. What effect does low gain have on a DAC output?

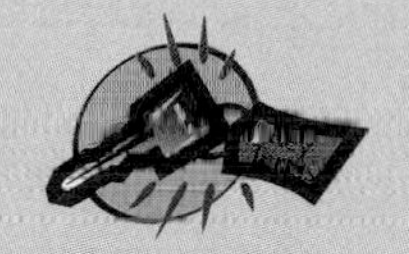

CHAPTER REVIEW

Key Terms

Aliasing The effect created when a signal is sampled at less than twice the Nyquist frequency. Aliasing creates unwanted frequencies that interfere with the signal frequency.

Analog-to-digital converter (ADC) A circuit used to convert an analog signal to digital form.

Decode A process of computer or DSP pipeline operation in which instructions are assigned to functional units and are decoded.

Digital-to-analog converter (DAC) A circuit used to convert the digital representation of an analog signal back to the analog signal.

DSP Digital signal processor; a special type of microprocessor that processes data in real time.

Execute A process of computer or DSP pipeline operation in which the decoded instructions are carried out.

Fetch A process of computer or DSP pipeline operation in which an instruction is obtained from the program memory.

Nyquist frequency The frequency that specifies the limit of the analog frequency to be sampled and defines it to be one-half of the sampling frequency.

Pipeline Part of the DSP architecture that allows multiple instructions to be processed simultaneously.

11. The output of the sampling circuit in Problem 10 is applied to a hold circuit. Show the output of the hold circuit.

12. If the output of the hold circuit in Problem 11 is quantized using two bits, what is the resulting sequence of binary codes?

13. Repeat Problem 12 using 4-bit quantization.

14. (a) Reconstruct the analog signal from the 2-bit quantization in Problem 12.

    (b) Reconstruct the analog signal from the 4-bit quantization in Problem 13.

15. Graph the analog function represented by the following sequence of binary numbers:

    1111, 1110, 1101, 1100, 1010, 1001, 1000, 0111, 0110, 0101, 0100, 0101, 0110, 0111, 1000, 1001, 1010, 1011, 1100, 1100, 1100, 1011, 1010, 1001.

16. The input voltage to a certain op-amp inverting amplifier is 10 mV, and the output is 2 V. What is the closed-loop voltage gain?

17. To achieve a closed-loop voltage gain of −330 with an inverting amplifier, what value of feedback resistor do you use if $R_i = 1.0\ \text{k}\Omega$?

18. Determine the binary output code of a 3-bit flash ADC for the analog input signal in Figure 12–46.

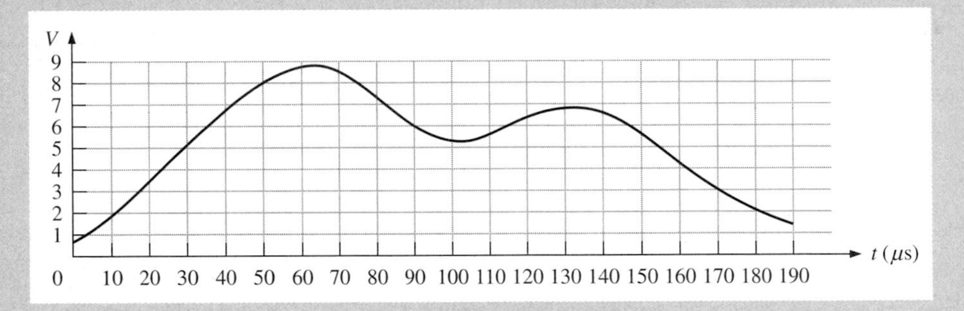

FIGURE 12–46

19. In the 4-bit DAC in Figure 12–35, the lowest-weighted resistor has a value of 10 kΩ. What should the values of the other input resistors be?

20. Determine the output of the DAC in Figure 12–47(a) if the sequence of 4-bit numbers in part (b) is applied to the inputs. The data inputs have a low value of 0 V and a high value of +5 V.

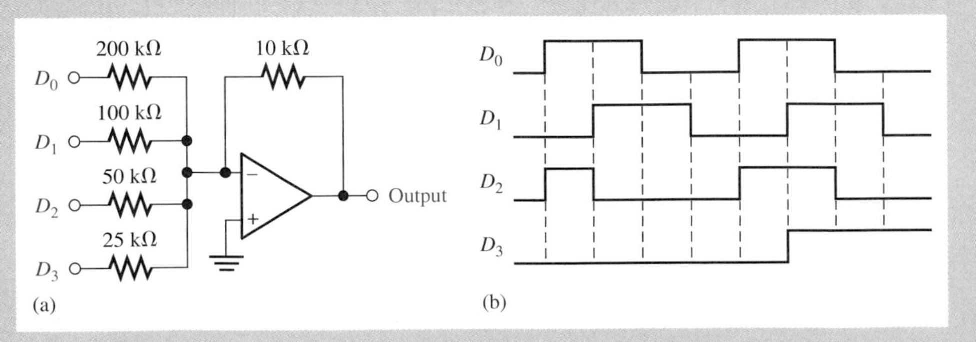

FIGURE 12–47

21. Repeat Problem 20 for the inputs in Figure 12–48.

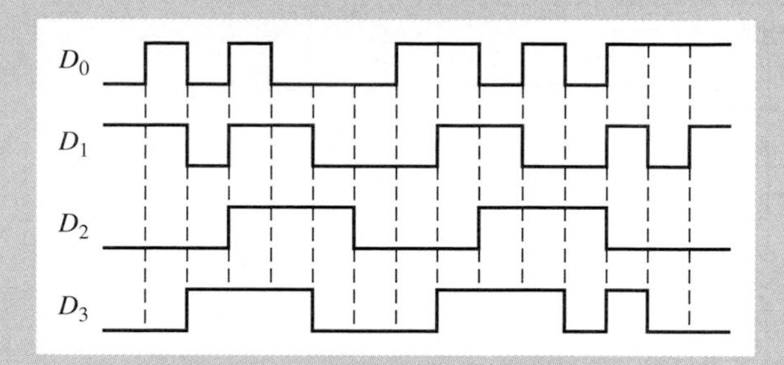

FIGURE 12–48

22. Determine the resolution expressed as a percentage, for each of the following DACs:
    (a) 3-bit
    (b) 10-bit
    (c) 18-bit
23. Determine the ratio of the largest to smallest resistor in a 12-bit binary-weighted input DAC.

Example Questions

ANSWERS

12–1: 100, 111, 100, 000, 011, 110. Yes, information is lost.

12–2: See Figure 12–49.

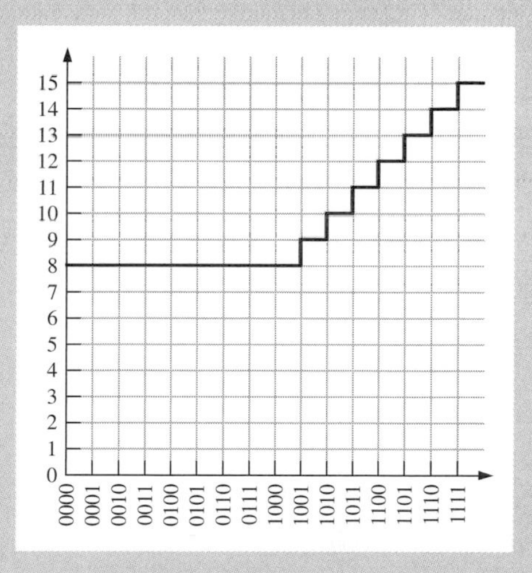

FIGURE 12–49

APPENDIX A

SUMMARY OF BOOLEAN ALGEBRA

Laws of Boolean Algebra

The basic laws of Boolean-algebra—the commutative laws for addition and multiplication, the associative laws for addition and multiplication, and the distributive law—are the same as in ordinary algebra. Each of the laws is illustrated with two or three variables, but the number of variables is not limited to this.

Commutative Laws

The *commutative law of addition* for two variables is written as

$$A + B = B + A$$

This law states that the order in which the variables are ORed makes no difference. Remember, in Boolean algebra as applied to logic circuits, addition and the OR operation are the same. Figure A–1 illustrates the commutative law as applied to the OR gate and shows that it doesn't matter to which input each variable is applied. (The symbol ≡ means "equivalent to.")

FIGURE A–1

Application of commutative law of addition.

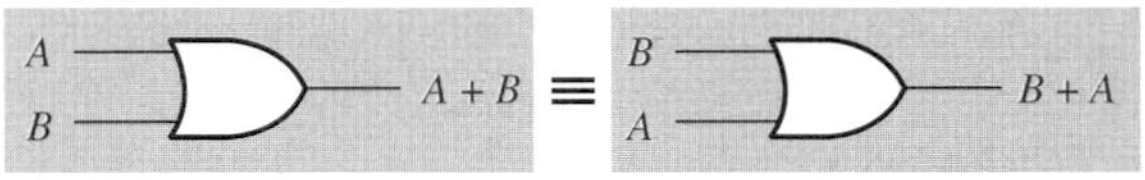

The *commutative law of multiplication* for two variables is

$$AB = BA$$

This law states that the order in which the variables are ANDed makes no difference. Figure A–2 illustrates this law as applied to the AND gate.

FIGURE A–2

Application of commutative law of multiplication.

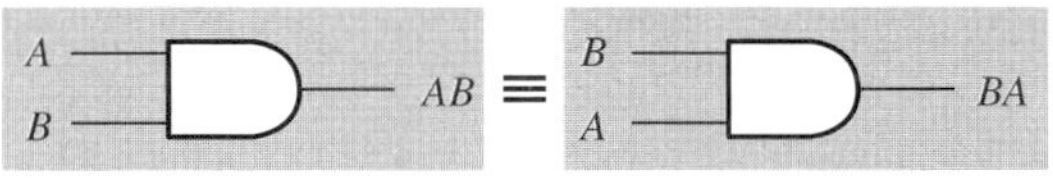

Associative Laws

The *associative law of addition* is written as follows for three variables:

$$A + (B + C) = (A + B) + C$$

This law states that when ORing more than two variables, the result is the same regardless of the grouping of the variables. Figure A–3 illustrates this law as applied to 2-input OR gates.

FIGURE A–3

Application of associative law of addition.

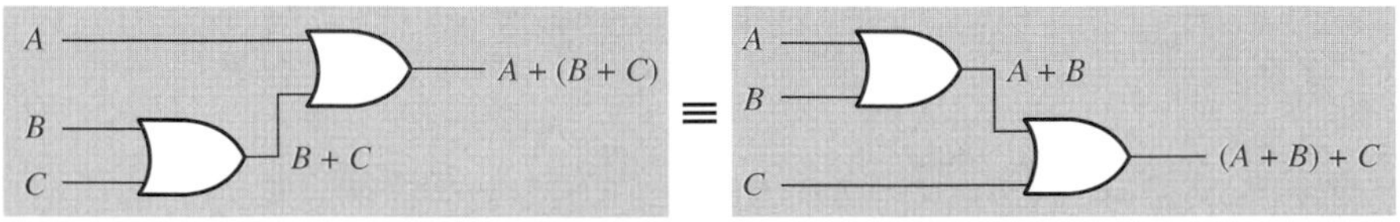

The *associative law of multiplication* is written as follows for three variables:

$$A(BC) = (AB)C$$

This law states that it makes no difference in what order the variables are grouped when ANDing more than two variables. Figure A–4 illustrates this law as applied to 2-input AND gates.

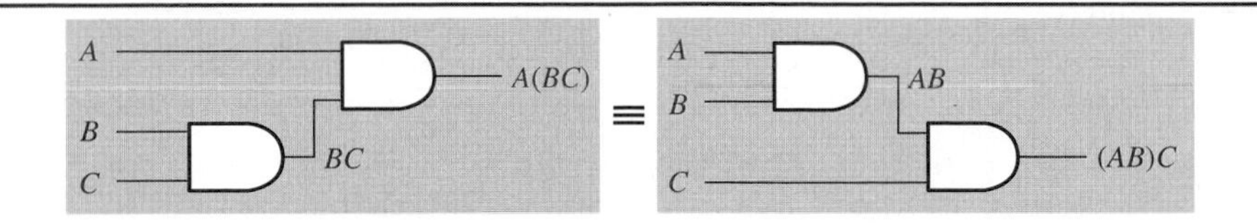

FIGURE A–4

Application of associative law of multiplication.

Distributive Law

The distributive law is written for three variables as follows:

$$A(B + C) = AB + AC$$

This law states that ORing two or more variables and then ANDing the result with a single variable is equivalent to ANDing the single variable with each of the two or more variables and then ORing the products. The distributive law also expresses the process of *factoring* in which the common variable A is factored out of the product terms, for example, $AB + AC = A(B + C)$. Figure A–5 illustrates the distributive law in terms of gate implementation.

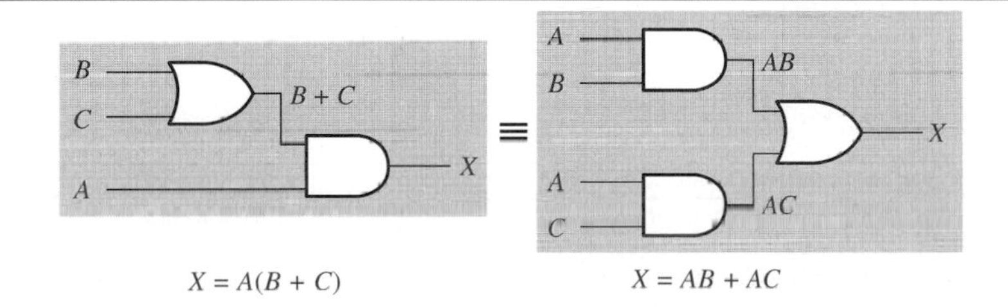

FIGURE A–5

Application of distributive law.

Rules of Boolean Algebra

Table A–1 lists 12 basic rules that are useful in manipulating and simplifying Boolean expressions. Rules 1 through 9 will be viewed in terms of their application to logic gates. Rules 10 through 12 will be derived in terms of the simpler rules and the laws previously discussed.

TABLE A–1

Basic rules of Boolean algebra.

1. $A + 0 = A$
2. $A + 1 = 1$
3. $A \cdot 0 = 0$
4. $A \cdot 1 = A$
5. $A + A = A$
6. $A + \overline{A} = 1$
7. $A \cdot A = A$
8. $A \cdot \overline{A} = 0$
9. $\overline{\overline{A}} = A$
10. $A + AB = A$
11. $A + \overline{A}B = A + B$
12. $(A + B)(A + C) = A + BC$

A, *B*, or *C* can represent a single variable or a combination of variables.

Rule 1. $A + 0 = A$ A variable ORed with 0 is always equal to the variable. If the input variable A is 1, the output variable X is 1, which is equal to A. If A is 0, the output is 0, which is also equal to A. This rule is illustrated in Figure A–6, where the lower input is fixed at 0.

FIGURE A–6

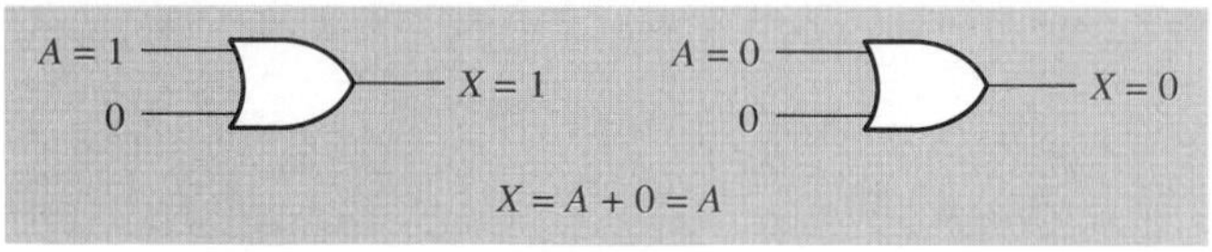

$X = A + 0 = A$

Rule 2. $A + 1 = 1$ A variable ORed with 1 is always equal to 1. A 1 on an input to an OR gate produces a 1 on the output, regardless of the value of the variable on the other input. This rule is illustrated in Figure A–7, where the lower input is fixed at 1.

FIGURE A–7

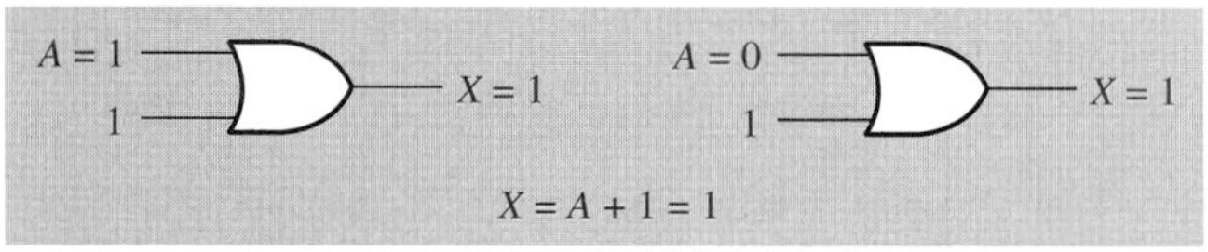

$X = A + 1 = 1$

Rule 3. $A \cdot 0 = 0$ A variable ANDed with 0 is always equal to 0. Any time one input to an AND gate is 0, the output is 0, regardless of the value of the variable on the other input. This rule is illustrated in Figure A–8, where the lower input is fixed at 0.

FIGURE A–8

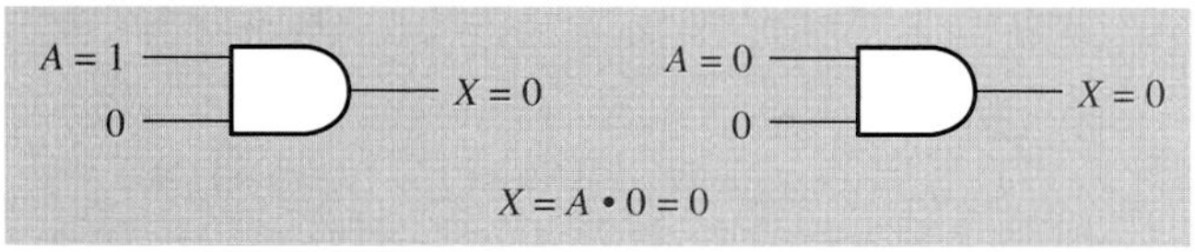

$X = A \cdot 0 = 0$

Rule 4. $A \cdot 1 = A$ A variable ANDed with 1 is always equal to the variable. If A is 0, the output of the AND gate is 0. If A is 1, the output of the AND gate is 1 because both inputs are now 1s. This rule is shown in Figure A–9, where the lower input is fixed at 1.

FIGURE A–9

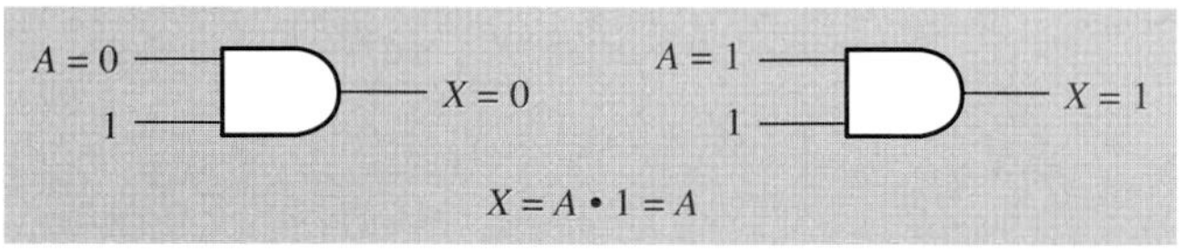

$X = A \cdot 1 = A$

Rule 5. $A + A = A$ A variable ORed with itself is always equal to the variable. If A is 0, then $0 + 0 = 0$; and if A is 1, then $1 + 1 = 1$. This is shown in Figure A–10, where both inputs are the same variable.

FIGURE A–10

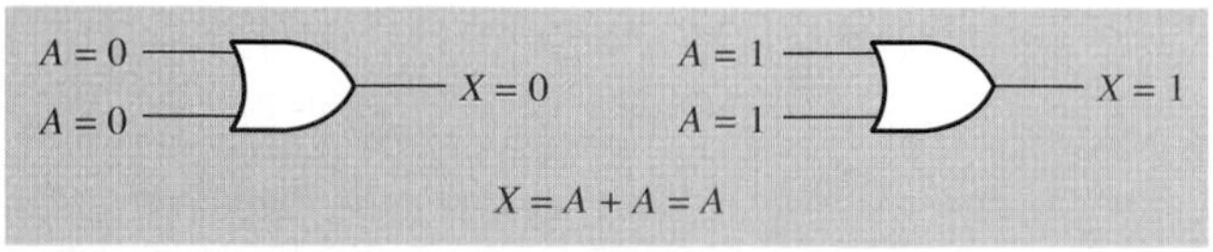

$X = A + A = A$

Rule 6. $A + \overline{A} = 1$ A variable ORed with its complement is always equal to 1. If A is 0, then $0 + \overline{0} = 0 + 1 = 1$. If A is 1, then $1 + \overline{1} = 1 + 0 = 1$. See Figure A–11, where one input is the complement of the other.

FIGURE A–11

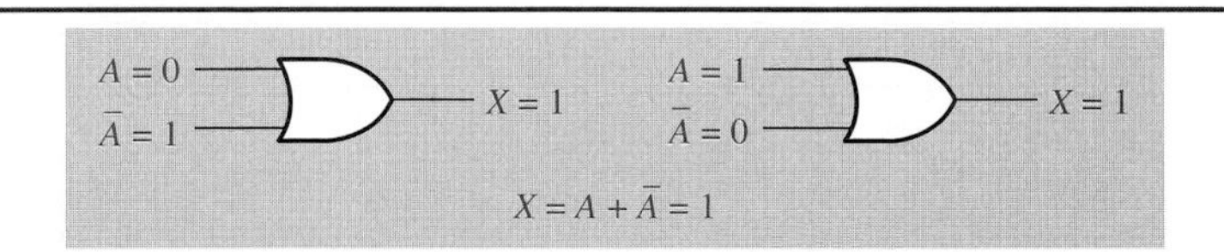

Rule 7. $A \cdot A = A$ A variable ANDed with itself is always equal to the variable. If $A = 0$, then $0 \cdot 0 = 0$; and if $A = 1$, then $1 \cdot 1 = 1$. Figure A–12 illustrates this rule.

FIGURE A–12

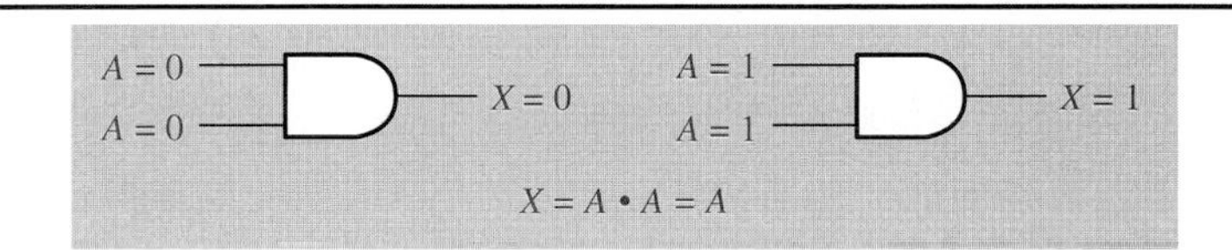

Rule 8. $A \cdot \overline{A} = 0$ A variable ANDed with its complement is always equal to 0. Either A or A will always be 0; and when a 0 is applied to the input of an AND gate, the output will be 0 also. Figure A–13 illustrates this rule.

FIGURE A–13

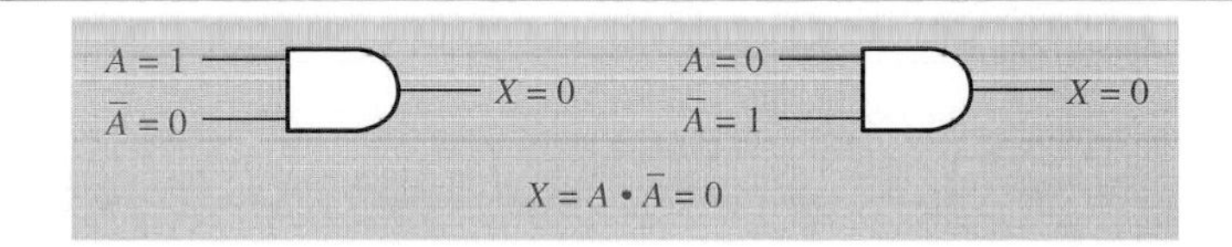

Rule 9. $\overline{\overline{A}} = A$ The double complement of a variable is always equal to the variable. If you start with the variable A and complement (invert) it once, you get $\overline{A}$. If you then take $\overline{A}$ and complement (invert) it, you get A, which is the original variable. This rule is shown in Figure A–14 using inverters.

FIGURE A–14

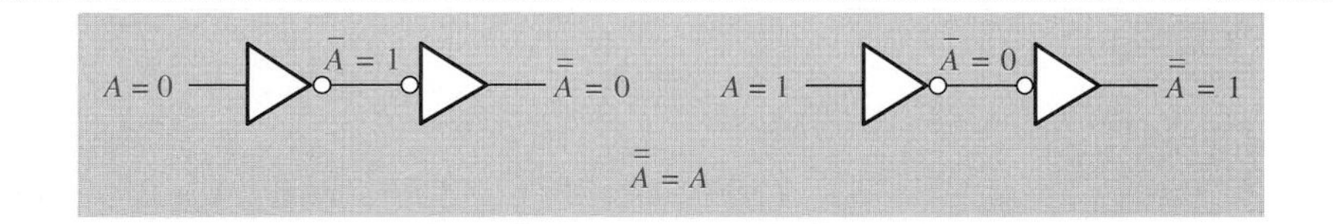

Rule 10. $A + AB = A$ This rule can be proved by applying the distributive law, rule 2, and rule 4 as follows:

$$A + AB = A(1 + B) \quad \text{Factoring (distributive law)}$$
$$= A \cdot 1 \quad \text{Rule 2: } (1 + B) = 1$$
$$= A \quad \text{Rule 4: } A \cdot 1 = A$$

The proof is shown in Table A–2, which shows the truth table and the resulting logic circuit simplification. The original circuit and the simplification are equivalent.

TABLE A–2

Rule 10: $A + AB = A$.

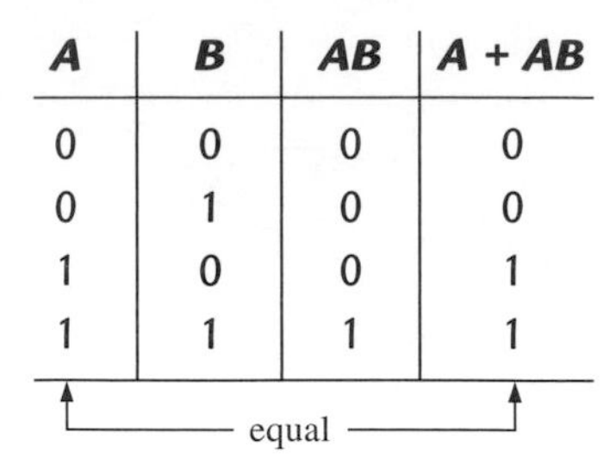

A	B	AB	$A + AB$
0	0	0	0
0	1	0	0
1	0	0	1
1	1	1	1

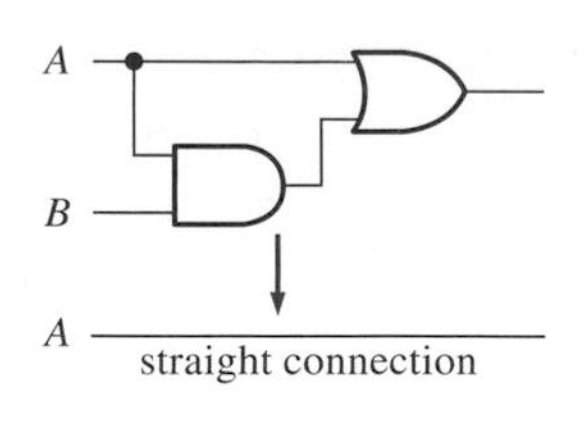

Rule 11. $\boldsymbol{A + \overline{A}B = A + B}$ This rule can be proved as follows:

$$
\begin{aligned}
A + \overline{A}B &= (A + AB) + \overline{A}B && \text{Rule 10: } A = A + AB \\
&= (AA + AB) + \overline{A}B && \text{Rule 7: } A = AA \\
&= AA + AB + A\overline{A} + \overline{A}B && \text{Rule 8: adding } A\overline{A} = 0 \\
&= (A + \overline{A})(A + B) && \text{Factoring} \\
&= 1 \cdot (A + B) && \text{Rule 6: } A + \overline{A} = 1 \\
&= A + B && \text{Rule 4: drop the 1}
\end{aligned}
$$

The proof is shown in Table A–3, which shows the truth table and the resulting logic circuit simplification, which are equivalent.

TABLE A–3

Rule 11: $A + \overline{A}B = A + B$.

A	B	$\overline{A}B$	$A + \overline{A}B$	$A + B$
0	0	0	0	0
0	1	1	1	1
1	0	0	1	1
1	1	0	1	1

equal

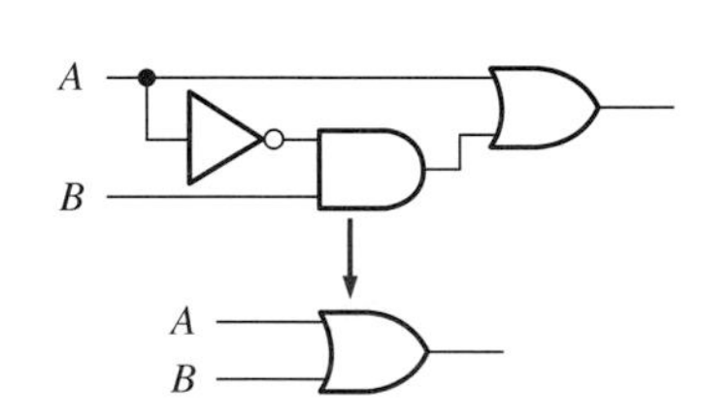

Rule 12. $\boldsymbol{(A + B)(A + C) = A + BC}$. This rule can be proved as follows:

$$
\begin{aligned}
(A + B)(A + C) &= AA + AC + AB + BC && \text{Distributive law} \\
&= A + AC + AB + BC && \text{Rule 7: } AA = A \\
&= A(1 + C) + AB + BC && \text{Factoring (distributive law)} \\
&= A \cdot 1 + AB + BC && \text{Rule 2: } 1 + C = 1 \\
&= A(1 + B) + BC && \text{Factoring (distributive law)} \\
&= A \cdot 1 + BC && \text{Rule 2: } 1 + B = 1 \\
&= A + BC && \text{Rule 4: } A \cdot 1 = A
\end{aligned}
$$

The proof is shown in Table A–4, which shows the truth table and the resulting logic circuit simplification, which are equivalent.

A	B	C	A + B	A + C	(A + B)(A + C)	BC	A + BC
0	0	0	0	0	0	0	0
0	0	1	0	1	0	0	0
0	1	0	1	0	0	0	0
0	1	1	1	1	1	1	1
1	0	0	1	1	1	0	1
1	0	1	1	1	1	0	1
1	1	0	1	1	1	0	1
1	1	1	1	1	1	1	1

(A + B)(A + C) and A + BC columns: equal

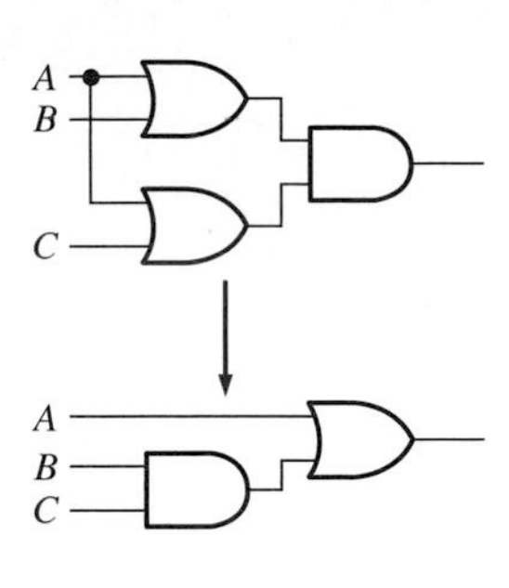

TABLE A–4

Rule 12: $(A + B)(A + C) = A + BC$.

DeMorgan's Theorems

One of DeMorgan's theorems is stated as follows:

> **The complement of a product of variables is equal to the sum of the complements of the variables.**

Stated another way,

> **The complement of two or more variables ANDed is equivalent to the OR of the complements of the individual variables.**

The formula for expressing this theorem for two variables is

$$\overline{XY} = \overline{X} + \overline{Y}$$

DeMorgan's second theorem is stated as follows:

> **The complement of a sum of variables is equal to the product of the complements of the variables.**

Stated another way,

> **The complement of two or more variables ORed is equivalent to the AND of the complements of the individual variables.**

The formula for expressing this theorem for two variables is

$$\overline{X + Y} = \overline{X}\,\overline{Y}$$

Figure A–15 shows the gate equivalencies and truth tables. As stated, DeMorgan's theorems also apply to expressions in which there are more than two variables.

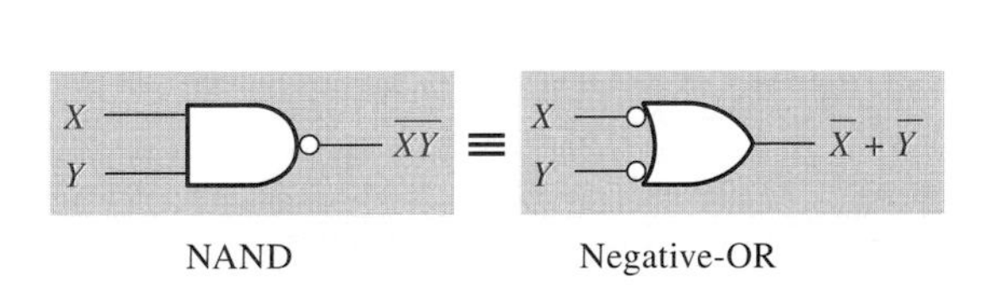

NAND　　Negative-OR

Inputs		Output	
X	Y	$\overline{XY}$	$\overline{X} + \overline{Y}$
0	0	1	1
0	1	1	1
1	0	1	1
1	1	0	0

X　Y　$\overline{X + Y}$ ≡ X　Y　$\overline{X}\,\overline{Y}$

NOR　　Negative-AND

Inputs		Output	
X	Y	$\overline{X + Y}$	$\overline{X}\,\overline{Y}$
0	0	1	1
0	1	0	0
1	0	0	0
1	1	0	0

FIGURE A–15

Gate equivalencies and the corresponding truth tables that illustrate DeMorgan's theorems. Notice the equality of the two output columns in each table. This shows that the equivalent gates perform the same logic function.

APPENDIX B

LAB MANUAL FIGURES

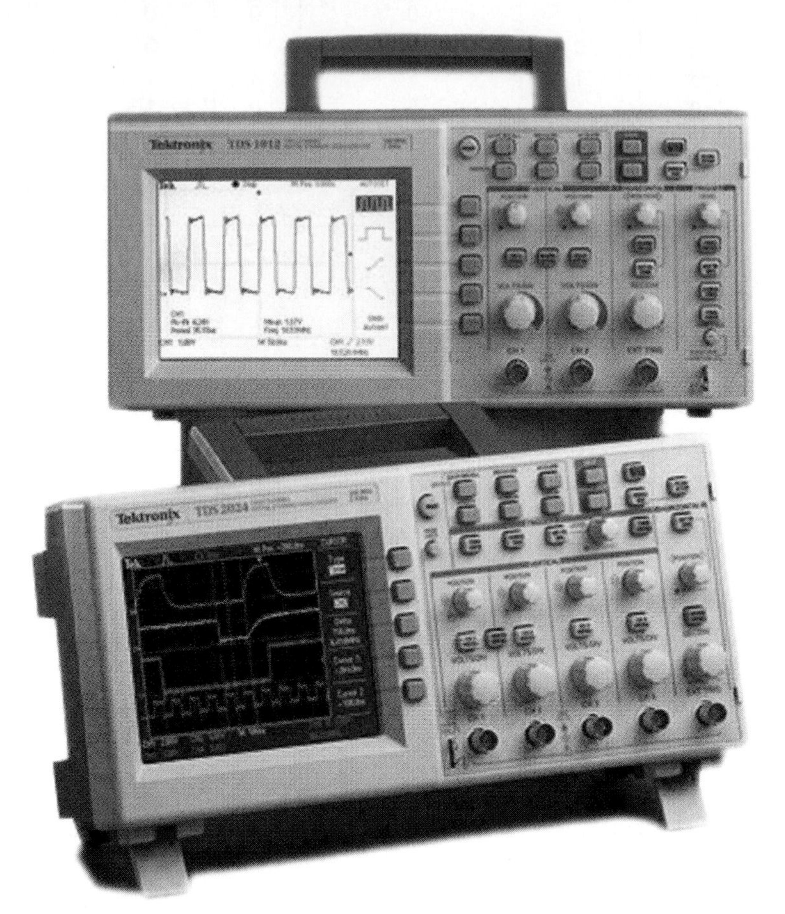

Figure I–13 Tektronix 1000 and 2000 series oscilloscopes. (Courtesy of Tektronix, Inc.)

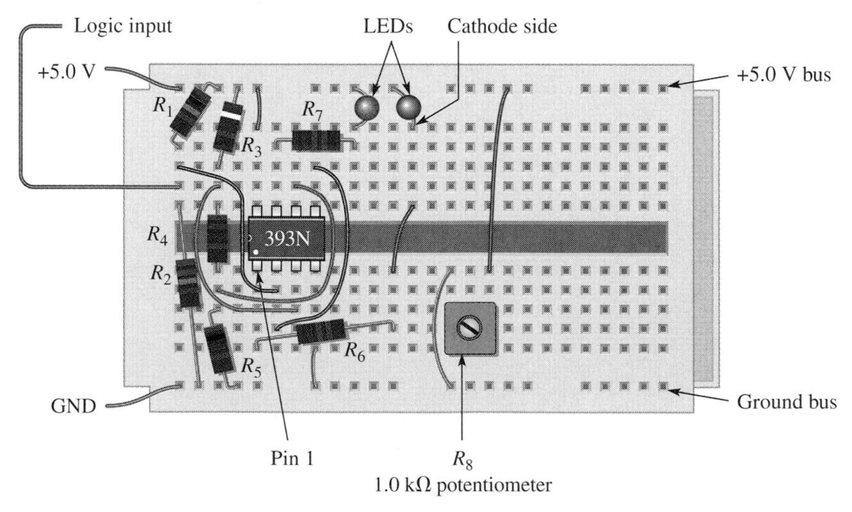

Figure 1–3

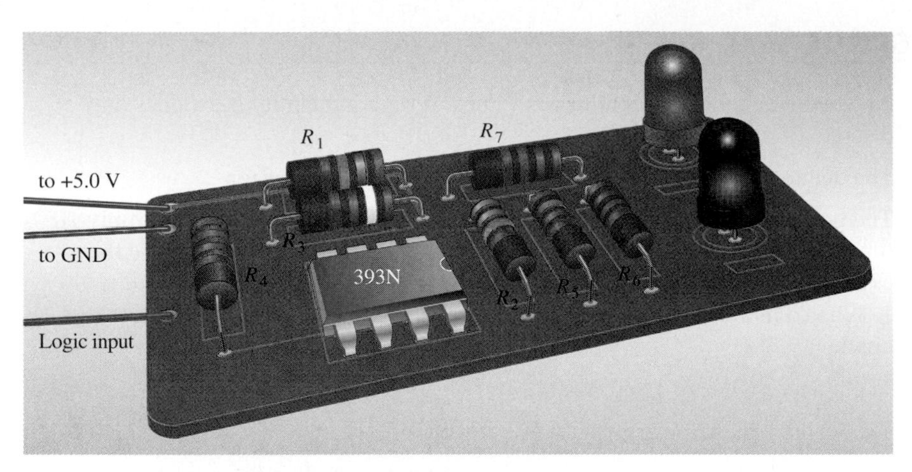

Figure 1–5 3-D view of logic probe (Picture adapted from the computer program Ultiboard).

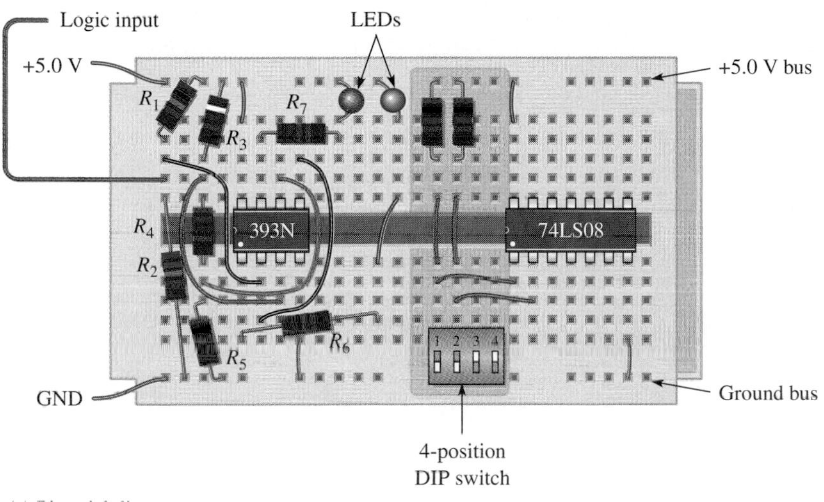

(a) Pictorial diagram

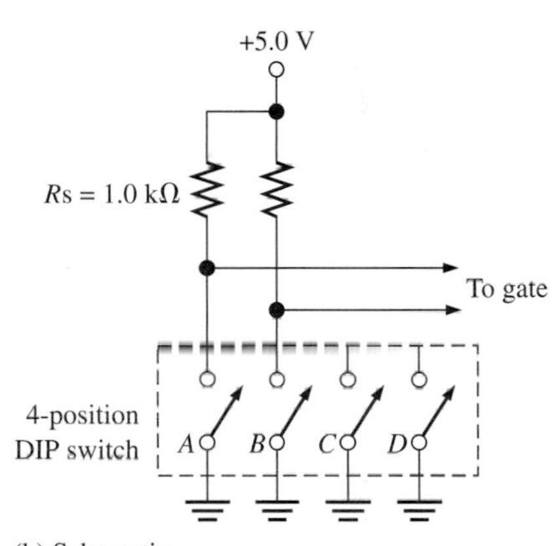

(b) Schematic

Figure 2–1

ANSWERS TO ODD-NUMBERED QUESTIONS

Chapter 1

1. An analog quantity has continuous values. A digital quantity has values only at discrete (individual) points.
3. The decimal system is based on weights that are powers-of-ten.
5. The power is the exponent. Raising 10 to a certain power means to multiply 10 times itself a number of times equal to the exponent. The exceptions are the exponents 0 and 1. When a number is raised to the power of 0, the result is always 1. When a number is raised to the power of 1, the result is always equal to the number.
7. The binary digits are 1 and 0.
9. The weights in the binary system are powers of 2 (1, 2, 4, 8, etc.).
11. LSB is the least significant bit in a binary number. MSB is the most significant bit.
13. Highest decimal number with 8 bits = 255. Highest decimal number with 10 bits = 1023.
15. Octal digits are 0, 1, 2, 3, 4, 5, 6, 7
17. The amplitude is the value of the maximum level measured from the baseline.
19. The period is 10 ms.
21. 10001011110010

Chapter 2

1. Both an AND gate and an OR gate have a HIGH output when all of their inputs are HIGH.
3. OR gate
5. $2^4 = 16$
7. Boolean variables are represented by letters.
9. OR
11. Commutative law of addition: The result of an OR operation does not depend on the order in which variables are ORed.
13. ULSI is ultra large-scale integration.
15. Three 3-input AND gates
17. LOW output
19. The output equals the OR of the nonshorted inputs.

Chapter 3

1. All inputs HIGH on either a NAND gate or a NOR gate produces a LOW output.
3. NAND gate
5. $2^8 = 256$
7. 0
9. The complement of a sum equals the product of the complements.
11. DeMorgan's second theorem: The complement of a sum (OR) of variables is equal to the product (AND) of the complemented variables.
13. A negative-AND operation is a NOR operation with asserted LOW inputs.

Chapter 4

1. XOR and XNOR can be made from combinations of other gates.
3. The inverters produce the complements of the input variables.
5. Any number
7. Product-of-sums (POS)
9. The distributive law
11. Groupings on a 4-variable Karnaugh map can contain 1, 2, 4, 8, or 16 cells.
13. The average power dissipation is $V_{CC}I_{CC}$.
15. A logic gate is limited to driving a number of gate inputs equal to the fan-out.

Chapter 5

1. A carry is produced when two 1s are added.
3. Half-adder and full-adder
5. A full-adder has three inputs.
7. Sixteen 2-bit adders
9. Eleven
11. Add 1 to the 1's complement to get the 2's complement.
13. The subtrahend is expressed as the 2's complement.
15. Multiplication is achieved by adding the multiplicand to itself a number of times equal to the multiplier.
17. A floating-point number is one expressed with a sign, mantissa (magnitude), and an exponent.

Chapter 6

1. 0010 and 0111 are valid BCD codes.
3. 39 = 0011 1001
5. ASCII for) = 0101001
7. Yes
9. A decoder accepts a binary input and produces the corresponding decimal output.
11. An encoder accepts a decimal input and produces the corresponding binary output.
13. The BCD code for 7: 0111
15. The S_2, S_1, S_0 inputs select the data input corresponding to the binary code applied to them.

Chapter 7

1. *S* input HIGH and *R* input LOW
3. A gated latch must be enabled before it can change state.
5. A flip-flop has a clock input and can change state only on the triggering edge of the clock pulse.
7. On the positive-going edge of the clock pulse

9. $Q = 0$
11. $Q = 0$
13. Q goes to the state opposite the one it had before the clock pulse.
15. The values of the external resistors and capacitor

Chapter 8

1. A counter is a digital circuit composed of flip-flops that is capable of progressing through a series of binary states when triggered by clock pulses.
3. Four flip-flops ($2^4 = 16$)
5. When a counter recycles, it goes from its terminal state back to the first state (usually 0).
7. 1111 (15)
9. 1010, 1011, 1100, 1101, 1110, 1111
11. The counter is in the 0 state when cleared.
13. 60
15. Clock frequency

Chapter 9

1. The shift register is a type of logic circuit consisting of flip-flops connected so as to pass data from one flip-flop to the next on a clock pulse.
3. Eight flip-flops
5. Bidirectional
7. Four
9. When data are shifted right from one flip-flop to the next in a shift register and from the last stage back to the first stage
11. A shift register counter produces special sequences of states.

Chapter 10

1. A CPLD contains multiple SPLDs with programmable interconnections.
3. Programmable gate arrays
5. HDL, hardware description language
7. An input buffer prevents loading.
9. A GAL is reprogrammable. A PAL is one-time programmable.
11. LAB: logic array block
13. Altera MAX 7000; Xilinx XC9500
15. JTAG: Joint Test Action Group
17. A target device is the PLD being programmed.
19. Entity and Architecture

Chapter 11

1. Desk top and lap top
3. Printer, modem, scanner, monitor, keyboard, mouse
5. The operating system software runs the computer. The application software performs various tasks for specific applications.
7. A data unit remains in the RAM after it has been read.
9. A flash memory is a high-density read/write memory and is often used for mass storage because it is nonvolatile.
11. An address is sent to memory on the address bus.
13. An assembler converts assembly language to machine language.
15. C++, BASIC, and COBOL

Chapter 12

1. The sample-and-hold process
3. Quantization is the representation of the stairstep levels as binary codes.
5. The Nyquist frequency specifies the limit of the analog signal frequency to be sampled and defines it to be one-half of the sampling frequency.
7. To give the ADC time to process the sampled value
9. Resolution is the number of bits used to represent a sampled value.
11. Sigma-delta is based on the difference between two successive sampled values.
13. Harvard architecture
15. Conversion of the digital voice signal to a coded form for secure transmission.
17. A type of digital-to-analog converter

ANSWERS TO ODD-NUMBERED PROBLEMS

Chapter 1

1. (a) $10_2 = 2_{10}$
   (b) $110_2 = 6_{10}$
   (c) $1010_2 = 10_{10}$
   (d) $11011_2 = 27_{10}$
3. (a) $25_{10} = 11001_2$
   (b) $76_{10} = 1001100_2$
   (c) $139_{10} = 10001011_2$
   (d) $245_{10} = 11110101_2$
5. (a) $2^3 - 1 = 7_{10}$
   (b) $2^5 - 1 = 31_{10}$
   (c) $2^7 - 1 = 127_{10}$
   (d) $2^9 - 1 = 511_{10}$
   (e) $2^{12} - 1 = 4095_{10}$
7. (a) $73_{16} = 0111\ 0011$
   (b) $A50_{16} = 1010\ 0101\ 0000$
   (c) $8B2C_{16} = 1000\ 1011\ 0010\ 1100$
9. (a) $67_8 = 110\ 111$
   (b) $146_8 = 001\ 100\ 110$
   (c) $3052_8 = 011\ 000\ 101\ 010$
11. (a) $2^3 = 8$
    (b) $2^8 = 256$
    (c) $2^{16} = 65{,}536$
    (d) $2^{32} = 4.294967296 \times 10^9$
13. (a) $18_{10} = 22_8$
    (b) $72_{10} = 110_8$
    (c) $555_{10} = 1053_8$
    (d) $1075_{10} = 2063_8$
15. (a) $27_8 = 23_{10}$
    (b) $670_8 = 440_{10}$
    (c) $5331_8 = 2777_{10}$
17. (a) $T = 4\ \mu s$
    (b) $f = 250$ kHz
    (c) Duty cycle = 0.25 or 25%
    (d) Amplitude = 3.3 V
19. (a) $295 = 2 \times 10^2 + 9 \times 10^1 + 5 \times 10^0$
    (b) $850 = 8 \times 10^2 + 5 \times 10^1 + 0 \times 10^0$
    (c) $1801 = 1 \times 10^3 + 8 \times 10^2 + 0 \times 10^1 + 1 \times 10^0$
    (d) $39{,}472 = 3 \times 10^4 + 9 \times 10^3 + 4 \times 10^2 + 7 \times 10^1 + 2 \times 10^0$

21. (a) $10^3 = 10 \times 10 \times 10 = 1000$
    (b) $10^5 = 10 \times 10 \times 10 \times 10 \times 10 = 100{,}000$
    (c) $10^8 = 10 \times 10 \times 10 \times 10 \times 10 \times 10 \times 10 \times 10 = 100{,}000{,}000$
23. (a) $110 = 1 \times 2^2 + 1 \times 2^1 + 0 \times 2^0$
    (b) $10010 = 1 \times 2^4 + 0 \times 2^3 + 0 \times 2^2 + 1 \times 2^1 + 0 \times 2^0$
    (c) $1110100 = 1 \times 2^6 + 1 \times 2^5 + 1 \times 2^4 + 0 \times 2^3 + 1 \times 2^2 + 0 \times 2^1 + 0 \times 2^0$
    (d) $1111000100 = 1 \times 2^9 + 1 \times 2^8 + 1 \times 2^7 + 1 \times 2^6 + 0 \times 2^5 + 0 \times 2^4 + 0 \times 2^3 + 1 \times 2^2 + 0 \times 2^1 + 0 \times 2^0$
25. 10001, 10010, 10011, 10100, 10101, 10110, 10111, 11000, 11001, 11010, 11011, 11100, 11101, 11110, 11111
27. These binary numbers are too large for the TI-36X. First manually convert binary to hex and then use the calculator to convert to decimal.
    (a) 100101011110 = 2,398
    (b) 11000011010111 = 12,503
    (c) 101000101111010 = 20,858
29. Duty cycle = 0.25 or 25%
31. $f = 70$ MHz
33. See Figure ANS–1.

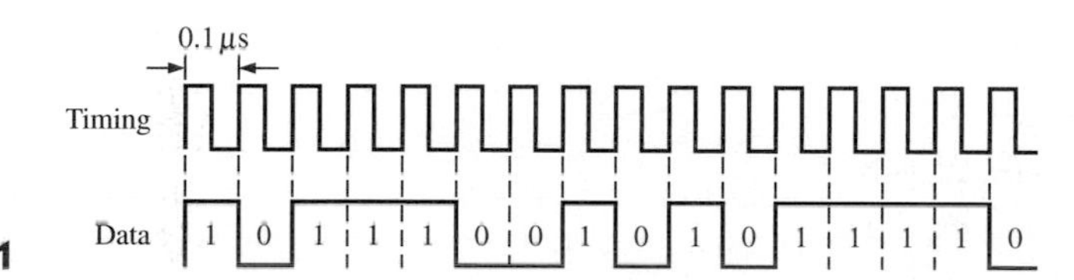

FIGURE ANS–1

35. See Figure ANS–2. MSB is first.

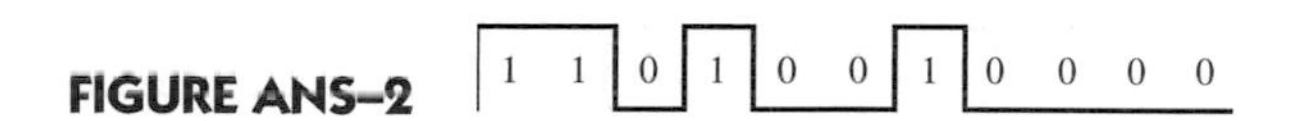

FIGURE ANS–2

Chapter 2

1. AND, OR, NOT
3. See Figure ANS–3.

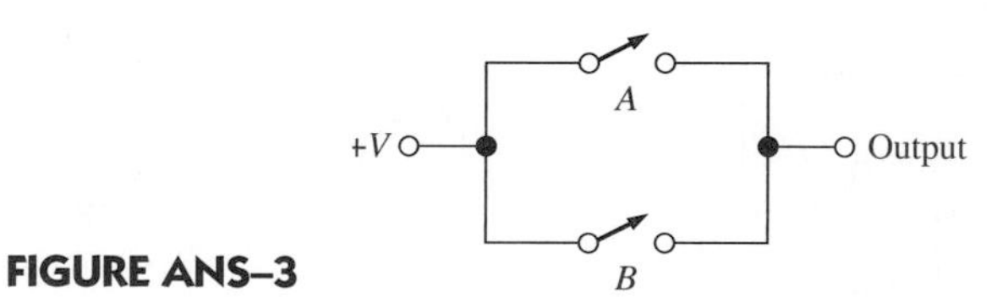

FIGURE ANS–3

5. (a) LOW
   (b) HIGH
   (c) HIGH
   (d) HIGH

7. (a) 2-input AND
   (b) Inverter (NOT)
   (c) 2-input OR
   (d) Inverter
   (e) 2-input OR
   (f) 4-input AND
9. (a) 2-variable AND
   (b) 3-variable OR
   (c) 4-variable AND
11. (a) $X = AB$
   (b) $X = ABC$
   (c) $X = ABCD$
13. (a) 0
   (b) 1
   (c) 0
   (d) 0
15. (a) 0
   (b) 1
   (c) 1
17. (a) $2^2 = 4$
   (b) $2^3 = 8$
   (c) $2^4 = 16$
   (d) $2^5 = 32$
19. (a) 3
   (b) 4
   (c) 5
   (d) 6
21. (a)

A	B	X
0	0	0
0	1	0
1	0	0
1	1	1

   (b)

A	B	C	X
0	0	0	0
0	0	1	0
0	1	0	0
0	1	1	0
1	0	0	0
1	0	1	0
1	1	0	0
1	1	1	1

(c)

A	*B*	*C*	*D*	*X*
0	0	0	0	0
0	0	0	1	0
0	0	1	0	0
0	0	1	1	0
0	1	0	0	0
0	1	0	1	0
0	1	1	0	0
0	1	1	1	0
1	0	0	0	0
1	0	0	1	0
1	0	1	0	0
1	0	1	1	0
1	1	0	0	0
1	1	0	1	0
1	1	1	0	0
1	1	1	1	1

23. See Figure ANS–4.

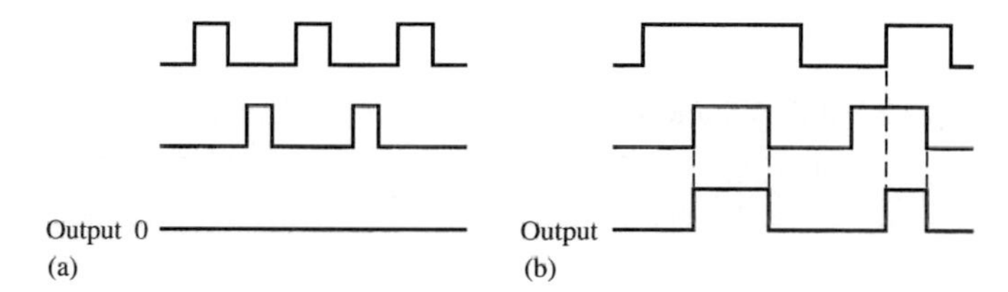

FIGURE ANS–4

25. See Figure ANS–5.

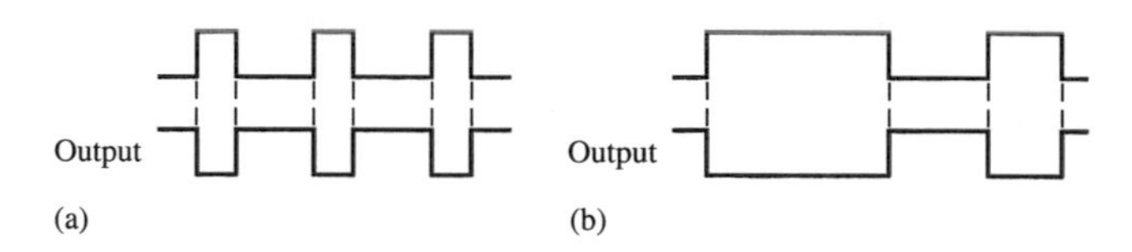

FIGURE ANS–5

27. (a) Upper input stuck HIGH
    (b) Output or one input stuck LOW
29. (a) No fault
    (b) One input or the output is stuck HIGH
31. The switch *A* inputs are shorted and always output a HIGH.
33. The output of switch *B* is open.

Chapter 3

1. (a) HIGH
   (b) HIGH
   (c) HIGH
   (d) LOW
3. (a) 2-input NAND
   (b) 2-input NOR
   (c) 2-input NOR
   (d) 4-input NAND
5. (a) 2-variable NAND
   (b) 3-variable NOR
   (c) 4-variable NAND
7. (a) $X = \overline{AB}$
   (b) $X = \overline{ABC}$
   (c) $X = \overline{ABCD}$
9. (a) $X = 1$
   (b) $X = 0$
   (c) $X = 1$
   (d) $X = 1$
11. (a) Commutative law of multiplication
    (b) Commutative law of addition
13. (a)

A	B	X
0	0	1
0	1	0
1	0	0
1	1	0

(b)

A	B	C	X
0	0	0	1
0	0	1	0
0	1	0	0
0	1	1	0
1	0	0	0
1	0	1	0
1	1	0	0
1	1	1	0

(c)

A	B	C	D	X
0	0	0	0	1
0	0	0	1	0
0	0	1	0	0
0	0	1	1	0
0	1	0	0	0
0	1	0	1	0
0	1	1	0	0
0	1	1	1	0
1	0	0	0	0
1	0	0	1	0
1	0	1	0	0
1	0	1	1	0
1	1	0	0	0
1	1	0	1	0
1	1	1	0	0
1	1	1	1	0

15. See Figure ANS–6.

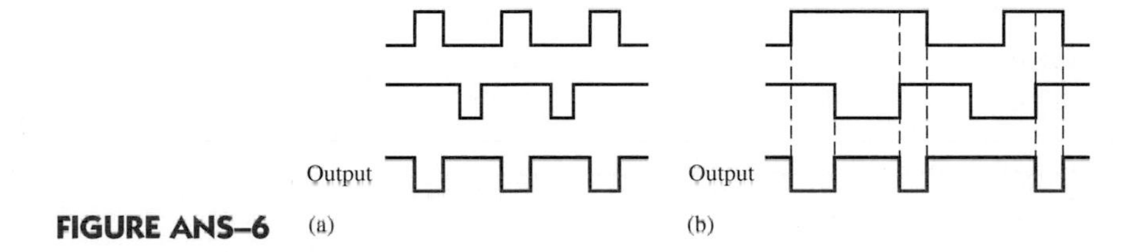

FIGURE ANS–6 (a) (b)

17. Channel 1: Amplitude = 2 V, $T = 50\ \mu s, f = 20$ kHz

    Channel 2: Amplitude = 2.5 V, $T = 30\ \mu s, f = 33.3$ kHz
19. From the truth table for a 2-input NOR gate, the output is HIGH only when both inputs are LOW. This is the same as for an AND gate with active-LOW inputs or a negative-AND.
21. (a) Input *B* stuck LOW

    (b) No apparent fault indicated
23. Switch *B* input to V_{CC} is open.
25. A switch input to NAND gate is open.

Chapter 4

1. (a) 0

   (b) 1

   (c) 1

   (d) 0

3. $A = 1, B = 1$ or $A = 0, B = 0$. See Figure ANS–7.

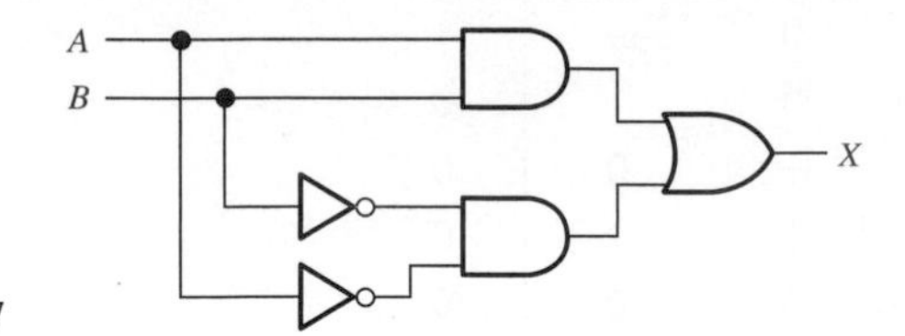

FIGURE ANS–7

5. $A = 0, B = 1, C = 0, D = 1$ or $A = 1, B = 0, C = 0, D = 1$. See Figure ANS–8.

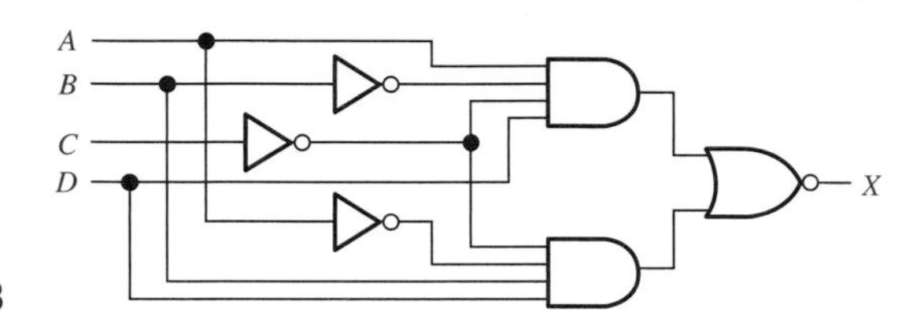

FIGURE ANS–8

7. (a) $(A + B) + C$
   (b) $A + (B + C)$
   (c) $(A + B + C) + D$ Other groupings are possible.
9. (a) 000, 101, 111; 3 inverters, three 3-input AND gates, and one 3-input OR gate
   (b) 110, 010, 011, 100, 001; 3 inverters, five 3-input AND gates, and one 5-input OR gate
11. (a) 4
    (b) 8
    (c) 16
13. 5 mW
15. 10
17. XNOR

A	B	X
0	0	1
0	1	0
1	0	0
1	1	1

19. See Figure ANS–9.

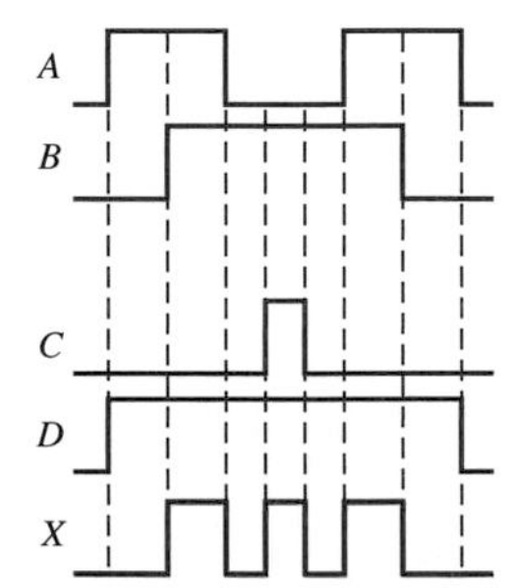

FIGURE ANS–9

21. See Figure ANS–10.

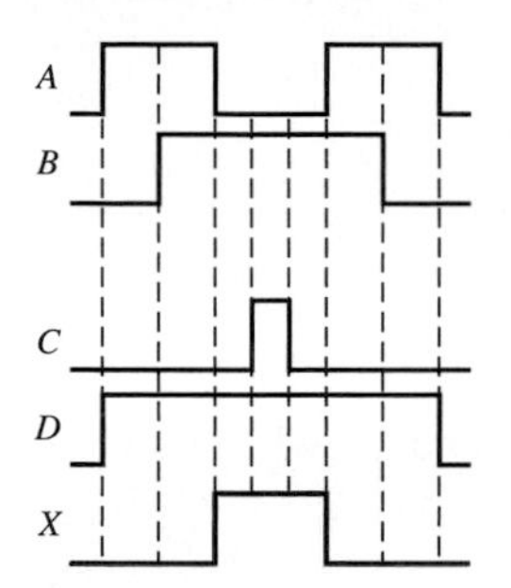

FIGURE ANS–10

23. See Figure ANS–11.

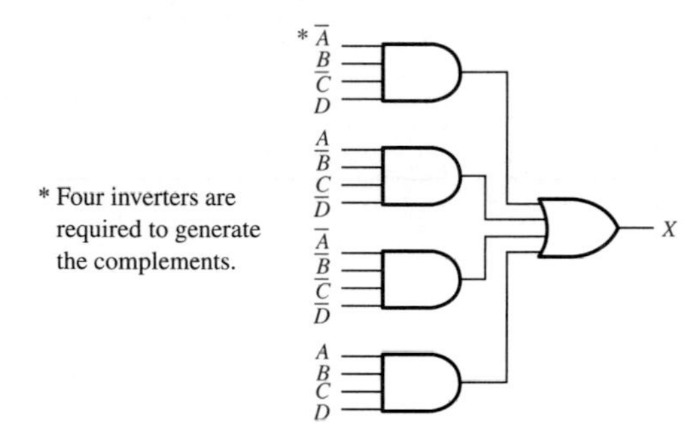

FIGURE ANS–11

25. (a) $X = \overline{A}\,\overline{B}\,\overline{C} + AC$
    (b) $\overline{A}C + \overline{A}B + A\overline{C}$
27. $X = ABD + A\overline{B}C$
29. No simplification is possible. Logic diagram is the same as Figure ANS–11.
31. 413 μW
33. The lower inverter output is open.
35. Output of G_2 is open.
37. The A and B inputs to gate G_0 are shorted.

Chapter 5

1. (a) 0
   (b) 1
   (c) 1
   (d) 10
3. (a) Sum = 0, $C_{out} = 0$
   (b) Sum = 1, $C_{out} = 0$
   (c) Sum = 0, $C_{out} = 1$
5. (a) $C_4\,\Sigma_4\Sigma_3\Sigma_2\Sigma_1 = 01111$
   (b) $C_4\,\Sigma_4\Sigma_3\Sigma_2\Sigma_1 = 01100$
   (c) $C_4\,\Sigma_4\Sigma_3\Sigma_2\Sigma_1 = 10000$

7. (a) 10110110
   (b) 00011010
   (c) 01110001
9. (a) 10000
   (b) 100100
   (c) 110111
11. An XOR gate produces a 1 when two bits are different and a 0 when two bits are the same.
13. See Figure ANS–12.

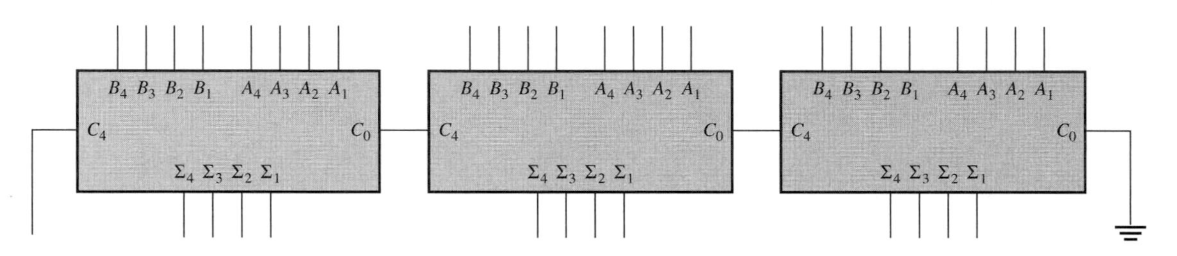

FIGURE ANS–12

15.

A	B	Σ	C_{out}
0	0	0	0
0	1	1	0
1	0	1	0
1	1	0	1

17. See Figure ANS–13.

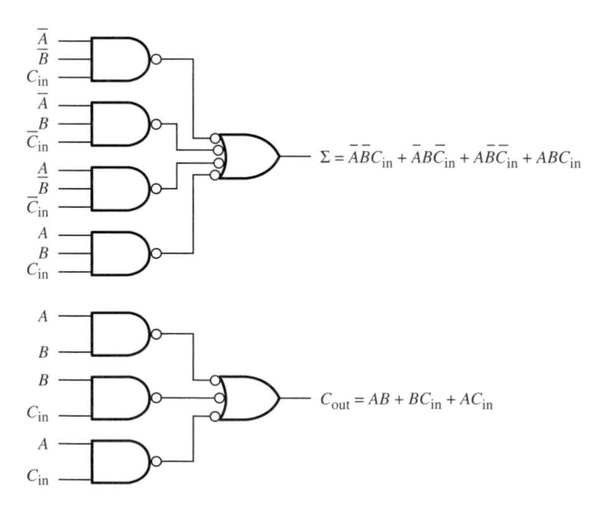

FIGURE ANS–13

19. (a) $C_8\,\Sigma_8\,\Sigma_7\,\Sigma_6\,\Sigma_5\,\Sigma_4\,\Sigma_3\,\Sigma_2\,\Sigma_1 = 011110011$
    (b) $C_8\,\Sigma_8\,\Sigma_7\,\Sigma_6\,\Sigma_5\,\Sigma_4\,\Sigma_3\,\Sigma_2\,\Sigma_1 = 011010001$
    (c) $C_8\,\Sigma_8\,\Sigma_7\,\Sigma_6\,\Sigma_5\,\Sigma_4\,\Sigma_3\,\Sigma_2\,\Sigma_1 = 110010001$

21. (a) 10100110
    (b) 01110111
    (c) 00111001
23. Because division can be done by a series of subtractions, it can be implemented using full-adders and 2's complement addition which is equivalent to subtraction.
25. The output of the comparator is 1.
27. See Figure ANS–14.

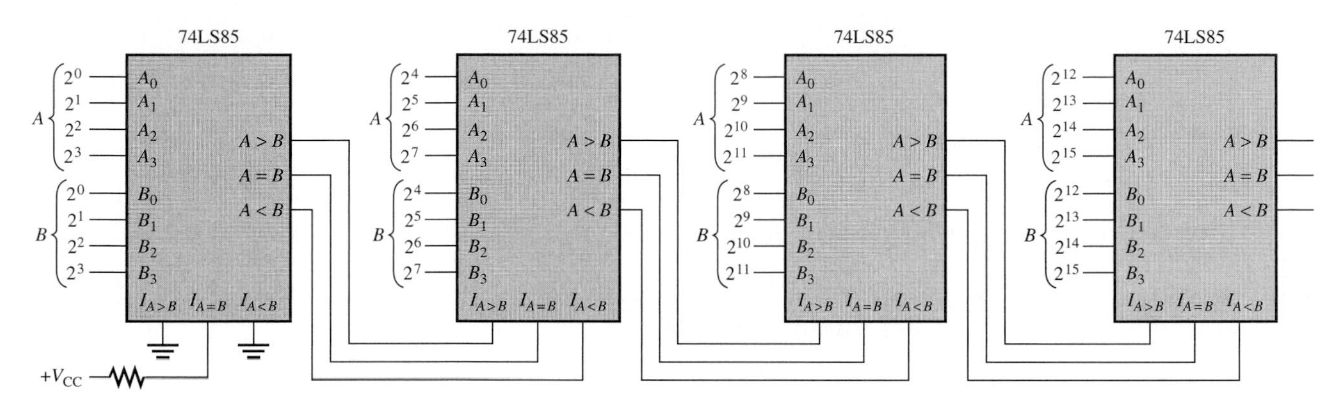

FIGURE ANS–14

29. The A_0 input is stuck LOW. Check the IC pin and if it is HIGH, the fault is internal.
31. Gate G_4 should be an OR gate.
33. The output of switch 6 is open.

Chapter 6

1. 0000, 0001, 0010, 0011, 0100, 0101, 0110, 0111, 1000, 1001
3. (a) 24 = 0010 0100
   (b) 39 = 0011 1001
   (c) 57 = 0101 0111
5. $2^7 = 128$
7. (a) 0; 0000
   (b) 4; 0100
   (c) 8; 1000
9. (a) **1**0110000
   (b) **0**0110100
   (c) **0**0111000
11. (a) 7
    (b) 5
    (c) 6
13. (a) *b, c*
    (b) *a, b, c, d, g*
    (c) *a, c, d, f, g*
    (d) *a, b, c, d, e, f, g*

15. (a) 2
    (b) 5
    (c) 8
    (d) 9
17. (a) 0011
    (b) 1001
    (c) 1000
19. (a) 1
    (b) 1
    (c) 0
21. (a) 4739 = 0100 0111 0011 1001
    (b) 7646 = 0111 0110 0100 0110
    (c) 1957 = 0001 1001 0101 0111
23. (a) **1**0001011101001001
    (b) **0**1000100100001000
    (c) **1**1001010001101001
25. Hello. How are you?
27. (a) 1001
    (b) 0010
    (c) 0100
    (d) 0111
29. $b = \overline{A}_3\overline{A}_2\overline{A}_1\overline{A}_0 + \overline{A}_3\overline{A}_2\overline{A}_1A_0 + \overline{A}_3\overline{A}_2A_1\overline{A}_0 + \overline{A}_3\overline{A}_2A_1A_0 + \overline{A}_3A_2\overline{A}_1\overline{A}_0 + \overline{A}_3A_2A_1A_0 + A_3\overline{A}_2\overline{A}_1\overline{A}_0 + A_3\overline{A}_2\overline{A}_1A_0$
31. See Figure ANS–15.

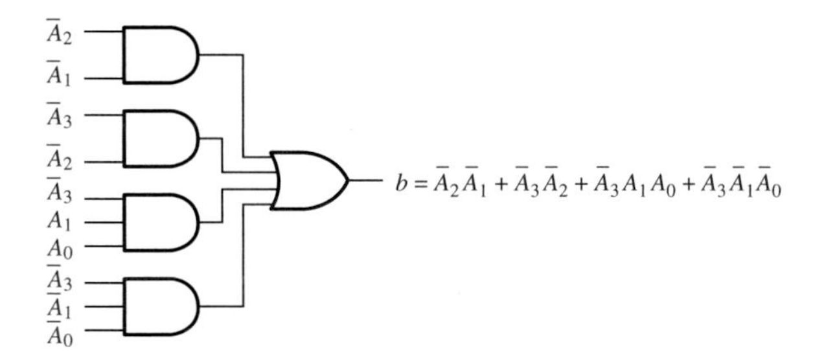

FIGURE ANS–15

33. $c = \overline{A}_3A_0 + \overline{A}_3A_2 + \overline{A}_2\overline{A}_1$
35. 1, 6, 9, 4, 4, 4, 8, 0
37. (a) $A_3A_2A_1A_0 = 1101$ (invalid)
    (b) $A_3A_2A_1A_0 = 0111$

39. See Figure ANS–16.

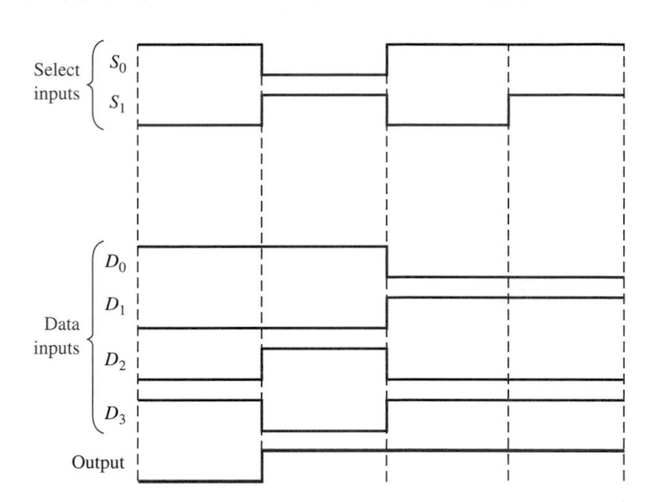

FIGURE ANS–16

41. See Figure ANS–17.

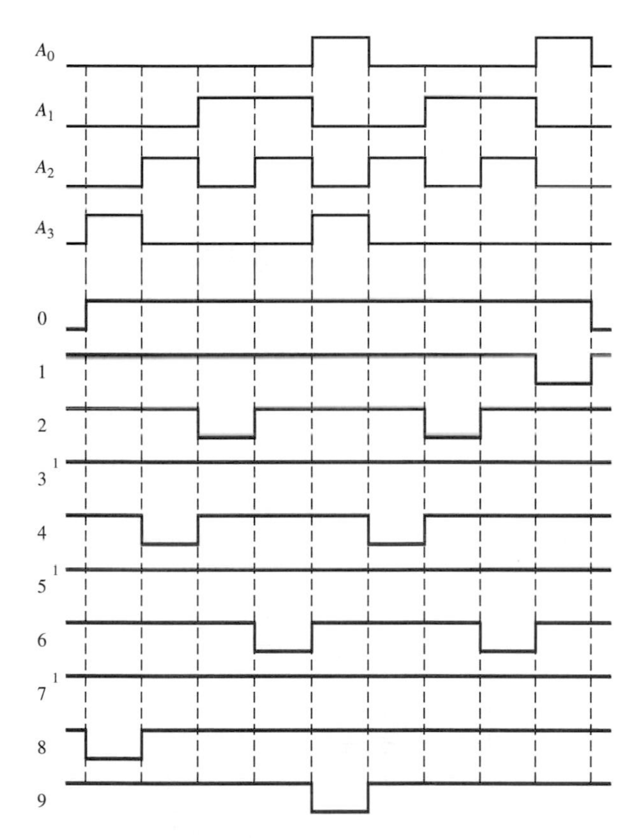

FIGURE ANS–17

43. A_2 stuck HIGH: 4, 5, 6, 7, 4, 5, 6, 7, no output active, no output active
A_1 stuck LOW: 0, 1, 0, 1, 4, 5, 4, 5, 8, 9

45. Switch output A is connected to output of B inverter in error. There should be no connection.

47. Inputs A and B of gate G_1 are shorted.

Chapter 7

1. (a) Q = HIGH
   (b) Q = LOW
   (c) Q = HIGH
3. (a) S-R latch
   (b) Positive edge-triggered J-K flip-flop
   (c) Gated D latch
   (d) Negative edge-triggered D flip-flop
5. (a) $\overline{Q} = 0$
   (b) $\overline{Q} = 1$
   (c) $\overline{Q} = 0$
   (d) $\overline{Q} = 1$
7. (a) $Q = 1$
   (b) $Q = 0$
   (c) $Q = 1$
9. 1.54 ms
11. 3.63 ms
13. See Figure ANS–18.

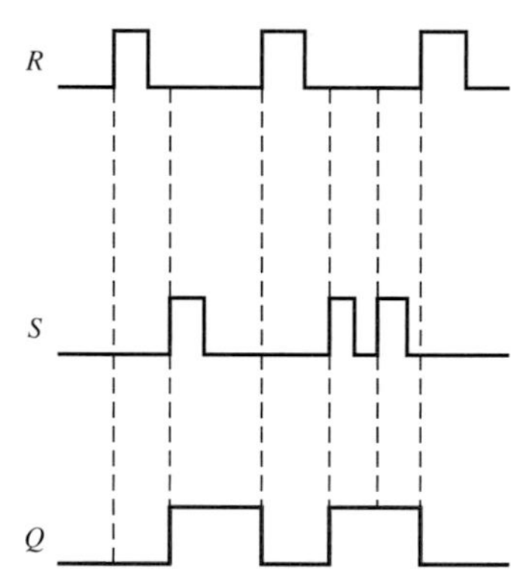

FIGURE ANS–18

15. See Figure ANS–19.

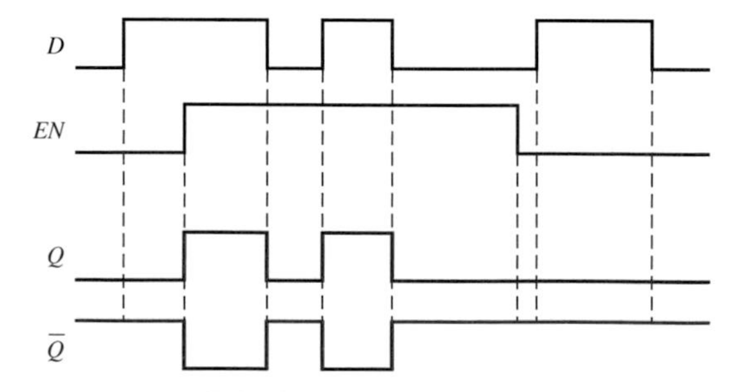

FIGURE ANS–19

17. See Figure ANS–20.

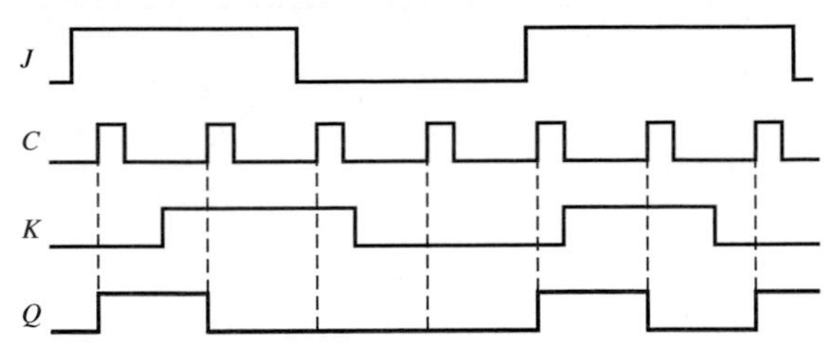

FIGURE ANS–20

19. See Figure ANS–21.

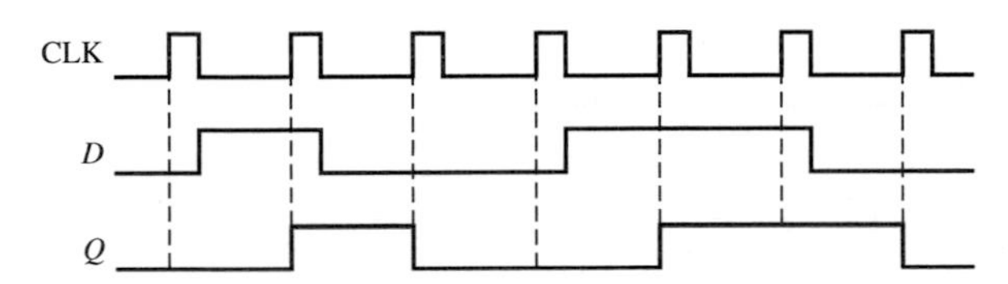

FIGURE ANS–21

21. See Figure ANS–22. $f_Q = (1/2) f_{CLK}$

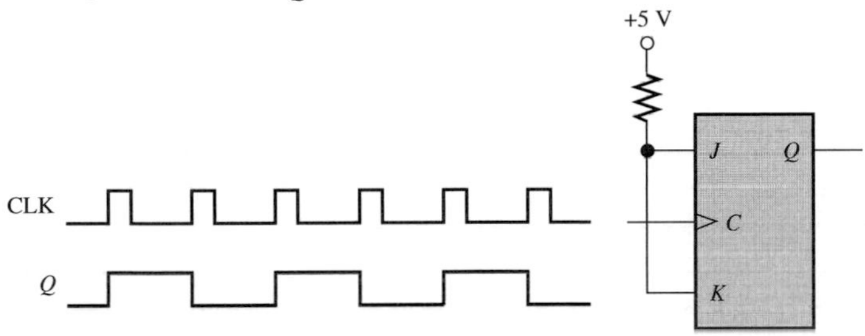

FIGURE ANS–22

23. See Figure ANS–23.

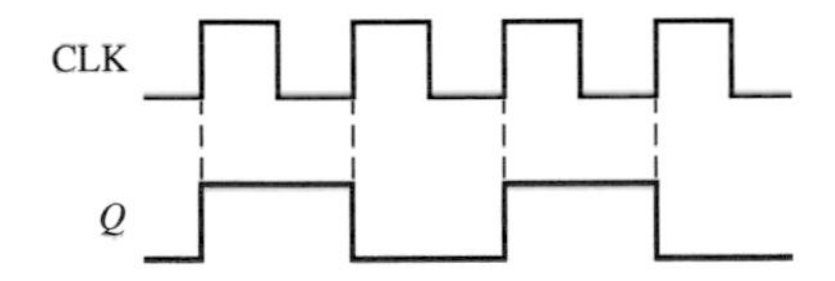

FIGURE ANS–23

25. 156 kΩ
27. 26.7 kHz
29. G_1 input from switch E is open.
31. $\overline{Q}$ is connected to D input in error. Should be no connection.
33. C_{EXT} is connected to the wrong side of R_{EXT2}, causing abnormal duty cycle.

Chapter 8

1. 000, 001, 010, 011, 100, 101, 110, 111
3. 0 through 31
5. Four
7. 1010, 1011, 1100, 1101, 1110, 1111 are invalid because they do not represent a decimal digit 0 through 9.

9. 512
11. $Q_3Q_2\overline{Q}_1Q_0$
13. See Figure ANS–24.

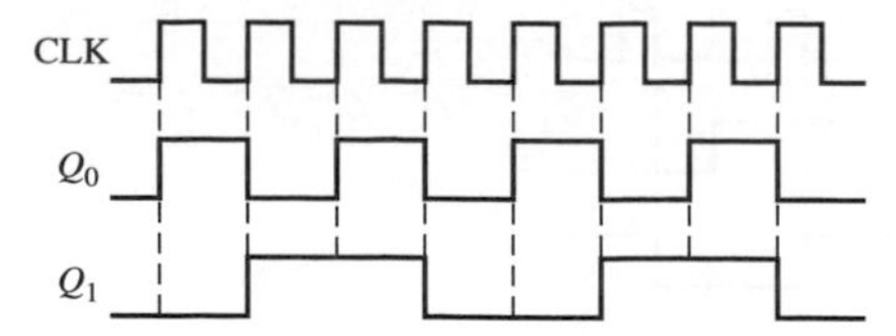

FIGURE ANS–24

15. See Figure ANS–25.

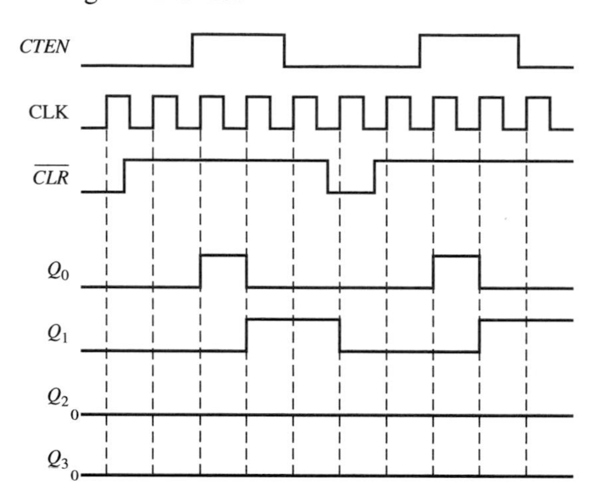

FIGURE ANS–25

17. See Figure ANS–26.

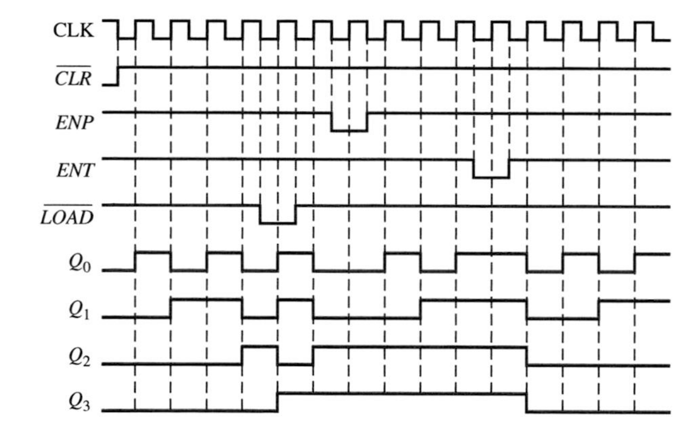

FIGURE ANS–26

19. See Figure ANS–27.

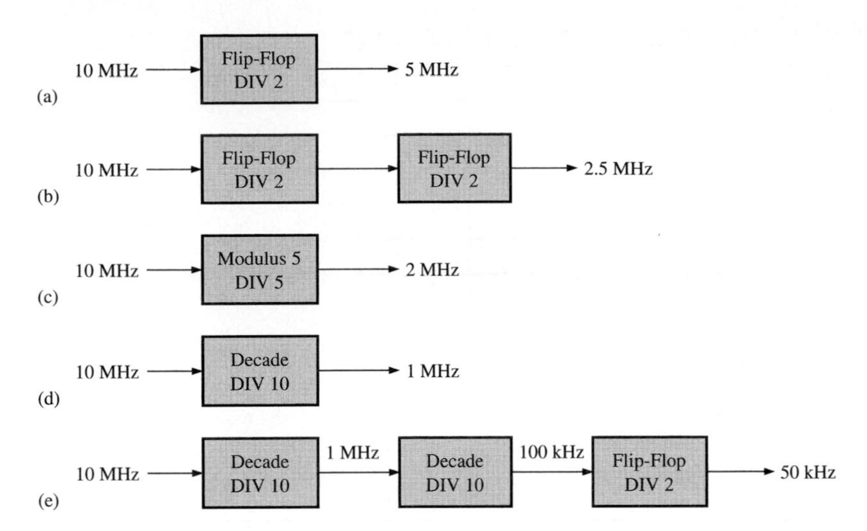

FIGURE ANS–27

21. See Figure ANS–28.

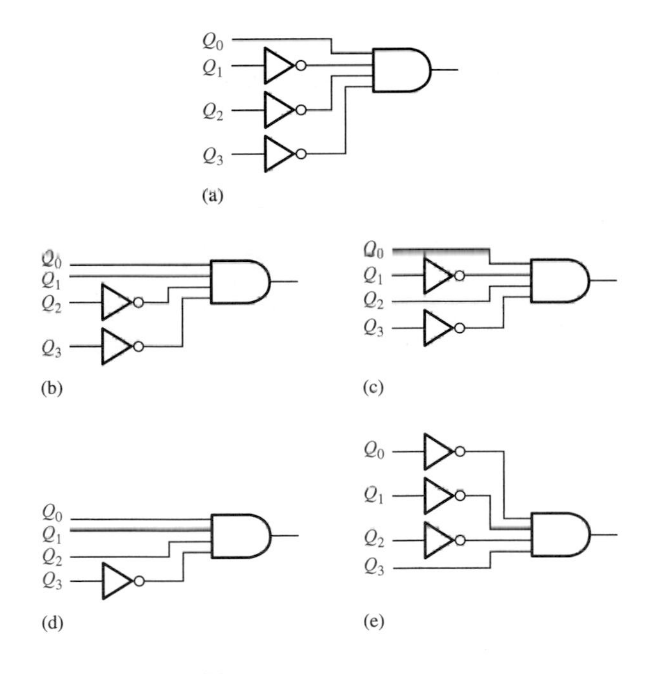

FIGURE ANS–28

23. See Figure ANS–29.

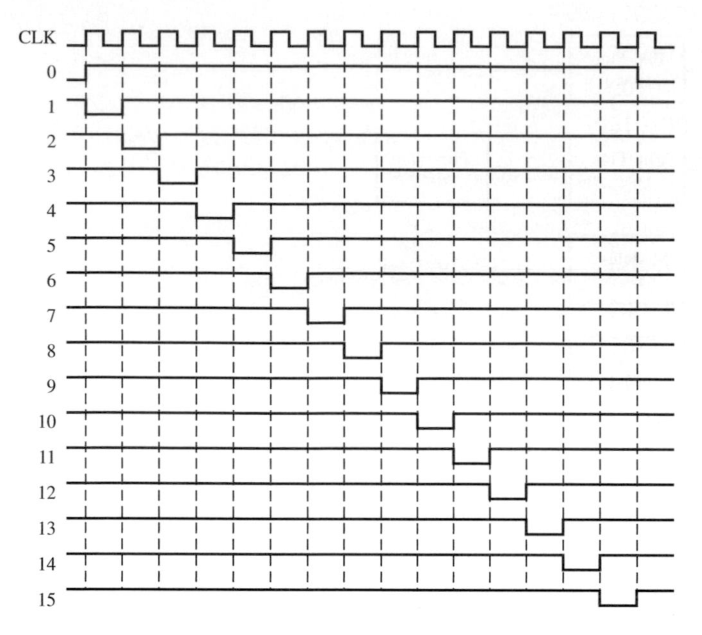

FIGURE ANS–29

25. AND gate output is stuck LOW, Q_2 output is stuck LOW, or clock input to FF2 is open.

27. G_2 output is open.

29. Gate G_2 has two inputs from Q_1. One should be moved to Q_2.

Chapter 9

1. Eight

3. 0.8 μs

5. 1,000,000

7. 100,000,000

9. 00010111

11. (a) 3 (b) 5 (c) 8

13. See Figure ANS–30.

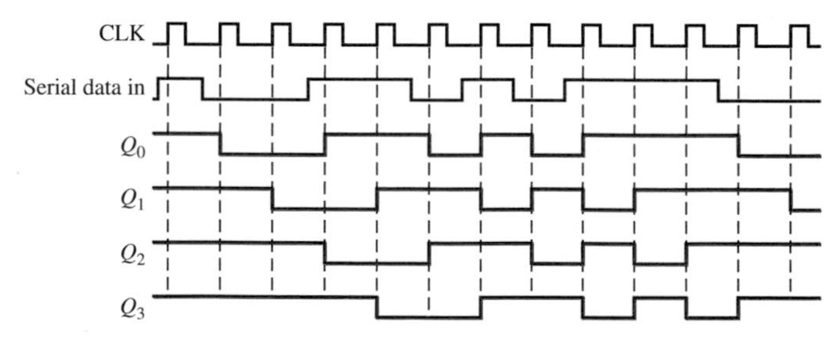

FIGURE ANS–30

15. Initially 101001111000
    CLK 1 010100111100
    CLK 2 001010011110
    CLK 3 000101001111
    CLK 4 000010100111
    CLK 5 100001010011
    CLK 6 110000101001
    CLK 7 111000010100
    CLK 8 011100001010
    CLK 9 001110000101
    CLK 10 000111000010
    CLK 11 100011100001
    CLK 12 110001110000

17. See Figure ANS–31.

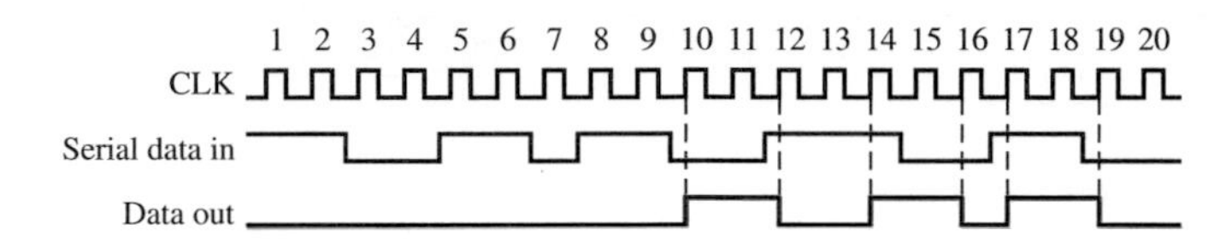

FIGURE ANS–31

19. See Figure ANS–32.

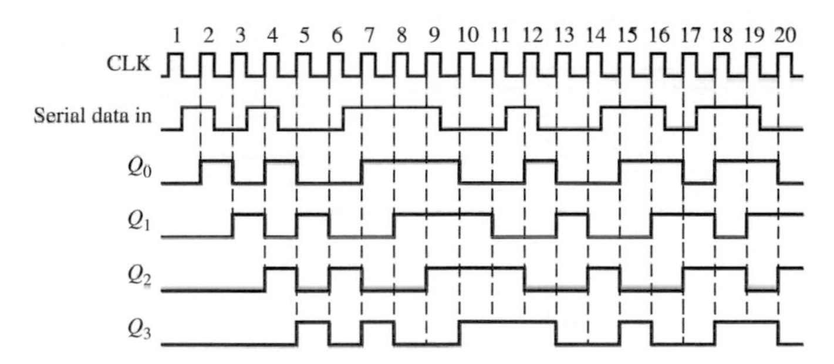

FIGURE ANS–32

21. See Figure ANS–33.

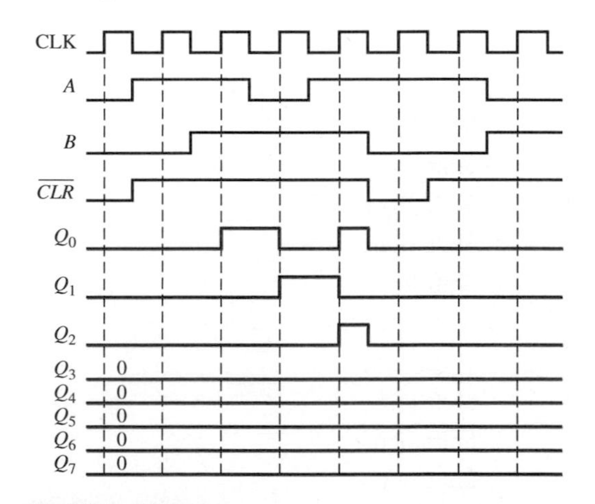

FIGURE ANS–33

23. See Figure ANS–34.

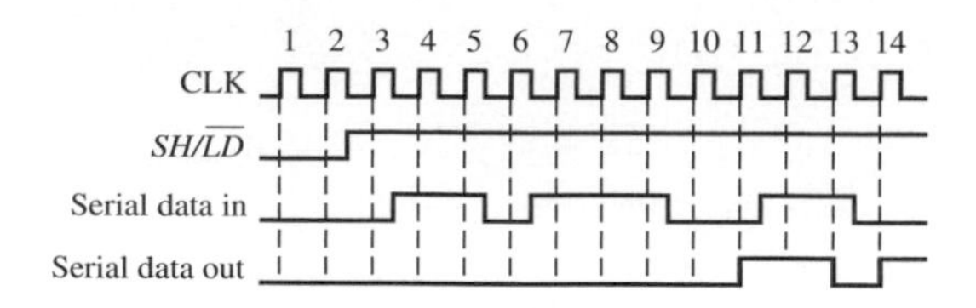

FIGURE ANS–34

25. See Figure ANS–35.

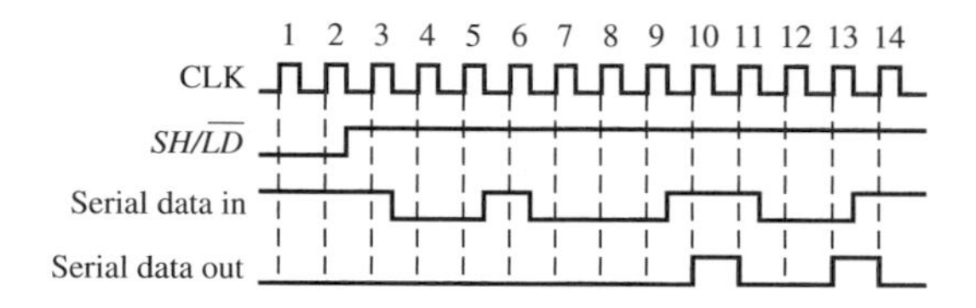

FIGURE ANS–35

27. See Figure ANS–36.

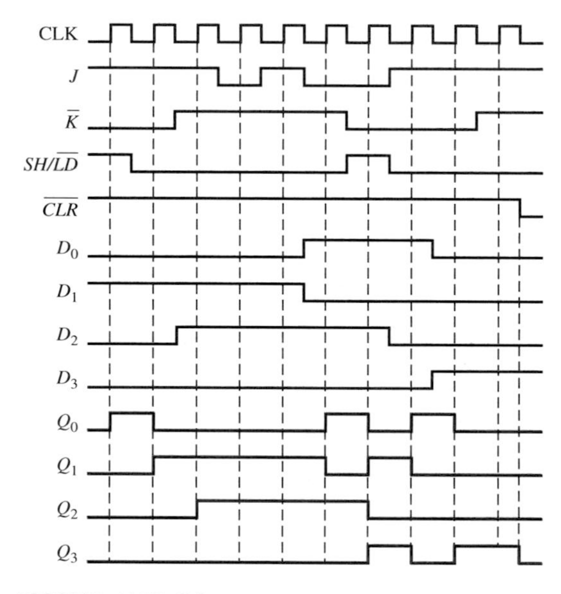

FIGURE ANS–36

29.

Initially (76)	01001100	
CLK 1	10011000	Shift left
CLK 2	01001100	Shift right
CLK 3	00100110	Shift right
CLK 4	00010011	Shift right
CLK 5	00100110	Shift left
CLK 6	01001100	Shift left
CLK 7	00100110	Shift right
CLK 8	01001100	Shift left
CLK 9	00100110	Shift right
CLK 10	01001100	Shift left
CLK 11	10011000	Shift left

31. See Figure ANS–37.

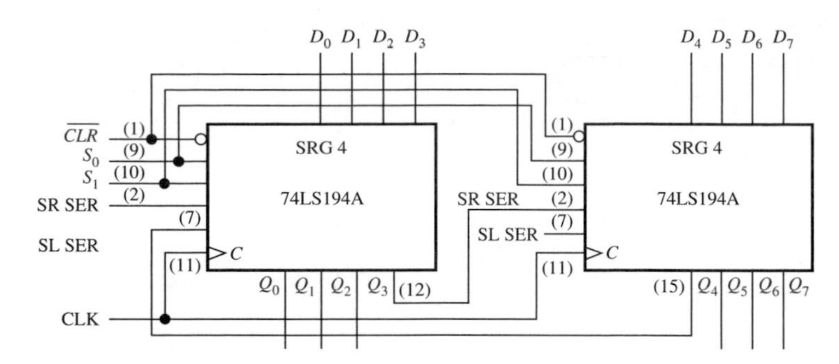

FIGURE ANS–37

33. Q_2 goes HIGH on the first clock pulse indicating that the D_2 input may be open.

35. *D* input to FF2 is open.

37. Output of OR gate G_8 is open.

Chapter 10

1. (a) Ball grid array (BGA)
   (b) Plastic leaded chip carrier (PLCC)

3. $X_1 = \overline{A}B, X_2 = \overline{A}\,\overline{B}, X_3 = A\overline{B}$

5. Very High Speed Integrated Circuit

7. Development software, computer, programming fixture.

9. (c) is correct.

11. 1:2, 2:3, 3:5, 4:1, 5:3, 6:6, 7:2, 8:3, 9:6

13. See Figure ANS–38.

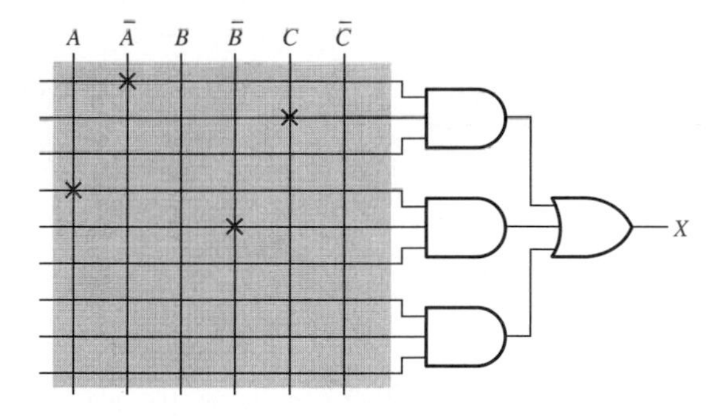

FIGURE ANS–38

15.
```
entity NAND_Gate4 is
   port (A, B, C, D: in bit; X: out bit);
end entity NAND_Gate4;
architecture NANDfunction of NAND_Gate4 is
begin
   X <= not (A and B and C and D);
end architecture NANDfunction;
```

17. See Figure ANS–39.

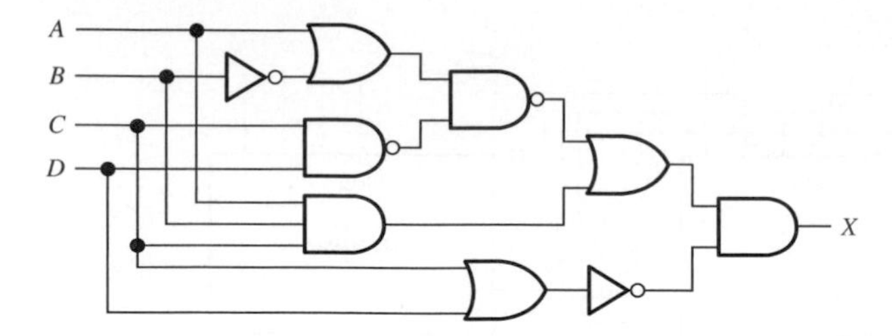

FIGURE ANS–39

19.
```
entity F10_45 is
    port (A, B, C: in bit; X: out bit);
end entity F10_45;
architecture Logic of F10_45 is
begin
    X <= (A and C) or B;
end architecture Logic;
```

21.
```
entity F10_47 is
    port (A, B, C, D: in bit; X: out bit);
end entity F10_47;
architecture Logic of F10_47 is
begin
    X <= (A or B) and (not B or C or D) and (not A or C);
end architecture Logic;
```

Chapter 11

1. During the fetch cycle, the CPU obtains an instruction from memory. During the execute cycle, the CPU carries out the instruction.
3. Bus
5. 8192 bits
7. Data are lost when new data are written in or when the power is removed.
9. Eight.
11. A rewritable compact disc in which data can be written to and read from.
13. See Figure ANS–40.

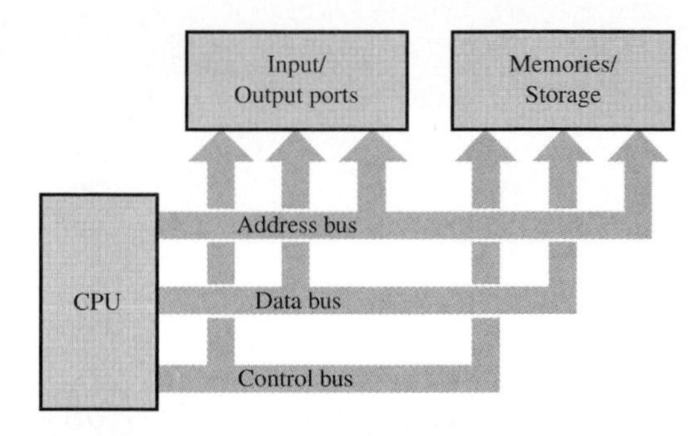

FIGURE ANS–40

15. 2,048 Mbits
17. (a) 10
    (b) 63
    (c) 205
    (d) 181
19. 262,144
21. The purpose of a refresh operation is to recharge the storage capacitors in a DRAM. A read operation automatically refreshes all addresses in a selected row.
23. Address multiplexing is a method for reducing the number of lines in the address bus. A row address is first placed on the bus followed by a column address.
25. The seek time is the average time for the read/write head to move to the desired track. The latency period is the time for the desired sector to spin under the read/write head once the head is positioned over the desired track.
27. Each number is two bytes and requires two addresses.
29. The ax register stores the sum of $3A_{16}$, 27_{16}, and $1B_{16}$, which is $7C_{16}$ after three times around the loop.
31. int total = 0;

Chapter 12

1. The purpose of analog-to-digital conversion is to change an analog signal into a sequence of digital codes that represent the amplitude of the analog signal with respect to time.
3. The purpose of digital-to-analog conversion is to change a sequence of digital codes into an analog signal represented by the digital codes.
5. 3200 million instructions (3200 MIPS)
7. 1. Program address generate (PG). The program address is generated by the CPU.
   2. Program address send (PS). The program address is sent to the memory.

3. Program access ready wait (PW). A memory read operation occurs.
4. Program fetch packet receive (PR). The CPU receives the packet of instructions.

9. 60 kHz

11. See Figure ANS–41.

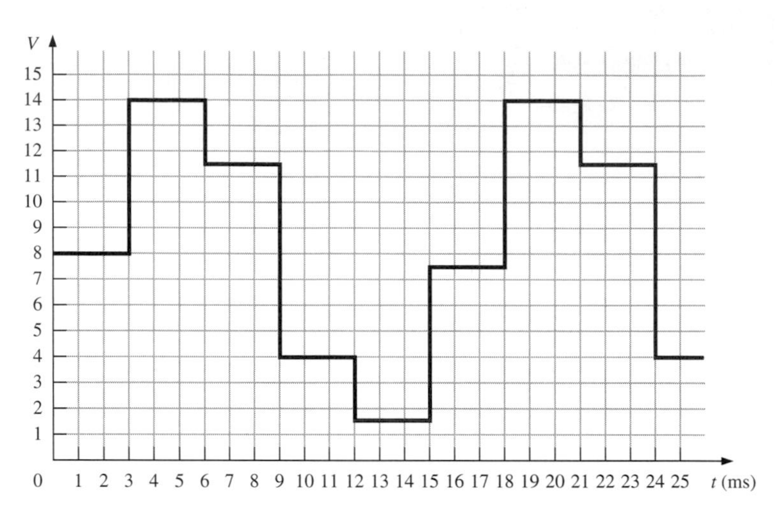

FIGURE ANS–41

13. 1000, 1110, 1011, 0100, 0001, 0111, 1110, 1011, 0100

15. See Figure ANS–42.

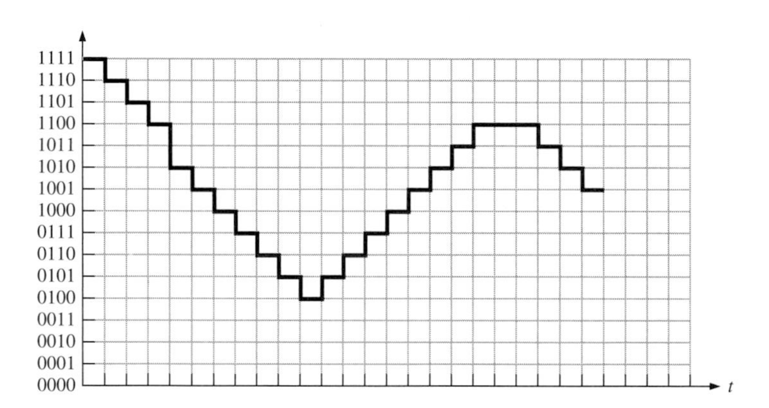

FIGURE ANS–42

17. 330 kΩ

19. $R_0 = 10\ \text{k}\Omega$, $R_1 = 5\ \text{k}\Omega$, $R_2 = 2.5\ \text{k}\Omega$, $R_3 = 1.25\ \text{k}\Omega$

21. See Figure ANS–43.

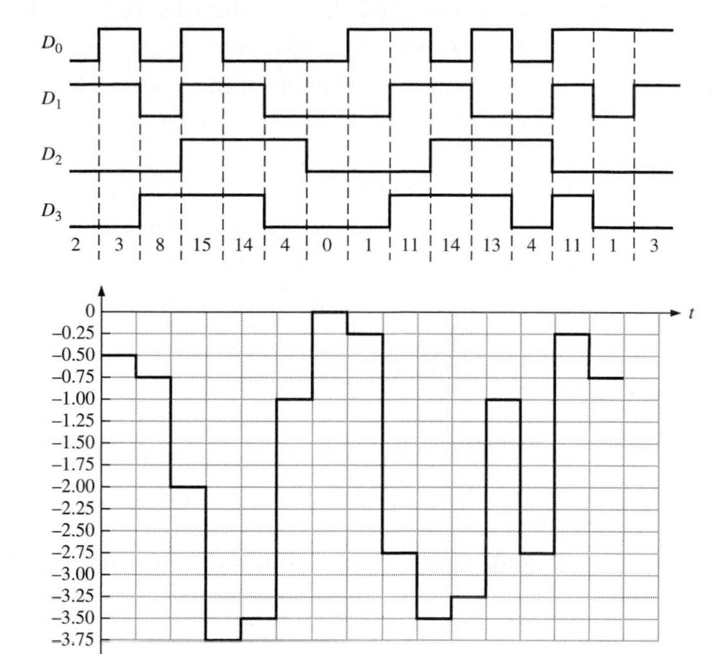

FIGURE ANS–43

23. $2^{11} = 2048$

GLOSSARY

Address A location in memory.

Address bus A one-way group of conductors from the microprocessor to memory, on which the address code is sent.

Algorithm The step-by-step process for achieving a given result.

Aliasing The effect created when a signal is sampled at less than twice the Nyquist frequency. Aliasing creates unwanted frequencies that interfere with the signal frequency.

Amplitude The "height" of a pulse measured from the baseline expressed in volts.

Analog Having a continuous set of values.

Analog-to-digital converter (ADC) A circuit used to convert an analog signal to digital form.

AND gate A digital circuit that has a HIGH ouput only when all of its inputs are HIGH.

AND-OR A logic circuit that consists of any number of AND gates with outputs connected to the inputs of a single OR gate.

AND-OR-Invert A logic circuit that consists of AND gates with outputs that are connected to a NOR gate.

Antifuse In programming a PLD, a one-time programmable link in which a normally open contact is shorted by melting the antifuse material to form a connection.

Architecture The VHDL unit that describes the internal operation of a logic function.

Arithmetic logic unit (ALU) The part of a microprocessor that performs arithmetic operations, logic operations, and comparison operations on binary data.

ASCII The American Standard Code for International Interchange is a 7-bit code that represents both numeric and alphabetic characters as well as symbols and commands. It is commonly used as a keyboard entry code for computers.

ASIC Application specific integrated circuit; a PLD preprogrammed by the manufacturer to perform a specified function.

Assembler A program that converts an assembly language program to machine language that is recognized by a computer.

Assembly language A programming language that uses English-like words and has a one-to-one correspondence to machine language.

Asserted state The logic state or level (HIGH or LOW) that brings about the action that a logic gate or inverter is intended to produce.

Associative law In addition (ORing) and multiplication (ANDing) of three or more variables, the order in which the variables are grouped makes no difference.

Astable Having no stable state.

Asynchronous Having no fixed time relationship.

Asynchronous counter A counter in which the flip-flops are not all clocked at the same time.

Biased exponent A number obtained by adding 127 to the actual exponent in floating-point format.

Bidirectional Having two directions. In a bidirectional shift register, the stored data can be shifted right or left.

Binary A number system with a base of 2 that has two digits, 1 and 0.

Binary addition The process of adding binary numbers.

Binary coded decimal Abbreviated BCD, this is a binary code that uses four bits to represent each of the ten decimal digits.

Binary counter A logic circuit made of flip-flops that goes through a sequence of binary states in response to clock pulses.

BIOS Basic Input/Output System; a set of programs in ROM that interfaces the I/O devices in a computer system.

Bistable Having two stable states.

Bit A binary digit.

Boolean addition The mathematical operation equivalent to the OR function.

Boolean multiplication The mathematical operation equivalent to the AND function.

Bus A standardized interconnection between major elements in a computer and between the computer and a peripheral.

Byte A group of eight bits.

Cache A small RAM used to store a limited amount of frequently used data. A cache memory can usually be accessed much faster than the main RAM.

Capacity The total number of data units that can be stored in a memory.

Cascade The connection of two or more counters such that the total modulus is increased.

Cell A fused cross point of a row and column in a PLD; a single storage element in a memory.

Clock A pulse waveform used as a basic timing signal in digital systems; also called a *timing waveform.*

Combinational logic Logic gates connected together to perform some specific function in which the output level is at all times dependent on the combination of input levels.

Commutative law In addition (ORing) and multiplication (ANDing) of two variables, the order in which the variables are ORed or ANDed makes no difference.

Comparator A logic circuit that compares the magnitudes of two binary numbers and produces an output, indicating whether or not the numbers are equal or which of the numbers is larger.

Compiler A program that functions as a programming language translator. A compiler translates high-level language to machine code.

Complement The inverse or opposite of a binary digit or Boolean variable. The complement of 1 is 0, and the complement of A is $\overline{A}$.

Control bus A set of conductive paths that connects the CPU to other parts of the computer to coordinate its operations and to communicate with external devices.

Counter decoding The process of detecting a specified state of a counter.

CPLD Complex programmable logic device; a type of PLD that can contain the equivalent of from two to sixty-four SPLDs.

CPU Central processing unit; the "brain" of a computer that processes the program instructions.

Cross-assembler A program that translates an assembly language program for one type of microprocessor to an assembly language for another type of microprocessor.

Data bus A bidirectional set of conductive paths on which data or instruction codes are transferred into a microprocessor or on which the result of an operation is sent out from a microprocessor.

Data waveform A digital waveform that contains binary information.

Decade counter A counter that has ten states.

Decode The process of converting coded information into a recognizable form; a process of computer or DSP pipeline operation.

Decoder A logic circuit that produces an output corresponding to the particular code on its inputs.

DeMorgan's theorems Two Boolean theorems that relate to the complementation of products and sums.

Demultiplexer A logic circuit (abbreviated DEMUX) that moves serial data from a single input line onto several parallel output lines.

D flip-flop A type of flip-flop that stores data based on the state of the D input at the triggering edge of a clock pulse.

Digital Having a series of individual values (discrete).

Digital-to-analog converter (DAC) A circuit used to convert the digital representation of an analog signal back to the analog signal.

DIP Dual in-line package; a type of integrated circuit package.

Distributive law ORing several variables and then ANDing the result with a single variable is equivalent to ANDing the single variable with each of the several variables and then ORing the products.

D latch A type of gated latch that stores the data bit on its D input when enabled.

DSP Digital signal processor; a special type of microprocessor that processes data in real time.

Duty cycle The ratio of the pulse width to the period of a digital waveform, usually expressed as a percentage.

Edge-triggering The method used to cause a flip-flop to store a data bit on the occurrence of a clock edge, based on the state of its inputs. If the flip-flop triggers on the rising edge of a clock pulse, it is called a positive edge-triggered flip-flop. If it triggers on the falling edge, it is called a negative edge-triggered flip-flop.

Enable An input on a logic circuit that prevents or allows response to another input, depending on the level of the Enable input.

Encoder A logic circuit that produces a code representing the particular input that is active.

Entity The VHDL unit that describes the inputs and outputs of a logic function.

Execute A process of computer or DSP operation in which the decoded instructions are carried out.

Exponent The part of a floating-point number that represents the number of places that the decimal point is to be moved.

Fan-out The maximum number of gate inputs of the same type that a single logic gate can drive.

Feedback The output voltage or a portion of it that is connected back to the input of a circuit.

Fetch A process of computer or DSP operation in which an instruction is obtained from the program.

Flash memory A high-density nonvolatile read/write memory.

Flip-flop A bistable logic circuit similar to a latch except that it can change state only on the occurrence of one edge of a clock pulse.

Floating-point A method commonly used in computers for representing numbers based on exponents.

FPGA Field programmable gate array; a type of PLD that generally has much more complexity than a CPLD and a different organization.

Frequency For a digital waveform, the number of pulses that occur in one second expressed in hertz (Hz).

Frequency divider A counter used to divide the input clock frequency by a specified amount.

Full-adder A logic circuit that performs addition on two bits and an input carry bit and produces a sum bit and an output carry bit.

Fuse A one-time programmable link in a PLD that is blown to eliminate selected variables from the output function.

GAL Generic array logic, a type of SPLD that is essentially a reprogrammable PAL.

Gated latch A type of latch with a gate that is used to enable or disable the *S, R,* or *D* inputs.

Half-adder A logic circuit that performs addition on two bits and produces a sum bit and an output carry bit.

Hard disk A large nonvolatile mass storage medium.

HDL Hardware description language used for programming PLDs.

Hexadecimal A number system with a base of 16 that has sixteen digits (0–9 and A–F).

High-level language A type of computer language closest to human language that is a level above assembly language.

Integrated circuit An electronic circuit made entirely on a tiny chip of semiconductive material, usually silicon.

Interrupt A computer signal or instruction that causes the current process to be temporarily stopped while a service routine is run.

Inversion The process of changing a HIGH (1) to a LOW (0) or vice versa; also known as complementation.

Inverter A logic circuit that performs inversion; also known as a NOT circuit.

ISP In-system programmable; a type of PLD that can be programmed after it has been installed on a circuit board.

J-K flip-flop A type of flip-flop that stores data based on the states of the *J* and *K* inputs at the triggering edge of a clock pulse.

Johnson counter A configuration of a shift register in which a specific prestored pattern of 1s and 0s is recirculated.

Karnaugh map A graphic tool used to simplify a Boolean expression. It is an arrangement of cells representing all product terms for a given number of Boolean variables.

Latch A bistable logic circuit that can store a binary 1 or 0.

Load To enter parallel data into a shift register.

Logic block A group of logic gates in a FPGA.

Logic level A voltage level that represents a binary digit.

LSB Least significant bit in a binary number.

LSI Large-scale integration; a fixed-function IC classification.

Machine language Computer instructions written in binary code that are understood by a computer; the lowest level of programming language.

Macrocell A section of logic in a CPLD.

Mantissa The magnitude of a floating-point number (excluding the exponent).

MFLOPS Million floating-point operations per second.

Microprocessor A large-scale digital integrated circuit that can be programmed to perform arithmetic, logic, and other operations; the CPU of a computer.

MIPS Million instructions per second.

MMACS Million multiply/accumulates per second.

Modulus The total number of unique states that a counter goes through.

Monostable Having only one stable state.

MSB Most significant bit in a binary number.

MSI Medium-scale integration; a fixed-function IC classification.

Multiplexer A logic circuit (abbreviated MUX) that moves data from several parallel input lines to a single output for serial transmission.

NAND gate A digital circuit that produces a LOW only when all of its inputs are HIGH.

Negative-AND The operation of a NOR gate when the asserted states of the inputs are LOW.

Negative-OR The operation of a NAND gate when the asserted states of the inputs are LOW.

Node A common electrical point in a circuit.

NOR gate A digital circuit that produces a LOW output when any of its inputs are HIGH.

NOT To invert or complement.

Nyquist frequency The frequency that specifies the limit of the analog signal frequency to be sampled and defines it to be one-half of the sampling frequency.

Object program A machine language translation of a high-level source program.

Octal A number system with a base of 8 that has eight digits (0–7).

OLMC Output logic macrocell; the part of a GAL that can be programmed for either combinational or registered outputs.

1's complement The result of inverting each bit in a binary number.

One-shot A type of triggered logic circuit that has only one stable state and is also referred to as a monostable multivibrator.

Op-code Operation code; the code representing a particular microprocessor instruction; a mnemonic.

OR-gate A digital circuit that has a HIGH output when one or more of its inputs are HIGH.

PAL Programmable array logic; a type of SPLD that is generally one-time programmable.

Parallel binary adder A logic circuit consisting of two or more full-adders that can add two binary numbers.

Parity bit A bit added to a given code to make the total number of bits either odd or even and used for error detection.

Period The time interval between pulses expressed in units of seconds.

Peripheral A device such as a printer or modem that provides communication with a computer.

Pipeline Part of the DSP architecture that allows multiple instructions to be processed simultaneously.

PLD Programmable logic device; an integrated circuit that can be programmed with any specified logic function.

Port In VHDL, the inputs and outputs of a logic function. In a computer, points at which data are taken in or out.

Power dissipation The amount of dc power used by a circuit; the product of V_{CC} and I_{CC}.

Power-of-two A number expressed as an exponent to the base of 2.

Priority encoder An encoder that ignores all active inputs except the one with the highest value.

Product-of-sums A standard form of Boolean expression that is the product (AND) of two or more sum terms (OR).

Product term A Boolean term that is the AND of two or more variables.

Program A list of instructions that a computer follows in order to achieve a specific result.

Programmable array An array of logic gates with a grid of programmable interconnections.

Propagation delay The time delay from a change in the input level of a gate to the corresponding change in output level.

Pulse width The time duration of a pulse expressed in units or seconds.

Quantization The process whereby a binary code is assigned to each sampled value during analog-to-digital conversion.

RAM Random-access memory; a type of volatile memory that can be written into or read from at any address selected randomly.

Read An operation that takes data out of a specified address in the memory.

Register An internal memory location within a microprocessor used as a temporary storage device.

Removable storage A part of a computer system such as a CD-ROM that allows for removable storage of data.

RESET One state of a latch or flip-flop in which the Q output is LOW and the device is effectively storing a 0.

Ring counter A configuration of a shift register in which a certain pattern of 1s and 0s is continuously recirculated.

ROM Read-only memory; a type of nonvolatile memory from which data can only be read.

Sampling The process of taking a sufficient number of discrete values at points on a waveform that will define the shape of waveform.

SET One state of a latch or flip-flop in which the Q output is HIGH and the device is effectively storing a 1.

7-segment display A display device that consists of seven segments that can be activated to form each of the ten decimal digits.

Shift To move binary data from stage to stage within a shift register or other storage device or to move binary data into or out of the device.

Shift register Two or more flip-flops connected to temporarily store and move binary data from one flip-flop to another.

Shift register counter A type of shift register with the serial output connected back to the serial input that exhibits a special sequence of states.

Sign bit The left-most bit of a binary number that designates whether the number is positive (0) or negative (1).

Signal tracing A method of troubleshooting that identifies the point in a circuit where the signal is first missing or incorrect.

SMT Surface-mount technology; a class of integrated circuit packages.

Software Programs that allow the computer hardware to perform specific tasks.

Source program A program written in either assembly or high-level language.

SPLD Simple programmable logic device; the least complex form of PLD. The PAL and GAL are types of SPLDs.

S-R latch A type of latch that is SET when its S input is active and is RESET when its R input is active.

SSI Small-scale integration; a fixed-function IC classification.

Stage One storage element (flip-flop) in a shift register.

Storage The property that allows a device to retain a 1 or 0 indefinitely.

Sum-of-products A standard form of Boolean expression that is the sum (OR) or two or more product terms (AND).

Sum term A Boolean term that is the OR of two or more variables.

Supply voltage The dc voltage that provides energy for a circuit to operate.

Synchronized A condition where two or more events occur with a fixed-time relationship.

Synchronous Having a fixed time relationship.

Synchronous counter A counter in which the flip-flops are all clocked simultaneously.

Target device A PLD that is being programmed.

Terminal count The final state in a counter's sequence.

Timer A term that refers to either a monostable or an astable multivibrator. An astable multivibrator is a type of oscillator that produces a continuous pulse waveform. A monostable multivibrator is a circuit that produces a single pulse when triggered.

Timing diagram A graph showing the time relationship of two or more waveforms.

Toggle The action of a flip-flop when it changes state on each clock pulse.

Truth table A list of all the combinations of inputs and the corresponding outputs of a logic circuit.

2's complement The result of adding 1 to the 1's complement of a binary number.

ULSI Ultra large-scale integration; a fixed-function IC classification.

Unit load A measure of fan-out. One gate input represents a unit load to the output of a gate within the same IC family.

Universal shift register A register that has both serial and parallel input and output capability.

Variable In Boolean algebra, a quantity represented by a letter that can take on a value of 1 or 0.

VHDL A standard hardware description language for programming PLDs.

VLSI Very large-scale integration; a fixed-function IC classification.

Waveform A pattern of the changes in voltage (or current) with time.

Weight The value of a digit in a number based on its position in the number.

Word A complete unit of digital information.

Write The operation that puts data into a specified address in the memory.

XNOR A logic gate that produces a 0 (LOW) if one and only one of its inputs is 1 (HIGH).

XOR A logic gate that produces a 1 (HIGH) if one and only one of its inputs is 1 (HIGH).

INDEX